Eco-Friendly Biobased Products Used in Microbial Diseases

Editors

Mahendra Rai

Department of Biotechnology
SGB Amravati University
Amravati, Maharashtra, India

Chistiane M. Feitosa

Department of Chemistry
Federal University of Piaui
Petronio Portela Campus, Teresina, Brazil

CRC Press
Taylor & Francis Group
Boca Raton London New York

CRC Press is an imprint of the
Taylor & Francis Group, an **informa** business
A SCIENCE PUBLISHERS BOOK

Cover illustration include free images partially obtained from rawpixel.com/Freepik and modified by the authors of Chapter 1 (*José Ribamar Costa Ferreira-Neto, Éderson Akio Kido, Flávia Figueira Aburjaile, Manassés Daniel da Silva, Marislane Carvalho Paz de Souza, Ana Maria Benko-Iseppon*).

First edition published 2022
by CRC Press
6000 Broken Sound Parkway NW, Suite 300, Boca Raton, FL 33487-2742

and by CRC Press
2 Park Square, Milton Park, Abingdon, Oxon, OX14 4RN

Library of Congress Cataloging-in-Publication Data (applied for)

ISBN: 978-1-032-15343-8 (hbk)
ISBN: 978-1-032-15344-5 (pbk)
ISBN: 978-1-003-24370-0 (ebk)

DOI: 10.1201/9781003243700

Typeset in Palatino
by Radiant Productions

Preface

The eco-friendly natural products have been used since ancient times. These originate from various natural sources such as plants, marine organisms, bacteria, fungi, actinomycetes, etc. At the beginning of 19th century, about 80% medicines came from plants. After the discovery of penicillin in 1928, the microorganisms attracted scientists to the discovery of new molecules to fight various microbial diseases and today, these microbes including bacteria, fungi and actinomycetes have become a major source of bioactive compounds. In late 1980s, combinatorial chemistry facilitated the discovery of bioactive compounds and lead molecules.

After the emergence of COVID-19 pandemic, researchers are looking towards a proper remedy to overcome the corona crisis. The bio-based products have great potential to fight with emerging microbial diseases including disease caused by coronaviruses. In the present scenario, there is a greater need to search for novel antibacterial, antifungal, antiviral and antiparasitic compounds from natural resources.

The proposed book is focused on the use of eco-friendly bio-based products of natural origin such as plants-derived products from the Amazon, the top hot spot of plants and other microbial flora, natural products as microbicidal agents against viruses, fungi, bacteria and microbial parasites. This also includes techniques for the identification of natural compounds present in extracts and essential oil, the role of terpenes and the bioactive antimicrobial agents of microbial origin. The authors participating in the book represent the major plant-rich areas/hotspots including Brazil, India, Argentina, Sri Lanka and Saudi Arabia, etc.

The book has been divided in three sections: Section I incorporates the bio-based natural products used in microbial diseases and role of omics in designing antimicrobial agents. Section II discusses the role of bio-based metabolites to overcome microbial diseases, and Section III deals with the antimicrobial metabolites of microbial and mushroom origin.

We are thankful to all the contributors for their sincere efforts to provide state-of-the-art information on the subject matter of their respective chapters. We also express our sincere thanks to the publisher and the authors of the chapters, whose research studies have been cited in the present book. Finally, we would like to thank the entire team at CRC Press for their generous cooperation and efforts in producing this book. MR wishes to thank the Polish National Agency for Academic Exchange (NAWA) No. PPN/ULM/2019/1/00117/U/00001 for financial support to visit the Department of Microbiology, Nicolaus Copernicus University, Toruń, Poland.

We hope that the proposed book would be an essential reading for postgraduate and research students of plant sciences, microbiology, chemistry and pharmacology. It will also enrich academicians and researchers of pharma industries.

Mahendra Rai
India
Chistiane M. Feitosa
Brazil

Contents

Section I

Natural Products Used in Microbial Diseases

1

The Journey through the Gene: a Focus on Plant Anti-pathogenic Agents Mining in the Omics Era

José Ribamar Costa Ferreira-Neto,[1,‡] *Éderson Akio Kido,*[1,‡]
Flávia Figueira Aburjaile,[2] *Manassés Daniel da Silva,*[1]
Marislane Carvalho Paz de Souza[1] and *Ana Maria Benko-Iseppon*[2,*]

Introduction

Plants are sessile beings. This way, they cannot escape from unfavorable situations. To combat or adapt to environmental oscillations, whether of biotic or abiotic kind, they have developed, through their evolution, the ability to synthesize a plethora of molecules. These important biological factors are primarily used to maintain/regulate the plant homeostasis.

Man has had close relationship with plants from immemorial times. This led to the discovery of many important medicinal applications. The following statements are some outstanding pharmacological insights:

- *'Let food be your medicine, and medicine be your food'* (Hippocrates; Greek physician; father of the medicine);
- *'There are no incurable diseases—only the lack of will. There are no worthless herbs—only the lack of knowledge'* (Avicenna; Persian polymath; wrote treatises on a variety of subjects, including medicine).

From the quotes of Avicenna and Hippocrates, two famous scientists in ancient times, it is evident that the therapeutic power of plants refers to the ancestry of modern humans. About 60% of the plants have medicinal use in post-Neolithic human history (Hao 2019). Fossils also revealed people from Mesopotamia (now Iraq; about 60.000 years ago) used a medicinal plant named Hollyhock (*Alcea rosea*) (Cowan 1999), this was, perhaps, the first recorded strategy used by man against illnesses. Plants are the largest biochemical and pharmaceutical 'stores' on Earth, and these 'living stores' help in understanding a myriad of biochemical compounds. The medicinal importance of plants has economic consequences.

[1] Laboratory of Molecular Genetics, Center of Biosciences, Genetics Department, Universidade Federal de Pernambuco, Av. Prof. Moraes Rego, 1235, Recife - PE, 50670-901, Brazil.
[2] Laboratory of Plant Genetics and Biotechnology, Genetics Department, Universidade Federal de Pernambuco, Av. Prof. Moraes Rego, 1235, Recife - PE, 50670-901, Brazil.
‡ Equal contribution
* Corresponding author: ana.iseppon@gmail.com

According to the British Chambers of Commerce, the market of plant-derived drugs and botanicals is projected to US$ 35.4 billion in 2020 (Sharma et al. 2019).

The medicinal essence from plants emanates from the versatility of compounds related to its secondary metabolism, besides its protein/peptides (Fig. 1.1). Secondary Metabolites (SMs) exist in plants with great structural diversity. It is generally accepted that SMs are made for a specific purpose. They are often found to be species-specific, and could be dispensable under many situations during a plant's life. In natural environments, they might be essential for plant survival and reproduction. Concerning the classes of SMs, many have medical impacts, acting as an anti-inflammatory, antioxidant or even anti-carcinogenic, while others have antibacterial, antifungal or antiviral activities. The SMs subset that are capable of restricting the spread of pathogens is also vast and diverse, being considered as 'Plant Antimicrobial Agents' (or PAAs). Didactically, PAAs derived from SMs (SM-PAAs) are classified (Tiku 2020) according to their presence in the plant (Fig. 1.1) in:

i. Phytoanticipins, which are defined as low-molecular-weight defense-related compounds present in plants even before the pathogen attack or produced from preexisting precursors (VanEtten et al. 1994);
ii. Phytoalexins, which are antimicrobial compounds, requiring *de novo* expression of enzymes related to biosynthetic pathways, induced only after pathogen elicitation (Bailey and Mansfield 1982). Phytoalexin production requires mechanisms of transcription and translation once the pathogen has been detected. Their up-regulation also involves the processing (trafficking and secretion) of phytoalexins at the infection site (Bednarek and Osbourn 2009).

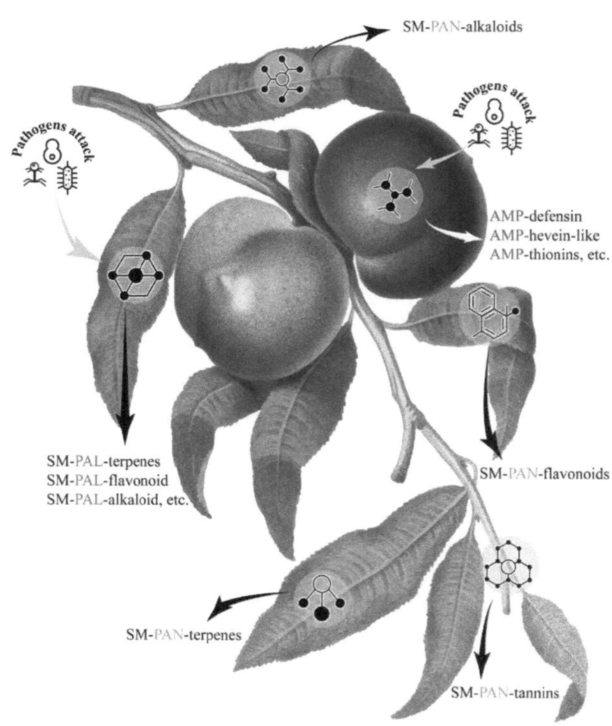

Figure 1.1. Scheme showing a hypothetical plant and the main classes of secondary metabolites and peptides (with anti-pathogenic activity), distributed in different plant tissues. Legend: SM (Secondary Metabolites); PAN (phytoanticipins); PAL (phytoalexins); AMP (antimicrobial peptides).

Additionally, some SMs could be classified as phytoanticipin or phytoalexins, depending on the plant strategy (Tiku 2020). The same SM may also function as phytoalexin or phytoanticipin in the same plant in different organs (for a review, see Tiku 2020; Vidhyasekaran 2004).

Regarding the anti-pathogenic (antimicrobial) effect, an irregular distribution of the scientific reports have covered distinct SMs. Some SMs, such as flavonoids, terpenes and alkaloids, give a of information in scientific literature, while saponins, coumarins, tannins, and quinones, among others, provide reduced information.

The above mentioned SMs classes represent a small fraction of those showing PAA activities. It is worth noting that these compounds may perform other physiological functions, sometimes associated with the plant clade. Some terpenoids provide plants odors (the real cause of the smell of cannabis; Mudge et al. 2019); tannins are plant pigments (Bindon et al. 2014); terpenes are decisive for plant flavor (e.g., sesquiterpenes valencene, alpha- and beta-sinensal play an important role in the orange fruit flavor; Maccarone et al. 1998; Vora et al. 1983) and some flavonoids determine the color of the flower (Zhu et al. 2012).

In turn, antimicrobial peptides (AMPs, Fig. 1.1) represent small chains of amino acids with anti-pathogenic activities. Unlike SMs, which present several functions, AMPs act mainly during plant defense. They are naturally occurring molecules produced as a front line of defense by multicellular organisms (Tam et al. 2015). These peptides can have a broad activity to kill bacteria, fungi and viruses. They act by exerting cytotoxicity on the invading pathogenic microorganisms, and serve as immune modulators in higher organisms (Santos-Silva et al. 2020).

Plant AMPs have been isolated from several tissues (roots, seeds, flowers, stems and leaves) from a broad number of species (Nawrot et al. 2014). They comprise different classes, including thionins, defensins, lipid transfer proteins, hevein-like peptides, knottins and cyclotides. Plant AMPs and SM-PAAs are promising antimicrobial agents with important biotechnological applications, such as drug development against human pathogenic organisms.

Next detailed information will be presented about the particular classes of AMPs and SM-PAAs, their anti-pathogenic mechanisms, and some current mining methods focusing on omics.

Getting to know PAAs better: a focus on their anti-pathogenic activities and mechanisms of action

Secondary metabolites with PAA activity

The plant kingdom constitutes an enormous reservoir of bioactive molecules, and as mentioned earlier, many plant SMs classes show anti-pathogenic activities. The complete study of all these classes would require an encyclopedic informational content. Therefore only some important representatives will be scrutinized. To this end, six large known SMs groups (Fig. 1.2) were data-mined, three of them for showing the lowest number of identified scientific manuscripts were selected, aiming to fill the information gap between the groups. The data-mining was carried out in the PubMed database (https://pubmed.ncbi.nlm.nih.gov/; June 2020), using a combination of keywords: 'SM group name AND antimicrobial activity AND plants' (e.g., flavonoid AND antimicrobial activity AND plants). The results are shown in Fig. 1.2, and the selected groups comprised coumarins, saponins and tannins.

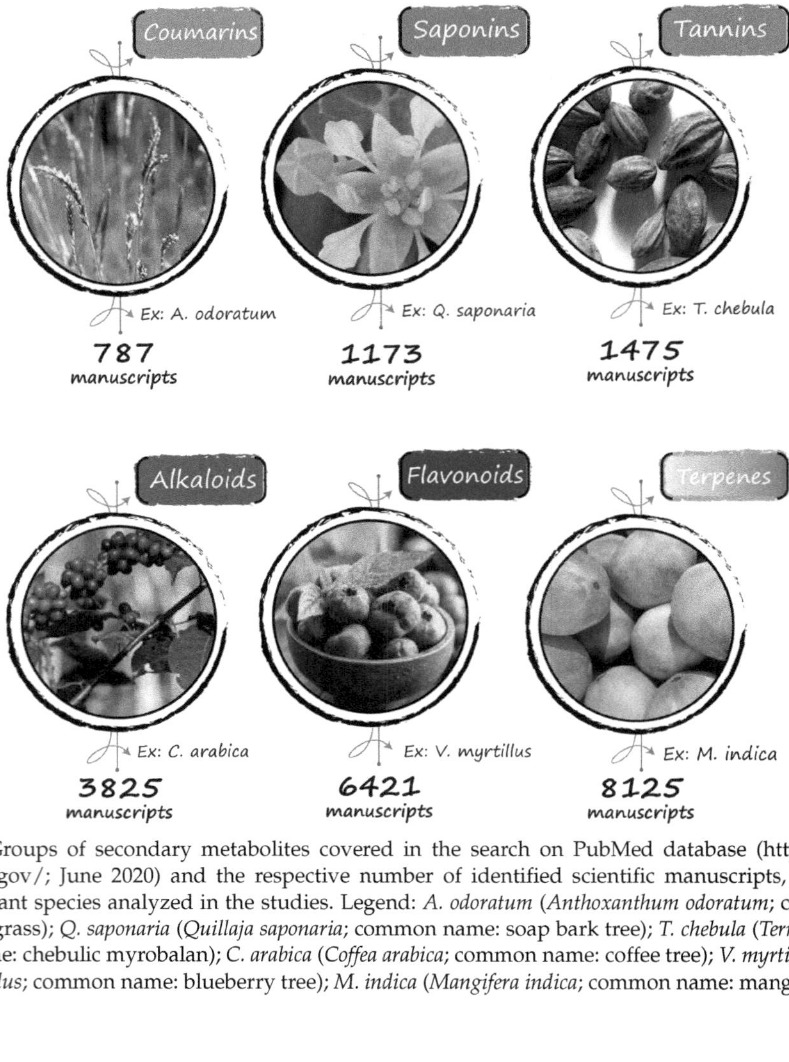

Figure 1.2. Groups of secondary metabolites covered in the search on PubMed database (https://pubmed.ncbi.nlm.nih.gov/; June 2020) and the respective number of identified scientific manuscripts, together with exemplary plant species analyzed in the studies. Legend: *A. odoratum* (*Anthoxanthum odoratum*; common name: sweet vernalgrass); *Q. saponaria* (*Quillaja saponaria*; common name: soap bark tree); *T. chebula* (*Terminalia chebula*; common name: chebulic myrobalan); *C. arabica* (*Coffea arabica*; common name: coffee tree); *V. myrtillus* (*Vaccinium myrtillus*; common name: blueberry tree); *M. indica* (*Mangifera indica*; common name: mango tree).

Coumarins

Coumarins, from the benzopyrones family (1,2-benzopyrones or 2H-1-benzopyran-2-ones), are widely distributed in natural environments. They present a benzene ring linked to a pyrone ring, and have been synthesized by the phenylpropanoid pathway (Bourgaud et al. 2006). The structural core of the coumarins is 2H-1-benzopyran-2-one or "fundamental coumarin". Based on modifications of this core (degree of oxygenation), coumarins are classified into simple coumarins, furocoumarins, dihydrofurocoumarins, pyranocoumarins (linear and angular), phenylcoumarins and biscoumarins (Borges et al. 2012; Venugopala et al. 2013).

The coumarin designation is after the species *Coumarouna odorata* (nowadays *Dipteryx odorata*), whose compound was initially isolated by Vogel in 1820 (Soine 1964). Coumarins are found in both mono- and dicots, with the richest sources being members of Rutaceae and Apiaceae families (Jain and Josh 2012; Stringlis et al. 2019), but also observed in Asteraceae, Leguminosae, Oleaceae, Moraceae, Thymelaeaceae and Poaceae (e.g., *Anthoxanthum odoratum*; Fig. 1.2) (Stringlis et al. 2019). While distributed throughout all parts of the plant, coumarins occur at the highest levels in fruits and flowers, followed by roots, stems and leaves (Jain and Josh 2012; Cuellar and Miranda 2001).

Coumarins have been tested *in vitro* for their activity against pathogens. Garcia et al. (1995) reported that scopoletin (an *Hevea brasiliensis* coumarin) showed *in vitro* antifungal activity against *Colletotrichum gloeosporioides*, *Corynespora cassiicola* and *Microcyclus ulei*. Antibacterial activity of coumarin was also reported. For instance, Tan et al. (2017) showed that root coumarins of *Prangos hulusii* exhibited antimicrobial activity against nine standards and six clinically isolated bacterial strains at concentrations ranging from five to 125 µg/mL. Other examples of coumarin antimicrobial activities are reported by Al-Majedy et al. (2017).

The antimicrobial activity of coumarins is dependent on the number and the polarity of the oxygen substituent(s) in the benzopyrone ring. Regarding scopoletin, the presence of a methoxy (-O-CH3) and a hydroxy group (-OH) in the benzopyrone ring explains its toxicity. According to Kayser and Kolodziej (1999), the structural requirements of simple coumarins for biological activity (antibacterial) against Gram-positive (e.g., *Staphylococcus aureus*, beta-hemolytic Streptococcus and *Streptococcus pneumoniae*) and Gram-negative bacteria (e.g., *Escherichia coli*, *Klebsiella pneumoniae*, *Pseudomonas aeruginosa*, *Proteus mirabilis* and *Haemophilus influenzae*), showed a methoxy function at C-7, and, if present, a hydroxyl group at either the C-6 or C-8 position, are invariably effective against the set of Gram-negative bacteria, and also *S. aureus* (Gram-positive). The presence of an aromatic dimethoxy arrangement, in turn, was favorable against beta-hemolytic *S. pneumoniae* and *H. influenza*. The authors suggested combining these structural features (observed in highly oxygenated coumarins) to identify promising candidates with broad-spectrum antibacterial activity.

Coumarins are also highlighted, concerning antiviral activity. There is much evidence showing their action against viruses, including the etiological agents of dengue, chikungunya and HIV (for a review, see Hassan et al. 2016; Mishra et al. 2020). Mechanisms in the antiviral activity comprise: inhibition of proteins essential for viral entry, replication, and infection; interference of NF-κB and Tat functions; anti-oxidative pathway, including NrF-2 (the nuclear factor erythroid 2 (NFE2)-related factor 2), among others (Hassan et al. 2016; Mishra et al. 2020).

Saponins

The term 'saponin' is derived from the Latin word 'sapo', which means soap. This is associated with foam formation when saponin-containing plants in an aqueous solution are agitated. Saponins are compounds of amphipathic nature (i.e., presenting a combination of polar and non-polar structural elements), which explains their soap-like behavior in water (Vincken et al. 2007). They are structurally diverse molecules and are classified as triterpene and steroid glycosides. The first group has three methyl groups removed (they have 27 C-atoms), whereas, in the second, all the 30 C-atoms are preserved (Vincken et al. 2007).

Saponins have been reported in more than 100 plant families (Kregiel et al. 2017). They are observed in seeds, roots, leaves and flowers (Kregiel et al. 2017). The primary sources of saponins in the human diet are legumes, mainly broad beans, kidney beans and lentils (Lásztity et al. 1998). The main plant sources are soap bark tree (*Quillaja saponaria*; Fig. 1.2), mojave yucca (*Yucca schidigera*) and ginseng (*Panax species*), among others which are concerned with medicinal use.

The results could be variable, when considering the saponin anti-pathogenic activity. Depending on the pathogen, this SM-PAA class may present higher or lesser antimicrobial properties. According to Kregiel et al. (2017), the saponins' antibacterial activity is often weak, while more significant antifungal activity has been observed. However, there are also good results against bacteria. Sonfack et al. (2019) reported a saponin isolated from roots of *Albizia adianthifolia* (Leguminosae), inhibiting the growth of 11 gram-negative multi-drug resistant bacterial strains tested. The results showed MICs (Minimum Inhibitory Concentrations) ranging from 16 to 128 µg/mL.

In turn, antifungal activities are widely reported (Barile et al. 2007; Porsche et al. 2017), and the antibiosis mechanism is clarified. The primary mode of saponin antifungal action involves pore formation and loss of membrane integrity, which cause fungal cell death. The deleterious effect is associated with a complex formation between saponin and ergosterol, the major fungal membrane sterol (Weete 1989; Bonanomi et al. 2009). As ergosterol is not present in plants, these complexes do not form, so phytotoxicity is avoided in plant cells.

The antiviral activities are also reported for saponins. Analyzing the extract of *Q. saponaria*, a complex mixture of triterpenoids (saponins), Roner et al. (2007) observed significant antiviral activity [against vaccinia virus, herpes simplex virus type 1, varicella-zoster virus, Human Immunodeficiency Viruses 1 and 2 (HIV-1, HIV-2), and reovirus], using concentrations more than 100-fold lower than those for cell cytotoxicity. The authors demonstrated that microgram amounts of the mentioned extract, while exhibiting no cell cytotoxicity or direct virucidal activity, prevented each of the six viruses tested from infecting their host cells. Several other reports are also available in the literature (Gosse et al. 2002; Pu et al. 2015). The potential action mechanism involved the inhibition of the virus nucleic acid synthesis or targeted the viral capsid protein synthesis (Simões et al. 1999).

Tannins

Tannins are a heterogeneous group of water-soluble, astringent polyphenolic compounds, classified into three main categories: hydrolysable, condensed and complex tannins. These SM-PAAs are thought to be partially responsible for the astringent taste of red wines and tea (Vidal et al. 2004; Bandyopadhyay et al. 2012). In plants, tannins were observed in seeds, fruits, flowers, leaves, roots and tree bark samples (Sieniawska and Baj 2017). The role of tannins is associated with plant defense against predation by herbivores and insects, resulting primarily in plant palatability reduction, particularly in young leaves (Barbehenn and Constabel 2011). In roots, tannins are chemical barriers against penetration and colonization by pathogens, while in seeds, they present bactericidal properties (Constabel et al. 2014).

Tannins exert biological effects according to their characteristics (Serrano et al. 2009), as follows:

i. As being unabsorbable, usually associated with tannins of complex structures and binding properties, which may produce local effects in the gastrointestinal tract (antioxidant, radical scavenging, antimicrobial, antiviral, antimutagenic and antinutritional effects);
ii. As being absorbable, usual tannins with low molecular weight structures, are easily absorbed and produce systemic effects in several organs.

From another point of view, the biological effects of tannins result from three main processes (Haslam 1996):

i. by chelation of metal ions;
ii. through antioxidant and reactive species scavenging activities;
iii. alternatively, through their ability to complex with other molecules, including macromolecules, such as protein and polysaccharides.

According to Sieniawska and Baj (2017), among the known mechanisms related to antimicrobial (antibacterial or antifungal) activities, some are highlighted: inhibition of extracellular microbial enzymes; deprivation of the substrates required for microbial growth; or direct action on microbial metabolism through inhibition of oxidative phosphorylation. Concerning the antiviral activity, Lin et al. (2013) observed that tannins (chebulagic acid and punicalagin) isolated from the fruits of *Terminalia chebula* Retz. (Fig. 1.2) were able to block attachment, penetration and the spread of human cytomegalovirus, hepatitis C virus and measles virus. The suggested mechanism implicated tannins mediating their antiviral activities by targeting HSV-1 viral glycoproteins that interact with cell surface heparin sulfate.

Peptides with PAAs activity

In a similar strategy applied to selected SMs groups reviewed in this work, a data-mining approach also allowed to select plant AMPs families. Thus, the keywords thionin, defensin, hevein-like, knottin, lipid transfer protein and snakin were data-mined in the PubMed database (https://pubmed.ncbi.nlm.nih.gov/; June 2020), following the combination: 'AMP group name AND plants' (e.g., defensin AND plants). The data-mining results are shown in Fig. 1.3. In general, manuscripts reporting AMPs were fewer than those reporting

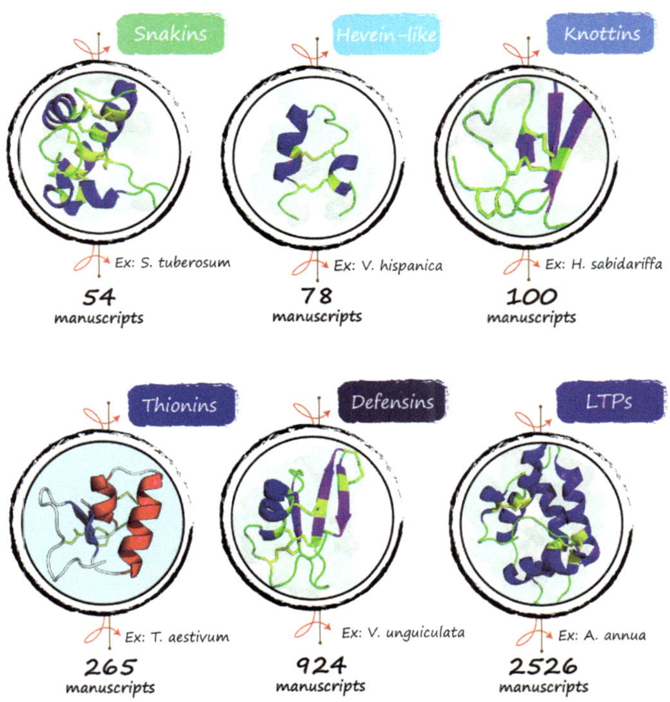

Figure 1.3. Schemes of the three-dimensional structure* of the antimicrobial peptide families searched in the PubMed database (https://pubmed.ncbi.nlm.nih.gov/; June 2020), with the respective number of scientific manuscripts identified, and some representative plant species. *All the three-dimensional structures of antimicrobial peptides, except for thionins (public domain), have been adapted from Finkina et al. (2019). Legend: *S. tuberosum* (*Solanum tuberosum*; common name: potato); *V. hispanica* (*Vaccaria hispanica*; common name: bladder soapwort); *H. sabidariffa* (*Hibiscus sabdariffa*; common name: roselle); *T. aestivum* (*Triticum aestivum*; common name: wheat); *V. unguiculata* (*Vigna unguiculata*; common name: cowpea); *A. annua* (*Artemisia annua*; common name: weet wormwood).

SM-PAAs, and the three AMP groups showing less information available were snakins, hevein-like and knottins. Before giving more details covering the selected AMP families, some general aspects of AMPs are presented.

General aspects of plant AMPs: a briefing guide

Usually, eukaryotes present AMPs, and are found in potentially all plants. They figure among the basic mechanisms of plant defense against pathogens, including bacteria, fungi and viruses (Nawrot et al. 2014). Every plant organ can also produce a complex blend of AMPs (Barbosa Pelegrini et al. 2011; Tam et al. 2015). Such peptides present a wide variation on their primary structure (the amino acid sequences), even when comparing representatives of the same class.

The AMPs classification is associated with aspects of its tridimensional structure (Campos et al. 2018), resulting from the high content of cysteine residues, whose number and arrangements are often conserved in each AMP group. The cysteine residues enable forming multiple disulfide bonds (usually two to six) that contribute to a compact fold and structural rigidity. Additionally, AMPs are characterized by their small size (15–100 amino acid residues) and typically by an overall positive charge (Nawrot et al. 2014). The mentioned characteristics confer AMPs high thermal, chemical and proteolytic stability that is crucial for these peptides to act as a chemical shield against multiple pathogens and some pests (Iqbal et al. 2019).

The mode of action of AMPs against pathogenic microbes is based on electrostatic interaction between the cationic nature of AMPs (i.e., positively charged molecules) and the anionic (i.e., negatively charged) cell surfaces of bacteria and fungi (Nawrot et al. 2014), provoking pores in membranes. Several models have been proposed to explain the pore formation: barrel-stave model, toroidal pore and carpet model (for further details, see Shai 1999).

Snakins

The snakin family (Fig. 1.3) figures among the less studied AMPs groups. These peptides have been described in literature as plant-exclusive. The term 'snakin' reflects their sequence similarity with snake hemotoxins (Berrocal-Lobo et al. 2002). Snakin AMPs were first described in potato (*Solanum tuberosum*; Fig. 1.3), showing 12 conserved cysteine residues, forming six disulfide bridges (Segura et al. 1999). These AMPs were also described in several other species, such as alfalfa (García et al. 2014), *Peltophorum dubium* (Rodríguez-Decuadro et al. 2018; 2020), and tomato (Herbel and Wink 2016), among others. Their expression was observed in plant tissues of inflorescences, tubers, leaves, stems, roots (Oliveira-Lima et al. 2017) and seedlings (Rodríguez-Decuadro et al. 2020).

Snakins can act against some fungi and bacteria at micromolar concentrations, as observed for snakin SN1 performance in potato plants (Segura et al. 1999), and Snakin-Z, from *Zizyphus jujube* (Daneshmand et al. 2013). Besides these results, only a few studies covered the snakin mechanism of action against pathogens. Herbel and Wink (2016) studied a Snakin-2 (SN2) from tomato (*Solanum lycopersicum*), showing a broad-spectrum antimicrobial activity against bacteria and fungi. The authors proposed the mode of action of SN2 regarding the specificity of the membrane, based on the cationic character of SN2, which allows diffusion through cell walls negatively charged from bacteria or fungi. Once it reaches the cell membrane, the formation of pores begins.

Hevein-like

The first representative of the hevein-like family was identified from the rubber tree (*Hevea brasiliensis*; Archer 1960), so the term 'hevein' was genus-derived. After that, the antifungal activity was attributed to a chitin-binding domain present in the peptide structure (Van Parijs et al. 1991). Thus, each peptide showing a similar domain is classified as a hevein-like peptide (Fig. 1.3). These peptides present three to five disulfide bridges, and consequently, six to ten cysteine residues (Slavokhotova et al. 2017).

Around 40 hevein-like peptides have been characterized from 14 plant species, covering 10 different plant families, including Amaranthaceae, Ginkgoaceae, Poaceae, and Moringaceae (Wong et al. 2017). Transcriptomic studies report the accumulation of hevein-like transcripts in leaves, stems and latex, but not in roots (Broekaert et al. 1990; Games et al. 2016).

Although heveins (and hevein-like peptides) have attracted the interest of scientists, with a growing number of studies in recent years, their mode of action remains unclear. They are reported to have antimicrobial activity against pathogens rich in chitin, such as fungi. These microorganisms may bind to the structural components of the chitin-binding site present on heveins (Slavokhotova et al. 2017). However, heveins can affect microorganisms devoid of these polysaccharides since their action against oomycetes and bacteria has also been reported (de Souza Cândido et al. 2014).

Knottins

Knottins are shorter AMPs in size (less than 50 amino acids in the mature form; Fig. 1.3), but with the most diverse functionally, according to Phazang et al. (2020). The knottin tridimensional structure contains six conserved cysteines in the C-terminal region, forming three disulfide bonds intertwined, giving rise to a unique structural scaffold (Molesini et al. 2017). They were identified in monocots (e.g., Poaceae) and dicots, as Solanaceae, Fabaceae, Rubiaceae, Violaceae, and Cucurbitaceae (Molesini et al. 2017; Phazang et al. 2020). They were also detected in seeds (He et al. 2013), leaves (Nguyen et al. 2014), fruit (Hass and Hermodson 1981) and flowers (Molesini et al. 2017).

Reported anti-pathogenic activities were associated with bacteria and fungi (Molesini et al. 2017). Concerning these activities, knottin-type peptides act as α-amylase or protease (carboxypeptidase A or trypsin) inhibitors (Konarev et al. 2002), metallo-carboxypeptidase inhibitor, and Serine protease inhibitor 1 (Molesini et al. 2017).

Brief contextualization on omics and bioinformatics

One of the most critical features in molecular genetics is the ability of DNA sequencing. At the beginning of the DNA sequencing era, the applied methodologies involved high cost/million bases and a long time/run necessary to obtain good quality sequence data. These factors became almost impossible to study large genomes. With the emergence of Second-Generation Sequencing (SGS; previously known as Next Generation Sequencing or NGS) technologies, this scenario started to be modified.

The SGS term describes platforms generating a huge amount (even billions of reads) of short nucleotide sequences (25 to 400 bp), increasing the high-throughput production, and reducing the cost/million bases by orders of magnitude (Mardis 2008). The pyrosequencing system developed by 454 Life Sciences, after acquired by Roche®, was the first successful NGS/SGS platforms. Other successful SGS platforms comprise the Illumina® Genome Analyzer (GA) II/IIx, the Applied Biosystems SOLiD™ (Sequencing by Oligo Ligation

Detection), the Helicos HeliScope™, among others (Mardis 2008). Further optimization has led to innovative third-generation (long-read sequencing) platforms as single-molecule real-time sequencing by PacBio®, nanopore sequencing, etc. (Bleidorn 2015; Xiao and Zhou 2020). The sequencing of entire genomes gives rise to a new genetics sub area, the genomics.

The term 'genomics' was proposed in 1986 by Thomas H. Roderick (Kuska 1998) to cover the mapping, sequencing and analysis of whole genomes. The suffix "omic" is currently placed after epithets of areas that already exist in life sciences or technologies, indicating the realization of global or high-scale (high throughput) approaches. In addition to genomics, transcriptomics (the study of all transcripts produced by an organism, tissue/organ, or even a cell at a given time, and under a specific stimulus or condition) has also taken advantage of SGS and third generation technologies. Additionally, the development of proteomics technologies, which globally study a set of proteins produced at a given time, under a specific stimulus in a specific tissue, completed the global analysis of the Central Dogma of Molecular Biology' biomolecules. Together, genomics, transcriptomics and proteomics are considered the first omics. However, there are more than a dozen omics approaches in the biology field of research, including:

i. metabolomics: the comprehensive analysis of metabolites in a biological specimen (Clish 2015);
ii. epigenomics: the systematic analysis of the global state of gene expression not attributable to mutational changes in the underlying DNA genome (Kalavacharla et al. 2017);
iii. phenomics: the acquisition of high-dimensional (multivariate) phenotypic data on an organism-wide scale (Houle et al. 2010).

As omics comprise large-scale generation and characterization of data (Argueso et al. 2019), sufficient computational power is necessary to extract the biological meaning from these 'large data', depending on the bioinformatics tools. Thus, as knowledge is central to life, bioinformatics is central to omics.

The knowledge generated by integrating omics data provides a better understanding of the organism's complexity and how they use their genetic resources in developmental, survival and adaptation processes. This also allows evolutionary and phylogenetic inferences and development in several areas, including agriculture, biotechnology, medicine, environment, among others (for a review, see VanEmon 2016; Tebani et al. 2016; Tanveer et al. 2018).

Exploiting plant biodiversity and genetic potential for PAAs mining in the omics era: the state of art

The plant genome encodes many fascinating features: their evolutionary history (plant origin and evolution), physiological potential (stress tolerance/resistance) and the ability to produce specific compounds, besides all the information necessary for development, survival and adaptation to new environmental conditions. The stated elements are the basis and foundation for deciphering genomic diversity and chemo-diversity at the molecular level, especially concerning PAAs. The application of high-throughput omics and bioinformatics approaches to plants omics is enlightening on defense elements and their regulation, playing a major role in the breeding of productive medicinal cultivars and transgenic crops.

Although the importance of plant omics is currently recognized, antimicrobial compounds and peptides omics are still scarcely explored. The performed data-mining using the combined keywords—'omics AND antimicrobial compounds AND plants' or 'omics AND antimicrobial peptides AND plants'—in the PubMed or Google Scholar databases (https://scholar.google.com.br/; June 2020), presented results indirectly related to the topic. Most of the identified manuscripts displayed the plant antimicrobial compounds and peptides acting inside the pathogens, which are their targets in microbial cells, or, approached how pathogens try to curb the PAAs' action. Thus, considering available information, the focus was the pathogenic organism, not the host plant. Examples of highly qualified and informative works on this subject can be seen in Dos Santos et al. (2016) and Chernov et al. (2019). These authors highlighted the advances in omics technologies (genomics, transcriptomics, proteomics and metabolomics) and bioinformatics, providing important insights into processes related to bacterial physiology, virulence, stress and the mechanisms of action of antimicrobial compounds/peptides derived from plants. The purpose of this chapter is to provide information on plants and their respective PAAs. Additionally, it approaches omics data and bioinformatic tools to identify plant genes, transcripts, proteins and metabolites with anti-pathogenic activity. Thus it can be divided into two branches: First, AMPs; then, SMs of interest. This stems from the fact that the strategies for AMPs mining are extremely different from those implemented for the analysis of SMs-synthesizing enzymes.

Antimicrobial peptides analysis by omics approaches

The recent increase in data derived from the sequencing of plant genomes and transcriptomes provides an opportunity for *in silico* prediction of novel AMPs. Due to the scarcity of new antibiotics and the increasing pathogen resistance to antibiotics currently in use, prospecting plant AMPs is a valuable new source to find antimicrobial molecules.

AMP prediction

From sequence-derived data, different strategies can be used for AMP mining. According to Santos-Silva et al. (2020), three basic approaches have been commonly used for AMPs identification: (i) local alignments, especially by using Basic Local Alignment Search Tool (BLAST); and by searching for specific patterns using (ii) REGEX (REgular EXpressions) or (iii) Hidden Markov Models (HMMs). For the first one, due to the high variability of AMPs, only a few highly conserved sequences can be identified using this type of inference.

REGEX and HMM strategies are used more effectively. The REGEX is a precise way of describing a pattern in a string where each regular expression position must be set, although ambiguous characters (or wildcards) can also be used. The HMMs are well-known for their effectiveness in modeling the correlations between adjacent symbols, domains or events, and they have been extensively employed in various fields of biological analysis.

As claimed by Santos-Junior et al. (2020), two significant computational difficulties affect AMPs prospection in sequence-derived data: (i) predicting small genes in DNA/RNA sequences; (ii) the prediction of AMP activity for small genes. Regarding the prediction of small genes in DNA- or RNA-encoding small peptides, most gene prediction approaches exclude small ORFs (Open Reading Frames). The regular tools preserve the longest ORF, conventionalizing it as the coding for a given protein. Thus, specific peptide mining approaches have been developed (e.g., Ramada et al. 2017; Lin et al. 2019). Additionally, a few recent surveys have shown that long ORFs identification methods can be used to detect small ORFs. In this way, the results are subsequently filtered while revealing that small ORFs are biologically active across a range of functions (Miravet-Verde et al. 2019; Sberro et al. 2019).

Plant AMP mining in sequence-derived data is a process currently in active progress. Several other computation strategies, besides those mentioned above, are available. Bhadra et al. (2018) proposed an accurate computational method for AMPs prediction by the random forest algorithm. The prediction model is based on the distribution patterns of amino acid properties along the sequence. Fallah Atanaki et al. (2020) developed a tool (BIPEP) for sequence-based prediction of biofilm inhibitory peptides based on combinations of descriptors, such as the Composition, Transition and Distribution (CTD) of different characteristics, besides the newly introduced NMR (Nuclear Magnetic Resonance spectroscopy)-based features, to map peptide sequences to numeric feature vectors as the input for the prediction method. Sharma et al. (2016) used machine learning tools to build six Support Vector Machine (SVM) and Weka-based models trained on 80 biofilm-active AMPs to identify anti-biofilm peptides based on their whole amino acid composition, selected residue features, and the positional preference of the residues.

While Yan et al. (2020) presented a short-length (≤ 30 amino acids) AMPs prediction method, Deep-AmPEP30, developed based on an optimal feature set of PseKRAAC reduced amino acids composition and convolutional neural network. This tool was capable of mining a powered peptide of 20 amino acids (P3, FWELWKFLKSLWSIFPRRRP) in the *Candida glabrata* genome. The potential AMP identified presents an antibiotic potential comparable to that of ampicillin. According to the authors, the Deep-AmPEP30 is a promising prediction tool to identify short-length AMPs from genomic sequences, aimed for drug discovery. Among other approaches, some authors also use efficient but less automatized analyses, as performed by Umadevi et al. (2018).

Recently, Shelenkov et al. (2020) reported a substantial effort for AMP prospection on plant sequence-derived data. The authors searched for AMPs in 1267 plant transcriptomes (downloaded from 1 kP website) using a Cysmotif searcher pipeline (Shelenkov et al. 2018). The pipeline performs cysteine motif searching in translated nucleotide sequences using regular expressions and then submit the identified motifs to several filtration steps to confirm that they are not by-products. On an average, 50–150 peptides were revealed in each transcriptome. According to Shelenkov et al. (2020), the datasets obtained will be useful for molecular biologists investigating AMPs in different organisms and for bioinformaticians developing novel algorithms of motif searching in nucleotides encoding peptide sequences.

From metabolites to the gene by omics approaches

Omics provides a powerful data source for understanding secondary metabolism. Approaches (genomics, transcriptomics, proteomics, metabolomics, etc.), datasets, bioinformatics tools for analysis, interpretation and data integration led to the comprehension of biosynthetic pathways of several SM-PAAs. When the metabolite and the pathway are known, the next step is to search the related genes, the substrate for biotechnological manipulations.

In the context 'from metabolites to the gene', medicinal plants are preferred targets for omics studies and SMs prospection due to the heterogeneity of its chemicals and pharmacological compounds and their beneficial effects (anti-inflammatory, anti-carcinogenic, etc.). Although omics studies covered secondary plant metabolism, only a few of them focused on the dynamics of SMs with recognized PAA' attributes. Such a gap is a valuable opportunity to study and bring the scientific community to work in this area. Despite valuable omics data available, the main challenges for studying secondary metabolism in plants involve the complexity of the plant genome, its large size, genetic redundancy, cellular compartmentalization and multilayered regulatory processes others (Rai et al. 2017).

Genomic information is an important resource for interpreting the biosynthetic pathways and the singularities of particular metabolic pathways across plant clades. Boutanaev et al. (2015) studied the diversification of terpene (a compound with antibiotic effects) in several plant genomes by performing a systematic genomic approach, dissecting genetic components of secondary metabolism, focusing on Terpenoid Synthases (TS) and cytochrome P450 (CYPs). They found evidence for distinct mechanisms of pathway assembly in eudicots and monocots. In eudicots, microsyntenic blocks of TS/CYP gene pairs were duplicates, thus providing templates for the evolution of new pathways. In monocots, new pathways emerged by mixing and matching individual TS and CYP genes through dynamic genome rearrangements. That was the first documented observation of TS and CYP assembly's unique pattern in eudicots and monocots.

Current studies on mining the genomes of diverse plant species have also shown Metabolic Gene Clusters (MGC) associated with secondary metabolites biosynthesis. Consequently, the involved genes were not randomly placed on plant chromosomes. MGC, a set of two or more non-homologous genes encoding enzymes from the same pathway, is a common prokaryotic genome characteristic. Considering the plant genomics studies, over 20 MGC has been reported (Nutzmann and Osbourn 2014; Nutzmann et al. 2016), covering the synthesis of compounds with antimicrobial activity, including terpenes (in *Solanum lycopersicum*; Falara et al. 2011; Matsuba et al. 2013), avenacin (in *Avena strigose*; Qi et al. 2004; Mylona et al. 2008), and cucurbitacin (in *Cucumis sativus*; Shang et al. 2014; Boutanaev et al. 2015). The above mentioned works applied genomic resources to gain insight into the biosynthesis and evolution of SMs, and they are crucial for the discovery of new SM-PAAs.

Functional genomics, such as transcriptomics, proteomics and metabolomics, also provide a powerful tool for understanding secondary metabolism. The *Medicago truncatula* transcriptome, providing ESTs (Expressed Sequence Tags) from a variety of plant tissues under (a)biotic treatments (Suzuki et al. 2002), allowed transcripts identification for three putative enzymes (squalene synthase, squalene epoxidase and beta-amyrin synthase) involved in the early steps of triterpene saponin biosynthesis.

Moreover, the co expression analysis of transcripts/genes allows a better understanding of unknown elements implicated in metabolic pathways. If an unknown gene is co expressed with a known gene component of a particular metabolic pathway, the unknown gene is probably connected to that pathway (Tohge et al. 2005; Wisecaver et al. 2017; Tai et al. 2018). Applying the transcriptome co expression analysis combined with a reverse genetics approach, Yonekura-Sakakibara et al. (2007) identified the gene encoding Flavonol 7-O-rhamnosyltransferase, a prominent enzyme involved in the flavonoid composition of Arabidopsis.

Studying the Triterpenic Acid (TA) metabolism in the mint family (Lamiaceae), Aminfar et al. (2019) conducted a gene expression tracking by transcriptome mining, alongwith with a co expression network analysis and tissue-specific metabolite-expression analysis, to identify potential genes involved in TAs biosynthetic pathways. Based on the results, genes encoding squalene epoxidase and oxidosqualene cyclases were proposed for boosting triterpene production. Some TAs present an anti-HIV activity (Cassels and Asensio 2010). According to Wisecaver et al. (2017), global gene co expression networks are a largely unexplored valuable resource for discovering the genetic basis and architecture of natural plant products.

Research efforts to identify candidate proteins/enzymes implicated in SMs biosynthesis have been successfully applied to proteomics data. Protein-based analysis tools have evolved very quickly in recent years. Some of the methods and references related to SMs, include Differential In-Gel Electrophoresis (DIGE; Champagne et al. 2012), Label-Free Protein Quantification (Oldham et al. 2010), and Isobaric Tags for Relative and Absolute Quantitation (iTRAQ; Martínez-Esteso et al. 2013).

Champagne et al. (2012) applied in-depth proteome analysis (2D-DIGE) on *Catharanthus roseus*. They identified 63 enzymes potentially implicated in secondary metabolism, 22 of them involved in terpenoid indole alkaloid biosynthesis and 16 predicted transporters associated with SMs. This plant species is considered a biochemical factory, producing more than 100 distinct terpenoid indole alkaloids, some of them showing powerful and important pharmacological attributes, including antimicrobial activity (Almagro et al. 2015).

Zhan et al. (2016) performed combined methods (RNA-Seq and iTRAQ analysis) to identify potential enzymes involved in the triterpenoid saponins biosynthetic pathway of *Anemone flaccida*. Among the annotated candidates, four were related to genes from the cytochrome P450 CYP716A subfamily, one of them encoding a key enzyme in the oleanane-type triterpenoid saponins biosynthetic pathway.

In turn, metabolomics has also been actively used to understand secondary plant metabolism and the discovery of SM-PAAs. French et al. (2018), using targeted and untargeted metabolite profiling by liquid-chromatography mass spectrometry, analyzed the metabolite profiles of 17 plant species, including wild plants (*Filipendula ulmaria*, *Prunella vulgaris*, *Centuarea nigra*, and *Rhinanthus minor*), two forage legumes (*Medicago sativa*, *Trifolium hybridium*) and 45 biomass samples from agricultural grasslands in the UK. The authors identified a range of SMs, including 32 compounds with known SM-PAA activities (antimicrobial/anthelmintic properties). The identified SM-PAAs targeted multiple aspects of pathogens metabolism and physiology, *in vitro* and *in vivo*, including inhibition of quorum sensing in bacteria and egg viability in nematodes. The most abundant bioactive compounds were benzoic acid, myricetin, p-coumaric acid, rhamnetin and rosmarinic acid.

Fudyma et al. (2019) generated a metabolite profiling of *Sphagnum fallax* using two kinds of mass spectrometry systems, 1H nuclear magnetic resonance spectroscopy and lipidome characterization by LC-MS/MS (liquid chromatography-mass spectrometry). They detected 655 metabolites, and 17 of them considered potential antimicrobial compounds. The metabolite profiling also highlighted acid-like and flavonoid glycoside compounds, two possible antimicrobial compounds.

Conclusions and perspectives

Plants and their compounds represent an extraordinary source of agents showing antimicrobial activity. Such organisms are traditional sources or substrates used in popular medicine. These days, the majority of pharmaceutically bioactive molecules are isolated from plants because their chemical synthesis is sometimes profitless. Omics and bioinformatics are currently the methods of choice for understanding plants' physiological dynamics, helping to identify relevant candidates for biotechnological applications. Thus, adopting new procedures/routines for PAAs mining is very necessary, meaningful and applicable.

The peptide nature of AMPs is fully associated with the omics strategies and the employment of bioinformatics tools. Thus, computational approaches, such as machine learning and random forest algorithms, have been applied, aiming to identify AMPs and their encoding genes. Concerning SM-PAAs, although SMs, in general, have significant omics data available, the specific SM-PAAs compounds do not have such abundant omics data. Such a gap represents a valuable opportunity to develop scientific works in this area.

Currently, the progress resulting from the omics applications leads to the integration of data. Such efforts already began involving several issues, and this is also the future for PAAs. It is desirable to integrate the different omics technologies to better understand biological systems and to uncover promising candidates. Therefore, this multidimensional integrated approach offers new roads for better comprehension of subjects regarding SM-PAAs and AMPs, attempting the exciting discovery of new genes.

Acknowledgments

The authors acknowledge the financial support of the Fundação de Amparo à Pesquisa do Estado de Pernambuco (FACEPE; grant number: BCT-0013-5.01/17), Conselho Nacional de Desenvolvimento Científico Tecnológico (CNPq) and Coordenação de Aperfeiçoamento de Pessoal de Nível Superior (CAPES) for fellowships and grants.

References

Almagro, L., Fernández-Pérez, F., Pedreño, M.A. (2015). Indole alkaloids from *Catharanthus roseus*: bioproduction and their effect on human health. Molecules 20: 2973–3000.

Al-Majedy, Y., Kadhum, A., Al-Amiery, A., Mohamad, A.B. (2017). Coumarins: the antimicrobial agents. Sys Rev Pharm 8: 62–70.

Aminfar, Z., Rabiei, B., Tohidfar, M., Mirjalili, M.H. (2019). Identification of key genes involved in the biosynthesis of triterpenic acids in the mint family. Sci Rep 9: 15826.

Archer, B.L. (1960). The proteins of *Hevea brasiliensis* latex: isolation and characterization of crystalline hevein. Biochem J 75: 236–240.

Argueso, C.T., Assmann, S.M., Birnbaum, K.D., Chen, S., Dinneny, J.R., Doherty, C.J., Eveland, A.L., Friesner, J., Greenlee, V.R., Law, J.A., Marshall-Colón, A., Mason, G.A., O'Lexy, R., Peck, S.C., Schmitz, R.J., Song, L., Stern, D., Varagona, M.J., Walley, J.W., Williams, C.M. (2019). Directions for research and training in plant omics: big questions and big data. Plant Direct 3: e00133.

Bailey, J.A., Mansfield, J.W. (1982). Phytoalexins. Wiley, New York.

Bandyopadhyay, P., Ghosh, A.K., Ghosh, C. (2012). Recent developments on polyphenol–protein interactions: effects on tea and coffee taste, antioxidant properties and the digestive system. Food Funct 3: 592–605.

Barbehenn, R.V. and Constabel, C.P. (2011). Tannins in plant–herbivore interactions. Phytochemistry 72: 1551–1565.

Barbosa Pelegrini, P., del Sarto, R.P., Silva, O.N., Franco, O.L., Grossi-de-Sa, M.F. (2011). Antibacterial peptides from plants: what they are and how they probably work. *In*: Biochem Res Int https://www.hindawi.com/journals/bri/2011/250349/. Accessed 10 Jun 2020.

Barile, E., Bonanomi, G., Antignani, V., Zolfaghari, B., Sajjadi, S.E., Scala, F., Lanzotti, V. (2007). Saponins from *Allium minutiflorum* with antifungal activity. Phytochemistry 68: 596–603.

Bednarek, P., Osbourn, A. (2009). Plant-microbe interactions: chemical diversity in plant defense. Science 324: 746–748.

Berrocal-Lobo, M., Segura, A., Moreno, M., López, G., García-Olmedo, F., Molina, A. (2002). Snakin-2, an antimicrobial peptide from potato whose gene is locally induced by wounding and responds to pathogen infection. Plant Physiol 128: 951–961.

Bhadra, P., Yan, J., Li, J., Fong, S., Siu, S.W.I. (2018). AmPEP: Sequence-based prediction of antimicrobial peptides using distribution patterns of amino acid properties and random forest. Sci Rep 8: 1697–1707.

Bindon, K., Kassara, S., Hayasaka, Y., Schulkin, A., Smith, P. (2014). Properties of wine polymeric pigments formed from anthocyanin and tannins differing in size distribution and subunit composition. J Agric Food Chem 62: 11582–11593.

Bleidorn, C. (2016). Third generation sequencing: technology and its potential impact on evolutionary biodiversity research. System Biodivers 14: 1–8.

Bonanomi, G., Vinale, F., Scala, F. (2009). The role of natural products in plant-microbe interactions. pp. 301–320. *In*: Osbourn, A.E., Lanzotti, V. (eds.). Plant-derived Natural Products: Synthesis, Function, and Application. Springer US, New York, NY.

Borges, M.F.M., Roleira, F.M.F., Milhazes, N.J.S.P., Villare, E.U. (2012). Simple coumarins: privileged scaffolds in medicinal chemistry. pp. 23–85. *In*: Atta-ur-Rahman, B., Reitz, A., Iqbal Choudhary, M. (eds.). Frontiers in Medicinal Chemistry - (Volume 4). Bentham Science Publishers.

Bourgaud, F., Hehn, A., Larbat, R., Doerper, S., Gontier, E., Kellner, S., Matern, U. (2006). Biosynthesis of coumarins in plants: a major pathway still to be unravelled for cytochrome P450 enzymes. Phytochem Rev 5: 293–308.

Boutanaev, A.M., Moses, T., Zi, J., Nelson, D.R., Mugford, S.T., Peters, R.J., Osbourn, A. (2015). Investigation of terpene diversification across multiple sequenced plant genomes. Proc Natl Acad Sci USA 112: E81–88.

Broekaert, I., Lee, H.I., Kush, A., Chua, N.H., Raikhel, N. (1990). Wound-induced accumulation of mRNA containing a hevein sequence in laticifers of rubber tree (*Hevea brasiliensis*). Proc Natl Acad Sci USA 87: 7633–7637.

Campos, M.L., Lião, L.M., Alves, E.S.F., Migliolo, L., Dias, S.C., Franco, O.L. (2018). A structural perspective of plant antimicrobial peptides. Biochem J 475: 3359–3375.

Cassels, B.K., Asencio, M. (2011). Anti-HIV activity of natural triterpenoids and hemisynthetic derivatives 2004–2009. Phytochem Rev 10: 545–564.

Champagne, A., Rischer, H., Oksman-Caldentey, K.-M., Boutry, M. (2012). In-depth proteome mining of cultured *Catharanthus roseus* cells identifies candidate proteins involved in the synthesis and transport of secondary metabolites. Proteomics 12: 3536–3547.

Chernov, V.M., Chernova, O.A., Mouzykantov, A.A., Lopukhov, L.L., Aminov, R.I. (2019). Omics of antimicrobials and antimicrobial resistance. Expert Opin Drug Discov 14: 455–468.

Clish, C.B. (2015). Metabolomics: an emerging but powerful tool for precision medicine. Cold Spring Harb Mol Case Stud 1: a000588.

Constabel, C.P., Yoshida, K., Walker, V. (2014). Diverse ecological roles of plant tannins: plant defense and beyond. Rec Adv Polyphen Res 4: 115–142.

Cowan, M.M. (1999). Plant products as antimicrobial agents. Clin Microbiol Rev 12: 564–582.

Cuellar, A., Miranda, M. (2001). Farmacognosia y productos naturales. La Habana: Editorial Félix Varela. pp. 207–213.

Daneshmand, F., Zare-Zardini, H., Ebrahimi, L. (2013). Investigation of the antimicrobial activities of Snakin-Z, a new cationic peptide derived from *Zizyphus jujuba* fruits. Nat Prod Res 27: 2292–2296.

de Souza Cândido, E., e Silva Cardoso, M.H., Sousa, D.A., Viana, J.C., de Oliveira-Júnior, N.G., Miranda, V., Franco, O.L. (2014). The use of versatile plant antimicrobial peptides in agribusiness and human health. Peptides 55: 65–78.

dos Santos, B.S., da Silva, L.C.N., da Silva, T.D., Rodrigues, J.F.S., Grisotto, M.A.G., Correia, M.T. dos S., Napoleão, T.H., da Silva, M.V., Paiva, P.M.G. (2016). Application of omics technologies for evaluation of antibacterial mechanisms of action of plant-derived products. Front Microbiol 7: 1466–1479.

Falara, V., Akhtar, T.A., Nguyen, T.T.H., Spyropoulou, E.A., Bleeker, P.M., Schauvinhold, I., Matsuba, Y., Bonini, M.E., Schilmiller, A.L., Last, R.L., Schuurink, R.C., Pichersky, E. (2011). The tomato terpene synthase gene family. Plant Physiol 157: 770–789.

Fallah Atanaki, F., Behrouzi, S., Ariaeenejad, S., Boroomand, A., Kavousi, K. (2020). BIPEP: Sequence-based prediction of biofilm inhibitory peptides using a combination of NMR and physicochemical descriptors. ACS Omega 5: 7290–7297.

Finkina, E.I., Melnikova, D.N., Bogdanov, I.V., Ovchinnikova, T.V. (2019). Peptides of the innate immune system of plants. Part I. structure, biological activity, and mechanisms of action. Russ J Bioorg Chem 44: 573–585.

French, K.E., Harvey, J., McCullagh, J.S.O. (2018). Targeted and untargeted metabolic profiling of wild grassland plants identifies antibiotic and anthelmintic compounds targeting pathogen physiology. Metabolism and Reproduction. Sci Rep 8: 1695–1705.

Fudyma, J.D., Lyon, J., AminiTabrizi, R., Gieschen, H., Chu, R.K., Hoyt, D.W., Kyle, J.E., Toyoda, J., Tolic, N., Heyman, H.M., Hess, N.J., Metz, T.O., Tfaily, M.M. (2019). Untargeted metabolomic profiling of *Sphagnum fallax* reveals novel antimicrobial metabolites. Plant Direct 3: e00179.

Games, P.D., da Silva, E.Q.G., Barbosa, M. de O., Almeida-Souza, H.O., Fontes, P.P., de Magalhães-Jr, M.J., Pereira, P.R.G., Prates, M.V., Franco, G.R., Faria-Campos, A., Campos, S.V.A., Baracat-Pereira, M.C. (2016). Computer aided identification of a Hevein-like antimicrobial peptide of bell pepper leaves for biotechnological use. BMC Genomics 17: 999–1012.

García, A.N., Ayub, N.D., Fox, A.R., Gómez, M.C., Diéguez, M.J., Pagano, E.M., Berini, C.A., Muschietti, J.P., Soto, G. (2014). Alfalfa snakin-1prevents fungal colonization and probably coevolved with rhizobia. BMC Plant Biol 14: 248.

Garcia, D., Sanier, C., Macheix, J.J., D'Auzac, J. (1995). Accumulation of scopoletin in *Hevea brasiliensis* infected by *Microcyclus ulei* (P. Henn.) V. ARX and evaluation of its fungitoxicity for three leaf pathogens of rubber tree. Physiol Mol Plant Path 47: 213–223.

Gosse, B., Gnabre, J., Bates, R.B., Dicus, C.W., Nakkiew, P., Huang, R.C.C. (2002). Antiviral saponins from *Tieghemella heckelii*. J Nat Prod 65: 1942–1944.

Hao, D.-C. (2019). Chapter 1 - Genomics and evolution of medicinal plants. pp. 1–33. *In*: Hao, D.-C. (ed.). Ranunculales Medicinal Plants. Academic Press.

Haslam, E. (1996). Natural polyphenols (vegetable tannins) as drugs: possible modes of action. J Nat Prod 59: 205–215.

Hass, G.M., Hermodson, M.A. (1981). Amino acid sequence of a carboxypeptidase inhibitor from tomato fruit. Biochemistry 20: 2256–2260.

Hassan Mohd, Z., Osman, H., Ali, M.A., Ahsan, M.J. (2016). Therapeutic potential of coumarins as antiviral agents. Eur J Med Chem 123: 236–255.

He, W.-J., Chan, L.Y., Clark, R.J., Tang, J., Zeng, G.-Z., Franco, O.L., Cantacessi, C., Craik, D.J., Daly, N.L., Tan, N.-H. (2013). Novel inhibitor cystine knot peptides from *Momordica charantia*. PLoS One 8: e75334.

Herbel, V., Wink, M. (2016). Mode of action and membrane specificity of the antimicrobial peptide snakin-2. PeerJ 4: e1987.

Houle, D., Govindaraju, D.R., Omholt, S. (2010). Phenomics: the next challenge. Nat Rev Genet 11: 855–866.

Iqbal, A., Khan, R.S., Shehryar, K., Imran, A., Ali, F., Attia, S., Shah, S., Mii, M. (2019). Antimicrobial peptides as effective tools for enhanced disease resistance in plants. Plant Cell Tiss Organ Cult 139: 1–15.

Jain, P.K., Joshi, H. (2012). Coumarin: chemical and pharmacological profile. J App Pharm Sci 2: 236–240.

Kalavacharla, V. (Kal), Subramani, M., Ayyappan, V., Dworkin, M.C., Hayford, R.K. (2017). Chapter 16 - Plant epigenomics. pp. 245–258. *In*: Tollefsbol, T.O. (ed.). Handbook of Epigenetics (Second Edition). Academic Press.

Kayser, O., Kolodziej, H. (1999). Antibacterial activity of simple coumarins: structural requirements for biological activity. Z Naturforsch, C, J Biosci 54: 169–174.

Konarev, A.V., Anisimova, I.N., Gavrilova, V.A., Vachrusheva, T.E., Konechnaya, G.Y., Lewis, M., Shewry, P.R. (2002). Serine proteinase inhibitors in the Compositae: distribution, polymorphism and properties. Phytochemistry 59: 279–291.

Kregiel, D., Berlowska, J., Witonska, I., Antolak, H., Proestos, C., Babic, M., Babic, L., Zhang, B. (2017). Saponin-based, biological-active surfactants from plants. Application and Characterization of Surfactants. DOI: 10.5772/68062.

Kuska, B. (1998). Beer, Bethesda, and biology: how "genomics" came into being. J Natl Cancer Inst 90: 93.

Lásztity, R., Hidvégi, M., Bata, Á. (1998). Saponins in food. Food Rev Int 14: 371–390.

Lin, L.-T., Chen, T.-Y., Lin, S.-C., Chung, C.-Y., Lin, T.-C., Wang, G.-H., Anderson, R., Lin, C.-C., Richardson, C.D. (2013). Broad-spectrum antiviral activity of chebulagic acid and punicalagin against viruses that use glycosaminoglycans for entry. BMC Microbiol 13: 187–202.

Lin, Y., Cai, Y., Liu, J., Lin, C., Liu, X. (2019). An advanced approach to identify antimicrobial peptides and their function types for penaeus through machine learning strategies. BMC Bioinformatics 20: 291–301.

Maccarone, E., Campisi, S., Fallico, B., Rapisarda, P., Sgarlata, R. (1998). Flavor components of italian orange juices. J Agric Food Chem 46: 2293–2298.

Mardis, E.R. (2008). Next-generation DNA sequencing methods. Annu Rev Genomics Hum Genet 9: 387–402.

Martínez-Esteso, M.J., Vilella-Antón, M.T., Pedreño, M.Á., Valero, M.L., Bru-Martínez, R. (2013). iTRAQ-based protein profiling provides insights into the central metabolism changes driving grape berry development and ripening. BMC Plant Biol 13: 167–187.

Matsuba, Y., Nguyen, T.T.H., Wiegert, K., Falara, V., Gonzales-Vigil, E., Leong, B., Schäfer, P., Kudrna, D., Wing, R.A., Bolger, A.M., Usadel, B., Tissier, A., Fernie, A.R., Barry, C.S., Pichersky, E. (2013). Evolution of a complex locus for terpene biosynthesis in Solanum. Plant Cell 25: 2022–2036.

Miravet-Verde, S., Ferrar, T., Espadas-García, G., Mazzolini, R., Gharrab, A., Sabido, E., Serrano, L., Lluch-Senar, M. (2019). Unraveling the hidden universe of small proteins in bacterial genomes. Mol Syst Biol 15: e8290.

Mishra, S., Pandey, A., Manvati, S. (2020). Coumarin: an emerging antiviral agent. Heliyon 6: e03217.

Molesini, B., Treggiari, D., Dalbeni, A., Minuz, P., Pandolfini, T. (2017). Plant cystine-knot peptides: pharmacological perspectives. Br J Clin Pharmacol 83: 63–70.

Mudge, E.M., Brown, P.N., Murch, S.J. (2019). The terroir of cannabis: terpene metabolomics as a tool to understand *Cannabis sativa* selections. Planta Med 85: 781–796.

Mylona, P., Owatworakit, A., Papadopoulou, K., Jenner, H., Qin, B., Findlay, K., Hill, L., Qi, X., Bakht, S., Melton, R., Osbourn, A. (2008). Sad3 and Sad4 are required for saponin biosynthesis and root development in oat. Plant Cell 20: 201–212.

Nawrot, R., Barylski, J., Nowicki, G., Broniarczyk, J., Buchwald, W., Goździcka-Józefiak, A. (2014). Plant antimicrobial peptides. Folia Microbiol (Praha) 59: 181–196.

Nguyen, P.Q.T., Wang, S., Kumar, A., Yap, L.J., Luu, T.T., Lescar, J., Tam, J.P. (2014). Discovery and characterization of pseudocyclic cystine-knot α-amylase inhibitors with high resistance to heat and proteolytic degradation. The FEBS Journal 281: 4351–4366.

Nutzmann, H.-W., Osbourn, A. (2014). Gene clustering in plant specialized metabolism. Curr Opin Biotechnol 26: 91–99.

Nutzmann, H.-W., Huang, A., Osbourn, A. (2016). Plant metabolic clusters—from genetics to genomics. The New Phytol 211: 771–789.

Oldham, J.T., Hincapie, M., Rejtar, T., Wall, P.K., Carlson, J.E., Lee-Parsons, C.W.T. (2010). Shotgun proteomic analysis of yeast-elicited California poppy (*Eschscholzia californica*) suspension cultures producing enhanced levels of benzophenanthridine alkaloids. J Proteome Res 9: 4337–4345.

Oliveira-Lima, M., Benko-Iseppon, A.M., Neto, J.R.C.F., Rodriguez-Decuadro, S., Kido, E.A., Crovella, S., Pandolfi, V. (2017). Snakin: structure, roles and applications of a plant antimicrobial peptide. Curr Protein Pept Sci 18: 368–374.

Phazang, P., Negi, N.P., Raina, M., Kumar, D. (2020). Plant antimicrobial peptides: next-generation bioactive molecules for plant protection. pp. 281–293. *In*: Kumar, M., Kumar, V., Prasad, R. (eds.). Phyto-microbiome in Stress Regulation. Springer, Singapore.

Porsche, F.M., Molitor, D., Beyer, M., Charton, S., André, C., Kollar, A. (2017). Antifungal activity of saponins from the fruit pericarp of *Sapindus mukorossi* against *Venturia inaequalis* and *Botrytis cinerea*. Plant Dis 102: 991–1000.

Pu, X., Ren, J., Ma, X., Liu, L., Yu, S., Li, X., Li, H. (2015). Polyphylla saponin I has antiviral activity against influenza A virus. Int J Clin Exp Med 8: 18963–18971.

Qi, X., Bakht, S., Leggett, M., Maxwell, C., Melton, R., Osbourn, A. (2004). A gene cluster for secondary metabolism in oat: implications for the evolution of metabolic diversity in plants. Proc Natl Acad Sci USA 101: 8233–8238.

Rai, A., Saito, K., Yamazaki, M. (2017). Integrated omics analysis of specialized metabolism in medicinal plants. Plant J 90: 764–787.

Ramada, M.H.S., Brand, G.D., Abrão, F.Y., Oliveira, M., Filho, J.L.C., Galbieri, R., Gramacho, K.P., Prates, M.V., Bloch, C. (2017). Encrypted antimicrobial peptides from plant proteins. Sci Rep 7: 13263.

Rodríguez-Decuadro, S., Barraco-Vega, M., Dans, P.D., Pandolfi, V., Benko-Iseppon, A.M., Cecchetto, G. (2018). Antimicrobial and structural insights of a new snakin-like peptide isolated from *Peltophorum dubium* (Fabaceae). Amino Acids 50: 1245–1259.

Rodríguez-Decuadro, S., Radío, S., da Rosa, G., Veja, M.B., Benko-Iseppon, A.M., Smircich, P., Cecchetto, G. (2020). Antimicrobial peptides in the seedling transcriptome of the tree legume *Peltophorum dubium*. Biochimie 180: 229–242.

Roner, M.R., Sprayberry, J., Spinks, M., Dhanji, S. (2007). Antiviral activity obtained from aqueous extracts of the Chilean soapbark tree (*Quillaja saponaria* Molina). J Gen Virol 88: 275–285.

Santos-Junior, C.D., Pan, S., Zhao, X.-M., Coelho, L.P. (2020). MACREL: antimicrobial peptide screening in genomes and metagenomes. bioRxiv 2019.12.17.880385.

Santos-Silva, C.A. dos, Zupin, L., Oliveira-Lima, M., Vilela, L.M.B., Bezerra-Neto, J.P., Ferreira-Neto, J.R., Ferreira, J.D.C., Oliveira-Silva, R.L. de, Pires, C. de J., Aburjaile, F.F., Oliveira, M.F. de, Kido, E.A., Crovella, S., Benko-Iseppon, A.M. (2020). Plant antimicrobial peptides: state of the art, *in silico* prediction and perspectives in the omics era. Bioinform Biol Insights 14: 1177932220952739.

Sberro, H., Fremin, B.J., Zlitni, S., Edfors, F., Greenfield, N., Snyder, M.P., Pavlopoulos, G.A., Kyrpides, N.C., Bhatt, A.S. (2019). Large-scale analyses of human microbiomes reveal thousands of small, novel genes. Cell 178: 1245–1259.e14.

Segura, A., Moreno, M., Madueño, F., Molina, A., García-Olmedo, F. (1999). Snakin-1, a peptide from potato that is active against plant pathogens. Mol Plant Microbe Interact 12: 16–23.

Serrano, J., Puupponen-Pimiä, R., Dauer, A., Aura, A.-M., Saura-Calixto, F. (2009). Tannins: current knowledge of food sources, intake, bioavailability and biological effects. Mol Nutr Food Res 53 Suppl 2: S310–329.

Shai, Y. (1999). Mechanism of the binding, insertion and destabilization of phospholipid bilayer membranes by α-helical antimicrobial and cell non-selective membrane-lytic peptides. Biochimica et Biophysica Acta (BBA) - Biomembranes 1462: 55–70.

Shang, Y., Ma, Y., Zhou, Y., Zhang, H., Duan, L., Chen, H., Zeng, J., Zhou, Q., Wang, S., Gu, W., Liu, M., Ren, J., Gu, X., Zhang, S., Wang, Y., Yasukawa, K., Bouwmeester, H.J., Qi, X., Zhang, Z., Lucas, W.J., Huang, S. (2014). Plant science. Biosynthesis, regulation, and domestication of bitterness in cucumber. Science 346: 1084–1088.

Sharma, A., Gupta, P., Kumar, R., Bhardwaj, A. (2016). dPABBs: A novel *in silico* approach for predicting and designing anti-biofilm peptides. Sci Rep 6: 21839.

Sharma, M., Koul, A., Sharma, D., Kaul, S., Swamy, M.K., Dhar, M.K. (2019). Metabolic engineering strategies for enhancing the production of bio-active compounds from medicinal plants. pp. 287–316. *In*: Akhtar, M.S., Swamy, M.K. (eds.). Natural Bio-Active Compounds: Volume 3: Biotechnology, Bioengineering, and Molecular Approaches. Springer, Singapore.

Shelenkov, A., Slavokhotova, A., Odintsova, T. (2020). Predicting antimicrobial and other cysteine-rich peptides in 1267 plant transcriptomes. Antibiotics (Basel) 9: 60–68.

Shelenkov, A.A., Slavokhotova, A.A., Odintsova, T.I. (2018). Cysmotif searcher pipeline for antimicrobial peptide identification in plant transcriptomes. Biochemistry Mosc 83: 1424–1432.

Sieniawska, E., Baj, T. (2017). Chapter 10 - Tannins. pp. 199–232. *In*: Badal, S., Delgoda, R. (eds.). Pharmacognosy. Academic Press, Boston.

Simões, C.M.O., Amoros, M., Girre, L. (1999). Mechanism of antiviral activity of triterpenoid saponins. Phytother Res 13: 323–328.

Slavokhotova, A.A., Shelenkov, A.A., Andreev, Y.A., Odintsova, T.I. (2017). Hevein-Like antimicrobial peptides of plants. Biochemistry Mosc 82: 1659–1674.

Soine, T.O. (1964). Naturally occurring coumarins and related physiological activities. J Pharm Sci 53: 231–264.

Sonfack, G., Tchinda, C.F., Simo, I.K., Bitchagno, G.T.M., Nganou, B.K., Çelik, İ., Tene, M., Görkem, S.F., Opatz, T., Beng, V.P., Kuete, V., Tane, P. (2019). Saponin with antibacterial activity from the roots of *Albizia adianthifolia*. Nat Prod Res 0: 1–9.

Stringlis, I.A., de Jonge, R., Pieterse, C.M.J. (2019). The age of coumarins in plant–microbe interactions. Plant Cell Physiol 60: 1405–1419.

Suzuki, H., Achnine, L., Xu, R., Matsuda, S.P.T., Dixon, R.A. (2002). A genomics approach to the early stages of triterpene saponin biosynthesis in *Medicago truncatula*. Plant J 32: 1033–1048.

Tai, Y., Liu, C., Yu, S., Yang, H., Sun, J., Guo, C., Huang, B., Liu, Z., Yuan, Y., Xia, E., Wei, C., Wan, X. (2018). Gene coexpression network analysis reveals coordinated regulation of three characteristic secondary biosynthetic pathways in tea plant (*Camellia sinensis*). BMC Genomics 19: 616–629.

Tam, J.P., Wang, S., Wong, K.H., Tan, W.L. (2015). Antimicrobial peptides from plants. Pharmaceuticals (Basel) 8: 711–757.

Tan, N., Yazıcı-Tütüniş, S., Bilgin, M., Tan, E., Miski, M. (2017). Antibacterial activities of pyrenylated coumarins from the roots of *Prangos hulusii*. Molecules 22.

Tanveer, T., Shaheen, K., Parveen, S., Misbah, Z.T., Babar, M.M., Gul, A. (2018). Chapter 19 - Omics-based bioengineering in environmental biotechnology. pp. 353–364. *In*: Barh, D., Azevedo, V. (eds.). Omics Technologies and Bio-Engineering. Academic Press.

Tebani, A., Afonso, C., Marret, S., Bekri, S. (2016). Omics-based strategies in precision medicine: toward a paradigm shift in inborn errors of metabolism investigations. Int J Mol Sci 17.

Tiku, A.R. (2020). Antimicrobial compounds (phytoanticipins and phytoalexins) and their role in plant defense. pp. 845–868. *In*: Mérillon, J.-M., Ramawat, K.G. (eds.). Co-evolution of Secondary Metabolites. Springer International Publishing, Cham.

Tohge, T., Nishiyama, Y., Hirai, M.Y., Yano, M., Nakajima, J., Awazuhara, M., Inoue, E., Takahashi, H., Goodenowe, D.B., Kitayama, M., Noji, M., Yamazaki, M., Saito, K. (2005). Functional genomics by integrated analysis of metabolome and transcriptome of Arabidopsis plants over-expressing an MYB transcription factor. Plant J 42: 218–235.

Umadevi, P., Soumya, M., George, J.K., Anandaraj, M. (2018). Proteomics assisted profiling of antimicrobial peptide signatures from black pepper (*Piper nigrum* L.). Physiol Mol Biol Plants 24: 379–387.

Van Parijs, J., Broekaert, W.F., Goldstein, I.J., Peumans, W.J. (1991). Hevein: an antifungal protein from rubber-tree (*Hevea brasiliensis*) latex. Planta 183: 258–264.

VanEmon, J.M. (2016). The omics revolution in agricultural research. J Agric Food Chem 64: 36–44.

VanEtten, H.D., Mansfield, J.W., Bailey, J.A., Farmer, E.E. (1994). Two classes of plant antibiotics: phytoalexins versus "phytoanticipins". Plant Cell 6: 1191–1192.

Venugopala, K.N., Rashmi, V., Odhav, B. (2013). Review on natural coumarin lead compounds for their pharmacological activity. Biomed Res Int 2013: 963248.

Vidal, S., Francis, L., Noble, A., Kwiatkowski, M., Cheynier, V., Waters, E. (2004). Taste and mouth-feel properties of different types of tannin-like polyphenolic compounds and anthocyanins in wine. Anal Chim Acta 513: 57–65.

Vidhyasekaran, P. (2004). Concise Encyclopedia of Plant Pathology. CRC Press.
Vincken, J.-P., Heng, L., de Groot, A., Gruppen, H. (2007). Saponins, classification and occurrence in the plant kingdom. Phytochemistry 68: 275–297.
Vora, J.D., Matthews, R.F., Crandall, P.G., Cook, R. (1983). Preparation and chemical composition of orange oil concentrates. J of Food Sci 48: 1197–1199.
Weete, J.D. (1989). Structure and function of sterols in fungi. pp. 115–167. *In*: Paoletti, R., Kritchevsky, D. (eds.). Advances in Lipid Research. Elsevier.
Wisecaver, J.H., Borowsky, A.T., Tzin, V., Jander, G., Kliebenstein, D.J., Rokas, A. (2017). A global coexpression network approach for connecting genes to specialized metabolic pathways in plants. Plant Cell 29: 944–959.
Wong, K.H., Tan, W.L., Kini, S.G., Xiao, T., Serra, A., Sze, S.K., Tam, J.P. (2017). Vaccatides: antifungal glutamine-rich hevein-like peptides from *Vaccaria hispanica*. Front Plant Sci 8: 1100–1014.
Xiao, T., Zhou, W. (2020). The third generation sequencing: the advanced approach to genetic diseases. Transl Pediatr 9: 163–173.
Yadav, S.P. (2007). The wholeness in suffix-omics, -omes, and the word om. J Biomol Tech 18: 277.
Yan, J., Bhadra, P., Li, A., Sethiya, P., Qin, L., Tai, H.K., Wong, K.H., Siu, S.W.I. (2020). Deep-AmPEP30: improve short antimicrobial peptides prediction with deep learning. Mol Ther Nucleic Acids 20: 882–894.
Yonekura-Sakakibara, K., Tohge, T., Niida, R., Saito, K. (2007). Identification of a flavonol 7-O-rhamnosyltransferase gene determining flavonoid pattern in Arabidopsis by transcriptome coexpression analysis and reverse genetics. J Biol Chem 282: 14932–14941.
Zhan, C., Li, X., Zhao, Z., Yang, T., Wang, X., Luo, B., Zhang, Q., Hu, Y., Hu, X. (2016). Comprehensive analysis of the triterpenoid saponins biosynthetic pathway in Anemone flaccida by transcriptome and proteome profiling. Front Plant Sci 7: 1094–1106.
Zhu, M., Zheng, X., Shu, Q., Li, H., Zhong, P., Zhang, H., Xu, Y., Wang, L., Wang, Liangsheng. (2012). Relationship between the composition of flavonoids and flower colors variation in tropical water Lily (Nymphaea) cultivars. Plos ONE 7: e34335.

2

Natural Products from the Amazon Region as Potential Antimicrobials

Josiane E. A. Silva,[1] *Iasmin L. D. Paranatinga,*[1] *Elaine C. P. Oliveira,*[2] *Silvia K. S. Escher,*[3] *Ananda S. Antonio,*[4] *Leandro S. Nascimento,*[5] *Patricia P. Orlandi*[6] and *Valdir F. Veiga-Júnior*[4,5,*]

Introduction

The identification of medicinal plants and use of natural products has been in existence since the beginning of human civilization. Humans have used and benefited of the natural products of plants with therapeutic potential, even without scientific understanding of the properties. According to the World Health Organization (WHO), at present both medicinal and aromatic plants form the basis of traditional or indigenous health systems (Bauri et al. 2015; Barata et al. 2016; Dar et al. 2017; Newman and Gragg 2020) in developing countries. The biological properties of medicinal plants are attributed to their chemical composition. Some of the activities already described in literature include anti-inflammatory, antinociceptive, hypoglycemic, antispasmodic, antiulcerogenic, antitumor, antiestrogenic, cytotoxic, antimicrobial and healing properties (Azab et al. 2016; Batista et al. 2016; Ugboko et al. 2020; Piriz et al. 2014). The use of herbal medicines has been growing across the world and consequently, there have been huge advancements in the studies of the pharmacological properties of plants used for therapeutic purposes in recent years (Silva et al. 2015).

Among the different diseases that affect humans, those caused by parasites, which, in 2010, resulted in 48.4 million cases and 8.78 million deaths have been highlighted here. Neurocysticercosis, toxoplasmosis, echinococcosis, malaria and schistosomiasis are the

[1] Biological Sciences Institute, Universidade Federal do Amazonas (UFAM), Av. General Rodrigo Octavio Jordão Ramos, 1200 - Coroado I | CEP 69067-005, Manaus-AM, Brazil.
[2] Departamento de Biotecnologia, Universidade Federal do Oeste do Pará, Rua Vera Paz, s/n (Unidade Tapajós) Bairro Salé | CEP 68040-000, Santarém-PA, Brazil.
[3] Departamento de Farmácia, Universidade Federal do Oeste do Pará, Rua Vera Paz, s/n (Unidade Tapajós) Bairro Salé | CEP 68040-000, Santarém-PA, Brazil.
[4] Chemistry Department, Universidade Federal do Amazonas (UFAM), Av. General Rodrigo Octávio Jordão Ramos, 1200, Coroado | 69067-005, Manaus-AM, Brazil.
[5] Chemical Engineering Section, Military Institute of Engineering (IME), Praça General Tibúrcio, 80, Praia Vermelha, Urca | CEP 22290-270, Rio de Janeiro-RJ, Brazil.
[6] Fundação Oswaldo Cruz (FIOCRUZ), Instituto Ageu Magalhães, Campus da UFPE - Av. Prof. Moraes Rego, s/n Cidade Universitária | CEP 50670-420, Recife-PE, Brazil.
* Corresponding author: valdir.veiga@gmail.com

greatest parasitic threats to the human population worldwide (Shih and Koeller 2015; Torgeson et al. 2015). In general, the spread of diseases caused by microorganisms, such as tuberculosis, typhoid and cholera, can be directly linked to a lack of basic sanitation, which in most cases is caused by inadequate and hazardous way of the population growth, in addition to a lack of drinking water and basic practices of hygiene (Shukla et al. 2020). Plants have been the main target of studies looking for therapeutic alternatives for these diseases as they represent a large source of bioactive substances (Fig. 2.1). *In vitro* studies have described the activity of plants and their products for the treatment of microbial diseases, which have reported antibacterial and antifungal activity. Essential Oils (EOs) and plant extracts are effective in controlling the growth of various microorganisms, including bacteria, filamentous fungi and yeasts through various mechanisms of action. Plant-based antibiotics have a chemical structure that differs from that of antibiotics derived from microorganisms, being able to regulate the intermediate metabolism of pathogens, activating or blocking reactions and enzymatic synthesis or even altering the membrane structure (Nazzaro 2013; Ramos et al. 2016; Silva et al. 2018). These mechanisms can also be related to the damage caused to the principle of the cell membrane by the lipophilic components of EOs. In the case of bacteria, Gram-negative bacteria are more resistant to EOs than Gram-positive ones, due to structural differences in the cell walls. Another advantage of the use of medicinal plants as a therapeutic source is the low risk of the pathogenic microorganisms developing resistance to vegetable oils and extracts, due to the synergism of antimicrobial chemicals that act through various mechanisms. All these aspects make the use of products derived from plants more advantageous and beneficial compared to other antimicrobial agents (Cleff 2012; Dahham et al. 2015; Nazzaro 2013).

The Amazon region has a rich biodiversity. Its plants have biologically active substances and therapeutic activities well known to traditional people for the treatment of their ailments. Several studies on the chemical composition of some of these plants have already been carried out, making it possible to identify the compounds responsible for biological actions, which include terpenoids, alkaloids, phenolic compounds and coumarins. Such compounds have high selectivity and specificity in their structures (Chen et al. 2015; Silva et al. 2015; Meneguetti et al. 2016; Fernandez et al. 2020).

The plants in this region have notable medicinal and food potential, such as copaíba, açaí, bacaba, buriti, buritirana, cupuaçu murici, piaçaba braba, pupunha, tucumã and tucumanzinho. Açai, an example of an Amazonian superfruit that recently captivated the

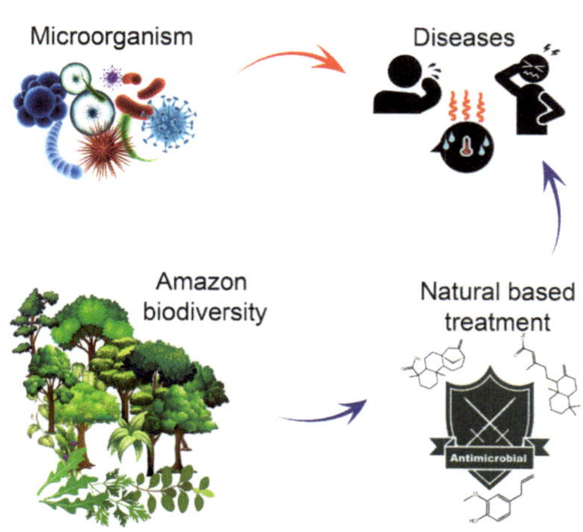

Figure 2.1. Antimicrobial activities of Amazon medicinal plants.

world, has immense bioactive potential with proven actions, such as neuroprotection, antioxidant and antimicrobial (Belda-Galbis et al. 2015; Peixoto et al. 2016; Tobouti et al. 2017; Pala et al. 2017). Fibigr et al. (2017) observed the presence of anthocyanins in açaí. According to the authors, in addition to anthocyanins, there are other bioactive substances, such as terpenes, flavonoids, phenolic compounds, lignoids, fatty acids, quinones and norisoprenoids. Several Amazonian plants are used to treat endemic diseases in the region, among them, the so-called Neglected Tropical Diseases (NTDs). Among the botanical families rich in specialized metabolites for use against these diseases, one can highlight the Anacardiaceae, Poaceae, Fabaceae, and Malpighiaceae (Lima et al. 2015), as well as bioactive alkaloids and pyrones of the *Lauraceae* family.

In Fabaceae, for example, studies performed with sesquiterpenes and diterpenes isolated from copaiba oils (*Copaifera* sp.) showed strong biological effects, whether isolated and in synergy, for leishmaniasis and the Chagas disease. However, despite the fact that the Amazon region has a great diversity of plants with biological actions, most of the species commonly used by traditional communities are poorly studied, delaying their potential use by the pharmaceutical industries (Santos et al. 2019). Understanding the possible therapeutic actions that Amazonian medicinal plants may have for the treatment of microbial diseases is fundamental for the development of new drugs.

In the largest tropical forest on the planet, the local population uses many plants to treat diseases, mainly for anti-inflammatory and antimicrobial purposes (Vargas et al. 2016; Valdivieso-Ugarte et al. 2019). Leaves, barks and roots are the main parts of the plants used for the preparation of these natural remedies, whether in the form of infusions, syrups, macerations or oils (Santos et al. 2014; Ellwanger et al. 2020). This empirical knowledge applied by the people of the Amazon rainforest is very important since infectious diseases are highly prevalent in this region and the compounds present in these botanical species are able to inhibit the development of these microorganisms (Jobim et al. 2014).

In traditional medicine, species of the genus *Lippia* are used as infusions or syrups for the treatment of bronchitis and other diseases, their EOs have excellent antimicrobial action. Species belonging to the genus *Eugenia* are also used to treat a wide variety of infectious, intestinal and gastrointestinal disorders, in addition to their use as a repellent (Fig. 2.2). Due to the high rates of leishmaniasis, the traditional Amazonian population has curated their knowledge of this disease and, for the purpose of treating this parasitosis, they typically employ the topical use of various plant extracts. *Nicotiana tabacum* is widely used for this purpose by the local population (Sarrazin et al. 2012; Odonne et al. 2017; Souza et al. 2018).

In the *Fabaceae* family, several species stand out for their traditional medicinal use and in treatment of parasitic diseases. In the *Copaifera* genus, not only are the extracts of the bark and leaves used, but also the oleoresins exudated from the trunk. Endemic species, such as *C. langsdorffii*, *C. officinalis*, *C. reticulata* and *C. multijuga* can exude more than 30 litres of resin, sought by animals to heal their wounds. The popular use of these resins includes healing, anti-inflammatory, antiseptic and anti-tumor activities, and as an agent to treat bronchitis, skin diseases, leishmaniasis, dermatoses and ulcers. "Barbatimão", *Stryphnodendron adstringens*, is another legume with traditional use for the treatment of wounds. Its most frequent use, as the name implies, is as an astringent, with these pharmacological actions confirmed scientifically, mainly due to its tannins (Leandro et al. 2012; Cavalcante et al. 2017; Ricardo et al. 2018; Mauro et al. 2019).

The monoterpenes present in the EOs of the leaves of *Lippia*, the sesquiterpenes and diterpenic acids of the oleoresins from the trunk of trees of various species of *Copaifera*, and

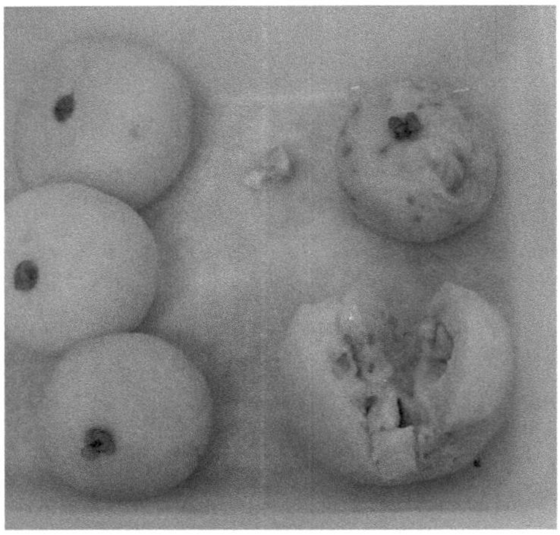

Figure 2.2. Fruits of *Eugenia stipitata*.

the tannins from the bark extracts of *Stryphnodendron* are some of the extremely bioactive substances of the Amazonian biodiversity. Traditional knowledge indicates how to use biodiversity and modern science confirms and isolates the bioactive substances.

Essential oils are natural aromatic mixtures, formed by volatile substances, that have different functions for the plants that produce them. Communication between species is the main function of the volatile constituents of plants, whether it is the attraction of pollinators, or to repel or deter predators. When the volatile constituents of plants are extracted by steam distillation techniques, it is said that an EO has been obtained. Otherwise, with the use of other technologies, such as infusion in water or maceration in organic solvents, a plant extract is obtained (sometimes, aromatic extracts). EOs are very lipophilic mixtures, often soluble in ethanol, but not in water. The lipophilic characteristic gives rise to its biological activities as it can interact with the lipid membrane of microorganisms.

The presence of EOs is quite common in angiosperms (the flowering plants) and especially in the so-called basal angiosperms. Grouped in the angiosperm phylogeny group APG IV (2016), the basal angiosperms are formed by the orders *Austrobaileyales*, *Nymphaeales*, *Amborellales* (which has only one species, in New Caledonia), *Chloranthales* and the *Magnoliids* clade. *Austrobaileyales* is made up of three families, *Austrobaileyaceae*, *Trimeniaceae*, and *Schisandraceae*. In the latter, the very common aromatic species are called Star Anise (*Illicium verum*), used as an aromatic tea with several biological properties described (Wang et al. 2011) (Fig. 2.3).

Nymphaeales consists of three botanical families: Cabombaceae, Hydatellaceae and Nymphaeaceae. The last of which has a cosmopolitan distribution and includes species such as *Nymphaea amazonum* (water lily), which has antimicrobial effects for the treatment of ulcers (Fenner et al. 2006). Phylogenetic studies showed that *Chloranthales* would be the sister order of the clade of *Magnoliids*, which brings together four orders: *Piperales*, *Canellales*, *Magnoliales* and *Laurales*. These orders contain the main botanical families with EOs, such as Piperaceae (in *Piperales*), Annonaceae and Myristicaceae (in *Magnoliales*) and Lauraceae (in *Laurales*). Avocado (*Persea americana*) is one of the most known *Lauraceae* fruits consumed worldwide, but it is not the only famous species in this family. *Laurus nobilis* (laurel) present several uses as a food ingredient, in alternative medicine, and also in ancient cultures, such as classical Greco-Roman for cultural purposes, its leaves were used to make wreaths given in honour to heroes and poets (Fig. 2.4).

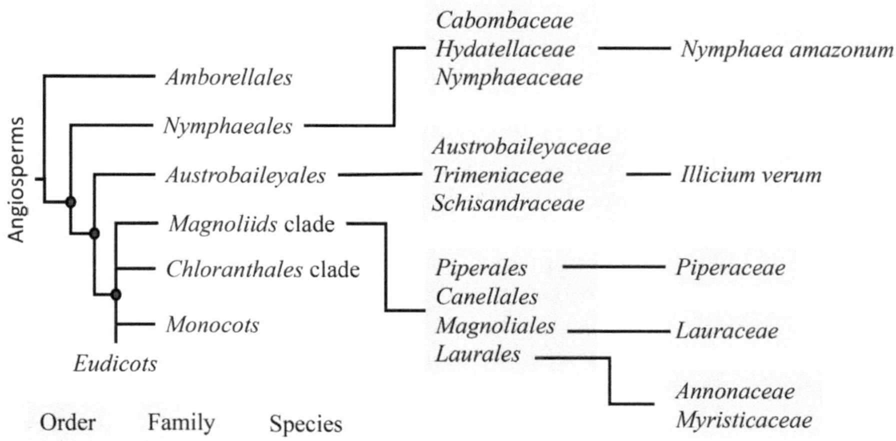

Figure 2.3. Basal angiosperms as EO producers.

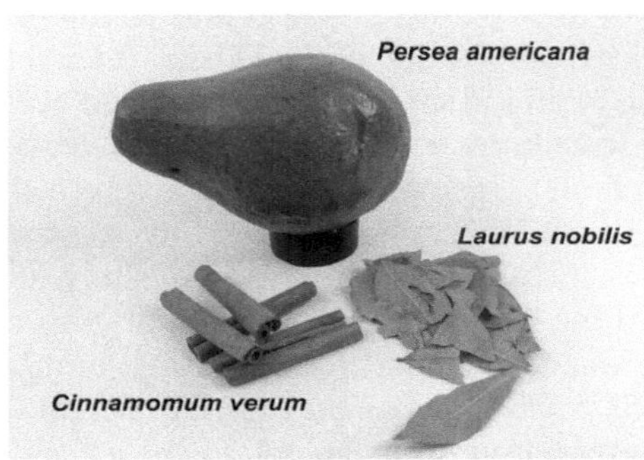

Figure 2.4. Some bioactive Lauraceae species.

The Great Age of Navigation was strongly stimulated by the search for extremely aromatic species such as cinnamon. The true cinnamon, *Cinnamomum zeylanicum* (or *Cinnamomum verum*), as well as *Cinnamomum camphora* and *Cinnamomum cassia* (Chinese cinnamon or cassia) are other Lauraceae aromatic species, which includes many others, such as the very aromatic sassafras (*Sassafras albidum*). In the Amazon region Lauraceae is not only endemic, but one of the most important and complex botanical families. *Aiouea, Endlicheria, Nectandra, Paraia, Pleurothyrium, Rhodostemonodaphne*, and *Sextonia* are some of the genera that occur together with the very common species of *Aniba* and *Ocotea* (Castro and Lorenzin 2005).

Aniba canelilla was one of the first species discovered in the Amazon region, when Spanish explorers sought to find Eldorado and valuable spices, especially cinnamon. With an aroma similar to real cinnamon (*Cinnamomum verum*), the EO of *Aniba canelilla* also contains cinnamaldehyde, but its main constituents are methyleugenol and 1-nitro-2-phenylethane, which was the first natural substance with the nitro group to be described in literature. The EO obtained from the wood of the *A. canelilla* tree showed a strong trypanocidal effect against *Trypanosoma evansi in vitro* at 0.5 to 2% concentrations (Giongo et al. 2017). Leishmanicidal (40 mg/mL) and antifungal (2 mg/mL to *Candida albicans*) activities were also observed for the EO obtained from the leaves of this species (Souza-Junior et al. 2020).

The species *Aniba parviflora* (synonymous with *Aniba fragrans*) is known in the Amazon as "macacaporanga" or laurel rose, being used in the preparation of teas, tinctures and poultices to treat persons of snake poisoning. The EOs obtained from leaves and branches of this species were tested for antimicrobial action against *Staphylococcus aureus* (ATCC-25923), *Enterococcus faecalis* (CCCD-E002), *Escherichia coli* (CCCD-E004), *Pseudomonas aeruginosa* (CCCD-P004) and *Candida albicans*. Both EOs showed similar activity in diffusion inhibition assays, with the same halos to those of the standard drug (gentamicin) for *Staphylococcus aureus* and *Enterococcus faecalis*. Using the Broth Microdilution technique, the *Aniba parviflora* EO showed a minimum inhibitory concentration (MIC) of 2.0 mg/mL for both bacteria (Batista et al. 2019). A study carried out with this species and another of the same genus, *Aniba rosaeodora*, evaluated the antimicrobial action of their EOs obtained only from the leaves, against bacteria including *Escherichia coli*, *Pseudomonas aeruginosa*, *Klebsiella pneumoniae*, *Staphylococcus aureus*, *Enterococcus faecalis*, *Streptococcus pyogenes* and *Staphylococcus epidermidis*. The results showed that the EOs were effective against these pathogenic bacteria, with the zone of inhibition values ranging from 8.8 ± 0.6 mm to 38.4 ± 1.4 mm (MIC, 1.3 to 10.0 µL/mL) for the *A. rosaeodora* EO and 9.2 ± 0.4 mm to 15.4 ± 0.9 mm for the *A. parviflora* EO. Both the bactericidal effect and its intensity were attributed to the linalool found in the chemical composition of the EOs (Sarrazin et al. 2016).

"Sacaca" was the name of an indigenous people located at the mouth of the Amazon River (Marajó Island) who were known for being very knowledgeable about the medicinal and "magical" actions of plants. This name was also given to shamanic practices, and is the popular name for an Amazonian Euphorbiaceae: *Croton cajucara*. This native shrub has several medicinal properties, including slimming and dyslipidemia actions. In the study of the antimicrobial activity of the EO from the leaves of varieties of this species, oils rich in 7-hydroxy-calamenene were active against methicillin-resistant *Mycobacterium tuberculosis*, *Mycobacterium smegmatis*, *Staphylococcus aureus* (*MRSA*), *Mucor circinelloides* and *Rhizopus oryzae*. The MICs of the oils were determined using the broth dilution assay, with results of 4.88 µg/mL for *M. tuberculosis*, 39.06 µg/mL for *M. smegmatis*, 4.76 ng/mL for *MRSA*, and 0.152 µg/mL for *R. oryzae*.

Oleoresins are materials that are naturally exuded from the tree trunk. Unlike EOs, they are not extracted by distillation methods, nor are they made up only of volatile substances. Oleoresins are always mixtures of liquid and light oily substances (10 to 15 carbons) that solubilize heavier resinous substances, containing 20 or 30 carbons. One of the most well-known oleoresins is that of pine trees (*Pinus* sp.), from which the pitch (the heavy part, with substances that have 20 carbons) and turpentine (lighter, composed of substances such as alpha-pinene, with 10 carbons) are obtained. Incense and myrrh are also oleoresins but with a large portion containing 30 carbons, all belonging to the Burseraceae botanical family. In the Amazon this family is quite common, particularly species of the genus *Protium*, but also other genera with fewer number of species, such as *Crepidospermum*, *Dacryodes*, *Tetragastris*, and *Trattinickia*. The antimicrobial action of EOs obtained from oleoresins of the genus *Protium* was studied against the fungi *Candida albicans* and *Aspergillus niger* and the bacteria *Staphylococcus aureus* and *Escherichia coli*. Oils rich in p-cymene (21.9 to 51.9%) showed a better inhibitory action against fungi (*Aspergillus niger* and *Candida albicans* at 2.5 mg/mL) compared to the results against the tested bacteria (*Staphylococcus aureus* and *Escherichia coli*), with MIC values ≥ 2.5 mg/mL (Silva et al. 2013).

Copaiba oils (*Copaifera* sp. - Fabaceae) are oleoresins with hundreds of pharmacological studies already carried out seeking to confirm and give scientific basis for traditional uses throughout Latin America, especially in the Amazon. Composed of sesquiterpene (the light constituents, with 15 carbons) and diterpenes (the resinous constituents, with 20 carbons), copaiba oils are sources of important bioactive molecules, such as beta-caryophyllene, an anti-inflammatory substance for which action on cannabinoid receptors has been recently identified. Among the several species of the genus *Copaifera* that produces this oleoresin, copaiba oils obtained from the most important species *Copaifera martii*, *C. officinalis*, *C. cearensis*, *C. multijuga*, *C. langsdorffii*, *C. lucens*, *C. paupera* and *C. reticulata* were chemically studied and their antimicrobial activities were evaluated against Gram-positive and Gram-negative bacteria, yeast, and dermatophytes (Almeida-Silva et al. 2020; Cavalcante et al. 2017; Ricardo et al. 2018). Copaiba oils showed no activity against Gram-negative bacteria and yeast and only moderate activity against dermatophyte fungi (*Trichophyton rubrum* and *Microsporum canis*). They showed strong bactericidal activity, decreasing the viability of the Gram-positive bacteria within 3 hours. The most active copaiba oils were those from *C. martii*, *C. officinalis* and *C. reticulata*, that were active against MRSA, *Staphylococcus aureus*, *Staphylococcus epidermidis*, *Bacillus subtilis* and *Enterococcus faecalis* (MIC from 31.3–62.5 µg/mL). The oleoresin from *C. martii* was studied further to identify the mechanism of action in *S. aureus*. Scanning electron microscopy revealed lysis of the bacteria, causing cellular agglomerates. Through transmission electron microscopy it was possible to identify a rupture and damage to the cell wall, which resulted in the release of cytoplasmic components. Other effects such as alterations in morphology, and a decrease in cell volume, were observed, indicating that copaiba oil may affect the cell wall (Santos et al. 2008). All these results show the strong effect that lipophilic substances can have on the cell walls of microorganism, one of the main mechanisms of action of EOs.

While oleoresins are specific and rare natural mixtures and EOs can only be obtained (industrially) by hydrodistillation, there are a large number of different ways to prepare organic extracts. In general, extracts are produced with organic solvents, but there are two main exceptions: extracts obtained in water and those obtained with fluids in their supercritical state. Thus, when it comes to the solvent, substances that have an affinity with the molecules of the plant to be extracted are usually applied. Water is often used with pH adjustments when the desired purpose is to extract acidic or basic substances. Hexane, ether, ethyl acetate, acetone, methanol and ethanol are the most common extracting solvents. Among the solvents that were frequent a few decades ago, but which are no longer used today because of their toxicity, are the aromatics (benzene and toluene) and those containing chlorine (carbon tetrachloride, chloroform, dichloromethane and dichloroethane). The solvents that have emerged in recent decades are supercritical fluids, ionic liquids and green solvents (limonene and glycerol, for example). As for the techniques for obtaining the extracts, the traditional infusion (in water) and maceration have been coupled with energy sources to assist the process of breaking down plant cell membranes, such as ultrasound and microwaves.

When looking for bioactive substances, the polarity of the extracting solvent is an important factor. Nonpolar substances are very commonly observed and not very active, such as fatty acids, waxes and sterols. On the other hand, very polar substances, extracted in water, are also sometimes less active and also easily observed very frequently among several species, such as polyphenols and sugars. Thus, medium polarity solvents are preferred when looking for new substances with biological activities that are unprecedented.

The use of a scan performed with several extracts (from several plant organs) prepared with a single solvent is a usual strategy in bioprospecting, in order to track plants that may have prominent biological activities. Sometimes, seeking to minimize the amounts of solvents used, a single hydroethanolic extract is prepared and then partitions are made with solvents of less polarity. In a study along these lines, 1,220 extracts were obtained from more than 350 species belonging to 71 different families of plants native to the Amazon rainforest and the Brazilian Atlantic forest. These extracts were tested for antibacterial activity against *Staphylococcus aureus*, *Enterococcus faecalis*, *Pseudomonas aeruginosa* and *Escherichia coli*. The extracts were obtained from leaves, wood or simply with all aerial organs, as in the case of herbs. Some results obtained for Amazonian plants are presented in Table 2.1 (Suffredini et al. 2006). The most active extracts (MIC 40.0 µg/mL) were observed to be from *Palicourea guianensis*, an alkaloid-rich Rubiaceae; and a leaf extract from *Virola theiodora*, another known toxic and hallucinogenic plant. The extracts from aerial parts of the Amazonian "araçá" tree, named *Psidium densicomum*, was also active at this same MIC.

A more common fruit in tropical regions, including Amazonia, is the cashew (*Anacardium occidentale* - Anacardiaceae). Several pharmacological activities have already been described, such as anti-inflammatory, antidiabetic and antimicrobial. Leaf plant extracts of the *Anacardium occidentale* L. species have shown to be potentially active against *Escherichia coli*, *Staphylococcus aureus* and *Candida albicans* strains, showing the antimicrobial potential of the cashew extract (Anand et al. 2015; Ajileye et al. 2015). The inhibition halo assay is another very useful method to analyze the potential antimicrobial activity. A paper disk containing an amount of a known antibiotic is put in a microorganism growth media, such as agar (Table 2.2). The zone of inhibition around the antibiotic-containing disk indicates the sensitivity of the bacteria to the antibiotic in the disk. With this assay, unknown substances or complex mixtures, such as extracts, can be evaluated in comparison to known antibiotics.

The pathogenic microorganisms *Staphylococcus aureus*, *Enterococcus faecalis*, *Salmonella enterica serovar Typhi*, *Acinetobacter baumannii* and *Candida albicans* were inhibited by the

Table 2.1. Antimicrobial potential of Amazon plants in MIC assays.

Species	Botanical Family	Microorganism	MIC µg/mL
Guatteria riparia	*Annonaceae*	*Staphylococcus aureus*	100.0
Mora paraensis	*Leguminosae*	*Staphylococcus aureus*	> 200.0
Virola theiodora	*Myristicaceae*	*Enterobacter faecalis*	40.0
Palicourea guianensis	*Rubiaceae*	*Staphylococcus aureus* *Enterobacter faecalis*	40.0 40.0
Piper arboreum	*Piperaceae*	*Staphylococcus aureus*	60.0
Psidium densicomum	*Myrtaceae*	*Staphylococcus aureus* *Enterobacter faecalis*	40.0 60.0

Table 2.2. Antimicrobial potential of Amazon plants in agar diffusion assays.

Species	Botanical Family	Microorganism	Inhibition halo (mm)
Hymenelobium petraeum	*Leguminoseae*	*Salmonella enterica* ser. Typhi *Staphylococcus aureus* *Enterococcus faecalis* *Candida albicans*	7.0 4.0 5.0 11.9
Vatairea guianensis	*Leguminoseae*	*Staphylococcus aureus*	4.0
Symphonia globulifera	*Clusiaceae*	*Staphylococcus aureus*	3.7
Pentaclethra macrolaba	*Leguminoseae*	*Acinetobacter baumannii* *Enterococcus faecalis*	5.8 4.0
Ptychopetalum olacoides	*Olacaceae*	*Acinetobacter baumannii* *Klebsiella ozanae*	4.9 9.2

aqueous extract of *Hymenolobium petraeum*, while the extracts of *Vatairea guianensis* and *Symphonia* and *guianensis globulifera* showed potential against *Staphylococcus aureus*, whereas a *Pentaclethra* macroloba type was able to inhibit the growth of *Acinetobacter baumannii* and *Klebsiella ozaenae*. These extracts had at least 40% antimicrobial activity when compared to the halo inhibition caused by the commercial antibiotic, ciprofloxacin, used as a control (Oliveira et al. 2010; Oliveira et al. 2013).

The MIC and disk tests are fast and reliable prospecting methods to screen various extracts in the search of bioactive substances. In addition, they speed up the discovery of new drugs using phytochemical methods. In these, the active extracts are fractionated and the fractions must be analyzed again, in a process called "bio-guided fractionation". Thus, the active fractions are fractionated and analyzed again and again. If the activity of the fraction decreases, it is a sign that the activity was lost during the fractionation process. On the other hand, if the activity increases as fractionation continues, it is a sign that the activity is being concentrated and the process continues until a pure bioactive substance is isolated.

When a bioassay is performed and an active substance is found, several aspects of the research must be evaluated next. The sustainability of the continuity of studies is one of these aspects. If the substance isolated from a natural product already has an economically viable synthetic route, it is likely that its price in the market will be low, making the choice to extract this substance from the plant in question unfeasible. Obviously, if the choice is to use a cheaper biological source, without needing isolation, then the isolated synthetic derivative, probably with more side effects, will not be the first choice. The isolated substance may be available not only by synthetic routes, but sometimes as a by-product of the bioindustry. In that case, they also become viable to being purchased commercially.

β-caryophyllene (BCP) and β-caryophyllene oxide (BCPO) are sesquiterpenes present in several plants. These compounds have anti-cancer, anti-inflammatory and antimicrobial activities, as well as other biological properties. BCP is a chemical compound found in several plants, such as oregano (*Origanum vulgare*), cinnamon (*Cinnamomum* spp.) and black pepper (*Piper nigrum*). This substance is observed in several Amazonian species, such as in the EOs of Lauraceae species, especially in the *Ocotea* genus (Yamaguchi et al. 2013) and also in the copaíba oleoresin, both from *Copaifera multijuga* and *Copaifera langsdorffii* (Oliveira et al. 2016). However, in Moroccan varieties of the species *Syzygium aromaticum* (cloves) the extraction yields of EOs are high (about 8.3%) and the percentage of BCP is also high, greater than 27%. Indonesia is the world's leading exporter of clove buds, the EOs of which are distilled to produce eugenol (present in quantities two to three times greater than BCP), a phenylpropanoid with strong antiseptic action, which will be discussed later. Thus, the BCP present in clove buds, after distillation to concentrate eugenol, becomes an extremely inexpensive residue from the extraction process.

Therefore, it is often interesting to analyze the biological actions of the constituents present in EOs, oleoresins and bioactive extracts from the Amazon, without having to isolate them, since they are substances present in plants from other parts of the Earth that have previously been isolated and tested.

Sesquiterpene is a chemical class comprising more than 300 compounds. They are terpenoids with skeletons containing 15 carbons, built from three isoprene units. Within this class there are several types of chemical structures, from linear to acyclic and cyclic carbon chains. The diversity of structures arises from the several biosynthetic forms in which sesquiterpenes can be produced and also by the diversity of possible chiral centers and stereospecific regions.

The β-caryophyllene (BCP) (Fig. 2.5) is a sesquiterpene commonly found in the Amazonian copaiba oils, such as that produced by *Copaifera reticulata*. This compound has several medicinal applications due to its low toxicity to humans and its remarkable biological activities including as an antioxidant, anti-inflammatory, antispasmodic, antihistaminic and antimicrobial (Francomano et al. 2019).

As an antimicrobial, BCP can inhibit aerobic Gram-positive and Gram-negative bacteria, such as *Staphylococcus aureus* and *Escherichia coli*, being more active against Gram-positive ones (Galdino et al. 2012; Yang et al. 2015; Dahham et al. 2015). It is estimated that BCP modifies the permeability of the bacterial membrane creating non-selective channels (Yang et al. 2019; Moo et al. 2020). Through these non-selective channels, the intracellular content leakage and the membrane is eventually disrupted. BCP also has inhibitory activity against *Bacillus cereus*, *Streptococcus mutans* and their biofilms (Yoo and Jwa 2018; Koyama et al. 2019).

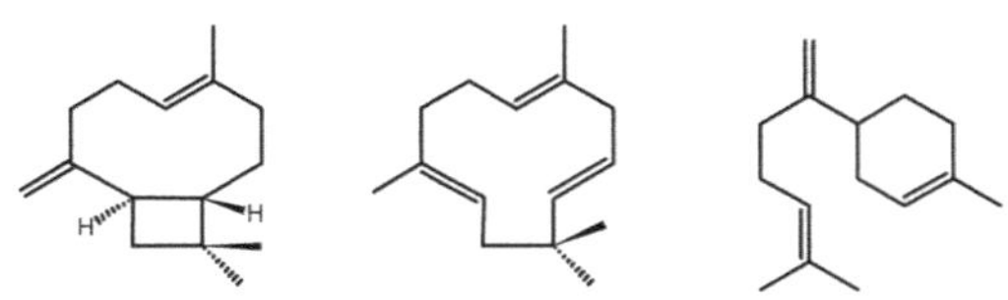

Figure 2.5. Examples of bioactive sesquiterpenes from Amazon plant species.

Another highly bioactive sesquiterpene is α-humulene (HML) (Fig. 2.5), an isomer of BCP. This compound can be isolated from the EOs of Amazonian plant species like *Eupatorium odoratum*, *Salvia officinalis*, *Copaifera reticulata*, *Mentha spicata* and *Myrica rubra*. HML has proven to be an anti-inflammatory, antibacterial, antidepressant, anticarcinogenic, anti-Alzheimer's agent, anesthetic and anti-hypertensive. The spectra of biological activities associated with HML is related to its interaction with several enzymes and neural receptors, such as prostaglandin G/H synthase 2, alcohol dehydrogenases, norepinephrine transporter and the 4 aminobutyric acid receptors.

HML was demonstrated to be a highly effective inhibitor of proliferation and biofilms of the Gram-negative bacteria, *Bacteroides fragilis*. This bacterium is common in the human colon flora, but it can cause blood infections and diarrheal disease. Violet crystal and MTT assays show that HML inhibits the proliferation of *Bacteroides fragilis* strains and biofilms by more than 90% at a concentration of 2 μg/mL. Even though the MIC90 of HML was 2 μg/mL, it is effective in both effects even at a concentration of 0.5 μg/mL. HML-enriched EOs also have activity against *Bacillus cereus*, *Staphylococcus aureus* and *Salmonella typhimurium* (Miladi et al. 2016; Jang et al. 2020; Zhang et al. 2020).

Different from BCP and HML, β-bisabolene (Fig. 2.5) is not a bicyclic sesquiterpene. This compound can be isolated from the EOs of *Eremanthus erythropappus*, *Copaifera* sp., *Commiphora guidottii* and *Piper* sp. β-bisabolene presents bioactivity including as an antimicrobial, anti-inflammatory, anti-leishmanial and being cytotoxic. In addition, it was also capable of restoring the activity of ampicillin against *Staphylococcus aureus* (Oliva et al. 2020).

Fernandes and coworkers (2020) demonstrated that isolated β-bisabolene had antimicrobial activity against eight standard pathogen strains. According to the microdilution method, β-bisabolene has a MIC (μg/mL) of 50 to *Streptococcus mutans*, 100 to *Streptococcus mitis* and *Streptococcus salivarius*, 250 to *Streptococcus sanguinis* and

Streptococcus sobrinus, 20 to *Mycobacterium tuberculosis* and 150 to *M. avium* and *M. kansasii*. The raw EO of *Eupatorium adenophorum* also demonstrated antimicrobial activity against these pathogens with an MIC ranging from 62.5 to 500 μg/mL. It is worth mentioning that β-bisabolene is in fact the major compound within the *E. adenophorum* EO. The microbial activity of *E. adenophorum* EO against *Arthrobacter protophormiae*, *Escherichia coli*, *Micrococcus luteus*, *Rhodococcus rhodochrous* and *Staphylococcus aureus* can be associated to the high concentrations of β-bisabolene (Kurade et al. 2010).

Phenylpropanoids are a simple but broad chemical class biosynthesized in plants. Compounds from this chemical class are produced in the shikimic acid pathway, being a derivate of phenylalanine or tyrosine. Phenylpropanoids contain a carbon backbone of C6–C3 and are also known as cinnamic acids. In Plantae kingdom this chemical class is responsible for protecting plants against herbivores, solar radiation, thermal stress, microorganisms and drought (Neelam and Sharma 2019). Within Amazon species specifically, eugenol and thymol are interesting compounds present in high concentration in EOs.

Eugenol (Fig. 2.6) can be obtained from the EO of clove, with concentrations ranging from 45 to 90% of the total EO composition (Arruda et al. 2019). This compound is known for its bioactivity as an anesthetic, anti-inflammatory, antioxidant, antifungal, antiviral and antimicrobial (Pramod et al. 2010; Bezerra et al. 2017; Barboza et al. 2018).

As an antimicrobial, eugenol had demonstrated activity against Gram-positive and Gram-negative bacteria such as, *Escherichia coli*, *Pseudomonas fluorescens*, *Staphylococcus aureus*, *Bacillus cereus* and *Listeria monocytogenes*. The mechanism of action of eugenol against microorganism is mainly through the increase of cellular membrane permeability, disruption of the cytoplasmic membrane and downregulation of gene transcription. By this mechanism, eugenol is also able to inhibit and disrupt the structure of *Acinetobacter baumannii* biofilms. In addition, eugenol also demonstrates synergic activity with penicillin and ampicillin against *Streptococcus pneumoniae* and *Streptococcus pyogenes* (Mak et al. 2019).

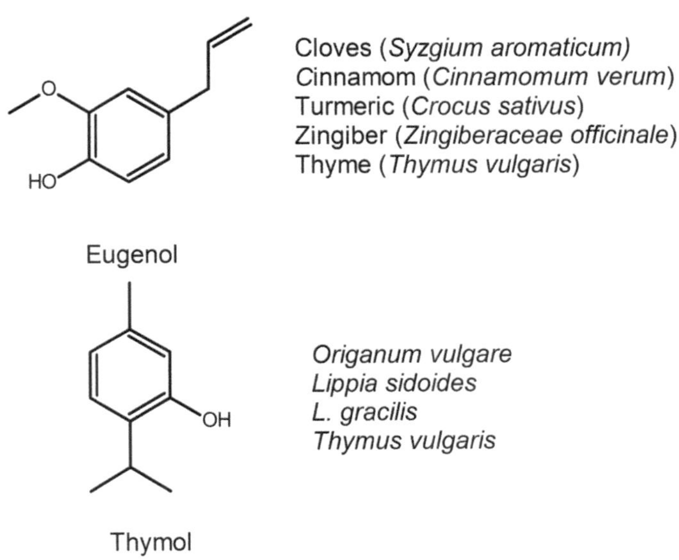

Figure 2.6. Eugenol and thymol chemical structures and their natural sources.

Thymol (Fig. 2.6) is one of major phenylpropanoids in the EOs of the *Lamiaceae* family, also found in *Verbenaceae*, *Scrophulariaceae*, *Ranunculaceae* and *Apiaceae* families. Thymol-enriched EOs have been used for many years in the food industry as a preservative and taste addictive. This compound is known for its biological properties as an antioxidant, antimicrobial, antifungal, anti-inflammatory, insect repellent and antiseptic (Zahran and Abdelgaleil 2011; Marchese et al. 2016; Islam et al. 2018; Silva et al. 2020; Arafa et al. 2020).

Regarding its antimicrobial and antifungal activity, thymol can inhibit Gram-positive and Gram-negative bacteria, such as *Escherichia coli*, *Proteus mirabilis*, *Proteus vulgaris*, *Salmonella typhimurium*, *Serratia marcescens*, *Yersinia enterocolitica*, *Pseudomonas fluorescens*, *Micrococcus* sp., *Sarcina flava*, *Staphylococcus aureus*, *Bacillus licheniformis*, *Bacillus thuringiensis* and *Listeria innocua*.

Thymol was demonstrated to be more inhibitory of 12 strains of *Salmonella typhimurium* than eugenol. By the microdilution method, the MIC of thymol ranged from 32 to 128 μg/mL, while eugenol ranged from 512 to 1024 μg/mL. Both eugenol and thymol were also capable of eradicating *Salmonella typhimurium* biofilms, with minimum biofilm eradication concentrations ranging from 10.50 to 697.6 and 15.3 to 934.4 μg/mL, respectively (Miladi et al. 2017). Hamzah and coworkers (2018) also demonstrated that by the crystal violet method thymol and eugenol are prominent inhibitors and antibiofilm for the polymicrobial biofilm of *Pseudomonas aeruginosa*, *Escherichia coli* and *Candida albicans*. Against these bacterial strains individually, thymol had an average inhibition of 84.60% and eugenol of 83.82%. In addition, the minimum biofilm inhibitor concentration to polymicrobial biofilms for both compounds (at a concentration of 1%) was 0.10 and 0.31%, respectively.

The monoterpenes are the simplest terpene compounds, being composed by two molecules of isoprene, with a carbon backbone of 10 atoms. Monoterpenes can be both, cyclic or acyclic structures, and can be present in concentrations as high as 90% of total composition of EOs in some higher plant species. They play a major role in plant metabolism but are also known for their biological activity as a hypotensive, anti-inflammatory, antipruritic, antigerminative, antiplasmodial, anticancer and food flavoring. Among the monoterpenes present in the EOs of Amazon plants, one needs to highlight limonene, linalool and geraniol.

Limonene (Fig. 2.7) is the major monoterpene found in citric plants such as oranges and lemons. Besides its characteristic citric aroma that made it famous in the food industry, its antimicrobial, antibacterial and antifungal properties enable it to be used as a food and cosmetic preservative. Limonene is more effective in the inhibition of Gram-positive bacterial strains, being capable of inhibiting *Salmonella senftenberg*, *Escherichia coli*, *Staphylococcus aureus*, *Streptococcus uberis*, *Listeria monocytogenes* and *Cronobacter sakazakii* (Espina et al. 2012; Montironi et al. 2016; Han et al. 2020).

The antibacterial mechanism of action of limonene is mainly based on its interaction with the outer membrane of Gram-positive bacteria. The presence of lipopolysaccharide in the outer membrane enables linkage of the limonene and further modification of its permeability and as a consequence, disruption. The inhibition efficacy of limonene can be as high as 92% against *Streptococcus uberis* strains known for biofilm formation. In addition, the MIC of limonene against Gram-positive bacteria, *Staphylococcus aureus*, *Bacillus subtilis*, and *Micrococcus luteus*, is respectively 0.625, 1.25, and 2.5 mg/mL. It is worth mentioning that limonene is considered a moderate antimicrobial, and not the main compound responsible for the bioactivity of citric plant EOs. In fact, raw EOs have lower MIC than isolated limonene, although this does not diminish all the potential of limonene (Espina et al. 2012; Hua-Li et al. 2019; Han et al. 2020).

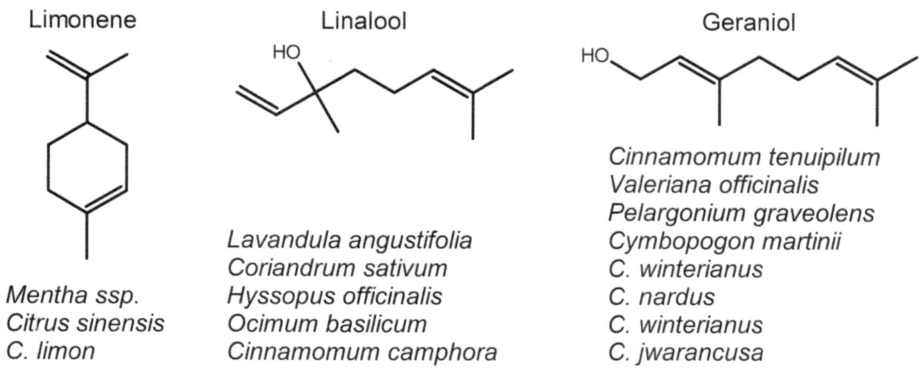

Figure 2.7. Bioactive monoterpenes from Amazon species and their natural sources.

Linalool (Fig. 2.7) is one of the bioactive monoterpenes of major occurrence in lavender and cinnamon EOs. It is largely used in the pharmaceutical and cosmetic industry due to its antibacterial, antifungal, antioxidant, anti-inflammatory and anticancer properties (Sabogal-Guáqueta et al. 2019). This compound is efficient against Gram-positive and Gram-negative bacteria. Several studies have already proven its inhibitory activity against *Staphylococcus aureus*, *Staphylococcus epidermidis*, *Pseudomonas aeruginosa*, *Escherichia coli*, *Candida albicans*, *Candida glabrata* and *Candida krusei*. In addition to the fact that the MIC of linalool (12.5 to 100 mM) is higher than eugenol (6.25 to 50 mM) in such cases, it is notable that isolated linalool is more active than that of raw EOs. In addition, linalool was also able to synergically increase the antimicrobial activity of *Juniperus communis*, *Citrus bergamia*, and *Pelargonium graveolens* EOs against *E. coli* (Park et al. 2012; Baky and Hashem 2016; Herman et al. 2016; Pereira et al. 2018; Aelenei et al. 2019).

Geraniol (Fig. 2.7) is a monoterpenic alcohol usually found in the EOs of cinnamon, pelargoniums (*Pelargonium*) and lemongrass (*Cymbopogon*). It is known for its fragrance of roses as much as for its biological activity as an antioxidant, anti-inflammatory, anticancer, antimicrobial and antinociceptive (Solórzano-Santos et al. 2012; Lei et al. 2018; Mączka et al. 2020).

The antimicrobial activity of geraniol has already been tested against 78 different microorganisms in total, with *Candida* and *Staphylococcus* strains being the most investigated (Lira et al. 2020). The antifungal activity of geraniol against *Candida* strains is usually performed by microdilution methods and presents an MIC ranging from 16 to 512 µg/mL. In addition, geraniol also modified the ATPase activity on the plasma membrane. Moreover, antibacterial activity against *Staphylococcus* strains can vary with an MIC as low as 2.44 µg/mL (brain heart infusion broth method) to 2773.6 µg/mL (nutrient broth method). Other microorganisms of interest that geraniol can inhibit include *Escherichia coli*, *Helicobacter pylori*, *Haemophilus influenzae* and *Mycobacterium tuberculosis* (Solórzano-Santos et al. 2012; Pereira et al. 2015; Singh et al. 2016; Singulani et al. 2018; Bhattamisra et al. 2019; Nee et al. 2019).

The diterpenoids are a chemical class comprising metabolites with a 20-carbon skeleton, formed by four isoprene units and derived from geranyl geranyl pyrophosphate. Their structure can be linear, bicyclic, tricyclic, tetracyclic, pentacyclic or macrocyclic with distinct functional groups. This chemical class is particularly interesting due to its antimicrobial activity.

For instance, the EO of *Copaifera reticulata* has as the major diterpenes, copalic and kaurenoic acids (Fig. 2.8). Both of these are known for their remarkable properties, such as *in vitro* antibacterial activity, antifungal activity against dermatophytes, as an antiallergic, an immunosuppressive, apoptosis-inducing activities and antitrypanosomal activity. In addition, kaurenoic acid is being suggested to be included in medicine for the treatment of breast cancer, hypertension, Alzheimer's, atherosclerosis and lung cancer, due to its interactions with the progesterone and mineralocorticoid receptors and myeloperoxidase enzyme (Katsuyama et al. 2013; Fidyt et al. 2016; Tobouti et al. 2017; Arruda et al. 2019).

Regarding the antimicrobial activity of the *C. reticulata* EO, copalic acid has an MIC ranging from 5 to 0.5 μg/mL for pathogens including *Kocuria rhizophila, Streptococcus pyogenes, Streptococcus pneumoniae, Bacillus subtillis, Streptococcus dysgalactiae, Streptococcus agalactiae* and *Staphylococcus epidermidis*. Meanwhile, the MIC of kaurenoic acid is approximately 5 μg/mL against *Streptococcus pneumoniae* and *Streptococcus pyogenes* (Mizuno et al. 2015; Arruda et al. 2019).

Dihydrocucurbitacin F-25-O-acetate (Fig. 2.8) is also an interesting compound in Amazon EOs. This substance is found in *Coutarea hexandra* (common name: murta-do-mato), a species traditionally used against fever, malaria and inflammations. Dihydrocucurbitacin F-25-O-acetate showed antimicrobial activity against *Staphylococcus aureus* with inhibitory results similar to the standard antibiotic vancomycin.

Lapachol, β-lapachone and α-lapachol are naphthoquinones commonly found in *Tabebuia* species (Bignoniaceae family), such as "Ipê roxo", a tree endemic to the Amazon rainforest, and. Ipê roxo was traditional known as the "miracle cure" tree for cancer and tumors. In the late 1960s, the miracle cure started to be attributed to lapachol, β-lapachone and α-lapachol. These compounds are known for their biological properties as antitumor, antivirals, antitrypanosomals and antimicrobials.

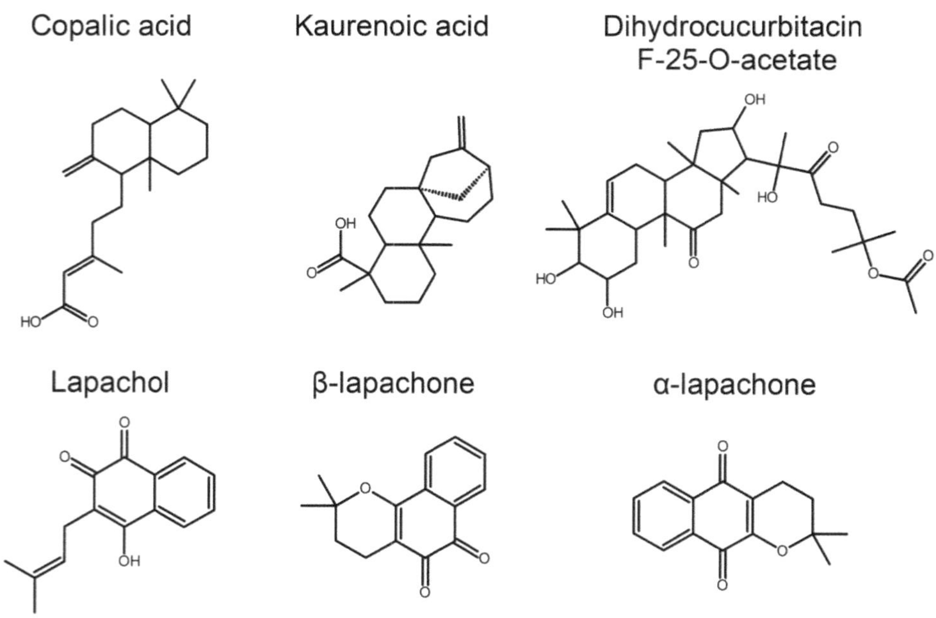

Figure 2.8. Diterpenes and non-volatile bioactive substances from Amazon species.

Lapachol can be isolated from *Tabebuia flavescens*, *Tabebuia guayacan*, *Kigelia pinnata*, *Phyllarthron comorense* and *Radermachera sinica*. As an antimicrobial, lapachol is an effective inhibitor of the *Brucella* genus, *Bacillus subtilis*, *Escherichia coli*, *Pseudomonas aeruginosa*, *Salmonella typhimurium*, *Enterococcus faecalis*, *Candida* sp., and *Micrococcus pyogenes*. It is particularly more effective against Gram-positive bacteria (Hussain and Green 2017). The MIC of lapachol against *Paracoccidioides brasiliensis* strains varied from 0.13 to 0.26 µmol/mL. Such a property is important as *Paracoccidioides* ringworm only occurs in Latin America (Souza et al. 2013).

B-lapachone (Fig. 2.8) can be isolated from the EOs of *Tabebuia impetiginosa*, *Tabebuia guayacan*, *Tabebuia pentaphylla*, *Hesperolinon adenophyllum*, *Phyllarthron comorense* and *Tectona grandis*. A remarkable fact about this compound is that it is more effective against MRSA than lapachol and α-lapachone. According to Macedo and coworkers (2013), the MIC of β-lapachone against this microorganism ranged from 8 to 16 µg/mL, while the other naphthoquinones presented MICs of 256 to 512 µg/mLand 64 to 256 µg/mL, respectively. Other microorganisms which β-lapachone can inhibit includes *Staphylococcus aureus* and *Cryptococcus* (Macedo et al. 2013; Cavalcanti et al. 2015).

Last but not least, α-lapachone (Fig. 2.8) can be obtained from species such *Catalpa ovata*, *Haplophragma adenophyllum*, *Tabebuia guayacan*, *Tabebuia impetiginosa*, *Tabebuia pentaphylla*, *Zeyhera tuberculosa* and *Ekmanianthe longiflora* (Epifano et al. 2014; Nepomuceno 2016; Hussain and Green 2017). Despite the fact that it is the least investigated of the naphthoquinones, this compound does have bioactivity against *Trypanosoma cruzi*, *Leishmania amazonensis* and *L. braziliensis*, the pathogens responsible for Chagas disease and leishmaniasis (Vieira et al. 2015; Hussain and Green 2017).

Conclusion

The natural products extracted from the Amazon region are widely documented, mainly as the biological properties of medicinal plants. Properties of these plants that can be highlighted are the anti-inflammatory, anticancer, antioxidant, antinociceptive and antimicrobial activities. The actions of natural products against microorganisms have been the subject of numerous studies, due to the increased resistance of microorganisms to existing antibiotics. However, there is still a lot unknown regarding the antimicrobial activity of vegetable oils and extracts, such as the mechanism of action, the synergistic effect between the compounds, the action of the isolated compounds and the pharmacological applications. Therefore, this chapter seeks to highlight medicinal plants in the Amazon with antimicrobial potential, as well as to describe their chemical compounds.

Acknowledgments

This study was financed in part by the Carlos Chagas Foundation for Research Support in the State of Rio de Janeiro (FAPERJ), Amazonas State Research Support Foundation (FAPEAM) and the National Council for Scientific and Technological Development (CNPq).

Abbreviations

APG	Angiosperm Phylogeny Group
BCP	β-Caryophyllene
BCPO	Caryophyllene oxide
NTDs	Neglected Tropical Diseases

EO	Essential Oil
HML	Humulene
MIC	Minimum Inhibitory Concentration
MRSA	Methicillin-Resistant *Staphylococcus aureus*
WHO	World Health Organization

References

Aelenei, P., Rimbu, C.M., Guguianu, E., Dimitriu, G., Aprotosoaie, A.C., Brebu, M., Horhogea, C.E., Miron, A. (2019). Coriander essential oil and linalool—interactions with antibiotics against Gram-positive and Gram-negative bacteria. Lett Appl Microbiol 68: 156–164.

Ajileye, O.O., Obuotor, E.M., Akinkunmi, E.O., Aderogba, M.A. (2015). Isolation and characterization of antioxidant and antimicrobial compounds from the leaf extract of *Anacardium occidentale* L. (*Anacardiaceae*). J King Saud Univers Scienc 27: 244–252.

Almeida, J.E., Andrade-Picanço, T., Paranatinga, I.L.D., Sousa Júnior, J.J., Silva-Escher, S.K., Oliveira, E.C.P. (2020). Evaluation of antimicrobial activity and determination of minimum inhibitory concentration (MIC) of copaiba oleoresin. Rev Uningá 57: 12–22.

Anand, G., Ravinanthan, M., Basaviah, R., Shetty, A.V. (2015). *In vitro* antimicrobial and cytotoxic effects of *Anacardium occidentale* and *Mangifera indica* in oral care. J Pharm Bioallied Sci 7: 69–74.

APG IV - The Angiosperm Phylogeny Group. (2016). An update of the Angiosperm Phylogeny Group classification for the orders and families of flowering plants: APG IV. Bot J Linn Soc 181: 01–20.

Arafa, W.M., Abolhadid, S.M., Moawad, A., Abdelaty, A.S., Moawad, U.K., Shokier, K.A.M., Shehata, O., Gadelhaq, S.M. (2020). Thymol efficacy against coccidiosis in pigeon (Columba livia domestica). Prev Vet Med 176: 01–08.

Arruda, C., Mejía, J.A.A., Ribeiro, V.P., Borges, C.H.G., Martins, C.H.G., Veneziani, R.C.S., Ambrósio, S.R., Bastos, J.K. (2019). Occurrence, chemical composition, biological activities and analytical methods on *Copaifera* genus—A review. Biomed Pharmacother 109: 01–20.

Azab, A., Nassar, A., Azab, A.N. (2016). Anti-inflammatory activity of natural products. Molecules 21: 01–19.

Baky, R.M.A.E., Hashem, Z.S. (2016). Eugenol and linalool: comparison of their antibacterial and antifungal activities. Afr J Microbiol Res 10: 1860–1872.

Barata, A.M., Rocha, F., Lopes, V.R., Carvalho, A.M. (2016). Conservation and sustainable use of genetic resources of medicinal and aromatic plants in the world for human well-being. Ind Crops Prod 88: 8–11.

Barboza, J.N., Silva, M.B.F.C., Silva, R.O., Medeiros, J.V.R., Sousa, D.P. (2018). An overview on the anti-inflammatory potential and antioxidant profile of eugenol. Oxid Med Cell Longev 2018: 01–09.

Batista, E.K.F., Trindade, H.I., Lira, S.R.S., Muller, J.B.B.S., Silva, L.L.B., Batista, M.D.C.D.S. (2016). Antinociceptive and anti-inflammatory activities of the ethanol extract of *Luehea divaricata*. Rev Bras Plant Medic 18: 433–441.

Batista, L.T., Sarrazin, S.L.F., Moura, V.M., Santos, I.G.C., Junior, S.D., Albuquerque, P.M. (2019). Chemical composition, antimicrobial and antioxidant activity of the essential oil of Aniba parviflora (Meisn) Mez. Rev Fitos 13: 181–191.

Bauri, R.K., Tigga, M.N., Kullu, S.S. (2015). A review of the use of medicinal plants to control parasites. Ind J Nat Prod Res 64: 268–277.

Belda-Galbis, C.M., Jiménez-Carretón, A., Pina-Pérez, M.C., Martínez, A., Rodrigo, D. (2015). Antimicrobial activity of açaí against *Listeria* innocua. Food Control 53: 212–216.

Bezerra, D.P., Militão, G.C.G., Morais, M.C., Sousa, D.P. (2017). The dual antioxidant/prooxidant effect of eugenol and its action in cancer development and treatment. Nutrients 9: 1367–1382.

Bhattamisra, S.K., Yan, V.L.Y., Lee, C.K., Kuean, C.H., Candasamy, M., Liew, Y.K., Sahu, O.S. (2019). Protective activity of geraniol against acetic acid and *Helicobacter pylori*-induced gastric ulcers in rats. J Tradit Compl Med 9: 206–214.

Castro, V., Lorenzi, H. (2005). Systematic botany: illustrated guide for the identification of Angiosperm families of Brazilian flora, based on APG II. Nova odessa: Instituto plamtarium.

Cavalcante, J.W., Cavalcante, V., Bieski, I. (2017). Traditional and ethnopharmacological knowledge of the Copaiba medicinal plant (*Copaifera langsdorffii* Desf.). Rev Biodiver 16: 01–10.

Cavalcanti, I.M.F., Pontes-Neto, J.G., Kocerginsky, P.O., Bezerra-Neto, A.M., Lima, J.L.C., Lira-Nogueira, M.C.B., Maciel, M.A.V., Neves, R.P., Pimentel, M.F., Santos-Magalhães, N.S. (2015). Antimicrobial activity of β-lapachone encapsulated into liposomes against meticillin-resistant *Staphylococcus aureus* and *Cryptococcus neoformans* clinical strains. J Glob Antimicrob Resist 3: 103–108.

Chen, J., Li, W., Yao, H., Xu, J. (2015). Insights on drug discovery of natural products by structural modification. Phytotherapy 1: 231–241.

Cleff, M.B. (2012). Susceptibility profile of *Candida* yeasts isolated from animals to the essential oil of *Rosmarinus officinalis* L. Rev Bras Pla Med 14: 43–49.

Dahham, S.S., Tabana, Y.M., Iqbal, M.A., Ahamed, M.B.K., Ezzat, M.O., Majid, A.S.A., Majid, A.M.S.A. (2015). The anticancer, antioxidant and antimicrobial properties of the sesquiterpene β-caryophyllene from the essential oil of *Aquilaria crassna*. Molecules 20: 11808–11829.

Dar, R.A., Shahnawaz, M., Qazi, P.H. (2017). Overview of medicinal plants: a review. J Phytopharmacol 6: 349–351.

Ellwanger, J.H., Kulmann-Leal, B., Kaminski, V.L., Valverde-Villegas, J.M., Veiga, A.B.G., Spilki, F.R., Fearnside, P.M., Caesar, L., Giatti, L.L., Wallau, G.L., Almeida, S.E.M., Borba, M.R., Hora, V.P., Chies, J.A.B. (2020). Beyond diversity loss and climate change: Impacts of Amazon deforestation on infectious diseases and public health. An Acad Bras Ciênc 92: 01–33.

Epifano, F., Genovese, S., Fiorito, S., Mathieu, M., Kiss, R. (2014). Lapachol and its congeners as anticancer agents: a review. Phytochem Rev 13: 37–49.

Espina, L., Gelaw, T.K., Lamo-Castellví, S., Pagán, R., García-Gonzalo, D. (2012). Mechanism of bacterial inactivation by (+)-limonene and its potential use in food preservation combined processes. Plos One 8: 01–10.

Fenner, R., Betti, A.H., Mentz, L.A., Rates, S.M.K. (2006). Plants with potential antifungal activity employed in Brazilian folk medicine. Rev Bras Ci Farm 42: 369–394.

Fernandes, C.C., Andrade, P.M., Santos, T.C.L., Santiago, M.B., Pagotti, M.C., Crotti, A.E.M., Martins, C.H.G., Magalhães, L.G., Miranda, M.L.D. (2020). '*In vitro*' evaluation of anti-caries, antimycobacterial, antileishmania and cytotoxic activities of the essential oils of '*Eremanthus erythropappus*' and of alpha-bisabolol, its main sesquiterpene. Austral J Crop Scienc 14: 236–243.

Fernandez, I.M., Chagas, E.A., Maldonado, S.A.S.M., Takahashi, J.A., Alemán, R.S., Melo Filho, A.A., Santos, R.C.S., Ribeiro, P.R.E., Fuentes, J.A.M., Chagas, P.C., Melo, A.C.G. (2020). Antimicrobial activity and acetilcolinesterase inhibition of oils and Amazon fruit extracts. J Med Plants Res 14: 88–97.

Fibigr, J., Satínský, D., Solich, P. (2017). A UHPLC method for the rapid separation and quantification of anthocyanins in acai berry and dry blueberry extracts. J Pharm Biomed Anal 143: 204–213.

Fidyt, K., Fiedorowicz, A., Strządała, L., Szumny, A. (2016). β-caryophyllene and β-caryophyllene oxide-natural compounds of anticancer and analgesic properties. Cancer Med 5: 3007–3017.

Francomano, F., Caruso, A., Barbarossa, A., Fazio, A., Torre, C.L., Ceramella, J., Mallamaci, R., Saturnino, C., Lacopetta, D., Sinicropi, M.S. (2019). β-caryophyllene: a sesquiterpene with countless biological properties. Appl Sci 9: 01–19.

Galdino, P.M., Nascimento, M.V., Florentino, I.F., Lino, R.C., Fajemiroye, J.O., Chaibub, B.A., Paula, J.R., Lima, T.C.M., Costa, E.A. (2012). The anxiolytic-like effect of an essential oil derived from *Spiranthera odoratissima* A. St. Hil. leaves and its major component, beta-caryophyllene, in male mice. Progr Neuropsychopharmacol Biol Psychiatry 38: 276–284.

Giongo, J.L., Vaucher, R.A., Silva, A.S., Oliveira, C.B., Mattos, C.B., Baldissera, M.D., Sagrillo, M.R., Monteiro, S.G., Custódio, D.L., Matos, M.S., Sampaio, P.T., Teixeira, H.F., Koester, L.S., Veiga Junior, V.F. (2017). Trypanocidal activity of the compounds present in *Aniba canelilla* oil against *Trypanosoma evansi* and its effects on viability of lymphocytes. Microb pathog 103: 13–18.

Hamzah, H., Pratiwi, S.U.T., Hertiani, T. (2018). Efficacy of thymol and eugenol against polymicrobial biofilm. Indonesian J Pharm 4: 214–221.

Han, Y., Sun, Z., Chen, W. (2020). Antimicrobial susceptibility and antibacterial mechanism of limonene against *Listeria monocytogenes*. Molecules 25: 01–15.

Herman, A., Tambor, K., Herman, A. (2016). Linalool affects the antimicrobial efficacy of essential oils. Curr Microbiol 72: 165–172.

Hua-Li, Z., Cai, M., Liu, Y.S., Sun, P.L., Luo, S.L. (2019). Antibacterial activity and mechanisms of essential oil from *Citrus medica* L. var. sarcodactylis. Molecules 24: 01–10.

Hussain, H., Green, I.R. (2017). Lapachol and lapachone analogs: a journey of two decades of patent research (1997–2016). Expert Opin Ther Pat 27: 1111–1121.

Islam, M.T., Khalipha, A.B.R., Bagchi, R., Mondal, M., Smrity, S.Z., Uddin, S.J., Shilpi, J.A., Rouf, R. (2018). Anticancer activity of thymol: A literature-based review and docking study with Emphasis on its anticancer mechanisms. IUBMB Life 71: 9–19.

Jang, H.I., Rhee, K.J., Eom, Y.B. (2020). Antibacterial and antibiofilm effects of α-humulene against bacteroides fragilis. Can J Microbiol 66: 389–399.

Jobim, M.L., Santos, R.C.V., Alves, C.F.S., Oliveira, R.M., Mostardeiro, C.P., Sagrillo, M.R., Souza Filho, O.C., Garcia, L.F.M., Manica-Cattani, M.F., Ribeiro, E.E., Cruz, I.B.M. (2014). Antimicrobial activity of extracts from Amazon *Astrocaryum aculeatum* and its association with oxidative metabolism. Microbiol Res 169: 314–323.

Katsuyama, S., Mizoguchi, H., Kuwahata, H., Komatsu, T., Nagaoka, K., Nakamura, H., Bagetta, G., Sakurada, T., Sakurada, S. (2013). Involvement of peripheral cannabinoid and opioid receptors in β-caryophyllene-induced antinociception. Eur J Pain 17: 664–675.

Koyama, S., Purk, A., Kaur, M., Soini, H.A., Novotny, M.V., Davis, K., Kao, C.C., Matsunami, H., Mescher, A. (2019). Beta-caryophyllene enhances wound healing through multiple routes. PLoS One 14: 01–21.

Kurade, N.P., Jaitak, V., Kaul, V.K., Sharma, O.P. (2010). Chemical composition and antibacterial activity of essential oils of *Lantana camara, Ageratum houstonianum* and *Eupatorium adenophorum*. Pharm Biol 48: 539–44.

Leandro, L.M., Sousa, F.V., Barbosa, P.C.S., Neves, J.K.O., Silva, J.Á., Veiga-Junior, V.F. (2012). Chemistry and biological activities of Copaíba terpenoids (*Copaifera* spp.) Oleoresins. Molecules 17: 3866–3889.

Lei, Y., Fu, P., Jun, X., Cheng, P. (2018). Pharmacological properties of geraniol—a review. Planta Med 85: 48–55.

Lima, G.S., Castro-Pinto, D.B., Machado, G.C., Maciel, M.A.M., Echevarria, A. (2015). Antileishmanial activity and trypanothione reductase effects of terpenes from the Amazonian species *Croton cajucara* Benth (Euphorbiaceae). Phytomedicine 22: 1133–1137.

Lira, M.H.P., Andrade Júnior, F.P., Moraes, G.F.Q., Macena, G.S., Pereira, F.O., Lima, I.O. (2020). Antimicrobial activity of geraniol: an integrative review. J Essent Oil Res 32: 01–12.

Macedo, L., Fernandes, T., Silveira, L., Mesquita, A., Franchitti, A.A., Ximenes, E.A. (2013). β-Lapachone activity in synergy with conventional antimicrobials against methicillin resistant *Staphylococcus aureus* strains. Phytomedicine 21: 25–29.

Mączka, W., Wińska, K., Grabarczyk, M. (2020). One hundred faces of Geraniol. Molecules 25: 3303.

Mak, K.K., Kamal, M., Ayuba, S., Sakirolla, R., Kang, Y.B., Mohandas, K., Pichika, M. (2019). A comprehensive review of eugenol's antimicrobial properties and industry applications: a transformation of ethnomedicine for industry. Pharmacogn Rev 13: 1–9.

Marchese, A., Orhan, I.E., Daglia, M., Barbieri, R., Di Lorenzo, A., Nabavi, S.F., Gortzi, O., Izadi, M., Nabavi, S.M. (2016). Antibacterial and antifungal activities of thymol: A brief review of the literature. Food Chem 210: 402–14.

Mauro, M., Grandis, R.A., Campos, M.L., Bauermeister, A., Peccinini, R.G., Pavan, F.R., Lopes, N.P., Moraes, N.V. (2019). Acid diterpenes from Copaiba oleoresin (*Copaifera langsdorffii*): Chemical and plasma stability and intestinal permeability using Caco-2 cells. J Ethnopharmacol 235: 183–189.

Meneguetti, D.U.O., Lima, R.A., Hurtado, F.B., Passarini, G.M., Macedo, S.R.A., Barros, N.B., Oliveira, F.A.Z., Medeiros, P.S.M., Militão, J.S.L.T., Nicolete, R., Facundo, V.A. (2016). Screening of the *in vitro* antileishmanial activities of compounds and secondary metabolites isolated from *Maytenus guianensis* Klotzsch ex Reissek (Celastraceae) chichuá Amazon. Rev Soc Bras Med Trop 49: 579–585.

Miladi, H., Mili, D., Slama, R.B., Zouari, S., Ammar, E., Bakhrouf, A. (2016). Antibiofilm formation and anti-adhesive property of three mediterranean essential oils against a foodborne pathogen *Salmonella strain*. Microb Pathog 93: 22–31.

Miladi, H., Zmantar, T., Kouidhi, B., Chaabouni, Y., Mahdouani, K., Bakhrouf, A., Chaieb, K. (2017). Use of carvacrol, thymol, and eugenol for biofilm eradication and resistance modifying susceptibility of *Salmonella enterica* serovar Typhimurium strains to nalidixic acid. Microb Pathog 104: 54–63.

Mizuno, C.S., Souza, A.B., Tekwani, B.L., Ambrósio, S.R., Veneziani, R.C.S. (2015). Synthesis and biological evaluation of polyalthic acid derivatives for the treatment of neglected diseases. Bioor Med Chem Lett 25: 5529–553.

Montironi, I.D., Cariddi, L.N., Reinoso, E.B. (2016). Evaluation of the antimicrobial efficacy of Minthostachys verticillata essential oil and limonene against Streptococcus uberis Taiwan. Nat Prod Commun 5: 1823–1828.

Moo, C.L., Yang, S.K., Osman, M.A., Yuswan, M.H., Loh, J.Y., Lim, W.M., Lim, S.H., Lai, K.S. (2020). Antibacterial activity and mode of action of β-caryophyllene on *Bacillus cereus*. Pol J Microbiol 69: 1–6.

Nazzaro, F. (2013). Effect of essential oils on pathogenic bacteria. Pharmaceuticals Basel 6: 1451–1474.

Nee, T.W., Ring, L.C., Arumugam, V., Sim, J.L.C., Hin, L.W., Yusof, F.A.M., Noor, M.A.M., Yenn, T.W. (2019). Sustained-release geraniol nanoparticles inhibit bacteria that cause human axillary odors. Arab J Sci Eng 44: 103–109.

Neelam, A.K., Sharma, K.K. (2019). Phenylpropanoids and its derivatives: biological activities and its role in food, pharmaceutical and cosmetic industries. Crit Rev Food Sci Nutr 60: 01–22.

Nepomuceno, J.C. (2016). Lapachol and its derivatives as potential drugs for cancer treatment. pp. 01–19. *In*: Breeding and Genetic Engineering: The Biology and Biotechnology. Research iConcept Press.

Newman, D.J., Cragg, G.M. (2020). Natural products as sources of new drugs over the nearly four decades from 01/1981 to 09/2019. Nat Prod 83: 770–803.

Odonne, G., Houël, E., Bourdy, G., Stien, D. (2017). Treating leishmaniasis in the Amazon: a review of ethnomedicinal concepts and pharmaco-chemical analyzes of traditional treatments to inspire modern herbal medicines. J ethnopharmacol 199: 211–230.

Oliva, A., Garzoli, S., Sabatino, M., Tadić, V., Costantini, S., Ragno, R., Božović, M. (2020). Chemical composition and antimicrobial activity of essential oil of *Helichrysum italicum* (Roth) G. Don fil. (Asteraceae) from Montenegro. Nat Prod Res 34: 445–448.

Oliveira, A.A., Segovia, J.F.O., Sousa, V.Y.K., Mata, E.C.G. (2013). Antimicrobial Activity of Amazonian Medicinal Plants. SpringerPlus 2: 01–06.

Oliveira, L.Z., Cesarino, F., Pantoja, T.F., Môro, F.V. (2010). Morphological aspects of fruits, seeds, germination and seedlings of *Hymenolobium petraeum*. Cienc Rural 40: 1732–1740.

Oliveira, L.G.S., Ribeiro, D.A., Saraiva, M.E., Macêdo, D.G., Macedo, J.G.F., Pinheiro, P.G., Costa, J.G.M., Solsa, M.M.A., Menezes, I.R.A. (2016). Chemical variability of essential oils of Copaifera langsdorffii Desf. in different phenological phases on a savannah in the Northeast, Ceará, Brazil. Ind Crop Prod 97: 455–464.

Pala, D., Barbosa, P.O., Silva, C.T., Souza, M.O., Freitas, F.R., Volp, A.C.P., Maranhão, R.C., Freitas, R.N. (2017). Açai (*Euterpe oleracea* Mart.) dietary intake affects plasma lipids, apolipoproteins, cholesteryl ester transfer to high-density lipoprotein and redox metabolism: A prospective study in women. Clin Nutr 37: 618–623.

Park, S.N., Lim, Y.K., Freire, M.O., Cho, E., Jin, D., Kook, J.K. (2012). Antimicrobial effect of linalool and α-terpineol against periodontopathic and cariogenic bacteria. Anaerobe 18: 369–72.

Peixoto, H., Roxo, M., Krstin, S., Wang, X., Wink, M. (2016). Anthocyanin-rich extract of Açai (*Euterpe precatoria* Mart.) mediates neuroprotective activities in *Caenorhabditis elegans*. J Funct Foods 26: 385–393.

Pereira, F.D.E.O., Mendes, J.M., Lima, I.O., Mota, K.S., Oliveira, W.A., Lima Ede, O. (2015). The antifungal activity of geraniol and citronelol, two monoterpene alcohols, against *Trichophyton rubrum* involves the inhibition of ergosterol biosynthesis. Pharm Biol 53: 228–234.

Pereira, I., Severino, P., Santos, A.C., Silva, A.M., Souto, E.B. (2018). Linalool bioactive properties and potential applicability in drug delivery systems. Colloids Surf 171: 566–578.

Piriz, M.A., Lima, C.A.B., Jardim, V.M.R., Mesquita, M.K., Souza, A.D.Z., Heck, R.M. (2014). Medicinal plants in the wound healing process: a literature review. Rev Bras Plantas 16: 628–636.

Pramod, K., Ansari, S.H., Ali, J. (2010). Eugenol: a natural compound with versatile pharmacological actions. Nat Prod Commun 5: 01–08.

Ramos, K., Andreani Junior, R., Kozusny-Andreani, D.I. (2016). Essential and vegetable oils in the *in vitro* control of *Colletotrichum gloeosporioides*. Rev Bras Pl Med 18: 605–612.

Ricardo, L.M., Dias, B.M., Müggea, F.L.B., Leite, V.V., Brandão, M.G.L. (2018). Evidence of the traditionality of Brazilian medicinal plants: The case studies of *Stryphnodendron adstringens* (Mart.) Coville (barbatimão) and *Copaíba* spp. (Copaíba) oleoresin in wound healing. J Ethnopharmacol 219: 319–336.

Sabogal-Guáqueta, A.M., Hobbie, F., Keerthi, A., Oun, A., Kortholt, A., Boddeke, E., Dolga, A. (2019). Linalool attenuates oxidative stress and mitochondrial dysfunction mediated by glutamate and NMDA toxicity. Biomed Pharmacother 118: 109295.

Santos, F.V., Santos, V.J.S.V., Farias, M.J., Mesquita, S.F.P., Maciel, M.A., Pinto, A.C., Cólus, I.M.S. (2008). Mutagenicity and antimutagenicity of Croton cajucara. Biologia 63: 327–331.

Santos, D.L., Moraes, J.S., Araújo, Z.T.S., Silva, I.R. (2019). Traditional knowledge about medicinal plants in the conservation of Amazonian biodiversity. Science in Focus 12: 86–96.

Santos, M.R.A., Lima, M.R., Oliveira, C.L.L.G. (2014). Medicinal plants used in Rondônia, Western Amazon, Brazil. Rev Bras Plantas Med 16: 707–720.

Sarrazin, S.L.F., Oliveira, R.B., Barata, L.E.S., Mourão, R.H.V. (2012). Chemical composition and antimicrobial activity of the essential oil of *Lippia grandis* Schauer (Verbenaceae) from the Western Amazon. Food Chem 134: 1474–1478.

Sarrazin, S.L.F., Oliveira, R.B., Maia, J.G.S., Mourão, R.H.V. (2016). Antibacterial activity of the rosewood (*Aniba rosaeodora* and *A. parviflora*) linalool-rich oils from the Amazon. Eur J Med Plants 12: 1–9.

Shih, R.Y., Koeller, K.K. (2015). Bacterial, fungal and parasitic infections of the central nervous system: radiological-pathological correlation and historical perspectives: from the archives of radiological pathology. Radiographics 35: 1141–1169.

Shukla, J.B., Naresh, R., Verma, S.R., Verma, S.R., Agarwal, M. (2020). Modeling the effect of sanitation in a human habitat to control the spread of bacterial diseases. Model Earth Syst Environ 6: 39–49.

Silva, E.F., Vieira, T.J.N., Gurgel, G.C.L. (2020). "*In vitro*" biological activity of the thymol phytoconstituent on species of the genus candida. Rev Pesq Interdisc 2: 677–685.

Silva, R.E., Oliveira, D.R., Leitão, S.G., Assis, I.M., Veiga-Junior, V.F., Lourenço, M.C., Alviano, D.S., Alviano, C.S., Bizzo, H.R. (2013). Essential oils of Protium spp. samples from Amazonian popular markets: chemical composition, physicochemical parameters and antimicrobial activity. J Essential Oil Res 25: 171–178.

Silva, L.E., Quadros, D.A., Alzino Neto, J.M. (2015). Ethnobotanical and ethnopharmacological study of medicinal plants used in the region of Matinhos – PR. Sci Nat 37: 266–276.

Silva, S.M.M., Costa, C.R.R., Gelfuso, G.M., Guerra, E.N.S., Nobrega, Y.K.M., Gomes, S.M., Pic-Taylor, A., Fonseca-Bazzo, Y.M., Silveira, D., Magalhães, P.O. (2018). Wound healing effect 429 of essential oil extracted from *Eugenia dysenterica* DC (Myrtaceae) leaves. Molecules 24: 01–16.

Singh, S., Fatima, Z., Hameed, S. (2016). Insights into the mode of action of the herbicide monoterpenoid geraniol anticandide reveal the disruption of multiple MDR mechanisms and virulence attributes in *Candida albicans*. Arch Microbiol 198: 459–472.

Singulani, J.L., Pedroso, R.S., Ribeiro, A.B., Nicolella, H.D., Freitas, K.S., Damasceno, J.L., Vieira, T.M., Crotti, A.E., Tavares, D.C., Martins, C.H., Mendes-Giannini, M.J., Pires, R.H. (2018). Geraniol and linalool anticandidal activity, genotoxic potential and embryotoxic effect on zebrafish. Future Microbiol 13: 1637–1646.

Solórzano-Santos, F., Miranda-Novales, M.G. (2012). Essential oils from aromatic herbs as antimicrobial agents. Curr Opin Biotechnol 23: 136–141.

Souza, M.A., Johann, S., Lima, L.A.R.S., Campos, F.F., Mendes, I.C., Beraldo, H., Souza-Fagundes, E.M., Cisalpino, P.S., Rosa, C.A., Alves, T.M.A., Sá, N.P., Zani, C.L. (2013). The antimicrobial activity of lapachol and its thiosemicarbazone and semicarbazone derivatives. Mem Inst Oswaldo Cruz 108: 342–351.

Souza, A.M., Oliveira, C.F., Oliveira, V.B., Betim, F.C.M., Miguel, O.G., Miguel, M.D. (2018). Traditional uses, phytochemistry and antimicrobial activities of Eugenia species—a review. Rev Plant Med 84: 1232–1248.

Souza-Junior, F.J.C., Luz-Moraes, D., Pereira, F.S., Barros, M.A., Fernandes, L.M.P., Queiroz, L.Y., Maia, C.F., Maia, J.G.S., Fontes-Junior, E.A. (2020). *Aniba canelilla* (Kunth) Mez (Lauraceae): A review of ethnobotany, phytochemical, antioxidant, anti-inflammatory, cardiovascular, and neurological properties. Frontiers in Pharmacol 11: 01–14.

Suffredini, I.B., Paciencia, M.L.B., Varella, A.D., Younes, R.N. (2006). Antibacterial activity of Brazilian Amazon plant extracts. BJID 10: 400–402.

Tobouti, P.L., Martins, T.C.A., Pereira, T.J., Mussi, M.C.M. (2017). Antimicrobial activity of Copaiba oil: A review and a call for further research. Biomed Pharmacother 94: 93–99.

Torgerson, P.R., Devleesschauwer, B., Praet, N., Speybroeck, N., Willingham, A.L., Kasuga, F., Rokni, M.B., Zhou, X.N., Fèvre, E.M., Sripa, B., Gargouri, N., Fürst, T., Budke, C.M., Carabin, H., Kirk, M.D., Ângulo, F.J., Havelaar, A., Silva, N. (2015). World Health Organization estimates of the global and regional disease burden of 11 foodborne parasitic diseases, 2010: a data synthesis. PLoS Med 12: 01–22.

Ugboko, H.U., Nwinyi, O.C., Oranusi, S.U., Fatoki, T.H., Omonhinmin, C.A. (2020). Antimicrobial importance of medicinal plants in Nigeria. The Scient Worl J 2020: 01–10.

Valdivieso-Ugarte, M., Gomez-Llorente, C., Plaza-Díaz, J., Gil, Á. (2019). Antimicrobial, antioxidant, and immunomodulatory properties of essential oils: a systematic. Rev Nutr 11: 01–46.

Vargas, F.S., Almeida, P.D.O., Boleti, A.P.A., Pereira, M.M., Souza, T.P., Vasconcellos, M.C., Nunez, C.V., Pohlit, A.M., Lima, E.S. (2016). Antioxidant activity and peroxidase inhibition of Amazonian plant extracts traditionally used as anti-inflammatories. BMC Complement Altern Med 16: 1–8.

Vieira, A.A., Brandão, I.R., Valença, W.O., Bruno, C.A.S., Cavalcanti, C., Pessoa, C., Carneiro, T.R., Braga, A.L., Silva-Júnior, E.N. (2015). Hybrid compounds with two redox centres: Modular synthesis of chalcogen-containing lapachones and studies on their antitumor activity. Eur J Med Chem 101: 254–265.

Wang, G.W., Hu, W.T., Huang, B.K., Qin, L.P. (2011). Illicium verum: a review on its botany, traditional use, chemistry and pharmacology. J Ethnopharmacol 136: 10–20.

Yamaguchi, K.K.D., Alcantara, J.M., Lima, E.S., Veiga Junior, V.F. (2013). Chemical composition and platelet aggregation activity of essential oils of two species of the *Genus Ocotea* (lauraceae). J Essent Oil Bear Plants 16: 518–523.

Yang, C., Hu, D.H., Feng, Y. (2015). Antibacterial activity and mode of action of the *Artemisia capillaris* essential oil and its constituents against respiratory tract infection-causing pathogens. Mol Med Rep 11: 2852–2860.

Yang, S.K., Yusoff, K., Ajat, M., Thomas, W., Abushelaibi, A., Akseer, R., Lim, S.H.E., Lai, K.S. (2019). Disruption of KPC-producing *Klebsiella pneumoniae* membrane via induction of oxidative stress by cinnamon bark (*Cinnamomum verum* J. Presl) essential oil. PLoS One 14: 01–20.

Yoo, H.J., Jwa, S.K. (2018). Inhibitory effects of β-caryophyllene on *Streptococcus mutans* biofilm. Arch Oral Biol 88: 42–46.

Zahran, H.E.D.M., Abdelgaleil, S.A. (2011). Insecticidal and developmental inhibitory properties of monoterpenes on *Culex pipiens* L. (Diptera: Culicidae). J Asia-Pacific Entomol 14: 46–51.

Zhang, C., Li, M., Zhao, G.R., Lu, W. (2020). Harnessing yeast peroxisomes and cytosol acetyl-CoA for sesquiterpene α-humulene production. J Agric Food Chem 68: 1382–1389.

3

Selective Antimicrobial Agents from Terrestrial Plants

A Hope in the Battle of Infection

Fadia S. Youssef [1,*] and *Mohamed L. Ashour*[2,*]

Introduction

Infection is a complex process that involves the attack of the living organism by disease-producing microbes and microorganisms with the subsequent multiplication within the hosting living organisms. This action stimulates the host organism's immune system to react to eliminate them or stop their harmful toxins. Infectious diseases are classified based on the causative organisms or the affected tissue(s). These infections are usually accompanied by a panel of side effects that range from pain of the organ, feeling persistent fatigue and the loss of appetite as well as fever. Additionally, some other symptoms are specific to the infected part, including cough, runny nose and body rashes may be noticed (Ryan and Ray 2004).

Despite the vast progress in drug discovery for curing illness, infectious diseases due to bacteria, fungi and viruses still represent a major obstacle to public health in the 21st century. This health problem is clear in many countries due to the weakness in the health care systems and the appearance of bacterial resistance to many antibiotics. Therefore, new classes of antibiotics are constantly demanded to overcome this problem. As a result, considerable attention has been given to the great therapeutic potential of herbs and medicinal plants that offer an acceptable model of cheap, substantially safe remedy that might be used as antimicrobial candidates for many developing countries (John 2010; Youssef et al. 2021).

The development of bacterial resistance against many antibiotics has been seriously reducing the success in curing such infections. There are many reported mechanisms involved in this resistance that include the inactivation of the antibiotic directly by changing important functional groups in the antimicrobial agents or even altering the dimensions or stereochemistry of the target enzymes. Moreover, the modification of the targeted site of action, the modification in the metabolic pathways, inhibition of drug uptake and

[1] Department of Pharmacognosy, Faculty of Pharmacy, Ain-Shams University, Abbasia, Cairo 11566, Egypt.
[2] Department of Pharmaceutical Sciences, Pharmacy Program, Batterjee Medical College, Jeddah 21442, Saudi Arabia.
* Corresponding author: fadiayoussef@pharma.asu.edu.eg; mohamed.ashour@bmc.edu.sa

reduction of the intracellular amount of the drug via its exist out of the infected cell by ABC transporters are other possible mechanisms (Wink et al. 2012). In addition, mechanisms for antimicrobial resistance comprise an effective efflux of the antimicrobial agent, reduction of drug entry into the microbial cells, degradation of antimicrobial drugs enzymatically in addition to the effective formation of biofilm (Varela et al. 2021).

Clerodendrum is a large genus with about 500 species belonging to the family Lamiaceae that broadly spread in subtropical and tropical areas of this world. Many Clerodendrum plants possess promising biological activities and have been used for ages in several folk medicine systems such as Japanese Kampo, Ayurveda, Traditional Chinese Medicine and Korean medicine for curing many ailments, i.e., syphilis, typhoid, cancer, jaundice and hypertension. Many studies have been conducted to assess their possible activities. In addition, phytochemical studies reported the isolation and chemical constituents' identification, steroids, diterpenoids, triterpenoids, flavonoids and phenolic compounds, as seen in Fig. 3.1 (Khan et al. 2017). Some popular *Clerodendrom* species that showed a potent antimicrobial activity are illustrated in Fig. 3.2.

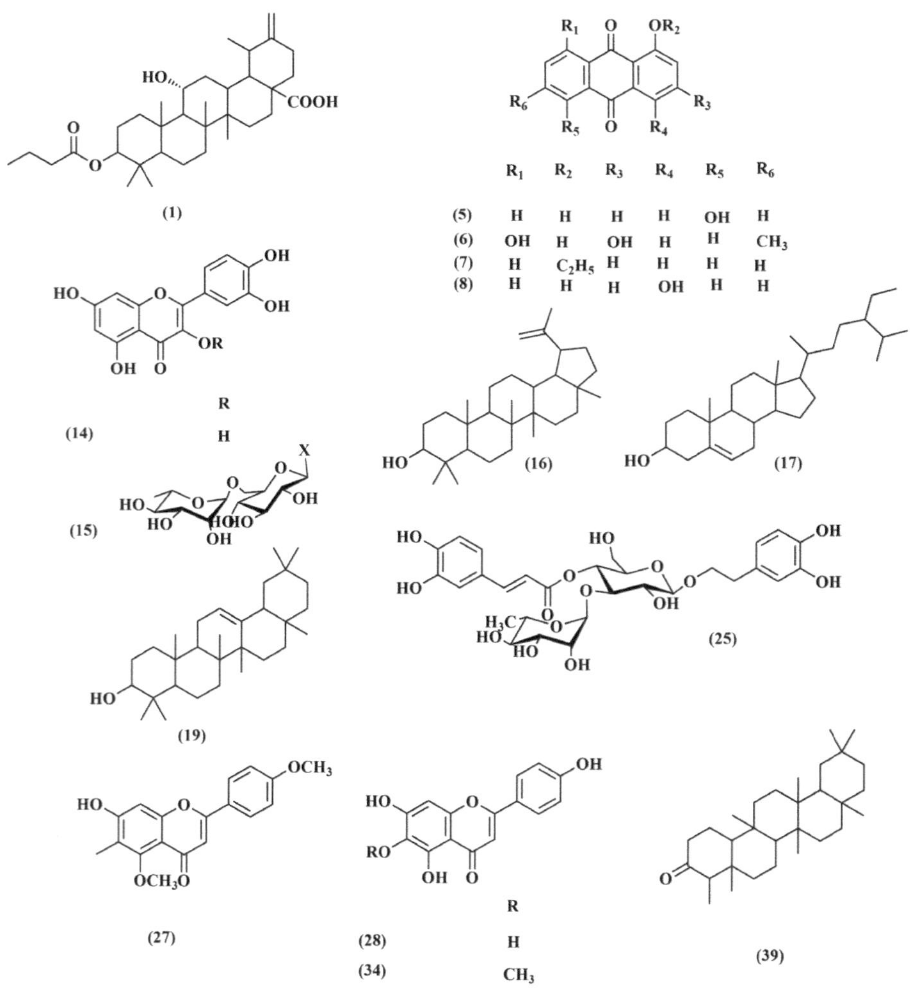

Figure 3.1. Some representative secondary metabolites identified in *Clerodendrum* species.

Figure 3.2. Some popular *Clerodendrom species* that showed a potent antimicrobial activity.

C. baronianum

The ethyl acetate extract of *C. baronianum* revealed antimicrobial activity against *B. subtilis, E. coli, P. aeruginosa* and *S. aureus* due to steroids (Borik 2013).

C. bungei

C. bungei roots (5–10 parts) have been included with many other plant ingredients as bacteriostatic powder in a chlorhexidine acetate. These herbal products act as an antimicrobial agent with no adverse or toxic signs on man after prolonged use. It also showed antiviral activity against hepatitis C. In addition, it displayed antiviral activity against hepatitis C (Shrivastava and Patel 2007). Its alcohol extract showed inhibition against both *S. aureus* and *E. coli* (IZDs = 11–20 mm and MIC/MBC = 256~ > 1024 mg/L) (Yang et al. 2012).

C. capitatum

Ethanol extract (70%) of *C. capitatum* leaves revealed a considerable antimicrobial activity, particularly against *Candida albicans*, as displayed at a dose of 80 mg/hole that was chosen as the highest dose for the susceptibility test using the disk diffusion technique showing a value of 20.33 ± 0.58 mm. This anti-candida activity may be attributed to the presence of flavonoids, tannins and cardiac glycoside in high percentages. However, ethanol extract of *C. capitatum* leaves showed the least effectiveness against *Staphylococcus aureus* with a minimum zone of inhibition equals to 7.33 ± 0.58 at a dose of 80 mg/hole (Hamidu Usman et al. 2017).

Furthermore, an additional study was carried on the different extracts of *C. capitatum* leaves, namely *n*-hexane, dichloromethane, ethyl acetate and methanol extracts, to evaluate their antimicrobial potential using the agar diffusion method against nine highly pathogenic organisms comprising bacteria and fungi. Bacteria include *Corynebacterium ulcerans, Klebsiella pneumoniae, Staphylococcus aureus,* Methicillin-resistant *Staphylococcus aureus, Streptococcus pneumoniae, Proteus mirabilis, Salmonella typhi* and *Shigella dysenteriae,* while fungi represented by *Candida albicans, C. tropicalis* and *C. krusei.* The *n*-hexane extract showed a mean zone of inhibition ranging from 17–26 mm, while dichloromethane, ethyl acetate and methanol showed 23–26, 26–29 and 20–23 mm as mean zones of inhibition, respectively against all tested organisms except for *Corynebacterium ulcerans*. The ethyl acetate extract showed a MIC value of 3.25 mg/mL against all the examined organisms except for *S. aureus* and *C. albicans*, where it displayed a MIC value of 7.5 mg/mL. However, *n*-hexane extract showed a MIC value of 7.5–15 mg/mL, whereas dichloromethane and methanol extracts showed a MIC of 7.5 mg/mL. Regarding MBC/MFC values, the ethyl acetate extract showed 7.5–15 mg/mL whereas dichloromethane, methanol and *n*-hexane extracts showed 15, 15–30 and 30 mg/mL, respectively as MBC/MFC value. Meanwhile, dichloromethane extract revealed the highest activity against *Mycobacterium bovis* with MIC value equals 0.675 mg/mL, thus allowing a new natural anti-TB drug (Habila et al. 2018).

C. chinense

C. chinense chloroform extract of the flower showed high activity against *Plasmodium falciparum* with IC_{50} < 10 μg/mL. Meanwhile, *C. chinense* chloroform extracts of the stem, flowers and leaves showed potent activity against *T. cruzi* with IC_{50} = 1.21, 1.12, and 3.39 μg/mL, respectively. In contrast, the tested samples showed no activity towards Gram-positive, Gram-negative bacteria or *Candida albicans*. Compounds isolated from *C. chinense* leaves such as Verbascoside and Rengyolone showed antimicrobial activity, the former towards *T. cruzi*. Nevertheless, the latter showed broad but non-specific behavior towards the examined organisms (AbouZid et al. 2013).

C. colebrookianum

The methanol extract obtained from the leaves of *C. colebrookianum* and its different fractions were obtained by solvent-solvent fractionation, namely *n*-hexane, ethyl acetate, *n*-butanol and the remaining aqueous fraction were evaluated for their antimicrobial activity. This activity was performed against *Staphylococcus aureus* and *Bacillus cereus* as Gram-positive bacteria and *Pseudomonas aeruginosa* and *Escherichia coli* as Gram-bacteria. The ethyl acetate fraction exerted a considerable activity against *Bacillus cereus* with a MIC value equals 1.5 mg/mL. This pronounced activity may be attributed to its richness by different metabolites comprising carbohydrates, flavonoids, phenols, saponins (Doley et al. 2016). Furthermore, the antimicrobial potential was significantly high for acetone extract displaying an obvious zone of inhibition of 14, 13 and 15 mm for *E. coli*, *S. marcescens* and *S. aureus*, respectively. This activity is undoubtedly attributed to its richness with phenol compounds, where it showed an elevated total phenol content that reaches 2.348 mg/mL (Mahesh et al. 2015).

C. cyrtophyllum

C. cyrtophyllum leaves have been included in *a* Chinese herbal preparation that showed a strong inhibitory potential against *S. aureus*, *E. coli*, and *B. subtilis*. Thus, it can act as safe, effective and relatively cheap. *C. cyrtophyllum* root was an ingredient in an herbal preparation that showed potent fungicidal and bactericidal activity with high water solubility, good stability and no irritation or side effects. Besides, a multifunctional traditional Chinese medicine comprising *C. cyrtophyllum* leaves as one of its ingredients showed an antimicrobial effect on *Staphylococcus aureus*, *Escherichia coli* and *Candida albicans* 99.9%, high stability with no adverse effect. Moreover, an additional herbal preparation comprising *C. cyrtophyllum* leaves as one of its components significantly reduced the hepatitis B virus (Huang 1998).

C. fragrans

The ethyl acetate extract of *C. fragrans* leaves that represents the semi-polar extract displayed a potent activity against Gram-positive bacteria like *S. aureus* showing 90% growth inhibition with a mean diameter of inhibition zone of 11.25 mm. In contrast, *n*-hexane fraction as a non-polar fraction showed the highest inhibition against *E. coli* with 77.21% inhibition in the microbial growth and mean diameter of inhibition zones of 10.50 mm. In addition, a preparation containing different herbal extracts with three-parts of *C. fragrans* roots as one of its ingredients showed a potent bacteriostatic effect (Simorangkir et al. 2019).

C. glabrum

The *n*-hexane extract of *C. glabrum* leaves showed a potent antimicrobial activity against *Candida albicans* and *Cryptococcus neoformans*, displaying a MIC of 0.10 mg/mL. Meanwhile, compounds isolated from the *n*-hexane extract, namely clerodendrumic acid **(1)** and heptadecanoic acid **(2)**, revealed mild antifungal activity with MIC value equals 125 and 188 μg/mL, respectively. They also showed mild antibacterial activity against Gram-positive and Gram-negative bacteria, with MIC values ranging between MIC values of average 157 and 172 μg/mL for Gram-positive and Gram-negative bacteria. In addition, clerodendrumic acid and heptadecanoic acid showed no toxicity *in vitro* against monkey kidney Vero cells displaying IC50s equal 202.6 and 108.4 μg/mL, respectively (Masevhe et al. 2013). However, *C. glabrum* exerted no antiviral activity against influenza A virus *in vitro* studies (Mehrbod et al. 2018).

C. inerme

C. inerme is a perennial plant belonging to the family Lamiaceae. Many studies showed that *C. inerme* is a rich source of many metabolites such as alkaloids, flavonoids, triterpenoids, tannins, diterpenoids, glycosides, phenols and volatile oils and steroids to which many of its biological activities comprising its antimicrobial activity were attributed (Al-Snafi 2016). Squalene **(3)** and linolenic acid methyl ester **(4)** are two compounds obtained from *C. inerme* leaves that showed *in vitro* antimicrobial activity that was validated using molecular modeling (Kar et al. 2020). *C. inerme* anthraquinones fraction of its leaves showed a considerable antimicrobial activity as evidenced by the diffusion agar method in which it significantly prohibits pathogenic bacteria comprising *Pseudomonas aeruginosa*. This activity is due to the presence of 1,5-dihydroxyanthrachinon **(5)**, emodin **(6)**, 1-ethoxyanthraquinone **(7)**, 1,4-dihydroxyanthraquinone **(8)**, 9,10-anthracenedione **(9)**, 2-ethyl anthraquinone **(10)**, 1,4-cyclohexanedione **(11)**, 1,4-dimethoxybenzene **(12)** and 2,3-dimethylquinone **(13)** in the anthraquinone fraction as displayed by GC-MS and FTIR analyses. Besides, gold (Au) and silver (Ag) nanoparticles (NPs) synthesized from *C. inerme* leaves extract showed considerable antimicrobial activity against a wide panel of organisms such as *B. subtilis*, *S. aureus*, *Klebsiella*, and *E. coli*, as well as *A. niger*, *T. harzianum* and *A. flavus* employing agar well diffusion method. The plant extract associated nanoparticles (CI-Ag and CI-Au NPs) displayed promising antifungal, antibacterial and perfect biofilm inhibitory potential against the studied pathogenic microorganisms compared to gold and silver NPs associated with dodecanethiol and polyvinylpyrrolidone, respectively. Furthermore, the biocompatibility test ascertained that CI-Ag and CI-Au NPs showed higher biocompatibility at 1–50 μM. Thus, this sheds light on the probable use of herbal drugs to synthesize nanomaterials with improved and powerful biomedical properties as a way of a new environmentally-friendly path direction. In addition, a preparation containing different herbal extracts with 0.1–0.3 parts of *C. inerme* roots as one of its ingredients showed a potent bacteriostatic effect (Khan et al. 2020).

An additional study was carried on *C. inerme* to assess its antimicrobial activity by agar as well as the diffusion method. The preliminary qualitative screening revealed the presence of alkaloids, tannins, flavonoids, saponins and phenols. The antimicrobial study revealed that *C. inerme* showed notable antimicrobial and antifungal activity against *S. aureus, E. coli,* and *A. niger* with high total phenol content was determined by Folin Ciocalteu reagent (Byadgi et al. 2017).

Furthermore, an additional study was carried on ZnO-NPs (zinc oxide nanoparticles) formulated from the aqueous and methanol leaves extracts of *C. inerme* to assess their antimicrobial activity against a wide panel of microorganisms comprising *Klebsiella pneumoniae, Staphylococcus aureus* and *Vibrio cholerae.* Results showed that the lowest concentration (10 mg/mL) showed no effect on the tested organisms. However, ZnO-NPs formulated from the methanol extract showed MIC equals 25 mg/mL against *Klebsiella pneumoniae, Staphylococcus aureus,* whereas it is 50 mg/mL for *Vibrio cholerae.* ZnO-NPs formulated from the aqueous extract showed MIC of 75 mg/mL for *Staphylococcus aureus* and 25 mg/mL for *Klebsiella pneumoniae* and *Vibrio cholerae* (Rad et al. 2019).

Moreover, un-doped tin oxide and Cobalt-doped tin oxide (Co-doped SnO_2) nanoparticles prepared from the aqueous extract of *C. inerme* revealed strong antibacterial and antifungal activities against *E. coli, B. subtilis, A. niger, A. flavus* and *C. albicans* with the zone of inhibitions of 30, 26, 17, 23 and 26, respectively. These findings were further supported by Scanning Electron Microscope (SEM), Transmission Electron Microscope (TEM), and confocal laser scanning microscopy (Khan et al. 2018). Besides, on testing the methanol, chloroform and ethyl acetate extracts of *C. inerme* leaves for their antimicrobial activities by the disk diffusion method against *Lactobacillus acidophilus, Rhizopus oryzae, Lactobacillus plantarum, Aspergillus niger, Staphylococcus aureus* and *Escherichia coli.* Amongst all the tested extracts, the methanol extract showed the highest effectiveness against all the test microorganisms, whereas the aqueous extract showed insignificant activity (Farooqui et al. 2010). In addition, a preparation containing different herbal extracts with 0.1–0.3 parts of *C. inerme* leaves as one of its ingredients showed a potent bacteriostatic effect. However, the aqueous extract of *C. inerme* in a pharmaceutically acceptable carrier was used as a bactericide in the ophthalmic diagnostic room that allows an effective and relatively low cost that fits the large scale ophthalmic diagnostic room in the hospital. Qualitative phytochemical analysis of *C. inerme* leaves ethanol extract revealed the presence of flavonoids, tannins, terpenoids and saponins in addition to 21 secondary metabolites. The ethanol extract showed promising antibacterial activity with a mean zone of inhibition as 0.30 and 1.23 against some pathogenic bacteria clarified by the agar well diffusion method (Anandhi and Ushadevi 2013).

C. inerme root extract showed potent antifungal activity against *Aspergillus niger, Monascus rubber, Candida albicans* and *Aspergillus fumigatus,* particularly the *Monascus ruber* comparable to the standard drug clotrimazole. In addition, the essential oil of *C. inerme* leaves showed significant antibacterial, antifungal and anti-MRSA activity that could be attributed to the prevalence of secondary metabolites represented by isoeugenol (35.50%), di-Bu phthalate (27.52%), eugenol (5.67%), α-pinene (5.54%) in addition to minor compounds as Benzoic acid, 4-formyl-benzoic acid Et ester (3.44%) and 2-Nitro-1-octanol (4.08%) as revealed by GC analysis (Kor-arnan and Thiraphibundet 2016).

C. infortunatum

A wet formulation of *C. infortunatum* displayed a potent fungicidal potential that could be adopted in agriculture to combat fungal attacks offering a relatively safe fungicide with reduced risk of environmental pollution in the agroecosystems. This formulation is free of petroleum-based solvents and thus shows no flammability risk, possible inhalation, skin toxicity or any harmful effect to the treated plants or animals (Saha et al. 2018).

In addition, *C. infortunatum* leaves extracted with different solvents like methanol, distilled water, acetone, *n*-hexane and dichloromethane are examined for their antimicrobial potential against *Staphylococcus pasteuri* as Gram-positive bacteria and *Pseudomonas aeruginosa* as a Gram-negative one. It showed a maximum inhibition zone at 5% against both tested organisms. Besides, methanol extract showed high content of phenols. Quantification of flavonoids and tannins revealed that *C. infortunatum* leaves possess high flavonoid content, which is 12.17 mg/g and total tannins equal to 13.87 mg/g in acetone extract. Purification of anthocyanin from an *in vitro* culture of *C. infortunatum* was done using various chromatographic techniques, which was further checked for purity using HPLC and the molar absorptivity assay. The antimicrobial activity was assessed by determining the Minimum Inhibitory Concentration (MIC) and Minimum Killing Concentration (MKC). The purified anthocyanins showed significant antimicrobial activity with different degrees with *C. albicans, S. aureus, P. aeruginosa* displayed significant values followed by MRSA, *E. coli* and *A. flavus*, approaching that of standard antibiotics. In contrast, *E. faecalis* showed a higher resistance. The mechanistic behavior shown was proved by analysis of intracellular potassium leakage and bacterial membrane integrity (Sivamaruthi et al. 2018).

Furthermore, green synthesis of silver nanoparticles (AgNPs) was done employing *C. infortunatum* alcohol extract that acts as a reducing and stabilizing agent that allows a simple and comparatively cheap biosynthetic method. AgNPs were examined using SEM, TEM, UV-visible spectroscopy as well as X-ray diffraction. The secondary metabolites that caused nano-transformation were mainly organic acids, tannins and phenolic compounds that predominate in the leaves. The antibacterial activity determination was performed using the agar diffusion method against *Staphylococcus aureus*, revealing a significant antibacterial activity (Jha and Prasad 2016).

The antimicrobial activity of *C. infortunatum* chloroform and ethanol extracts obtained from different organs comprising leaves, stems and roots was assessed *in vitro* using micro broth dilution, disk diffusion and serial dilution assays. All the tested samples revealed considerable antibacterial and antifungal activity approaching the standard drugs tetracycline and fluconazole. The ethanol leaf extract showed the maximum inhibition with MIC values equal 64 µg/mL to *Klebsiella pneumoniae, Salmonella typhi* and *Bacillus megaterium*. In contrast, it is 128 µg/mL for *Escherichia coli, Streptococcus-β-haemolyticus* and *Staphylococcus aureus* (Waliullah et al. 2014).

C. myricoides

C. myricoides was investigated for its antimicrobial activity against *Mycobacterium tuberculosis, Klebsiella pneumoniae, Escherichia coli, Candida albicans, Staphylococcus aureus, Cryptococcus neoformans, Salmonella typhi, Pseudomonas aeruginosa, Methicillin-resistant S. aureus* and *Shigella sonnei* employing disk diffusion as well as *microdilution* assays. The methanol extract exhibited a broad spectrum potential causing inhibition in 75% of test pathogens with the highest zone of inhibition of 20.3 mm for *S. aureus*, revealing the lowest

MIC value of 90 μg/mL. The methanol extract also showed a pronounced effect against *Methicillin-resistant S. aureus, Staphylococcus aureus, Escherichia coli, Shigella sonnei* and *Candida albicans*. Additionally, it revealed a high content of terpenoids, anthraquinones, flavonoids, alkaloids and phenols (Njeru et al. 2016).

C. paniculatum

C. paniculatum is one of the popular members in the genus *Clerodendrum*, widely known for its ornamental purposes. It showed a wide array of biological activities comprising antimicrobial activity that could be attributed to the abundance of phytoconstituents in its extract represented by quercetin **(14)**, rutin **(15)**, lupeol **(16)**, β-sitosterol **(17)**, stigmasta-4,25-dien-3-one **(18)**, β-amyrin **(19)**, oleanolic aldehyde acetate **(20)** and (3β)-stigmasta-4,22,25-trien-3-ol **(21)** (Kekuda and Sudharshan 2018). In addition, *C. paniculatum* showed considerable activity against multidrug-resistant bacteria such as *E. coli* and *K. pneumoniae* that are highly resistant to streptomycin and cephalosporin. *C. paniculatum* showed greater antibacterial activity against *E. coli* superior to ampicillin and comparable effect against *K. pneumoniae* as ampicillin showing diameter of zone inhibitions equal 14 and 15 mm, respectively (Shruthi et al. 2017).

C. phlomidis

C. phlomidis n-hexane and dichloromethane have been examined for their antibacterial effect against *Bacillus licheniformis* using the diffusion assay. The extracts showed zones of inhibition of 19 and 15.5 mm, respectively. The exact mode of action for the inhibition of *Bacillus licheniformis* was determined using *in vitro* studies represented by SEM (Scanning Electron Microscopy), DNA (Deoxyribonucleic acid) fragmentation, and FACS (Fluorescence-Activated Cell Sorting) analysis. Scanning Electron Microscopy analysis revealed the existence of irregularity in the cell wall, damage of its membrane and the leakage and protrusion of the cellular components. Moreover, DNA analysis showed the apoptosis of bacterial cellular DNA. Besides, FACS analysis revealed the potential disruption of the membrane potential triggered by the extracts where dichloromethane showed higher potency when compared to the *n*-hexane fraction. FTIR spectroscopic analysis was performed to determine the predominant functional groups in the extract that revealed the richness of the bioactive extract by halogen, alcohol and aldehyde groups. Thus, it can be concluded that *C. phlomidis* dichloromethane extract exerts its significant antimicrobial activity by membrane disruption with concomitant cellular contents leakage in addition to bacterial DNA degradation (Sirohi et al. 2019).

In addition, the antimicrobial effect of *C. phlomidis* was tested using the agar diffusion method against a panel of infectious microorganisms represented by *S. aureus, B. subtilis, E. coli* and *P. aeruginosa*. The ethanol extract of the shade-dried leaves showed the highest activity exerting the mean diameter of zone inhibition ranging between 13–14 mm that could be attributed to its richness with different secondary metabolites (Latha et al. 2018). In addition, the ethyl acetate extract of *C. phlomidis* formulated in the form of Electrospun Poly Caprolactone (PCL) nanofibrous may revealed potent antibacterial activity against *S. aureus, S. typhi, E. coli* and *P. aeruginosa* with good releasing properties that was evidenced using a period of 60 hours for examining the related studies. Results of the releasing studies showed the initial release started in the first 8 hours and continued for a long time. It is noteworthy to highlight that the extract is rich in secondary metabolites, as revealed by GC-MS analysis that showed the presence of 40 compounds in the non-polar solvent extract mainly belonging to terpenoids, flavonoids with phytol, hexadecanoic acid and palmitic acid representing the predominant compounds. Thus it can be used as an antibacterial wound dressing (Ravichandran et al. 2019).

An additional study was performed on petroleum ether, ethyl acetate, acetone and methanol extracts of the dried powdered leaves of *C. phlomidis* to examine their antimicrobial activities against 20 bacterial strains and 15 fungal strains. Results revealed that ethyl acetate and petroleum ether extracts showed antimicrobial activity against four bacteria and seven fungal strains. The ethyl acetate extract exerted the highest zone of inhibition estimated by 18 mm *versus Staphylococcus epidermidis*; meanwhile, the petroleum ether extract showed a mean diameter for the zone of inhibition estimated by 13 mm against *Trichophyton rubrum* (Hepsibah et al. 2017).

Besides, the antimicrobial activity was additionally assessed on different extracts of *C. phlomidis* leaves, namely petroleum ether, ethyl acetate, acetone and methanol, against multidrug-resistant clinical pathogens comprising 20 bacteria and 15 fungi using the disk diffusion method. Petroleum ether and ethyl acetate revealed a promising activity against four bacteria and seven fungal strains. The greatest zone of inhibition, 18 mm, was reported by the ethyl acetate extract against *Staphylococcus epidermidis*. In contrast, the petroleum ether extract displayed the highest activity against *Trichophyton rubrum*, with a mean of zone inhibition equals 13 mm (Hepsibah et al. 2017).

In addition, the crude flavonoid-rich fractions of *C. phlomidis* were evaluated against some pathogenic organisms. Both plants parts and callus culture displayed high effectiveness against both Gram-positive and Gram-negative bacteria. The former was represented by *Staphylococcus aureus* and *Bacillus subtilis*, while the latter was represented by *Pseudomonas aeruginosa* and *Escherichia coli*. Additionally, they revealed a potent activity against fungal strains represented by *Aspergillus flavus*, *Aspergillus niger* and *Candida albicans*. In contrast, aqueous extracts exerted no inhibition against all the tested pathogenic organisms (Kaur and Renu 2012).

In vitro evaluation of the antimicrobial activity was performed on *C. phlomidis* root extracts from different solvents (petroleum ether, chloroform, ethyl acetate and ethanol) as well as its isolated compounds, which are ethyl-2-hydroxy-4-Me benzoate **(22)**, phenylacetic acid **(23)** and 3,6,7-trihydroxy-2-(3-methoxyphenyl)-4H-chromen-4-one **(24)**. These were tested against *Streptococcus pyogenes*, *Staphylococcus aureus*, *Pseudomonas aeruginosa*, *Escherichia coli*, *Aspergillus niger* and *Candida albicans* by the disk diffusion method, and amikacin and ketoconazole were used as standards. Results revealed that ethanol extract at a concentration of 106.66 µg/mL showed a mean zone of inhibition of 15.33 mm against *Escherichia coli*, whereas the chloroform extract was effective against *Staphylococcus aureus* showing a mean zone of inhibition of 14.67 mm. In addition, the isolated compound ethyl-2-hydroxy-4-Me benzoate showed promising antimicrobial activity against *Staphylococcus pyogenes*, and *Candida albicans* reporting a mean zone of inhibition of 9 mm at 60 µg/mL (Sathish et al. 2013).

C. trichotomum

C. trichotomum herbal extract showed potent antiviral activity against the Respiratory Syncytial Virus (RSV) as experienced in both *in vitro* and *in vivo* models. The extract effectively decreased RSV division RSV-stimulated cell death in addition to the prohibition of RSV-gene transcription as well as RSV protein synthesis and syncytia production. This activity may be attributed to acteoside **(25)** that constitutes the predominant secondary metabolite in the extract as revealed by the HPLC, where acteoside showed similar results as the extract towards RSV (Chathuranga et al. 2019).

Besides, the antibacterial activity of *C. trichotomum n*-hexane, dichloromethane, ethyl acetate and *n*-butanol fractions was assessed against *Staphylococcus aureus*, *Escherichia coli* and *Helicobacter pylori*. *n*-Hexane and dichloromethane fractions displayed antimicrobial activity against *H. pylori* at a concentration of 1.7 mg/mL and displayed mean zones of inhibition of 10 and 11 mm in the disk technique, respectively. 22-Dehydroclerosterol **(26)** and *β*-amyrin **(19)**, compounds isolated from dichloromethane fraction, displayed moderate antibacterial potential against *S. aureus*, *E. coli* and *H. pylori*, at a concentration of 3.4 mg/mL. β-amyrin in particular displayed clear zones of 12 and 13 mm against *E. coli* and *H. pylori*, respectively, showing its potential as an antibacterial agent (Choi et al. 2012).

C. serratum

The antifungal activity of *C. serratum* was performed against some yeast pathogens, namely *Candida glabrata*, *Candida tropicalis* and *Candida albicans*. *C. serratum* showed a pronounced effect against all the tested organisms with a MIC value of 50 µg/mL (Vidya et al. 2019). Methanol extract exhibited the highest activity against a panel of pathogens with inhibition zone ranging from 14 to 32 mm, and MIC equals 0.5–10 mg/mL showing superior activity compared to the aqueous extract activity (Mahajan et al. 2019).

Phytochemical studies of *C. serratum* leaf ethanol extract revealed the presence of alkaloids, phenolics, carbohydrates, proteins, carbonate and glycoside. The antimicrobial potential was determined *in vitro* by using the disk diffusion technique. This was evaluated against *Pseudomonas putida*, *Escherichia coli*, *Staphylococcus hominis*, *Bacillus subtilis* and *Proteus vulgaris*. The results revealed that *Pseudomonas putida* and *Escherichia coli* as Gram-negative in addition to *Staphylococcus hominis*, *Proteus vulgaris* and *Bacillus subtilis* as Gram-positive bacteria are highly susceptible to the antimicrobial effect of the extract displaying zones of inhibition of 10 and 23 mm against the Gram-negative, respectively and ranging 10–20 mm *versus* Gram-positive bacteria that is concentration dependant (Kajaria et al. 2012). Furthermore, different extracts of *C. serratum* displayed antifungal, antibacterial and anti-tuberculosis activity due to various metabolites, which are alkaloids, flavonoids, and phenol in both chloroform and methanol extracts. While terpenoids and saponins are present in methanol extract, steroids were predominant in chloroform extract only. Cleroflavone **(27)**, scutellarein **(28)**, queretaroic acid **(29)**, apigenin **(30)**, serratagenic acid **(31)**, *β*-sitosterol **(17)** as well as D-mannitol **(32)**, verbascoside **(25)**, oleanolic acid **(33)**, hispidulin **(34)**, cholestanol **(35)**, clerosterol **(36)**, campesterol **(37)** and 24-ethyl cholesterol **(38)** are among the predominant components in *C. serratum*. An additional study was conducted to ascertain the antimicrobial effect of different extracts of *C. serratum* against pathogenic organisms listed above. *n*-Hexane leaf extract at a 100 mg/mL concentration exerts a promising antimicrobial against all the examined organisms activity, particularly *Bacillus subtilis* showing a zone of inhibition of 20 mm. In contrast, *Pseudomonas putida* was the least susceptible to the antimicrobial effect of the extract, with a zone of inhibition of 17.1 mm. In comparison, at a concentration of 25 mg/mL, *n*-hexane leaf extract showed the higher inhibition towards *Escherichia coli*, recording a zone of inhibition of 3.62 mm, whereas it recorded the minimum inhibition against *Pseudomonas putida* with a zone of inhibition of 2.5 mm (Singh et al. 2012; Patel et al. 2014).

Silver nanoparticles were synthesized from *C. serratum* leaf extract. It acts as a reducing agent on exposure to the leaf extract; the aqueous silver ions were reduced with the concomitant formation of silver nanoparticles with 35 nm as an average size. The silver nanoparticles were further characterized using UV-Visible as well as transmission electron microscopy (TEM) techniques. The silver nanoparticles formed from *C. serratum* extract were highly effective against human pathogenic bacteria represented by *Escherichia coli* and *Staphylococcus aureus* (Patil and Mengane 2016).

The essential oil obtained from *C. serratum* bark was evaluated for its antimicrobial activity and its phytochemical composition using GC-MS that leads to the identification of 48 compounds represented mainly by linoleic acid, α-zingiberene, oleic acid, palmitic acid and β-sesquiphellandrene. The antimicrobial potential of oil was determined against four pathogenic strains, viz. *Escherichia coli*, methicillin-sensitive *Staphylococcus aureus*, methicillin-resistance *Staphylococcus aureus* and *Proteus* spp. The results showed that the efficacy of oil follows the principle of dose-dependency, displaying a zone of inhibition of 38, 34, 31 and 30 mm for *Proteus, E. coli*, MRSA and MSSA at 200 µL (Noreen et al. 2018).

C. splendens

C. splendens is an African plant that grows as a climbing shrub in West Africa, it has been adopted traditionally to alleviate wounds and infectious disorders. The petroleum ether, ethyl acetate and 70% ethanol extracts cause a pronounced inhibition in the growth of *Staphylococcus aureus, Pseudomonas aeruginosa, Escherichia coli, Bacillus subtilis* and *Candida albicans* (Fleischer et al. 2011). Moreover, silver nanoparticles (AgNPs) were formulated from *C. splendens* leaf extract that permitted an environment-friendly technique where the phytoconstituents predominating in the extract successfully caused a bio-reduction in silver ions concomitant stabilization in AgNPs. The AgNPs displayed a pronounced antimicrobial behavior against airborne pathogens such as *Bacillus subtilis* and *Pseudomonas aeruginosa* (Jayakumar and Vedhaiyan 2019). *C. splendens* chloroform extract of the flower showed high activity *versus Plasmodium falciparum* with IC_{50} < 10 µg/mL. Meanwhile, *C. splendens* chloroform extract of the leaf showed high activity against *T. cruzi* with IC_{50} = 1.98 µg/mL (AbouZid et al. 2013).

In addition, a study was conducted to assess the MICs of some antibiotics represented by ciprofloxacin, amoxicillin, tetracycline and erythromycin in the presence of sub-inhibitory concentrations of the methanol extract of *C. splendens* against some infectious organisms, namely, *Bacillus subtilis, Staphylococcus aureus, Pseudomonas aeruginosa* and *Escherichia coli*. The extracts were found to have different degrees of antimicrobial activity with MICs between 4 and 50 mg/mL. In most cases, the combination between the extract and the antibiotics results in activity reduction (Adu et al. 2019).

C. viscosum

Silver nanoparticles (AgNPs) synthesized from *C. viscosum* water extract displayed a pronounced antiplasmodial activity against the 3D7 sensitive strain of *Plasmodium falciparum* exhibiting IC_{50} = 2.30 µg/mL (Sahoo et al. 2020). Besides, the antimicrobial activity of *C. viscosum* different leaves extracts, namely petroleum ether, chloroform and methanol, were investigated against *Escherichia coli, Shigella flexneri, Proteus vulgaris, Bacillus licheniformis, Staphylococcus epidermidis, Pseudomonas aeruginosa, Staphylococcus aureus* and *Bacillus subtilis*. MIC value of the methanol extract was found to be in the range of 100 µg/mL. In addition, phytochemical screening revealed the presence of alkaloids, flavonoids and reducing sugars in the extract. Another study was conducted on *C. viscosum* root to investigate the antimicrobial activity of the methanol extract as well as its fractions comprising ethyl acetate, carbon tetrachloride and chloroform. Results showed that the ethyl acetate fraction revealed more activity against Gram-negative bacteria than the Gram-positive ones. This activity was undoubtedly attributed to the significant amount of flavonoids and tannins in *C. viscosum* root (Nayak et al. 2016).

C. volubile

The different extracts of *C. volubile* (ethanol, cold and hot aqueous extracts) were examined for their antifungal activity *in vitro* against toxigenic *Aspergillus flavus* using the agar well diffusion method. The ethanol showed strong antifungal activity. As revealed by the phytochemical investigations, saponins and tannins were predominant in the ethanol extract (Jeff-Agboola and Awe 2016).

Role of different *Clerodendrum* species in combating COVID-19

In December 2019, the COVID-19 pandemic causing severe pneumonia was triggered by SARS-CoV-2, a novel coronavirus, has threatened peoples' health all over the globe. By October 2020, SARS-CoV-2 resulted in more than 33 million infected people and over 1 million deaths according to the World Health Organization (Alesawy et al. 2021). This acts as a driving force encouraging scientists to search for therapeutic weapons to combat this lethal virus. Thus, an *in silico* study has recently been conducted on the different metabolites identified by the GC-MS analysis of 12 *Clerodendrum* species within the active sites of RNA-dependent RNA polymerase, the imperative spike protein and main protease enzyme Mpro of SARS-CoV-2. An in-depth molecular docking study revealed that some metabolites such as friedelin **(39)**, taraxerol **(40)**, as well as stigmasterol **(41)** exerted a strong inhibitory potential with the active sites of the targeted proteins. Physicochemical attributes of the compounds are following Lipinski's rule of five as well as Veber's rule. The estimation of Molecular Mechanics-Generalized Born Surface Area (MM-GBSA) binding free energy showed that taraxerol exerted the firm binding within the active sites of all the targeted SARS-CoV-2 proteins. This observation was further ascertained by robust molecular dynamics simulations of the complexes of the viral proteins with taraxerol. It is concluded that taraxerol displayed better binding energy scores with the targeted viral proteins than the drugs specifically targeted against them, thus serving as a promising tool against SARS-CoV-2 infection that needs further *in vivo* studies (Kar et al. 2020).

Conclusion and future perspectives

Members of the genus *Clerodendrom* are known for possessing many biological activities comprising antibacterial potential. They showed considerable antimicrobial activity against a wide array of Gram-positive, Gram-negative, fungi and viruses. This significant anti-infective potential is taken by the plant species either in the crude extracts form or in their nano-formulations. This activity has ultimately relied on their richness in secondary metabolites, particularly squalene, linolenic acid methyl ester, anthraquinones and ethyl-2-hydroxy-4-Me benzoate. Moreover, acteoside, isolated from *C. trichotomum* herbal extract, showed potent antiviral activity against the Respiratory Syncytial Virus (RSV) as experienced in both *in vitro* and *in vivo* models. Thus, genus *Clerodendrom* could lead to promising drugs of natural origin to combat infectious agents.

These therapeutic agents solve the problem of the COVID-19 pandemic that has attacked the world since December 2019, resulting in severe pneumonia that, consequently, threatened human life and caused many deaths. Unfortunately, no specific antiviral drug is available until now to completely cure COVID-19 infection. Meanwhile, antiviral agents as remdesivir displayed mild effect by reducing the rate of mortality and the duration of treatment.

Hence, various extracts and isolated metabolites from *Clerodendrom* species act as an appealing source of drug entities that may encourage scientists to search among them for therapeutic weapons to combat this lethal virus. This could be done by a preliminary screening of the bioactive metabolites using *in silico* studies for their pronounced ability to bind with the active center of SARS-COV-2 crucial proteins such as main protease (M^{Pro}), papain-like protease (PL^{pro}) as well as 3-chymotrypsin-like protease ($3CL^{pro}$), Angiotensin-Converting Enzyme 2 (ACE2) and the RNA-dependent RNA polymerase (RdRp).

Components that affect SARS CoV-2 crucial proteins serve as a perfect strategy for discovering lead drug entities combating COVID-19 infection. Those exerting high binding affinities and their original extracts should be further subjected *in vitro* and *in vivo* assays to confirm their activities. Moreover, clinical trials based on the formulated dosage should be conducted on the bioactive ones to ensure their efficacy. Besides, *Clerodendrom* species should be further examined for combating many other infectious agents, particularly those that cause antimicrobial resistance. They should also be assessed to inhibit bacterial biofilm formation that greatly attributes to their resistance.

References

AbouZid, S.F., Wahba, H.M., Elshamy, A., Cos, P., Maes, L., Apers, S., Pieters, L., Shahat, A.A. (2013). Antimicrobial activity of some *Clerodendrum* species from Egypt. Nat Prod Res 27: 1032–1036.

Adu, F., Yaw, D.B., Agyare, C., Sam, G.H., Etsiapa, V. (2019). Antibacterial resistance modulatory properties of selected medicinal plants from Ghana. African J Pharm Pharmacol 13: 57–69.

Al-Snafi, A.E. (2016). Chemical constituents and pharmacological effects of *Clerodendrum inerme*—A review. SMU Med J 1 3: 129–153.

Alesawy, M.S., Abdallah, A.E., Taghour, M.S., Elkaeed, E.B., Eissa, H.I., Metwaly, A.M. (2021). *In silico* studies of some isoflavonoids as potential candidates against COVID-19 targeting human ACE2 (hACE2) and viral main protease (Mpro). Molecules 26: 2806, doi.org/10.3390/molecules26092806.

Anandhi, K., Ushadevi, T. (2013). Analysis of phytochemical constituents and antibacterial activities of *Clerodendrum inerme* L. against some selected pathogens. Int J Biotechnol Allied Fields 1: 387–393.

Borik, R.M. (2013). Isolation and structural characterization of a steroidal antimicrobial agent from *Clerodendrum baronianum*. World J Chem 8: 48–54.

Byadgi, S., Kulloli, S., Venugopa, C. (2017). Phytochemical screening and antimicrobial activity of plant extracts for textile applications. International J of Biochem Res Rev 20: 1–10.

Chathuranga, K., Kim, M.S., Lee, H.-C., Kim, T.-H., Kim, J.-H., Gayan Chathuranga, W., Ekanayaka, P., Wijerathne, H., Cho, W.-K., Kim, H.I. (2019). Anti-respiratory syncytial virus activity of Plantago asiatica and *Clerodendrum trichotomum* extracts *in vitro* and *in vivo*. Viruses 11: 604, doi.org/10.3390/v11070604.

Choi, J.W., Cho, E.J., Lee, D.G., Choi, K., Ku, J., Park, K.-W., Lee, S. (2012). Antibacterial activity of triterpenoids from *Clerodendron trichotomum*. J App Biol Chem 55: 169–172.

Doley, P., Singh, A., Devi, N.M., Singh, C.B., Thokchom, A. (2016). Reverse phase HPLC estimation of antioxidants and antimicrobial activities of *Clerodendrum colebrookianum* Walp. J of Pharmacog Phytochem 5: 199–205.

Farooqui, M.A., Chauhan, P.S., Krishnamoorthy, P., Shaik, J. (2010). Extraction of silver nanoparticles from the leaf extracts of *Clerodendrum inerme*. Digest J Nanomat and Biostr 5: 43–49.

Fleischer, T., Mensah, A., Oppong, A., Mensah, M., Dickson, R., Annan, K. (2011). Antimicrobial and anti-inflammatory activities of the leaves of *Clerodendrum splendens* leaves. Pharmacog Comm 1: 85–89.

Habila, J.H., Isah, Y., Adeyemi, M., Ayo, R., Ndukwe, G., Momoh, H.M. (2018). Antimicrobial and antitubercular activities of bioactive triterpenes from *Clerodendrum capitatum* (Willd) leaves. J Chem Soci Nigeria 43, doi: www.journals.chemsociety.org.ng/index.php/jcsn/article/view/152.

Hamidu Usman, V.I., Suyi, H.E., Umar, H.A. (2017). Comparative phytochemistry and *in vitro* antimicrobial effectiveness of the leaf extracts of *Clerodendrum capitatum* (Verbenaceae) and *Ficus glumosa* (Moraceae). J Pharmacog Phytochem 6: 945–948.

Hepsibah, A.H., Mala, M., Jothi, G.J. (2017). Antimicrobial activity and TLC profiling of *Clerodendrum phlomidis* Linn. F. leaf extract against multidrug resistant clinical pathogens. Int J Pharm Pharm Sci 9: 275–278.

Huang, K.C. (1998). The Pharmacology of Chinese Herbs. CRC Press, Boca Raton.

Jayakumar, A., Vedhaiyan, R.K. (2019). Rapid synthesis of phytogenic silver nanoparticles using *Clerodendrum splendens*: its antibacterial and antioxidant activities. Korean J Chem Eng 36: 1869–1881.

Jeff-Agboola, Y., Awe, L. (2016). Antifungal and phytochemical screening of some Nigerian medicinal plant extracts against toxigenic *Aspergillus flavus*. Cogent Food Agri 2: 1210556.

Jha, A.K., Prasad, K. (2016). Green synthesis and antimicrobial activity of silver nanoparticles onto cotton fabric: An amenable option for textile industries. Adv Mat Lett 7: 42–46.

John, J.E. (2010). Natural products as lead-structures: a role for biotechnology. Drug Disc Today 15: 409–410.

Kajaria, D., Gangwar, M., Kumar, D., Sharma, A.K., Nath, G., Tilak, R., Tripathi, Y.B., Tripathi, J., Tiwari, S. (2012). Qualitative phytochemical characterization and antimicrobial evaluation of a polyherbal compound-Bharangyadi. J Pharm Res 5: 416–419.

Kar, P., Mishra, D.K., Roy, A., Chakraborty, A.K., Sinha, B., Sen, A. (2020). Elucidation of phytomedicinal efficacies of *Clerodendrum inerme* (L.) Gaertn. (Wild Jasmine). South Afr J Bot 1–9.

Kar, P., Sharma, N.R., Singh, B., Sen, A., Roy, A. (2020). Natural compounds from *Clerodendrum* spp. as possible therapeutic candidates against SARS-CoV-2: An *in silico* investigation. J Biomol Str Dynam 1–12, doi.org/10.1080/07391102.2020.1780947.

Kaur, C., Renu, S. (2012). Evaluation of antimicrobial activity of crude flavonoids in medicinally important arid zone plant *Clerodendrum phlomidis* Linn. Int Res J Pharm 3: 1835–1843.

Kekuda, T.P., Sudharshan, S. (2018). Ethnobotanical uses, phytochemistry and biological activities of *Clerodendrum paniculatum* L. (Lamiaceae): A comprehensive review. J Drug Delivery Ther 8: 28–34.

Khan, S.A., Shahid, S., Ahmad, W., Ullah, S. (2017). Pharmacological importance of *Clerodendrum* genus: a current review. Int J Pharm Sci Res 2: 22–30.

Khan, S.A., Kanwal, S., Rizwan, K., Shahid, S. (2018). Enhanced antimicrobial, antioxidant, *in vivo* antitumor and *in vitro* anticancer effects against breast cancer cell line by green synthesized un-doped SnO_2 and Co-doped SnO_2 nanoparticles from *Clerodendrum inerme*. Microb Path 125: 366–384.

Khan, S.A., Shahid, S., Lee, C.-S. (2020). Green synthesis of gold and silver nanoparticles using leaf extract of *Clerodendrum inerme*; characterization, antimicrobial, and antioxidant activities. Biomolecules 10: 835, doi.org/10.3390/biom10060835.

Kor-arnan, S., Thiraphibundet, P. (2016). Antibacterial and antifungal activities of isolated compounds from *Clerodendrum inerme*. J Sci Technol MSU 35: 60–66.

Latha, R., Fathima, M., Dhanraj, K. (2018). *In vitro* studies on the antimicrobial effect of different solvent extracts of *Dodonaea aungustifolia* L. and *Clerodendrum phlomidis* L. against some pathogenic bacteria. Int J Drug Develop Res 10: 13, 188: 8–18.

Mahajan, H., Singh Arora, D., Singh, H., Jain, S.K., Namarta, K., Singh, J. (2019). Scientific validation of the antimicrobial and antiproliferative potential of *Clerodendrum serratum* (L.) Moon, its phytoconstituents and their biosafety by acute oral toxicity study. Drug Chem Toxicol 1–11, doi.org/10.1080/01480545.2019.1643875.

Mahesh, M., Preenon, B., Lalthanzama, V., Somashkar, R., Ravishankara, B.E., Benaka, S.B., Richard, S.A., Dhananjaya, B.L. (2015). The antioxidant and antimicrobial activity of the leaves extract of *Clerodendrum colebrookianum* Walp. (Fam: Verbenaceae). Int J Pharm Pharm Sci 7: 96–99.

Masevhe, N.A., Awouafack, M.D., Ahmed, A.S., McGaw, L.J., Eloff, J.N. (2013). Clerodendrumic acid, a new triterpenoid from *Clerodendrum glabrum* (Verbenaceae), and antimicrobial activities of fractions and constituents. Helv Chimica Acta 96: 1693–1703.

Mehrbod, P., Abdalla, M.A., Njoya, E.M., Ahmed, A.S., Fotouhi, F., Farahmand, B., Gado, D.A., Tabatabaian, M., Fasanmi, O.G., Eloff, J.N. (2018). South African medicinal plant extracts active against influenza A virus. BMC Comp Alt Med 18: 112, doi.org/10.1186/s12906-018-2184-y.

Nayak, S.K., Das, D., Tripathy, N.K., Sarangi, R.R. (2016). Comparative *in vitro* antimicrobial activity studies of *Clerodendrum viscosum* Vent leave extracts. Int J Pharma Res Rev 5: 1–6.

Njeru, S.N., Obonyo, M., Nyambati, S., Ngari, S., Mwakubambanya, R., Mavura, H. (2016). Antimicrobial and cytotoxicity properties of the organic solvent fractions of *Clerodendrum myricoides* (Hochst.) R. Br. ex Vatke: Kenyan traditional medicinal plant. J Intercult Ethnopharmacol 5: 226–232.

Noreen, R., Intisar, A., Ghaffar, A., Jabeen, F., Abid, M.A., Din, M.I., Irfan, M., Faiz, F., Sattar, T. (2018). Constituents of volatile oil from bark of *Clerodendrum serratum* (L.) and its antibacterial activity. J Essent Oil Bear Plants 21: 198–205.

Patel, J.J., Acharya, S.R., Acharya, N.S. (2014). *Clerodendrum serratum* (L.) Moon.—A review on traditional uses, phytochemistry and pharmacological activities. J Ethnopharmacol 154: 268–285.

Patil, A., Mengane, S. (2016). Green synthesis of silver nano particles from *Clerodendrum serratum* and antimicrobial activity against human pathogens. J Bionanosci 10: 491–494.

Rad, S.S., Sani, A.M., Mohseni, S. (2019). Biosynthesis, characterization and antimicrobial activities of zinc oxide nanoparticles from leaf extract of *Mentha pulegium* (L.). Micro Patho 131: 239–245.

Ravichandran, S., Radhakrishnan, J., Jayabal, P., Venkatasubbu, G.D. (2019). Antibacterial screening studies of electrospun Polycaprolactone nano fibrous mat containing *Clerodendrum phlomidis* leaves extract. App Sur Sci 484: 676–687.

Ryan, K.J., Ray, C.G. (2004). Medical Microbiology. McGraw Hill, New York 4: 370.

Saha, S., Mukherjee, A., Biswas, S., Choudhury, D., Saha, J., Pal, S., Sarkar, M., Kole, R. (2018). Formulation and chemical characterization of *Clerodendrum infortunatum* leaf extract in relation to antifungal activity. Heliyon 4: e01047, doi.org/10.1016/j.heliyon.2018.e01047.

Sahoo, R.K., Tamuli, K.J., Narzary, B., Bordoloi, M., Sharma, H.K., Gogoi, K., Bhattacharyya, D.R. (2020). *Clerodendrum viscosum* Vent leaf extract supported nanosilver particles: Characterization, antiplasmodial and anticancer activity. Chem Phys Lett 738: 136893, doi.org/10.1016/j.cplett.2019.136893.

Sathish, M., Priyadarsini, R., Sunitha, P., Saraswathy, T. (2013). Antimicrobial activity of the extracts and isolated compounds of *Clerodendrum phlomidis*. Int J Pharm Pharma Sci 5: 362–366.

Shrivastava, N., Patel, T. (2007). Clerodendrum and healthcare: an overview. Med Arom Plant Sci Biotechnol 1: 142–150.

Shruthi, G., Joyappa, M.P., Chandrashekhar, S., Kollur, S.P., Prasad, M., Prasad, A., Shivamallu, C. (2017). Bactericidal property of *Clerodendrum paniculatum* and Saraca asoka against multidrug resistant bacteria, restoring the faith in herbal medicine. Int J Pharm Pharma Sci 8: 3863–3871.

Simorangkir, M., Hutabarat, W., Nainggolan, B., Silaban, S. (2019). Antioxidant and antibacterial activities of nonplolar to polar solvent extracts Sarang Banua (*Clerodenrum fragrans* Vent Wild) leaves. J Chem [RJC] 12: 959–965.

Singh, M.K., Khare, G., Iyer, S.K., Sharwan, G., Tripathi, D. (2012). *Clerodendrum serratum*: A clinical approach. J App Pharma Sci 2: 11–15.

Sirohi, P., Yadav, A.K., Singh, N.K. (2019). Inhibitory mechanism of *n*-hexane and dichloromethane leaf extracts of *Clerodenrum phlomidis*. J Plant Develop Sci 11: 501–510.

Sivamaruthi, B.S., Kesika, P., Subasankari, K., Chaiyasut, C. (2018). Beneficial effects of anthocyanins against diabetes mellitus associated consequences—A mini review. Asian Pacific J Trop Biomed 8: 471–477.

Varela, M.F., Stephen, J., Lekshmi, M., Ojha, M., Wenzel, N., Sanford, L.M., Hernandez, A.J., Parvathi, A., Kumar, S.H. (2021). Bacterial resistance to antimicrobial agents. Antibiotics 10: 593; doi.org/10.3390/antibiotics10050593.

Vidya, A.G., Vijayan, A., Jyothis, L., Nair, R., Suja, K. (2019). Evaluation of antifungal efficacy of some medicinal plants on *Candida* spp. causing vulvovaginitis. Ind J Exp Biol 57: 297–301.

Waliullah, T.M., Yeasmin, A.M., Wahedul, I.M., Parvez, H. (2014). Evaluation of antimicrobial study *in vitro* application of *Clerodendrum infortunatum* Linn. Asian Pacific J Trop Dis 4: 484–488.

Wink, M., Ashour, M.L., El-Readi, M.Z. (2012). Secondary metabolites from plants inhibiting ABC transporters and reversing resistance of cancer cells and microbes to cytotoxic and antimicrobial agents. Front Microbiol 3: 130–145.

Yang, C.-X., Zuo, G.-Y., Han, J., Wang, G.-C, Bian, Z-Q. (2012). *In vitro* antimicrobial activities of 26 Yunnan plant extracts against multidrug resistant pathogens. Afr J Microbiol Res 6: 4132–4137.

Youssef, F.S., Altyar, A.E., Omar, A.M., Ashour, M.L. (2021). Phytoconstituents, *in vitro*, anti-infective activity of *Buddleja indica* Lam., and *in silico* evaluation of its SARS-CoV-2 inhibitory potential. Front Pharmacol 12: 298, doi.org/10.3389/fphar.2021.619373.

4

Antimicrobial Activity of Extracts and Fractions of *Ximenia americana* L. (Olacaceae)

Valdiléia Teixeira Uchôa,[1,*] *Mahendra Rai,*[2,3]
Gilmânia Francisca Sousa Carvalho,[1] *Herbert Gonzaga Sousa,*[1]
Patrícia e Silva Alves[1] and *Renata da Silva Carneiro*[1]

Introduction

Since the early days of civilization, popular medicine has been present in our lives, through the ingestion of herbs and part of plants for diagnostic purposes, prophylaxis, treatment and cure of diseases (Falcão et al. 2017). The first herbal records date back to the period 2838–2698 BC when the Chinese Emperor Shen Nung catalogued 365 medicinal herbs and poisons that were used under the Taoist inspiration of Pan Ku, considered a God of creation (França et al. 2008). Medicinal plants are well known natural sources of therapeutic agents used in the treatment of various diseases, based on the empirical knowledge of indigenous, African and European cultures, historical, social and religious factors. Despite the great technological advances in pharmacology and health, natural remedies are widely prescribed, as they are recognized as a safe material with less side effects than synthetic drugs, becoming the most frequent use in therapeutic medicine and as supplementary material (Cioch et al. 2017; Ginovyan et al. 2017).

Medicinal plants present a source of active natural products with pharmacological action, which differs directly between the plant species, the structural variability of chemical compounds and their biological properties. Since 1978, the World Health Organization has recognized phytotherapy as an official practice and recommended the dissemination of the knowledge necessary for its use (Souza et al. 2016; Oliveira et al. 2018). Medicinal plants are over-the-counter products, commercialized in the world as herbal medicines or as active ingredients. Besides their popular use to treat and cure human diseases, natural products derived from plants can play an important role as a source of pharmacological tools for the understanding of biochemical routes and mechanisms of disease (Amparo et al. 2018).

[1] PPGQ-GERATEC, Department of Chemistry-State University of Piauí, Teresina, Piauí, Brazil.
[2] Nanobiotechnology Lab., Department of Biotechnology, Sant Gadge Baba Amravati University, Amravati-444602, Maharashtra.
[3] Department of Microbiology, Nicolaus Copernicus University, Torun, Poland.
* Corresponding author: valdileiatexeira@ccn.uespi.br

Many researches have attempted to develop and strengthen pharmaceutical properties in popular medicine, allowing significant discoveries of natural drugs containing antibiotic, antineoplastic, analgesic, cardioprotective effect, among others. According to the World Health Organization, approximately four billion people (80% of the population in developing countries) depend on medicinal plants for their primary health care (De Andrade et al. 2019). However, the safety of its use should be questioned due to reports of toxicity or disease. Before being used with therapeutic purposes, medicinal plant extracts should be evaluated for their exact chemical composition, the mechanism of action and their therapeutic doses. High levels of toxic agents can occur in medicinal plants when they are in places contaminated with microorganisms or metals, such as near roads, metal mining or landfills (Abdallah and Koko 2017).

Even with the expansion of technology and the advancement of medicine, infectious diseases are increasing and continue to be a great health problem around the world (Poulios et al. 2020). Among these are pathologies caused by bacteria, unicellular beings and procreation of the Monera kingdom, most times associated with the occurrence and proliferation of several pathologies and processes of decomposition of organic matter. The prevalence of microbial infectious diseases and their complications have increased in recent years, especially the resistance of microorganisms due to the indiscriminate use of multiple antimicrobial drugs commonly used (Francisco et al. 2020).

However, due to the rapid development of resistance against drugs and pharmaceuticals in general, there is a need for the development of new active antimicrobial compounds, which have an efficacy against resistant microbes (De Andrade et al. 2019). According to Borges et al. (2016), the plant materials are demonstrated as one of the most promising sources for the development of new approaches to the discovery of antimicrobial compounds. As the use of traditional medicines clearly shows that bacteria are unable to develop resistance to multiple phytochemicals present in plant extracts (Santos et al. 2018).

The secondary metabolites of the plants are mainly responsible for their antimicrobial activity in the various types of extracts, this occurs due to the compounds presenting the main phenolic and polyphenols groups (flavonoids, quinones, tannins, coumarins), terpenoids, alkaloids, lectins and polypeptides. Among the several types of reaction of the compounds with antimicrobial action, metabolites can act in the destruction of the microbial membrane (carvacrol, thymol, eugenol, etc.), inhibition of the cellular metabolism (cinnamaldehyde), control of the formation of biofilms, inhibition of bacterial capsule production and reduction of microbial toxin production (Poulios et al. 2020).

Ximenia americana L.

In various parts of the world, the population uses different plants as a therapeutic resource to fight against diseases. The species *Ximenia americana* L. is recognized as a medicinal plant with ethnopharmacological relevance.

The genus *Ximenia* is a representative of the Olacaceae group being grown as shrubs and distributed in tropical and subtropical regions (Sobeh et al. 2017).

The species *Ximenia americana* (Linn) has common names such as sour plum, sea lemon (Agyigra et al. 2017), wild plum and thorn plum (De Menezes et al. 2019). It is a small tree of up to 6 meters, its branches have straight spines, the leaves are alternated or grouped and its flowers are greenish-white and scented (Feyssa et al. 2012), as seen in Fig. 4.1.

It is a pantropical plant, native to the region of Central and South America and quite common in all intertropical Africa (Soro et al. 2009). In Brazil, *X. americana* L. is present in the Caatinga biome and in the dry period, when most of the species in this region lose

(a) (b)

Figure 4.1. Morphological aspects of *X. americana*. (a) young shrub; (b) leaves and flowers.

their leaves, this plant stands out for its green leaves, which characterizes a plant resistant to lack of water (Silva et al. 2008).

The *X. americana* is characterized by the presence of an axillary stem on branches, by the non-added chalice on fruits and by petals with an inner surface densely covered by sinuous and transparent trichomes (Rocha et al. 2017).

For medicinal purposes, the most used parts of this plant are the roots and the stem bark (Le et al. 2012). The roots can treat rheumatism, diarrhea (Koné et al. 2004), leprosy and syphilis (Shettar et al. 2015).

Phytochemical investigations of the *X. americana* stem bark extract demonstrated the presence of bioactive compounds such as flavonoids, saponins, tannins and terpenoids (Maikai et al. 2009), widely used in the treatment of ulcers, presenting a purifying action (Júnior et al. 2019), wound healing, well-known menstrual regulator (De Araújo et al. 2008). Several studies support the anti-inflammatory and antioxidant action of the stem bark of this species (Shettar et al. 2015; Dias et al. 2018). The presence of flavonoids identified in the aqueous extract of the barks suggested the stagnation of the pulmonary inflammatory process (Carvalho et al. 2020). When tested on animals undergoing tendon injury treatment combined with pulsed ultrasound, the *X. americana* stem bark gel caused a significant improvement in edema (Leal et al. 2016).

In the chemical composition of the leaves, flavonoids and tannins can be identified (Brasileiro et al. 2008). They are used in the treatment of measles, toothache and also as a laxative (Mevy et al. 2006).

The fruits are similar in shape and size to a plum, dark reddish-brown to bright orange-scarlet. The edible pulp clings to a stone, which contains an oil-rich, soft cream-colored seeds (Rezanka and Sigler 2007). In phytochemical research, the green, yellow and red fruits of *X. americana* had a higher content of flavonoids, anthocyanins, polyphenols and starch (Almeida et al. 2016; Bazezew et al. 2021).

Bioactive metabolites from *X. americana*

Preliminary and quantitative phytochemical studies of *X. americana* indicate the presence of several classes of secondary metabolites in this species. Shettar et al. (2015) identified phenolic, flavonoid, tannin, saponin and steroid compounds in a study developed with the chloroform, ethyl acetate, methanolic, ethanolic and hydroalcoholic fractions of the leaves of *X. americana*. The results of their study corroborate with the previous findings of many researchers (Onguleye and Ibitoye 2003; James et al. 2007; Palma et al. 2020) when investigating alcoholic and hydroalcoholic extracts from the leaves and stem bark of the same species also by qualitative techniques.

A phytochemical study by HPLC-MS developed by Maia (2018) found the presence of polyphenolic compounds quercetin and (–)-epicatechin in the methanolic extracts of the roots, leaves and stem bark of *X. americana*. Similar results were found by da Silva et al. (2018) in a study developed with hydroalcoholic extract from the stem bark of this plant. Uchôa et al. (2016) isolated the flavonoids quercetin (**1**) from the leaf and epicatechin (**2**) from the stem bark ethyl acetate fraction (Fig. 4.2).

All the phytochemicals found in *X. americana* are considered bioactive, with biological and pharmacological activities of broad interest to the scientific community, which gives this plant great potential for the development of herbal medicines. These constituents are attributed to their antioxidant, anti-inflammatory, antinociceptive, molluscicidal, gastroprotective, hepatoprotective and antimicrobial activities (Costa et al. 2010; Almeida et al. 2016; Da Silva et al. 2018).

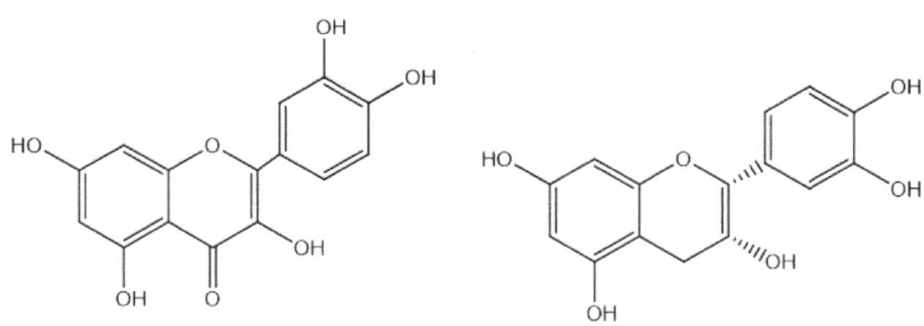

Figure 4.2. Structure of quercetin (1) and epicapicatechin (2) isolated from *X. americana*.

Antimicrobial activity of *Ximenia americana*

Several researchers have studied the biological properties of different parts of *Ximenia americana* L. and their antimicrobial action (antibacterial, antifungal and antiviral effect). They carried out tests by means of broth microdilution, disk diffusion, plate diffusion and cup-plate method using extracts of leaves, stems, roots and barks with solvents of different polarities such as butanol, petroleum ether, chloroform, ethanol, methanol and water, which can be seen in Table 4.1.

Omer and Elnima (2003) investigated the antimicrobial activity of *X. americana* against bacteria and fungi using extracts from the bark, leaves, roots and stem of this species. Among the solvents used (chloroform, methanol and water), it was observed that methanolic extracts were more active, followed by aqueous extracts. *Staphylococcus aureus*, among the microorganisms used, showed greater sensitivity to the extracts of *X. americana*.

The leaves of *X. americana* showed antimicrobial activity when investigating organic (ethanol/chloroform) and aqueous extracts by the cup-plate method (Onguleye and Ibitoye 2003). The organic extracts were shown to be active against *Escherichia coli*, NCTC 10418; *Pseudomonas aeruginosa*, NCTC 6750; *Proteus vulgaris*, NCTC 4636; *Bacillus subtilis*, NCTC 8241; *S. aureus*, NCTC 6571; and *Candida albicans*, AATC 10231. Greater inhibition zones were revealed by the ethanol/chloroform extracts against *P. aeruginosa*, with bactericidal action similar to penicillin 2 µg, a reference antibiotic sold commercially.

Table 4.1. Antibacterial, antifungal and antiviral activity of *Ximenia americana* L.

Antimicrobial method	Used part/solvents	Antimicrobial activity	Paper	Reference
Plate diffusion	Bark, leaves, roots and stem/ Chloroform, methanol and water	Methanolic and aqueous extracts were the most active, with *Staphylococcus aureus* being the most susceptible bacterium among the tested organisms. *Candida albicans* was the least suceptible among the tested fungi.	Antimicrobial activity of *Ximenia americana*	(Omer and Elnima 2003)
Cup-plate method	Leaves/Ethanol/ Chloroform, Water	The results indicated that the ethanol/chloroform extract of *X. americana* was active against all test isolates. The inhibition zone recorded for the various organisms was higher for *Pseudomonas aeruginosa*	Studies of antimicrobial activity and chemical constituents of *Ximenia americana*	(Ogunleye and Ibitoye 2003)
Plate diffusion and microdilution	Roots/Ethanol	Ethanolic extract from the roots showed antimicrobial activity against Gram positive bacteria (*S. aureus, Enterococcus faecalis, Bacillus subtilis* and *Streptococcus pyogenes*) with IC_{50} over 47 μg/mL	Traditional medicine in North Côte-d'Ivoire: screening of 50 medicinal plants for antibacterial activity	(Koné et al. 2004)
Broth microdilution	Leaves, stem barks and roots/ Methanol, Water	The extracts inhibited *S. aureus* and *Klebsiella pneumoniae* while *Shingella flexneri* was inhibited only by the leaves methanolic extracts, aqueous stem bark and aqueous leaves. *Salmonella typhi* and *Escherichia coli* were not affected by these extracts	Phytochemical and Antimicrobial Investigation of the Aqueous and Methanolic Extracts of *Ximenia americana*	(James et al. 2007)
Disk diffusion	Stem bark/ Petroleum ether, chloroform, butanol, methanol and water	Methanolic and aqueous extracts showed ample activity in inhibiting the growth of *E. coli, P. vulgaris, S. aureus, P. aeruginosa* and *B. subtilis*	Antimicrobial Properties of Stem Bark Extracts of *Ximenia americana*	(Maikai et al. 2009)
Disk diffusion	Leaves/Methanol, water	There was no activity of both extracts against *S. aureus, P. aeruginosa, Proteus* sp., *S. pneumoniae* and *E. coli*. Minimum inhibitory concentrations (MIC) above 1000 μg/mL	Studies on the antibacterial activity and chemical constituents of Khaya senegalensis and *Ximenia americana* leaf extracts	(Kawo et al. 2011)

Table 4.1 contd. ...

...*Table 4.1 contd.*

Antimicrobial method	Used part/solvents	Antimicrobial activity	Paper	Reference
Plate diffusion/ Microdilution	Stem bark/ Ethanol/water	Inhibition of *S. aureus* (MIC = 0.250–0.063 μL/μL), *E. faecalis* (MIC = 1.000 μL/μL) and *S. oralis* (MIC = 1.000 μL/μL)	Study bioprospecting of medicinal plant extracts of the semiarid northeast: contribution to the control of oral microorganisms	(Silva et al. 2012)
Microdilution	Leaves and bark/ Ethanol	The extracts have potential as bactericidal agents and modulators of bacterial resistance, and can be used as adjuvants in the treatment of infections by *S. aureus* resistant to erythromycin	Modulation of the erythromycin resistance in *Staphylococcus aureus* by ethanolic extracts of *Ximenia americana* L. and *Schinopsis brasiliensis* Engl	(Silva et al. 2015)
Broth microdilution	Barks	The extracts showed inhibitory activity angainst *C. albicans* in the concentration range between 125 and 250 μg/mL. The MIC of 125 μg/mL was presented by extracts of ethanol 70% and 90% for maceration, 50% and 70% for turbolysis and 50% for ultrasound	Tablet of *Ximenia americana* L. developed from mucoadhesive polymers for future use in oral treatment of fungal infections	(Almeida et al. 2019)
Microdilution	Stem barks/ Ethanol/Water	Hydroalcoholic extract showed MIC ≥ 1024 μg/mL against all tested bacteria and synergistic effect against strains of *S. aureus* when associated with norfloxacin	*Ximenia americana* L. enhances the antibiotic activity and inhibit the development of kinetoplastid parasites	(De Menezes et al. 2019)
Evaluation of cell death caused by extract toxicity and inhibition of viral cytopathic effect	Stem-bark	The hydroalcoholic extract inhibited the replication of HIV-2 at concentrations that were non-toxic to the host cells (EC_{50}: 27.1 mg/mL)	Antiviral Activity Against Human Immunodeficiency Virus Type 1 (HIV-1) and Type 2 (HIV-2) of Ethnobotanically Selected Ethiopian Medicinal Plants	(Asres et al. 2001)

Koné et al. (2004) investigated the antimicrobial activity of ethanolic extracts from 50 medicinal plants used by communities in the North of Côte d'Ivoire as treatment for bacterial diseases. The root extract of *X. americana* was shown to be active against Gram positive bacteria *S. aureus*, *B. subtilis*, *Enterococcus faecalis* and *Streptococcus pyogenes*, the last two being the most sensitive to the ethanolic extract of the roots of this species.

The study developed by James et al. (2007) evaluated the antibacterial activity of methanolic and aqueous extracts from the bark, leaves and roots of *X. americana* against five bacteria, using the broth microdilution method. It was observed that, among the extracts used, the methanolic extract of the roots showed greater bactericidal activity against *Klebsiella pneumoniae*, also standing out with greater significant inhibition ($p < 0.05$) on *S. aureus* when compared to the aqueous extracts of the leaves and stem bark. The Minimum Inhibitory Concentration (MIC) was only evident for methanolic extracts at 1.25×10^4 µg/mL against *S. aureus* while the Minimal Bactericidal Concentration (MBC) of these extracts in *S. aureus* was obtained at 2.50×10^4 µg/mL.

In another study by James et al. (2007), the aqueous and methanolic extracts of the leaves and aqueous extracts of the barks were able to inhibit *Shigella flexneri*. *Salmonella typhi* and *Escherichia coli* were not susceptible to the extracts investigated. The authors justify this behavior because Gram-negative bacteria have more complex cell structures. According to Silver and Bostian (1993), the same extract can present different inhibition halos for different strains, and this is combined with the specific resistance of each microorganism. The antimicrobial activity of *X. americana* in this study is directly related to its chemical composition, mainly regarding the presence of tannins and flavonoids.

The antimicrobial activity of *X. americana* was also investigated by Maikai et al. (2009) who tested the inhibitory capacity of stem bark extracts of the species against the microorganisms *E. coli*, *P. aeruginosa*, *S. aureus*, *P. vulgaris*, *B. subtilis* and *C. albicans*. The methanolic and aqueous extracts showed greater zones of growth inhibition of the tested organisms when compared to the butanolic extract. The results of this study indicate that the greater the polarity of the solvent, the greater the bactericidal and fungicidal activity of the studied plant (Migliato et al. 2011).

In a study developed by Kawo et al. (2011), aqueous and methanolic extracts from the leaves of *X. americana* were not active against *S. aureus*, *P. aeruginosa*, *Proteus* sp., *Streptococcus pneumoniae* and *E. coli* when executed by the method of disk diffusion. The results show minimal inhibitory concentrations above 1000 µg/mL, indicating that the extracts examined were not able to inhibit the growth of microorganisms. The authors relate these results to the possible absence or low concentration of bioactive compounds with bactericidal activity. In addition to this, other factors may interfere with the inhibition of the growth of microorganisms, which include the composition of the plant extract, extractive method, solvent polarity, as well as operational care and collection period (Tortora et al. 2005; Ayres et al. 2008).

Silva et al. (2012) evaluated the antimicrobial activity of 30 plant species against 10 oral pathogenic microorganisms. In that study, hydroalcoholic extracts 10, 20, 30, 50 and 70% of *X. americana* stem barks proved effective against *S. aureus* (MIC = 0.250–0.063 µL/µL), *E. faecalis* (MIC = 1,000 µL/µL) and *Streptococcus oralis* (MIC = 1,000 µL/µL). These results indicate that *X. americana* exhibits great potential in the formulation of bioproducts with bactericidal actions for use in the treatment of oral problems.

Regarding the modulating activity, Silva et al. (2015) studied it using ethanol extracts from the leaves and barks of *X. americana* against strains of *S. aureus* resistant to erythromycin. It was observed by the microdilution method that the barks ethanolic extracts showed antimicrobial potential against most of the tested organisms in concentrations lower than the reference antibiotic. Even when combined with erythromycin, the bark ethanolic extracts still showed a good synergistic effect, decreasing MIC. However, the leaf extracts showed bactericidal activity only when tested alone, showing no synergistic effect with the antibiotic.

Almeida et al. (2019), researched the antifungal activity of hydroalcoholic extracts from the barks of *X. americana* obtained by different extraction methods, which were later incorporated into polymeric structures for the production of tablets. Hydroalcoholic extracts at concentrations of 50, 70 and 90% were tested against the fungus *C. albicans* and the minimum inhibitory concentrations (MIC) found varied between 125 and 250 µg/mL. The MIC of 125 µg/mL was found in hydroalcoholic extracts 50 and 70% by the turbolysis method, 50% in the ultrasonic and 70% and 90% for the maceration method.

The study also suggests that the secondary metabolites found in plants, especially in the species studied, are responsible for antifungal activity. Through synergism, these biocomposites can act to inhibit the formation of the fungal cell appearance of microorganisms, causing disruptions and preventing cell division (Zaynab et al. 2018). Earlier ethnopharmacological investigations revealed the antifungal activity against *C. albicans* of several species of the genus *Ximenia*, indicating the potential of these plants in the treatment of oral candidiasis (Runyoro et al. 2016; Maroyi 2016). In addition, the leaves of *X. americana* showed antifungal activity (IC_{50} 8.12 µg/mL) against *C. albicans* when investigating methanol extracts in the study developed by Traoré et al. (2015).

A recent study developed by De Menezes et al. (2019), investigated the bactericidal and modulatory activity of hydroalcoholic extracts from *X. americana* stem bark against the bacteria *E. coli*, *S. aureus* and *P. aeruginosa* and multidrug-resistant strains of these microorganisms. Through the microdilution method, MIC values ≥ 1024 µg/mL were found against all bacteria used. The MIC values were reduced by 50% when the modulating activity of these extracts combined with norfloxacim against *S. aureus* was evaluated, indicating that the bioactive compounds present in the *X. americana* have the capacity to act synergistically with antibiotics in the fight against pathogenic microorganisms.

X. americana exhibited antiviral activity in a study developed by Asres et al. (2001). In this study, hydroalcoholic extracts from the stem bark of this species were tested against HIV-1 (IIIB) and HIV-2 (ROD), showing inhibitory activity on replication of these viruses at concentrations ranging from 8.3 to 27.7 µg/mL.

By the studies analyzed here, it was possible to verify a good antimicrobial and modulator activity of *X. americana* against different bacterial strains, which can be attributed, according to literature, to the properties exposed by compounds synthesized in secondary metabolism. Secondary metabolites such as alkaloids and phenolic substances such as simple phenols, phenolic acids, quinones and flavones are the main groups of compounds extracted from plants with antimicrobial properties (Gotlieb 1981; Aderinokun et al. 1999; Souza et al. 2000; Esquenazi et al. 2002; Singh et al. 2002; Hatano et al. 2005). These classes of metabolites are widely reported in phytochemical studies of the species *X. americana* (Shettar et al. 2015; da Silva et al. 2018).

Conclusion

The effectiveness of *Ximenia americana* as an antimicrobial agent has been proven, since extracts of this species have shown satisfactory results in controlling the growth of a wide variety of microorganisms, which include bacteria, yeasts, filamentous fungi and viruses. Moreover, considering the antimicrobial activity, tablets of *X. americana* have been developed for the treatment of oral fungal infections. However, further studies are needed in the search for new antimicrobial agents from natural products, since plant extracts can slow or prevent the growth of these resistant pathogens, as these plant species have several chemical constituents that can potentiate the antimicrobial effect.

References

Abdallah, E.M., Koko, W.S. (2017). Medicinal plants of antimicrobial and immunomodulating properties. Antimicrobial Research: Novelbioknowledge and Educational Programs. Formatex Research Center, Spain, 127–139.

Aderinokun, G.A., Lawoyin, J.O., Onyeaso, C.O. (1999). Effect of two common Nigerian chewing sticks on gingival health and oral hygiene. Odont Trop 22: 13.

Agyigra, I.A., Ejiofor, J.I., Magaji, M.G. (2017). Avaliação da toxicidade aguda e subcrônica do extrato metanólico da casca do caule de *Ximenia americana* Linn (Olacaceae) em ratos Wistar. Bol Fac Farm 55(2): 263–267.

Almeida, L., Oshiro Junior, J.A., Silva, M., Nóbrega, F., Andrade, J., Santos, W., Ribeiro, A., Conceição, M., Veras, G., Medeiros, A.C. (2019). Tablet of *Ximenia americana* L. developed from mucoadhesive polymers for future use in oral treatment of fungal infections. Polymers 11(2): 379. DOI: https://doi.org/10.3390/polym11020379.

Almeida, M.L.B., Freitas, W.E.S., Morais, P.L.D., Sarmento, J.D.A., Alves, R.E. (2016). Bioactive compounds and antioxidant potential fruit of *Ximenia americana* L. Food Chem 192: 1078–1082. DOI: https://doi.org/10.1016/j.foodchem.2015.07.129.

Amparo, T.R., Braga, V.C.D.C., Seibert, J.B., Souza, G.H.B.D., Teixeira, L.F.D.M. (2018). Methods for *in vitro* evaluating antimicrobial activity of medicinal plants: the need for standardization. Infarma - Ciências Farmacêuticas 30(1): 50–59. DOI: http://dx.doi.org/10.14450/2318-9312.v30.e1.a2018.pp50-59.

Asres, K., Bucar, F., Kartnig, T., Witvrouw, M., Pannecouque, C., De Clercq, E. (2001). Antiviral activity against human immunodeficiency virus type 1 (HIV-1) and type 2 (HIV-2) of ethnobotanically selected Ethiopian medicinal plants. Phytother Res 15(1): 62–9. doi: 10.1002/1099-1573(200102)15:1<62::aid-ptr956>3.0.co;2-x. PMID: 11180526.

Ayres, M.C.C., Brandão, M.S., Vieira-Junior, G.M., Menor, J.C.A.S., Silva, H.B., Soares, M.J.S., Chaves, M.H. (2008). Antibacterial activity of useful plants and chemical constituents of the roots of *Copernicia prunifera*. Rev Bras Farmacogn 18(1): 90–97. DOI: http://dx.doi.org/10.1590/S0102-695X2008000100017.

Bazezew, A.M., Emire, S.A., Sisay, M.T. (2021). Bioactive composition, free radical scavenging and fatty acid profile of *Ximenia americana* grown in Ethiopia. Heliyon, p. e07187. DOI: https://doi.org/10.1016/j.heliyon.2021.e07187.

Borges, A., Abreu, A.C., Dias, C., Saavedra, M.J., Borges, F., Simões, M. (2016). New perspectives on the use of phytochemicals as an emergent strategy to control bacterial infections including biofilms. Molecules 21(7): 877. DOI: http://dx.doi.org/10.3390/molecules21070877.

Brasileiro, M.T., Egito, A.A., Lima, J.R., Randau, K.P., Pereira, G.C., Rolim-Neto, P.J. (2008). *Ximenia americana* L.: botânica, química e farmacologia no interesse da tecnologia farmacêutica. Rev Bras Farm 89(2): 164–167.

Carvalho, G.F.S., Marques, L.K., Sousa, H.G., Silva, L.R., Ferreira, D.C.L., Amaral, F.P.M., Maia Filho, A.L.M., Silva, J.F., Alves, W.S., Oliveira, M.D.A., Costa Júnior, J.S., Costa Junior, F.L.C., Ramos, R.M., Rai, M., Uchôa, V.T. (2020). Phytochemical study, molecular docking, genotoxicity and therapeutic efficacy of the aqueous extract of the stem bark of *Ximenia americana* L. in the treatment of experimental COPD in rats. J Ethnopharmacol 247: 112259. DOI: https://doi.org/10.1016/j.jep.2019.112259.

Cioch, M., Satora, P.A.W.E.Ł., Skotniczny, M., Semik-Szczurak, D., Tarko, T. (2017). Characterisation of antimicrobial properties of extracts of selected medicinal plants. Pol J Microbiol 66(4): 463–472. DOI: https://doi.org/10.5604/01.3001.0010.7002.

Costa, E.M.M.B., Barbosa, A.S., Arruda, T.A., Oliveira, P.T., Dametto, F.R., Carvalho, R.A., Melo, M.D. (2010). *In vitro* study of the antimicrobial action of plant extracts against *Enterococcus faecalis*. J Bra Pathol and Med Lab 46: 175.

Da Silva, B.A.F., Costa, R.H.S., Fernandes, C.N., Leite, L.H.I., Ribeiro-Filho, J., Garcia, T.R., Coutinho, H.D.M., Wanderley, A.G., Menezes, I.R.A. (2018). HPLC profile and antiedematogenic activity of *Ximenia americana* L. (Olacaceae) in mice models of skin inflammation. Food Chem Toxicol 119: 199–205. DOI: https://doi.org/10.1016/j.fct.2018.04.041.

De Andrade, A.P.C., Magalhães, N.P., Da Silva, A.S.A., De Oliveira, D.R., De Farias, A.S., Da Silva Rios, D.A. (2019). A ação antimicrobiana dos extratos alcoólicos e aquosos da folha da goiabeira (*Psidium guajava* L.) no controle de *Staphylococcus aureus* ATTCC 27922, *Escherichia coli* ATTCC 25922 e *Listeria monocytogenes* SCOTT A. Segur. Aliment Nut 26: 1–7. DOI: https://doi.org/10.20396/san.v26i0.8655619.

De Araújo, M.R.S., Assunção, J.C.C., Dantas, I.N.F., Costa-Lotufo, L.V., Monte, F.J.Q. (2008). Chemical constituents of *Ximenia americana*. Nat Prod Commun 3(6): 1934578X0800300605. DOI: htttps://10.1177/1934578X0800300605.

De Menezes, I.R.A., Costa, R.H.S., Boligon, A.A., Rolón, M., Coronel, C., Veja, C., Coutinho, H.D.M., Costa, M.S., Tintino, S.R., Pereira, R.L.S., Albuquerque, T.R., Almeida, J.R.G.S., Quintáns-Júnior, L.J. (2019). *Ximenia americana* L. enhances the antibiotic activity and inhibit the development of kinetoplastid parasites. Comp Immunol Microbiol Infect Dis 64: 40–46. DOI: https://doi.org/10.1016/j.cimid.2019.02.007.

Dias, T.L.M.F., Melo, G.M.A., da Silva, Y.K.C., Queiroz, A.C., Goulart, H.F., Alexandre-Moreira, M.S., Santana, A.E.G., Uchôa, V.T. (2018). Antinociceptive and anti-inflammatory activities of the ethanolic extract, of fractions and of epicatechin isolated from the stem bark of *Ximenia americana* L. (Oleacaceae). Rev Virtual Quim 10(1): 86–101. DOI: http://dx.doi.org/10.21577/1984-6835.20180009.

Esquenazi, D., Wigg, M.D., Miranda, M.M.F.S., Rodrigues, H.M., Tostes, J.B.F., Rozental, S., Silva, A.J.R., Alviano, C.S. (2002). Antimicrobial and antiviral activities of polyphenolics from *Cocos nucifera* Linn. (Palmae) husk fiber extract. Res Microbiol 53: 647–52.

Falcão, L., Roman, S.S., Zakrzevski, S.B., Pereira, A., Paroul, N., Cansian, R.L. (2017). Antimicrobial action of leaves essential oil of *Casearia sylvestris swartz*. Erechim Perspect 41(153): 115–123.

Feyssa, D.H., Njoka, J.T., Asfaw, Z., Nyangito, M.M. (2012). Uses and management of *Ximenia americana*, Olacaceae in semi-arid East Shewa, Ethiopia. Pak J Bot 44(4): 1177–1184.

França, I.S.X.D., Souza, J.A., Baptista, R.S., Britto, V.R.S. (2008). Popular medicine: benefits and drawbacks of medicinal plant. Rev Bras Enferm 61(2): 201–208. DOI: https://doi.org/10.1590/S0034-71672008000200009.

Francisco, I.P.B., De Jesus, L.F., Borges, D.A. (2020). Métodos da técnica do DNA recombinante feitos na *Escherichia coli*/Recombinant DNA technique methods made in the *Escherichia coli*. Id on Line Rev Mult Psic 14(51): 467–474. DOI: https://doi.org/10.14295/idonline.v14i51.2147.

Ginovyan, M., Petrosyan, M., trchounian, A. (2017). Antimicrobial activity of some plant materials used in Armenian traditional medicine. BMC Complement Altern Med 17(1): 50. DOI: https://doi.org/10.1186/s12906-017-1573-y.

Gotlieb, O. (1981). New and underutilized plants in the Américas: solution to problems of inventory through systematics. Interciência 6: 22–9.

Hatano, T., Kusuda, M., Inada, K., Ogawa, T., Shiota, S., Tsuchiya, T., Yoshida, T. (2005). Effects of tannins and related polyphenols on methicillin-resistant *Staphylococcus aureus*. Phytochemistry 66: 2047–55. DOI: https://doi.org/10.1016/j.phytochem.2005.01.013.

James, D.B., Abu, E.A., Wurochekke, A.U., Orji, G.N. (2007). Phytochemical and antimicrobial investigation of the aqueous and methanolic extracts of *Ximenia americana*. J Med Sci 7(2).

Júnior, J.C.S., Estevão, L.R.M., Ferraz, A.A., Simões, R.S., Vieira, M.G.F., Evêncio-Neto, J. (2019). Pomada de *Ximenes americana* promove aceleração da cicatrização de feridas em ratos1. Acta Cir Bras 34(3): e201900307. DOI:https://10.1590/s0102-865020190030000007.

Kawo, A.H., Suleiman, Z.A., Yusha'u, M. (2011). Studies on the antibacterial activity and chemical constituents of Khaya senegalensis and *Ximenia americana* leaf extracts. Afr J Microbiol Res 5(26): 4562–68. DOI: https://doi.org/10.5897/AJMR11.597.

KONÉ, W.M., Atindehou, K.K., Terreaux, C., Hostettmann, K., Traoré, D., Dosso, M. (2004). Traditional medicine in North Côte-d'Ivoire: screening of 50 medicinal plants for antibacterial activity. J Ethnopharmacol 93(1): 43–49. DOI: https://doi.org/10.1016/j.jep.2004.03.006.

Le, N.H.T., Malterud, K.E., Diallo, D., Paulsen, B.S., Nergård, C.S., Wangensteen. H. (2012). Bioactive polyphenols in *Ximenia americana* and the traditional use among Malian healers. J Ethnopharmacol 139(3): 858–862. DOI: https://doi.org/10.1016/j.jep.2011.12.031.

Leal, S.S., Uchôa, V.T., Figuerêdo-Silva, J., Soares, R.B., Mota, D.M., Alencar, R.C., Maia Filho, A.L.M., Sant'Ana, A.E.G., Beltrame Junior, M. (2016). Eficácia da fonoforese com *Ximenia americana* L. na inflamação de tendão de ratos. Rev Bras Med Esporte 22(5): 355–360. DOI: https://doi.org/10.1590/1517-869220162205156899.

Maia, C.E.G. (2018). Analysis of the phytochemical potential of *Ximenia americana* and antiglicante capacity of mangiferin and methanolic extracts obtained from coffee leaves. Master's thesis. Federal University of Ceara.

Maikai, V., Maikai, B., Kobo, P. (2009). Antimicrobial properties of stem bark extracts of *Ximenia americana*. J Agr Sci 1(2): 30.

Maroyi, A. (2016). *Ximenia caffra* Sond. (Ximeniaceae) in sub-Saharan Africa: A synthesis and review of its medicinal potential. J Ethnopharmacol 184: 81–100. DOI: https://doi.org/10.1016/j.jep.2016.02.052.

Mevy, J.P., Bessiere, J.M., Greff, S., Zombre, G., Viano, J. (2006). Composition of the volatile oil from the leaves of *Ximenia americana* L. Biochem Syst Ecol 34(7): 549–553. DOI: https://doi.org/10.1016/j.bse.2006.01.007.

Migliato, K.F., Corrêa, M.A.I., Salgado, H.R.N., Tognolli, J.O., Sacramento, L.V.S., Mello, J.C.P., Giannini, M.J.S.M., Almeida, A.M.F., Pizzolitto, A.C. (2011). Factorial design of the optimization extraction of *Syzygium cumini* (L.) Skeels fruits. Quím Nova 34(4): 695–699. DOI: https://doi.org/10.1590/S0100-40422011000400024.

Ogunleye, D., Ibitoye, S. (2003). Studies of antimicrobial activity and chemical constituents of *Ximenia americana*. Trop J Pharm Res 2(2): 239–241. DOI: https://doi.org/10.4314/tjpr.v2i2.14606.

Oliveira, V.B.D., Mezzomo, T.R., Moraes, E.F.D. (2018). The knowledge and use of medicinal plants by users of primary healthcare facilities in Colombo, PR. Rev Bras Ciênc Saúde 22(1): 57–64. DOI: https://doi.org/10.4034/RBCS.2018.22.01.08.

Omer, M., Elnima, E. (2003). Antimicrobial activity of *Ximenia americana*. Phytotherapy 74: 122–126. DOI: https://doi.org/10.1016/S0367-326X(02)00302-7.

Palma, A.F.M., Marques, L.K.M., Carneiro, R.S., Carvalho, G.F.S., Ferreira, D.C.L., Sant'Ana, A.E.G., Maia-Filho, A.L.M., Marques, R.B., Alves, W.S., Uchôa, V.T. (2020). Avaliação dos Extratos Hidroalcoólicos do Caule e Folhas da *Ximenia americana* L. na Cicatrização de Feridas Excisionais Aguda em Pele de Camundongos, Rev Virtual Quim 12: 37–50. http://dx.doi.org/10.21577/1984-6835.20200005.

Poulios, E., Vasios, G.K., Psara, E., Giaginis, C. (2020). Medicinal plants consumption against urinary tract infections: a narrative review of the current evidence. Expert Rev Ant Infect Ther 5: 1–10. DOI: https://doi.org/10.1080/ 14787210.2021.1828061.

Řezanka, T., Sigler, K. (2007). Identification of very long chain unsaturated fatty acids from Ximenia oil by atmospheric pressure chemical ionization liquid chromatography–mass spectroscopy. Phytochemistry 68(6): 925–934. DOI: https://doi.org/10.1016/j.phytochem.2006.11.034.

Rocha, K.C.J., Goldenberg, R., Meirelles, J., Viana, P.L. (2017). Flora das cangas da Serra dos Carajás, Pará, Brasil: Olacaceae. Rodriguésia 68(3): 1053–1057. DOI: https://doi.org/10.1590/2175-7860201768336.

Runyoro, D.K.B., Ngassapa, O.D., Matee, M.I.N., Joseph, C.C., Moshi, M.J. (2006). Medicinal plants used by Tanzanian traditional healers in the management of Candida infections. J Ethnopharmacol 106: 158–165. DOI: https://doi.org/10.1016/j.jep.2005.12.010.

Santos, J.J.F., Coelho-Ferreira, M., Lima, P.G.C. (2018). Ethnobotany of medicinal plants on public markets in the metropolitan region of Belém do Pará, Brazil. Biota Amazônia 8(1): 1–9. DOI: http://dx.doi.org/10.18561/2179-5746/biotaamazonia.v8n1p1-9.

Souza, L.F., Dias, R.F., Guilherme, F.A.G., Coelho, C.P. (2016). Medicinal plants referenced by "raizeiros" from Jataí county, Goiás state. Rev Bras Plantas Med 18(2): 451–461. DOI: https://doi.org/10.1590/1983-084X/15_173.

Shettar, A.K., Kotresha, K., Kaliwal, B.B., Vedamurthy, A.B. (2015). Evaluation of *in vitro* antioxidant and anti-inflammatory activities of *Ximenia americana* extracts. Asian Pac J Trop Dis 5(11): 918–923. DOI: https://doi.org/10.1016/S2222-1808(15)60957-4.

Silva, G.G., Souza, P.A., Morais, P.L.D., Santos, E.C., Moura, R.D., Menezes, J.B. (2008). Caracterização do fruto de ameixa silvestre (*Ximenia americana* L.). Rev Bras Frutic 30(2): 311–314. DOI: https://doi.org/10.1590/S0100-29452008000200007.

Silva, M.S.P., Brandão, D.O., Chaves, T.P., Formiga Filho, AL.L.N., Costa, E.M.M.B., Santos, V.L., Medeiro, A.C.D. (2012). Study bioprospecting of medicinal plant extracts of the semiarid northeast: contribution to the control of oral microorganisms. Evid Based Complement Alternat M. DOI: https://doi.org/10.1155/2012/681207.

Silva, K.M.A., Chaves, T.P., Santos, R.L., Brandão, D.O., Fernandes, H.A., Ramos Júnior, F.J.L., Santos, V.L., Felismino, D.C., Medeiro, A.C.D. (2015). Modulation of the erythromycin resistance in *Staphylococcus aureus* by ethanolic extracts of *Ximenia americana* L. and *Schinopsis brasiliensis* Engl. Bol latinoam Caribe Plantas Med Aromát 14(2): 92–98. http://www.redalyc.org/articulo.oa?id=85636183003.

Silver, L.L., Bostian, K.A. (1993). Discovery and development of new antibiotics: the problem of antibiotic resistance. Antimicrob Agents Chemoter 37(3): 377–383. DOI: https://dx.doi.org/10.1128%2Faac.37.3.377.

Singh, B., Sahu, P.M., Singh, S. (2002). Antimicrobial activity of pyrrolizidine alkaloids from *Heliotropium subulatum*. Fitoterapia 73: 153–155. DOI: https://doi.org/10.1016/S0367-326X(02)00016-3.

Sobeh, M., Mahmoud, M.F., Abdelfattah, M.A.O., El-Beshbishy, H.A., El-Shazly, A.M., Wink, M. (2017). Hepatoprotective and hypoglycemic effects of a tannin rich extract from *Ximenia americana* var. caffra root. Phytomedicine 33: 36–42. DOI: https://doi.org/10.1016/j.phymed.2017.07.003.

Soro, T.Y., Traore, F., Sakande, J. (2009). Analgesic activity of the aqueous extract from *Ximenia americana*. CR Biol 332(4): 371–377. DOI: https://10.1016/j.crvi.2008.08.022.

Souza, C.A.S., Avancini, C.A.M., Wiest, J.M. (2000). Antimicrobial activity of *Tagetes minuta* L. - *Compositae* (Chinchilho) against Gram-positive and Gram-negative bacteria. Braz J Vet Res Anim Sci 37(6). DOI: https://doi.org/10.1590/S1413-95962000000600001.

Tortora, G., Funke, B., Case, C. (2005). Microbiologia. 8 ed. Porto Alegre: Artes Médicas Sul, p.872.

Traoré, M.S., Baldé, M.A., Camara, A., Baldé, E.S., Diané, S., Diallo, M.S.T., Keita, A., Cos, P., Maes, L., Pieters, L. (2015). The malaria co-infection challenge: An investigation into the antimicrobial activity of selected Guinean medicinal plants. J Ethnopharmacol 174: 576–581. DOI: https://doi.org/10.1016/j.jep.2015.03.008.

Uchôa, V.T., Sousa, C.M.M., Carvalho, A.A., Santana, A.E.G., Chaves, M.H. (2016). Free radical scavenging ability of *Ximenia americana* L. stem bark and leaf extracts. J Appl Pharma Sci 6: 91.

Zaynab, M., Fatima, M., Abbas, S., Sharif, Y., Umair, M., Zafar, M.H., Bahadar, K. (2018). Role of secondary metabolites in plant defense against pathogens. Microb Pathog 124: 198–202. DOI: https://doi.org/10.1016/j.micpath.2018.08.034.

5

Spices as Eco-friendly Microbicides:

From Kitchen to Clinic

Sahadeo D. Patil

Introduction

Development of resistance to antimicrobial agents in pathogenic microbes poses a serious risk to public health and hampers the economy of any nation. With major pathogens becoming resistant to currently available antibiotics there is an urgent need for research of a newer class of antimicrobial agents.

Since time immemorial, people have gathered plant and animal resources for their need. Among these uses, medicinal plants play a central role, not only as traditional medicine used in many cultures, but also as trade commodities, meeting the demand of distant markets. About 80% of the population in low and middle income countries rely on traditional medicines and up to 65% of the population in high income countries use complementary/alternative medicines (WHO 2002). Several hundred genera are used medicinally, mainly as herbal preparations in the indigenous system of medicine in different countries and are sources of very potent and powerful drugs (Ahmad et al. 1998). China by itself uses 4,941 (18.9%) of 26,092 native species as drugs in Chinese traditional medicine (Duke and Ayensu 1985). India exhibits a remarkable outlook in modern medicines that are based on natural products, besides the traditional system of Indian medicines. Almost, 70% modern medicines in India are derived from natural products. The world average stands at 12.5% while India has 20% plant species of medicinal value.

Plants as a source of medicine have a special importance in countries like India, Pakistan, Bangladesh and Sri Lanka which have well developed traditional systems of medicine called Ayurveda, Sidha and Unani, all of which derived more than 90% of medicaments from higher plants. There are around 25,000 effective plant-based formulations, used in folk medicine and known to rural communities in India. The derivatives of medicinal plants are non-narcotic with little or no side effects (Wakdikar 2004). India stands second in the 12 leading countries of export of medicinal and aromatic plant material from 1991–1998 (Lange 2002). It is well known that developed countries, are also using medicinal system

Department of Microbiology and Biotechnology, Shri Shivaji Science College, Amravati-444603, Maharashtra, India.
Email: sdpatil20957@yahoo.co.in

that involves the use of herbal drugs and remedies. Thus the demand for plant-derived products has increased worldwide and will continue to grow in the years to come.

Plants produce a large number of secondary metabolites that shows inhibitory effects on pathogenic microbes. Some of these are used as drugs, flavors, fragrances, insecticides and dyes, and thus have a great economic value. Major groups of plant secondary metabolites comprise terpenes, phenolic compounds, flavonoids and alkaloids. Antimicrobial agents used for fighting infectious diseases are currently becoming ineffective and its misuse and overuse leads to the development of antimicrobial resistance. Multidrug-resistant tuberculosis alone is responsible for 230,000 deaths per annum globally. Presently problems of antimicrobial resistance are becoming more complex and serious. Plant derived antimicrobials particularly plant secondary metabolites either in purified form or plant part preparation as crude extract appears to be a promising tool to combat pathogenic microbes.

This chapter aims to discuss antibacterial and antifungal potential of spices commonly used in kitchen and the role of spice extracts and spice essential oils highlighting the promising use in clinics to combat infections. It also aims to highlight antimicrobial principles of spices with mechanism of action and combination effects for improved antimicrobial activity with a possible solution to antimicrobial resistance.

Spices

Spices may be defined as natural plant products used to improve the flavour, aroma, taste and color of food products. They are also used in beverages, liquors, pharmaceutical, cosmetic and perfumery products. According to Pruthi 1976; the term 'spices and condiments' applies to "such natural plant or vegetable products or mixtures thereof, in whole or ground form, as are used for flavoring, seasoning and imparting aroma or flavour to foods". Spices are vital culinary addendums enhancing organoleptic characteristics of food and are used in medicines and perfumes, in oils for religious ceremonies and in aphrodisiacs, and also in the flavor and preserve food. India is known as 'The Home of Spices' and often considered as the 'Spice bowl' of the world. Spices constitute an important group of agricultural commodities, which are indispensable in culinary art.

Many of the spices that are popular today are indigenous to India. Some of the spices grown in India today such as long pepper, black pepper, cinnamon, turmeric and cardamom have been known for thousands of years. Cardamom and turmeric indigenous to India were cultivated as early as the 8th century B.C. in the gardens of Babylon (Rosengarten 1969). India is also the world's largest exporter of spices and has also emerged as a major supplier of spice products such as ground spices, spice mixes, spice pastes and curry powders (Subbulakshmi and Naik 2002). Cardamom, ginger, black pepper, cumin and mustard seeds are included in herbal medicines, prescribed to remove fats and to cure urinary complaints, piles and jaundice. In Ayurvedic medicine, spices such as cloves and cardamom were wrapped in betel-nut leaves and chewed after meals to increase the flow of saliva and aid digestion. Throughout the world, spices and herbs are frequently used in cuisine, to improve flavor, and to provide new tastes.

Spices vary considerably in shape, size, color, flavor and texture. According to the International Organization for Standardization (ISO), there is no clear cut division between 'Spices' and 'Condiments' and as such they have been clubbed together. Different parts of the plants used as spices include seeds (coriander, mustard, cumin, celery, aniseed, caraway), fruit (cardamom, chilly), leaves (basil), bark (cinnamon), bulb (garlic), rhizome (ginger), berries (black pepper), bud (clove). Before the early 1800s, spices were available only in the whole form. Today one has whole spices, ground spices, seasoning blends and spice extractives. Spice extractives include essential oils, oleoresins, liquid soluble (oleoresins plus solubilizing agents to create a liquid seasoning), dry soluble (oleoresins plated on a dry carrier), encapsulated spices, etc. India, considered as the Land of Spices is one of the major spice producing and exporting country of the world, contributing about 20–25% of the world trade in spices. Among 109 spices recognized by the International Organization for Standardization (ISO) for the world more than 52–60 spice crops are grown in India. Agro-climatic conditions in India, have a large scope for the cultivation of majority of the spices.

Classification of spices

Different approaches are commonly used to classify spices (Subbulakshmi and Naik 2002; Bakhru 2004). These include:

Plant parts used: Different plant parts like leaf (mint, curry leaf), root, bulb (onion, garlic, etc.), fruit (chilli, cardamom, allspice), seed (cumin, fenugreek, coriander, fennel, bishop's weed, poppy, aniseed), bark (cinnamon and cassia), rhizome (turmeric, ginger), kernel (nutmeg), bud (clove), berry: black pepper, juniper and allspice.

Economic importance: Spices are grouped as major (small cardamom, black pepper, chilli, turmeric and ginger) and minor spices (all spices excluding major spices).

Climate requirement: Spices are grouped into three categories as tropical spices (ginger, turmeric, black pepper, cinnamon, small cardamom and clove), subtropical spices (cumin, fennel, coriander, fenugreek, onion and garlic) and temperate spices (thymes, saffron, caraway seed).

Origin and flavor: Spices can be classified as aromatic spices (cardamon, aniseed, celery, cumin, coriander, fenugreek and cinnamon), pungent spices (ginger, chilli, black pepper), phenolic spices (clove and allspice), colored spices (turmeric, saffron).

Season of growth: Accordingly these are annual spices (coriander, cumin, fennel, fenugreek, bishop's weed), biennial spices (onion and parsley) and perennial spices (black pepper, saffron, clove, nutmeg and cinnamon).

Botanical description: This includes the scientific name of the spice plant and family.

In India, pepper, cardamom, chilli, ginger, turmeric, coriander, cumin, fennel, fenugreek, celery, saffron, tamarind and garlic are produced on large scale while aniseed, bishop's weed, dill seed, poppy seed, bay leaf, curry leaves and cinnamon are produced and exported in small quantities. Due to Covid-19 pandemic during 2019–20 foreign trade of many countries registered a negative growth. However, despite the COVID pandemic, the export of spices from India has increased during 2019–20 and has crossed the US$ 3 billion mark for the first time in the history of the export of spices. The export during 2019–20 attained an all-time record in terms of both volume and value.

Spice extracts as antimicrobial agents

Extracts from many types of plants are used as flavoring and seasoning agents in foods and beverages. Due to antimicrobial properties they have been used therapeutically for centuries. Antimicrobial properties of spices are attributed to secondary metabolites that include essential oils, alkaloids, terpenoids, phenolics, etc. More than 100,000 plant secondary metabolites have been identified and the majority of them possess antimicrobial potentials. Such plant secondary metabolites can be obtained using solvent extraction processes. Soxhlet extraction is normally a method of choice. Solvents commonly used include water, methanol, chloroform, ethanol, petroleum ether, etc. Not a single solvent is helpful in extracting all antimicrobial principles present. For example, terpenoids, flavonoids and alkaloids are extracted in chloroform; flavonols and alkaloids in acetone; tannins, alkaloids, terpenoids, flavonol in ethanol; and saponins, tannins, flavones, terpenoides in methanol. Water is used for extracting saponins, tannins and terpenoids.

It is evident (Elgayyar et al. 2001), that some spices and herbs are potential sources of antibacterial and antifungal agents. The antimicrobial activity of a large number of extracts of spices has been extensively studied since the end of the last century, and the active components in these herbs are also known today.

Antibacterial activities

Spices and aromatic herbs are well known for their role as a 'natural' method of food preservation. A lot of work has been done on garlic extract including inhibition of *Staphylococcus aureus, Salmonalla typhi, Shigella sonnei, Shigella dysenteriae* and lactic acid bacterium (*Lactobacillus plantarum*). Garlic also possesses greater antimicrobial activity than ginger and onion. Aqueous extract of freeze-dried garlic inhibit bacteria and other microorganisms while fresh garlic extract (10%) is known to exhibit maximum germicidal activity towards *Bacillus cereus* and *B. brevis*. The bactericidal effect of garlic extract with 93% killing of *Staphylococcus epidermidis* and *Salmonella typhi* has been reported by Arora and Kaur (1999). Ginger extract effectively inhibit *E. coli, E. aerogenes, Proteus* sp., Staphylococci, Streptococci, *B. subtilis, Salmonella* and *P. aeruginosa* (Kamble and Patil 2007). However, different extracts of coriander normally do not show inhibitory activity towards bacterial species.

Allicin is the active inhibitory principle in garlic which is enzymatically released from precursor after the garlic bulb is crushed. Aqueous garlic extract brings about oxidation of thiol (L-cysteine glutathione 2-mercaptoethanol) groups present in essential proteins, causing enzyme inactivation and subsequent inhibition of microbial growth. Bulb extract of garlic is effective in inhibiting growth of food poisoning bacteria and also blocks the growth respiration in pathogenic bacteria. Several studies showed sensitivity of many bacterial species including *B. sphaericus, S. aureus, S. epidermidis, E. coli, E. aerogenes, P. aeruginosa, S. flexneri, S. typhi, S. typhimurium*, and *Yersinia enterocolitica* to garlic (Arora and Kaur 1999; Elizabeth 2001). Chloroform extract of garlic, cinnamon, thyme, oregano clove, saffron, nutmeg, coriander, pepper, pimento and juniper berry exhibit bacteriostatic action against *S. aureus* and *Vibrio* spp., while extract of garlic, cinnamon, thyme, oregano, pepper, pimento and juniper berry also demonstrate genotoxic effect.

Cardamom (*Elettaria cardamomum*), the 'Queen of spices' is a perennial aromatic plant traditionally used as a culinary ingredient, commonly cultivated in southern India, Sri Lanka, Tanzania and Guatemala. This plant is considered as an important source of flavonoids, alkaloids, terpenoids, anthocyanins and phenolic compounds. Cardamom has been used for the treatment of several disorders, including asthma, indigestion and congestive jaundice. It possesses various pharmacological properties such as antioxidant, anti-inflammatory, anti-cancer and antimicrobial activities. Its fruit and seed extracts prepared by supercritical fluid extraction with carbon dioxide exert an antibacterial effect against major periodontal pathogens *Aggregatibacter actinomycetemcomitans, Fusobacterium nucleatum, Porphyromonas gingivalis*, and *Prevotella intermedia* (Souissi et al. 2020).

Kamble and Patil (2007) used aqueous extracts of a large number of spices including garlic, clove, star-anise, tamarind, cinnamon, ginger, rosemary, turmeric, asafoetida, cardamom (small and large), cumin, basil, parsley, mustard, nutmeg, black pepper, caraway, fenugreek, fennel, coriander, aniseed, Gracinia Indica, curry leaf, pepper long, poppy seed, mace, bishop's weed, dill, mint, pomegranate and saffron are used for testing antimicrobial activity against *B. cereus, B. thuringiensis, B. subtilis, E. coli, K. pneumoniae, S. pyogenes, S. paratyphi* A, *S. typhi, P. vulgaris, P. mirabilis, P. aeruginosa* and *E. aerogenes*. Garlic and clove possess excellent antimicrobial activity against both Gram-positive and Gram-negative bacteria with 15–40 mm inhibition zones. A study conducted by Mostafa et al. (2018) examined antibacterial efficacy of cumin, clove, thyme and ginger ethanolic extracts against *Bacillus cereus, Staphylococcus aureus, Escherichia coli, Pseudomonas aeruginosa* and *Salmonella typhi*. The results of this study showed that the clove ethanolic extract was the most effective plant extract and showed bacteriostatic and bactericidal activities against *S. aureus* and *P. aeruginosa* with MIC's from 2.5 to 5.0 mg/ml but no activity against *S. typhi*.

Nair et al. (1996) proved bacteriostatic properties of black pepper on *E. coli, Salmonella virochow* and *Staphylococcus aureus* at the concentration used in meat preparation. Black and white pepper also inhibits *Clostridium botulinum* (Huhtanen 1980). Sagdic et al. (2003) evaluated antibacterial effects of 18 extracts of spices commonly consumed worldwide and grown naturally in Turkey against 23 bacterial strains to compare their antibacterial effects with 11 antibiotics and concluded higher inhibitory effects of hop extracts than erythromycin against *S. aureus*.

Plants and plant products have been used extensively throughout history to treat medical problems. Infections with methicillin resistant *S. aureus*, multi drug resistant *P. aeruginosa* and enterohemorrhagic *E. coli* O157 results in life threatening complications. Ethanol and methanol extract of clove is highly effective in inhibiting the growth of multi drug resistant *P. aeruginosa* and enterohemorrhagic *E. coli* O157 respectively, while both extracts are highly effective against methicillin resistant *S. aureus*. The antibacterial enhancement (synergistic effect) by the combinations of extracts is promising in designing a potentially active antibacterial agent of plant origin (Bassam 2004). Huhtanen 1980 studied inhibition of *Clostridium botulinum* by alcoholic extracts of mace, bay leaf, nutmeg, white and black pepper, rosemary, clove, turmeric, thyme, paprika and oregano and concluded the best inhibitory activity of mace against *C. botulinum*. Generally aliphatic straight chain alcohols with chain length of C_{14} or C_{16} are most inhibitory towards *C. botulinum*.

De et al. (1999) reported potent antimicrobial activity of celery against *B. subtilis* and *E. coli*. Leaf extract of celery is an effective inhibitor of *S. aureus, E. coli, B. cereus, E. faecalis* and *E. aerogenes*, however the extracts isolated from the roots exhibit less activity. Aktug and Karapinar (1986) reported inhibitory effect of mint (*Menthapiperita*) ground leaves and their extract on *S. aureus, S. typhimurium* and *V. parahaemolyticus*. Alcohol, ethyl acetate, acetone and chloroform extract of aniseed have no inhibitory activity against *Bacillus cereus, B. subtilis, K. pneumoniae, S. aureus* and *P. aeruginosa*.

A solvent used for preparing plant extracts play a major role in sensitivity of microorganisms to extracts. For example, water extracts have less or no antimicrobial activity but other extracts exhibit good antimicrobial activity as bioactive aromatic or saturated organic compounds are generally obtained through initial ethanol or methanol extraction (Bassam et al. 2004; Cowan 1999). Many studies have concentrated on antimicrobial effects of spice extracts, some of which are given in Table 5.1.

Evaluation of antimicrobial properties of water, ethanol and hexane extracts of 30 types of commercial herbs and spices commonly used in the production of ready meals showed that the choice of the solvent used to extract herbs and spices significantly influence

Table 5.1. Antimicrobial activity of Spice extracts.

Spice extract	Method	Sensitive organisms	Reference
Syzygium aromaticum	Disk diffusion	*Bacillus cereus, Staphylococcus aureus, Escherichia coli* & *Pseudomonas aeruginosa*	Mostafa et al. 2018
Zingiber officinales (ethanolic)		*Bacillus cereus, Staphylococcus aureus* & *Pseudomocas aeruginosa*	
Thymus vulgaris		*Staphylococcus aureus* & *Pseudomonas aeruginosa*	
Clove, cumin, carom (ethanolic)	Disk diffusion	*Staphylococcus aureus, Escherichia coli*	Garg and Giri 2020
cinnamon and black pepper			
Aniseed (*Pimpinell aanisum* L.), basil (*Ocimum basilicum* L.), bay (*Laurus nobilis* L.), black pepper (*Piper nigrum* L.), cayenne pepper (*Capsicum annuum* L.), celery (*Apium graveolens* L.), chili (*Capsicum annuum* L.), cinnamon (*Cinnamomum zeylanicum* L.), clove (*Syzygium aromaticum* L.), coriander (*Coriandrum sativum* L.), cumin (*Cuminum cyminum*L.), dill (*Anethum graveolens* L.), fennel (*Foeniculum vulgare* Mill.), French onion (*Allium cepa* L.), garlic (*Allium sativum* L.), ginger (*Zingiber officinale* Rosc.), mace (*Myristica fragrans* Houtt.), marjoram (*Origanum majorana* L.), mustard (*Sinapis alba* L.), nutmeg (*Myristica fragrans* Houtt.), onion (*Allium cepa* L.), oregano (*Origanum vulgare* L.), paprika (*Capsicum annuum* L.), parsley (*Petroselinum crispum* L.), pimento (*Pimenta dioica* L.), rosemary (*Rosmarinus officinalis* L.), sage (*Salvia officinalis* L.), thyme (*Thymus vulgaris* L.), turmeric (*Curcuma longa* L.) and white pepper (*Piper nigrum* L.).	Disk diffusion	*Listeria innocua, Escherichia coli, Staphylococcus aureus* & *Pseudomonas fluorescens*	Witkowska et al. 2013
Garlic, Turmeric, Ginger (Aqueous)	Disk diffusion	*Escherichia coli, Salmonella* spp., *Pseudomonas* spp. & *Staphylococcus aureus*	Sah et al. 2020

the antimicrobial activities. Generally, water extracts display little or no antimicrobial activity, but hexane and ethanol extracts display good and excellent antimicrobial activity respectively. Further, Gram-positive bacteria are slightly more susceptible to the action of herb and spice extracts than Gram-negative bacteria (Witkowska et al. 2013). Nanasombat and Lohasupthawee (2005), examined antibacterial activity of crude ethanolic extracts of different spices including cardamom, cinnamon, clove, coriander, cumin, garlic, ginger, holy basil, kaffir lime leaves and peels, lemongrass, mace, nutmeg, black and white pepper, and turmeric against *Salmonella* (20 serotypes), *Citrobacter freundii, Enterobacter aerogenes, Escherichia coli, Klebsiella pneumoniae* and *Serratia marcescens* using a disk diffusion test.

Clove, cumin, and kaffir lime peels showed the broadest antibacterial activity by inhibiting growth of all bacterial strains with 8–22 mm inhibition zone. The sensitivity of various microorganisms to ethanol or methanol extracts of spices appears to be more than aqueous extract (Bassam et al. 2004) that could be due to the extraction of aromatic and saturated organic compounds of plants by these solvents (Cowan 1999). However, the unique advantage of aqueous extracts is that these are comparatively cheap and less cumbersome for preparation and thus form a strong basis for further exploration in antimicrobial testing systems.

Antifungal activities

Antifungal properties of plant secondary metabolites are of great importance to humans and plants as well. Most of the work is focused on human fungal infections including candidiasis, aspergillosis, cryptococcosis due to development of drug resistance in the causative agents of these infections. As such the research community is now looking for spice plants as an alternative to antifungal drugs.

Elizabeth (2001) reported the inhibitory activity of *Allium sativum* against *Candida albicans*. Spices such as clove and oregano significantly delay the growth of molds like *Cladosporium herbarum*, *Eurotium repens*, *Penicillium expansum* and *Rhizopus stolonifer*. Plant juices are more active against yeast than bacteria. The juice of onion, garlic and dill were reported to be highly effective towards yeast. In another study, Thyagaraja and Hosono (1996) reported promising result of asafoetida extract in inhibiting the growth of food spoilage molds. Ethanolic extract show antifungal property, while aqueous extract remains ineffective towards fungi. Ginger inhibits aflatoxin producing *Aspergillus*. Besides, fresh ginger juice also shows inhibitory action towards *A. niger*, *S. cerevisiae*, *Mycoderma* spp. and *L. acidophilus*.

Azzouz and Bullerman (1982) studied the antimycotic effects of ground herbs and spices and other plant materials and commercial antifungal agents against the mycotoxin producing molds and showed clove as the strongest antifungal spice followed by cinnamon. Mustard, garlic, allspice and oregano produce various degrees of inhibition. Other spices like white pepper, black pepper, rosemary, thyme, turmeric, anise, and onion shown relatively minor effects on growth of the *Aspergillus* and *Penicillium* strains, while sage, red pepper and paprika possess no antimycotic effects. Inhibition of mold growth and aflatoxin production by mycotoxigenic strains of *Aspergillus* and *Penicillium* is best explained in yeast extract sucrose broth (Hitokoto et al. 1980).

Spice essential oils as antimicrobial agents

Essential oils are known as man's first medicine. Priests and physicians were using essential oils thousands of years before the time of Christ. The fragrance of plants is carried in essential oil fraction. Spice essential oils are the volatile aromatic extracts prepared by steam distillation of the ground spices. These oils are secondary metabolites that are highly enriched in compounds called terpenes and terpenoides, which are active against bacteria, fungi and protozoa. On the other hand, cellulose is the bulk of all spices but do not contribute to the aroma and flavor of the spice. The aromatic profile is largely determined by the essential oil content, quantitatively as a measure of aromatic strength and qualitatively as a determinant of aromatic character.

Historically, Egyptians cultivated plants for their oils about 3,500 BC and used them extensively in their religion, in cosmetics as well as for medicinal purposes. At the same time, China and India were also exploring herbs and aromatic plants, which would become an integral part of the Indian Ayurvedic medical system. Ayurveda, has a 3,000-year history of incorporating essential oils into healing elixirs. Vedic literature lists more than 700 substances including cinnamon and ginger. In 1937 French chemist and perfumer Rene Maurice Gattefosse coined the term aromatherapy. During the Second World War, doctors successfully treated many injured soldiers with essential oils. In fact, virtually all essential oils possess antiseptic, antifungal, antiviral, antioxidant and anti-inflammatory properties. Due to wide array of properties, essential oils are used in perfumes, cosmetics, soaps and other products; as well as for flavoring food and drinks, and for adding scent to incense and household cleaning products.

Spice oils are volatile substances which are mostly terpenic in nature and are obtained by steam or hydrodistillation methods. The oil is made up of hydrocarbons like terpenes, sesquiterpenes and their oxygenated derivatives. Oleoresins are obtained by extraction of the powdered spices using organic solvent like alcohol, acetone, hexane, methylene chloride, ethylene dichloride and ethyl acetate and contain essential oils fraction, bite principles, color and resinous matter. Spice oils only give the aroma of the spice, whereas spice oleoresins represent the total flavor of the spice.

The mechanisms of action of the essential oils (Nazzaro et al. 2013) include:

1. Cell wall degradation.
2. Cytoplasmic membrane damage.
3. Cytoplasm coagulation.
4. Membrane proteins damage.
5. Leakage of the cell contents.
6. Proton motive force reduction.
7. Intracellular ATP pool reduction via decreased ATP synthesis and augmented hydrolysis and
8. Membrane potential reduction via increased membrane permeability.

Disk diffusion and broth microdilution methods are commonly used for determining antimicrobial activity of spice essential oils. The former method describes the zone of inhibition while the latter estimates the quantity of essential oil as MIC. Spice essential oils inhibit the growth of pathogens within MIC range of 5–900 μg/ml or more (Table 5.2).

Table 5.2. Antimicrobial activity of Spice essential oils.

Spice EO	Method	Sensitive organisms (ZOI/MIC)	Reference
Cinnamon EO	Disk diffusion/ Microdilution	*Escherichia coli* (29.0 ± 0.7 mm/4.88 μg/ml)	El Atki et al. 2019
		Staphylococcus aureus (40 ± 0.5 mm/4.88 μg/ml)	
		Pseudomonas aeruginosa (30.5 ± 1.0 mm/19.53 μg/ml)	
Clove EO	Microdilution	*Candida tropicalis* (500 μg/ml)	Souza et al. 2016
Thyme EO	Microdilution	*Malassezia furfur* (920 μg/ml)	Vinciguerra et al. 2018
Cumin EO	Microdilution	*Staphylococcus aureus* (125 μg/ml)	Tanhaeian et al. 2020
		Klebsiella pneumoniae (500 μg/ml)	
		Pseudomonas aeruginosa (500 μg/ml)	
		Escherichia coli (125 μg/ml)	
		Salmonella typhimurium (500 μg/ml)	

Antibacterial activities

The emergence of multidrug resistant and disinfectant resistant bacteria like *Staphylococcus aureus, Escherichia coli* and *Pseudomonas aeruginosa* has increased rapidly, causing the increase of morbidity and mortality. This generates the need to develop safe and effective natural plant-based products such as essential oils. Farag et al. (1989) tested the inhibitory effects of some Egyptian spice essential oils from sage leaves, rosemary leaves, caraway fruits, cumin fruits, clove flower buds and thyme leaves and their basic constituents against Gram-negative bacteria (*Pseudomonas fluorescens, Escherichia coli* and *Serritia marcescens*), Gram-positive bacteria (*Staphylococcus aureus, Micrococcus* spp., *Sarcina* spp., and *Bacillus subtilis*), and an acid-fast bacterium (*Mycobacterium phlei*). The Gram-positive bacteria were reported to be more sensitive to the essential oils than were Gram-negative bacteria. The inhibition zones for different microorganisms produced by essential oils were smaller than those produced by their basic constituents. Thyme and cumin oil had stronger antimicrobial activity than other oils. They also reported that there is a relationship between the chemical structures of the most abundant compounds in the oil and their antimicrobial activities.

Thymol and eugenol, major compounds of thyme and clove essential oil are effective towards oral bacteria (Shapiro et al. 1994). Essential oil of cumin exhibit marked inhibitory activity at a low concentration (800–1200 ppm), while coriander and fennel oil at a higher concentration (1600–2400 ppm) against *Staphylococcus aureus, E. coli, Salmonella typhi, Shigella dysenteriae* and *Vibrio cholerae* (Syed et al. 1986). Chao et al. (1998) tested the antibacterial activity of proprietary blend of oils (named thieves) containing cinnamon, rosemary, clove, eucalyptus, and lemon against airborne *Micrococcus luteus, Pseudomonas aeruginosa* and *Staphylococcus aureus*. The bacterial cultures were sprayed in an enclosed area and thieves was diffused for a given amount of time. Such treatment effectively caused 82% reduction in *M. luteus* bioaerosol, a 96% reduction in the *P. aeruginosa* bioaerosol and 44% reduction in the *S. aureus* bioaerosols after 10 minutes exposure. Thyme, clove, cinnamon, pimento, bay, almond, marjoram, geranium, lovage and angelica were reported to be the most effective essential oils against 25 genera of bacteria (Deans and Ritchie 1987). Similarly Beuchat (1994) reported the inhibitory activity of thyme, oregano, savory, sage, cinnamon, clove, vanilla and other spices against Gram-positive and Gram-negative bacteria.

Essential oils of ginger, cumin, bishop's weed, coriander, basil, clove and mustard as well as eugenol exhibit variable degrees of inhibition against *L. acidophilus* and *B. cereus* as determined by the paper disk agar diffusion method (Meena 1994). Dorman and Deans (2000) reported considerable antibacterial activity of volatile oils of black pepper, clove, geranium, oregano and thyme against 25 genera of bacteria including animal and plant pathogens, food poisoning and spoilage bacteria. Burt and Reinders (2003) quantified the antibacterial properties essential oils from bay, clove bud, oregano and red and light thyme on a non-toxigenic strain of *Escherichia coli* O157:H7 and reported significant colicidal and colistatic properties of oregano and light thyme oil. The addition of 0.05% (w/v) agar as a stabilizer helps in substantial improvement in the antibacterial activity. Nguefack et al. (2004) reported the highest antibacterial activity of essential oil extracted from *Thymus vulgaris*, against strains of *Listeria monocytogenes, L. innocua* and *Staphylococcus aureus*. The antibacterial effect was due to permeabilization of the cytoplasmic membrane.

The complex bioactive composition of essential oils interacts with multiple bacterial cellular targets instead of a particular single mode of action, thus inhibiting pathogens from acquiring resistance. Bioactive monoterpenes such as thymol and carvacrol found mainly in thyme and oregano essential oil and phenylpropanoid such as eugenol found mainly in the essential oil of clove contribute to inhibition of both Gram-positive and negative bacteria. Lambert and colleagues (2001) reported that carvacol and thymol possess the ability to disintegrate the outer membrane of Gram-negative bacteria and disrupt the cell membrane of Gram-positive bacteria, thus counteracting the 'reduced membrane permeability' mechanism employed by multidrug resistant strains. The disturbances in the outer membrane cause conformational changes, leading to leakage of ions across the cell membrane. Besides that, disruption of the cell membrane also increases susceptibility of the bacteria towards the actions of other bioactive compounds. Yap et al. (2013) demonstrated complete annihilation of multidrug-resistant bacteria when treated with a particular combination of essential oil and antibiotic at a fairly low concentration. The significant reduction in the minimum inhibitory concentration obtained for both essential oils and antibiotics implicate that compounds such as carvacol and thymol within essential oils permeate the cell membrane of the bacteria to the actions of antibiotics. In addition, phenylpropanoids such as eugenol and *trans*-cinnamaldehyde, commonly found in cinnamon essential oil are able to inactivate microbial enzymes by forming linkages via hydrogen, ionic and hydrophobic interactions, eventually leading to bacterial cell death. Other monoterpenes such as carvone, limonene and 1, 8-cineole found commonly in eucalyptus and rosemary essential oil are known to affect the metabolism and growth rate of bacteria.

Phenolics are compounds which are structurally comprised of one or more aromatic rings with hydroxyl groups and can be classified into several groups such as phenolic acids, flavonoid, stilbenes, tannins, coumarin, etc. Flavonoids, one of the natural phenolics, have been intensively studied for their antimicrobial potentials. Most of the antimicrobial drugs available these days are ineffective due to development of resistance to them. Flavonoids have the potential to overcome this emerging crisis as this class of natural products showed the antimicrobial activity by different mechanisms than those of conventional drugs, so flavonoid could be an effective treatment of pathogenic infections (Biharee et al. 2020). Tannins are water-soluble polyphenols that are known for disrupting the peptidoglycan cell wall and cytoplasmic membrane of drug-resistant strain of *Helicobacter pylori* and *Candida albicans*, leading to the leakage of cellular content and cell death. Alkaloids are not widely used for disease therapy because most of the alkaloids exhibit carcinogenic properties and may cause severe toxicity to humans.

Antifungal activities

Essential oils of spices contribute to aroma and flavoring properties of spices and the aromatic profile is largely determined by the essential oil content. Hitokoto et al. (1980) reported that eugenol extracted from cloves and thymol from thyme cause complete inhibition of the growth of *Aspergillus flavus* and *A. versicolor* at 0.4 mg/ml or less. Anethol extracted from star-anise inhibit growth of *Aspergillus* at a concentration of 2 mg/ml. Conner and Beachut (1984) studied the effect of essential oils of spices including cinnamon, clove, garlic, onion, savory and thyme and showed that garlic oil acts as a potent inhibitor of yeast growth at concentrations as low as 25 ppm. On the other hand, cinnamon and clove oils stimulates pseudomycelium production by *Saccharomyces cerevisiae*.

Essential oil of clove exhibit marked germicidal activity in a broth containing higher percentage of sugar. Sugar is not necessary for antimicrobial action but the concentration of sugar solution provides a good vehicle for obtaining oil dispersion. The essential oil of aniseed exhibit fungistatic properties to *Aspergillus flavus* and *Penicillium italicum* at 1000 ppm, as well as other fungal species from 12 different genera (Shukla and Tripathi 1987). They identified anethole as the active constituent of aniseed oil having fungistatic action at 600 ppm and thus are of potential use to protect food against fungal spoilage during storage. Among the most active components found in spices, eugenol is the most effective followed by thymol, anethol and menthol, against food borne pathogens (Karapinar and Aktug 1987).

Some spice essential oils like thyme, cumin, clove, caraway, rosemary and sage completely inhibit growth of fungal mycelium and aflatoxin production by *Aspergillus parasiticus*. Similarly, essential oils from celery and cumin inhibit aflatoxin producing strains of *Aspergillus flavus* and *A. parasiticus*. The essential oils are fungistatic at lower doses but turn fungicidal at higher doses. Fungal and yeast cultures are more sensitive to cumin oil and cuminaldehyde than bacteria. Thyme, oregano, savory, sage, cinnamon, clove, vanilla and other spices also inhibit yeasts and molds. Essential oils from spices viz. ginger, cumin, bishop's weed, coriander, basil, clove and mustard as well as eugenol exhibit variable degrees of inhibition against *A. niger*, *S. cerevisiae*, *Mycoderma* spp., as determined by the paper disk agar diffusion method.

In a separate study Elgayyar et al. (2001) examined the effectiveness of the cardamom, anise, basil, coriander, rosemary, dill and angelica essential oil for controlling the growth and survival of pathogenic and saprophytic microorganisms and showed inhibitory property for oregano, basil and coriander essential oil, with minimum lethal concentration ranging between 8 and 10 ppm. The methanol and ethanol oleoresins of cardamom have also been shown to possess antifungal activities against *Aspergillus terreus*, *Penicillium purpurogenum*, *Fusarium graminearum* and *Penicillium madriti* given by the poison food method (Singh et al. 2008). Cinnamon and clove oil exert fungicidal and fungistatic properties against the *Colletotrichum musae*, *Lasiodiplodia theobromiae* and *Fusarium proliferatum* that causes anthracnose and crown rot disease in banana.

Medicinal importance of spices

Spices have been used historically to increase metabolism, raise body heat (thermogenesis), improve digestion and assimilation and potentiate the effect of other substances. For this reason, in many herbal traditions, small amounts of hot pungent spices are added to many traditional compounds. For example, cinnamon (*Cinnamomum aromaticum*) directly stimulates cellular glucose metabolism as cinnamon powder reduces blood glucose level.

Leja and Czaczyk (2016) reviewed industrial uses of plant secondary compounds, such as food production (to improve taste, to provide vitamins and macro and microelements, and also to inhibit food spoilage caused by food borne bacteria), in medicine (in the treatment of various diseases; in chemoprevention and cancer therapy; as a source of natural antimicrobials for the treatment of infectious disease), and in pharmacology and cosmetology (in dietary supplements, and as a result of the demand for preservative-free cosmetics, to reduce the risk of methylparaben allergies).

China, India, Greece, Rome and Egypt have a rich history in the medicinal uses of spices and herbs. In the Indian system of medicine, spices and aromatics have a vital role as spices and condiments possess medicinal properties and have a profound effect on human health. Spices and condiments are the inevitable ingredients in most of the indigenous medicines and possess the property of curing several common ailments. Some of them can serve as powerful natural drugs like antibiotics, carminatives, antidepressants, analgesics, cholesterol reducers, sex stimulants, antihypertensive, diuretics, anti-inflammatory agents, blood vessel dilators and so on. Aromatherapy, using essential oils, relaxes or stimulates the body, creates positive moods, relieves cold symptoms and respiratory problems and eases muscle pain. The vapors are inhaled to release neurochemicals in the brain through receptors in the mouth and nose, which then cause the desired effects. Spices are also used as balms or massage oil and applied on the skin, joints and muscles to relieve stress and pain.

Many spices have anticholesterolenic and antitumor properties, which hold lot of promises in the management of health problems. They also increase bile secretion and favorably enhance the absorption of fat soluble vitamins. Turmeric contains curcumin, which is an antibacterial and also lowers blood and liver cholesterol along with appropriate changes in serum lipoprotein. Curcumin is also known for its antitumor activity and the action is mediated through the induction of apoptosis in AK-5 tumor cells.

In traditional medicine, ginger is used to improve the flow of body fluids. It stimulates blood circulation throughout the body by powerful stimulatory effect on the heart muscles and by diluting blood. Essential oils from cumin, poppy, basil and rosemary components, carsinol, carnosol and urosolic acid possess anticarcinocenic activity.

Capsicum is another spice most commonly used in our diet. Capsaicin, the pungent principle of red hot chilli often serve as analgesic, anti-inflammatory and anti-obesity agent. The role of red pepper and turmeric has been studied for treatment of arthritis. In Ayurveda, the indigenous system of medicine, spices have an indispensable position in the formulation of various drugs.

The basic problem in using the essential oils in clinics is the high variability of their chemical composition, which depends on many factors, both biotic and abiotic. A need to find a way to standardize the chemical composition of essential oils is necessary in order to have a stable mixture (Oliveira et al. 2020). Hassan et al. 2018 used the experimental model of *Candida albicans* keratitis in rabbits giving 50 µL of 4 mg/mL eugenol in propylene glycol drops hourly for 15 days for determining the efficacy of eugenol for the treatment of *Candida* keratitis in an experimental model. At least 75% of all eugenol-treated eyes recovered from keratitis, with improvement in the remaining 25% of the eyes thereby suggesting eugenol as a natural, safe and effective treatment for fungal keratitis.

Essential oils and its individual compounds are widely used in cosmetic products due to analgesic, antiseptic, antimicrobial, carminative, diuretic, spasmolytic to hyperaemic and stimulatory activities. Creams, gels and ointments containing essential oils possess antibacterial and antifungal activities and do not require any additional chemical preservative. For example, thymol is used in small concentrations (less than 1 mM concentrations) in

a) Cosmetic formulations, as denaturant and fragrance ingredient,
b) Foods, as preservative against food-borne microorganisms and synthetic flavoring,
c) External drugs for its preservative properties, and
d) Products such as insecticides, fungicides, other biocidal products and as a room deodorant. The use of thymol in drugs is limited by its moderate cytotoxicity as evident from *in vitro* human and animal cells, and in experimental animal studies.

Antimicrobial principles of Spices

Spice plants synthesize a wide spectrum of many compounds with complex molecular structures by secondary metabolism and are present in spice essential oils and oleoresins. These secondary metabolites are found as alkaloids, flavonoids, isoflavonoids, tanins, cumarins, glycosides, terpens and phenolic compounds.

Although each spice essential oil contains many chemical components, major components participate in antimicrobial activities (Table 5.3).

Table 5.3. Major antimicrobial components of spice essential oils.

Spice	**EO major antimicrobial component**
Celery leaf	Limonene (40.5%)
Clove	Eugenol (85–92%)
Basil	Methyl chavicol (74–87%)
Cinnamon	Cinnamaldehyde (42–75%)
Coriander	Linalool (43.67%)
Star anise	*Trans-anethole* (82.7%)
Black pepper	Piperine (35–60%)
Cumin	Cuminaldehyde (35–62%)
Garlic	Allicin (0.4%)
Ginger	Zingerone (25–30%)
Red pepper (chilli)	Capsaicin (2–4%)
Turmeric	Curcumin (20–30%)
Oregano	Carvacrol
Fenugreek	Saponin

Both oils and oleoresins are widely used in pharmaceutical products, to provide either a pleasant taste or aroma to make them pleasing and easy to use, which would otherwise be difficult to accept. These include medications, skin creams, cold remedies, etc. Most staple foods are bland, although they provide important sources of nutrition and are essential for growth and maintenance. Food is made palatable and enjoyable by using spices and herbs which are valued for their flavour and color. The flavor moiety is well protected from evaporators and in whole spices by the cells. Whole spices release flavor slowly; but ground spices release the flavor more quickly. Characteristic flavor compounds are unique for different spices (Table 5.4).

Total polyphenols in the non-volatile parts of *C. cassia* branches and leaves exhibit antibacterial activity *in vitro* against Gram-positive (*Staphylococcus aureus* and *Streptococcus pneumoniae*) and Gram-negative (*Escherichia coli* and *Pseudomonas aeruginosa*) bacteria. Acetone extract of *C. cassia* contain 88.39% cinnamaldehyde and is most sensitive to the plant pathogen *Alteraria solani* (Lu et al. 2019).

Allicin is a chemically oxygenated sulfur amino acid that occurs as a major component of garlic extract and has potent *in vitro* antimicrobial activity. Bacterial species like methicillin-resistant *Staphylococcus aureus, Escherichia coli, Enterococcus, Shigella dysenteriae, S. flexneri,* and *S. sonnei* are sensitive while *Pseudomonas aeruginosa, Streptococcus β hemolyticus* and *Enterococcus faecium*, are resistant to allicin (Ankri and Mirelman 1999). According to Yamada and Azuma (1977) pure allicin possess fungistatic and fungicidal activity against *Candida, Cryptococcus, Trichophyton, Epidermophyton,* and *Microsporum* at low concentration with MIC between 1.57 and 6.25 µg/mL. Allicin reacts with thiol compound (L-cysteine) to form the S-thiolation product, S-allylmercaptocysteine. However the main antimicrobial effect of allicin is due to its interaction with important thiol-containing enzymes. Allicin also specifically inhibits other bacterial enzymes such as the acetyl-CoA-forming system, consisting of acetate kinase and phosphotransacetyl-CoA synthetase. The wide spectrum antimicrobial effects of allicin are due to the multiple inhibitory effects they may have on various thiol-dependent enzymatic systems.

Table 5.4. Important flavor compounds in Spices.

Spice	:	**Flavour compounds**
Allspice	:	Eugenol, β-caryophyllene
Anise	:	(E)-anethole, methyle, chavicol
Bay laurel	:	1.8-cineole
Black pepper	:	Piperine, S-3-Carene, β-caryophyllene
Caraway	:	δ-carvone, caronederivati yes
Cardamom	:	α-terpinyl acetate, 1-8-cineule, linalool
Cinnamon, cassia	:	Cinnamaldehyde, eugenol
Chilli	:	Capsaicin, dihydro capsaicin
Clove	:	Eugenol, eugeneyl acetate
Coriander	:	δ-linalool, C10-C 14-2-alkenals
Cumin	:	Cuminaldehyde, p-l.3-mentha-dienal
Dill	:	δ-carvone
Fennel	:	(E)-anethole, fenchone
Ginger	:	Gingerol, Shogaol, neral, geranial
Mace	:	α-pinene, sabinene, 1-terpenin-4-ol
Mustard	:	Ally isothiocynate
Nutmeg	:	Sabinine, a-pinene, myristicin
Parsley	:	Apiol
Saffron	:	Safranol
Turmeric	:	Turmerone, Zingeberene, 1,8-cineole
Vanilla	:	Vanillin, p-OH-benzyl-methyl ether
Important flavour compounds in a few culinary herbal Spices		
Herbal spices	:	**Flavour compounds**
Basil, Sweet	:	Methylchavicol, linalool, methyl eugenol
Marjoram	:	e- and t-sabinene hydrates, terpinen-4-ol
Oregano	:	Carvacrol, thymol
Rosemary	:	Verbenone, 1-8-cineole, camphor, linanool
Sage, Clary	:	Salvial-4(14)-en-l-one, linalool
Sage, Dalmation	:	Thujone, 1,8-cineole, camphor
Sage, Spanish	:	e- and t-sabinylacetate, 1,8-cineole, camphor
Savory	:	Carvacrol
Tarragon	:	Methyl chavicol, anethole
Thyme	:	Thymol, carvacrol
Peppermint	:	I-menthol, menthone, menthfuran
Spear mint	:	I-carvone, carvone derivatives

Source: Spices Board, India. Available at https://www.indianspices.com/spices-development/properties/flavor-profile-spices.html.

Thymol is a phenol derivative of terpenes. Thymol (2-isopropyl-5-methylphenol) is the main monoterpene phenol found in the essential oils extracted from plants belonging to the genus Thymus. Due to the lack of health risk to the consumer, it is registered by the European Commission for use as flavorings in foodstuffs. Thymol possesses antiseptic, antibacterial, antifungal, anthelmintic, antiviral, antioxidant, expectorant, antispasmodic, carminative, diaphoretic, sedative, anti-rheumatic, even anti-cancer, anti-hyperlipidemic and anti-hyperglycemic activity. Thymol is the most effective *S. typhimurium, E. coli, Staphylococcus aureus, Staphylococcus epidermidis*, and methicillin-resistant *S. aureus, Klebsiella pneumoniae, Yersinia enterolitica O9, Brucella melitensis, P. aeruginosa* and *Proteus* spp. Furthermore, 2.0 mg/ml concentration is sufficient to inhibit *E. coli*.

Bacterial and fungal resistance to antibiotics is related to their ability of forming biofilm that prevents the penetration of the antibacterial agent at the site of infection. It is estimated that 65–80% of all infections are caused by biofilm-related microorganisms, which is the main reason for the ineffectiveness of treatment. Thymol has shown the ability to inhibit biofilms of *E. coli, Listeria monocytogenes, Pseudomonas putida* and *S. aureus*. Besides antibacterial action, thymol also exerts antifungal activity towards *Candida albicans*, and *Cryptococcus neoformans*.

Carvacrol, a phenolic monoterpene is abundant in essential oils of Origanum, and Thymus. It affects periplasmic enzymes and membrane proteins in many bacteria and makes *E. coli* nonmotile by inhibiting flagellin synthesis. The mechanism of action of carvacrol against microorganisms is associated with the presence of a hydroxyl group and an aromatic ring.

Jeyakumar and Lawrence (2021) studied antibacterial activity of the eugenol against multi-drug resistant clinical isolates of Gram positive (*Listeria monocytogenes, Staphylococcus aureus, Streptococcus pyogenes, Clostridium perfringens, Bacillus cereus, Bacillus subtilis*) and Gram negative (*Escherichia coli, Salmonella typhi, Shigella dysenteriae, Campylobacter jejuni, Helicobacter pylori, Vibrio cholerae, Pseudomonas aeruginosa*). Their results showed highest antibacterial activity with maximum zones of inhibition against *Campylobacter jejuni* (36.33 $\pm$ 1.53 mm) and *Helicobacter pylori* (34.00 $\pm$ 1.00 mm) and the lowest activity against *Pseudomonas aeruginosa* (10.00 $\pm$ 1.00 mm) followed by *Listeria monocytogenes, Bacillus subtilis* and *Streptococcus pyogenes* with zones of inhibition corresponding to 17.33, 21.00 and 20.33 mm, respectively. Furthermore eugenol causes extensive disruption of the *E. coli* bacterial membrane, with the drainage of cytoplasmic content. The data on mechanism of eugenol suggest disturbance of lipid fraction of the bacterial membrane which alters membrane permeability resulting in leakage of intracellular contents. de Oliveira Pereira and coworkers (2013), investigated antifungal activity of eugenol against 13 clinical isolates of dermatophytic fungus *Trichophyton rubrum* using broth microdilution assay. MIC values of eugenol against *T. rubrum* strains varied from 64–512 µg/ml, with 50% of the strains tested inhibited at 256 µg/ml and 90% at 512 µg/ml. Conidia formation and ergosterol production by fungal cells was also decreased.

Curcumin is a natural pigment with a characteristic yellow–orange color, predominantly found in the rhizomes of turmeric (*Curcuma longa* L.) from the ginger family, Zingiberaceae, native to tropical South Asia. Together with essential oils and other curcuminoids, curcumin is a principal bioactive compound of turmeric powder. It is very popular in South Asian and Middle Eastern cuisines, especially for preparing curry dishes. It blocks bacterial growth owing to its structural characteristics and the generation of antioxidation products. Curcumin can inhibit bacterial virulence factors, inhibit bacterial biofilm formation and prevent bacterial adhesion to host receptors through the bacterial quorum sensing regulation system. Adamczak et al. 2020, reported broad spectrum of antimicrobial activity of curcumin against six species of Gram-positive bacteria (*Enterococcus faecalis, Staphylococcus aureus, S. epidermidis, S. haemolyticus, Streptococcus agalactiae, S. pyogenes*), nine Gram-negative bacteria (*Acinetobacter baumannii, A. lwoffii, Escherichia coli, Klebsiella oxytoca, K. pneumoniae, Proteus mirabilis, Pseudomonas aeruginosa, Serratia marcescens, Stenotrophomonas maltophilia*), and four species of yeast-like fungi (*Candida albicans, C. glabrata, C. tropicalis, Saccharomyces cerevisiae*) and confirmed the much greater sensitivity of Gram-positive than Gram-negative bacteria except *Pseudomonas aeruginosa* (MIC = 62.5 μg/mL), to curcumin.

Cinnamaldehyde, the main bioactive compound of cinnamon inhibits Extended-Spectrum-Beta-Lactamase (ESBL) producing bacteria *E. coli* and *Klebsiella pneumoniae* at MICs of 122 μg/mL and 245 μg/mL respectively (Dhara and Tripathi 2013). This bioactive compound was also found to have greater antimicrobial activity than eugenol, the main bioactive compound of clove. Resistance to third generation cephalosporins is associated with production of Extended-Spectrum-Beta-Lactamase (ESBL) by pathogenic bacteria, and such findings certainly help the replacement of cephalosporins by plant derived natural product, cinnamaldehyde.

Mechanism of antimicrobial activity of essential oils

Mechanism studies of potential bioactive compounds of plant origin and their interaction aims at a better understanding of activity, toxicity and the possibility as a candidate for the design of second-generation inhibitors. Although the antimicrobial effects of essential oils are well established, the mechanism of action of such compounds is poorly understood. Research on the antimicrobial action of phenolic compounds focuses on their effects on cellular membranes (Ultee et al. 1999). Plant essential oils contain such compounds in their composition are known to cause structural and functional damage to plasma membranes.

The mechanism responsible for phenolic toxicity to microorganisms includes enzymes inhibition by the oxidized compounds, possibly through reaction with sulfydryl groups or through more specific interaction with the proteins (Mason 1987). The site(s) and number of hydroxyl groups on the phenolic ring are thought to be associated with their relative toxicity to microorganisms. According to Conner and Beuchat (1984), the antimicrobial action of essential oils may be due to impairment of a variety of enzymes systems including those involved in energy production and structural component synthesis. Various preservatives including essential oils, phenols and bacteriocins acts on the permeability barrier of cytoplasmic membrane and causes leakage of various substances such as ions, ATP, nucleic acid and amino acids.

Lambert et al. (2001) studied the mechanism of the action of oregano essential oils and two of its principle components namely thymol and carvacrol against *Pseudomonas aeruginosa* and *Staphylococcus aureus*. They reported that mixing of carvacrol and thymol at a correct amount may exert the total inhibition, which is similar to oregano essential oil. Such inhibition is due to damage in membrane integrity, which further affects pH homeostasis and equilibrium of inorganic ions.

In general, studies on the mechanism of action of essential oils have shown their deleterious effects on cellular membrane including permeability and protons motive force, destruction of electron transport systems, hydrogen bonding of phenolic compounds to membrane proteins causing partition in the lipid bilayer.

Generally, Gram-negative bacteria have been reported to be more resistant than Gram-positive to essential oils because of their cell wall lipopolysaccharide (Russel 1991). Cell wall lipopolysaccharide may prevent essential oils active compounds from reaching the cytoplasmic membrane of Gram-negative bacteria. Inhibitory action of natural products on the mold involves cytoplasm granulation, cytoplasmic membrane rupture and inactivation and/or inhibition of intracellular and extracellular enzymes. These biological events could take place separately or concomitantly causing inhibition of mycelium germination (Cowan 1999).

Cinnamaldehyde is an electronegative molecule that interferes with the cellular biological process, particularly nitrogen containing substances such as proteins and nucleic acids. Cinnamon EO has been reported to inhibit bacteria through antiquorum sensing effects; inhibiting cell division, ATPase, biofilm formation membrane porin and mobility; altering the lipid profile and thereby acting cell membrane producing lumps and auto aggregation (Vasconcelos et al. 2018; Clemente et al. 2016). In a separate study, He et al. 2018 showed that cinnamaldehyde has potential antimicrobial abilities, including alteration and destruction of cell membrane and interaction with genomic DNA. It changes the secondary structure and morphology of genomic DNA by binding to the minor groove and changes morphology of *E. coli* cells to wrinkled, distorted and even lysed cells as a result of decrease in cell viability. Thus cinnamaldehyde has an important impact on the *E. coli* cell membrane in order to enhance its permeability so that it can enter and spread throughout the interior of the cell. Generally Gram-positive bacteria are more sensitive to essential oils than Gram-negative strains. This is due to differences of the membranes between Gram-positive and Gram-negative bacteria. The cell envelope of Gram-positive strains is composed of glycopolymers such as teichoic acid and several layers of peptidoglycan, making the envelope thicker than the one of Gram-negative strains, which is composed of a thin layer of peptidoglycan. Lipoproteins and lipopolysaccharides in the outer membrane of Gram-negative strains reduces the entrance of certain drugs such as lipophilic molecules and makes the organism more resistant.

According to literature, essential oils exert their activities through different mechanisms, such as membrane destabilization or inhibition of ATPase activity, cell division or efflux pumps activity. The activity of essential oils depends on their chemical composition, but mostly studies indicate the membrane as the first target independent of the active compound or the bacterial strain. For example, the E-cinnamaldehyde present at 82.05% in Chinese cinnamon, thymol present at 45.17% in bishop's weed and carvacrol present at 57.21% in oregano interferes with membrane integrity as the first target. These three compounds can perturb *in vitro* the lipid membrane of the bacteria causing changes in the permeability, causing leakage of the intracellular material and therefore the death of the bacteria (Oliveira et al. 2020). The mode of action of thymol has been associated with its interaction with membrane proteins, modification of the membrane permeability and intracellular targets.

The fungal cell wall is a dynamic structure that protects fungal protoplasts from external osmotic shocks and defines fungal morphogenesis. Thus, changes in the organization or functional disruption of the cell wall induced by antifungal agents are involved in fungal death. The fungal cell membrane is a dynamic structure composed of a lipid bilayer where enzymes and transport proteins are embedded. Ergosterol is a unique sterol found only in the cell membrane of fungi, important for their proper growth and functioning and also acts as an important regulator of membrane fluidity. Thymol have been shown to decrease ergosterol in cell membranes of *Candida* and *Cryptococcus*, thereby causing disruption of membrane integrity, membrane-associated enzyme disturbances, extensive damage and, finally cell death (Kowalczyk et al. 2020). Similarly eugenol exerts its antifungal activity on the cell wall and cell membrane of *Trichophyton rubrum* by disrupting ergosterol biosynthesis and such a result may serve as a guide for future *in vivo* studies of clinical use of eugenol in treating dermatophyte infections (de Oliveira Pereira et al. 2013).

Combination efficacy of spices

Synergism comes from the Greek word 'Synergos' meaning working together. Synergism is a positive interaction. It refers to the interaction between two or more antimicrobial agents against the pathogen, when the combined effect of antimicrobials is significantly greater than the sum of inhibitory effects of the individual antimicrobial agent when they are tested separately. Plant derived natural products potentiate the activity of antibiotics through synergistic effect and such a combination approach for solving the problem of antimicrobial resistance. Combination therapy of known drugs began to be considered as an alternative to the existing ones since the discovery of the excellent synergistic effect of 5-flucytosine and amphotericin B (Zacchino et al. 2003).

Antimicrobial drug resistance is of two types, 'microbiological' and 'clinical'. When a strain is not susceptible to a dosage above an established breakpoint level, based on standardized CLSI protocols, is called microbiological resistance. It can be 'primary' if the pathogen is innately resistant without any prior exposure to the drug and 'secondary' if the resistance develops after repeated exposure to a drug. Chemosensitization by natural compounds to increase effectiveness of commercial antimycotics is a somewhat new approach to deal with drug resistant organisms. Chemosensitizing agents possess antifungal activity, but at insufficient levels to serve as antimycotics, alone. Their main function is to disrupt fungal stress response, destabilize the structural integrity of cellular and vacuolar membranes or stimulate production of reactive oxygen species, augmenting oxidative stress and apoptosis. These agents function mainly as synergists that affect the target pathogen in such a way that it becomes more vulnerable (chemosensitized) to the commercial antimycotic agent (Campbell et al. 2012).

The main purpose of using combinations of plant secondary metabolites with antibiotics is to increase the antimicrobial effect against the target pathogen. Although different methods are available for detecting the interactions between plant extracts/ essential oils, the checkerboard technique is a method of choice. In this *in vitro* method concentrations of one substance are arranged horizontally and concentrations of another substance are arranged vertically in 96-well broth microdilution tray. Fractional Inhibitory Concentration Index (FICI) is then calculated and interaction of two antimicrobials is interpreted as synergistic, additive and antagonistic with FICI values of ≤ 0.5, > 0.5 and ≤ 2 and > 2 respectively (Table 5.5).

Thymol, was used as a traditional, natural antifungal agent in Chinese medicine, but has far lower antifungal activity than commercially available azoles. To increase its activity, Faria et al. (2011) used the combination approach by mixing thymol with known anticandidal drugs amphotericin B, fluconazole and itraconazole. The combination antifungal actions were determined as fractional inhibitory concentration index ($FICI_{90}$) and classified synergistic if $FICI_{90}$ values ≤ 0.5. The results of this study showed thymol synergism with amphotericin B, fluconazole and itraconazole with $FICI_{90}$ values 0.50, 0.37 and 0.21 respectively. Synergism by thymol may be associated with the disruption of the fungal cell wall/membrane.

Synergistic interactions are also well documented for pathogenic bacteria as reviewed by Vasconcelos et al. 2018. Mixing antibiotics with cinnamon essential oil reduce the MICs, 2 fold for *P. aeruginosa*, 2–8 fold for *S. aureus*, and 2–4 fold for *E. coli*. The synergistic effect was obtained by the combination of cinnamon essential oil with ampicillin or chloramphenicol

Table 5.5. Combination effects of Spices.

Spice EO/ Component	Combination with	Pathogen (Interaction)	Reference
Allicin	Nystatin	*Candida albicans* (Synergistic)	Zainal 2021
	CHX	*Staphylococcus aureus* (Synergistic)	
Cinnamon EO	Chloramphenicol	*Staphylococcus aureus, Escherichia coli* (Synergistic)	El Atki et al. 2019
	Streptomycin	*Staphylococcus aureus, Escherichia coli* (Additive)	
	Ampicillin	*Staphylococcus aureus* (Synergistic)	
Carvacrol	Eugenol	*Listeria monocytogenes, Enterobacter aerogenes, Escherichia coli* (Synergistic)	Bassole 2010
	Thymol	*Listeria monocytogenes, Enterobacter aerogenes, Escherichia coli, Pseudomonas aeruginosa* (Synergistic)	
	Linalool	*Enterobacter aerogenes, Escherichia coli* (Synergistic)	
Thyme EO	Cinnamon EO	*Listeria monocytogenes, Escherichia coli* (Synergistic)	García-Díez et al. 2017
	Cumin EO	*Listeria monocytogenes, Salmonella* (Synergistic)	
	Parsley	*Listeria monocytogenes* (Indifferent)	
	Garlic EO	*Staphylococcus aureus* (Additive)	

against *S. aureus* and by the combination of essential oil and chloramphenicol against *E. coli* and an additive effect was found when essential oil was combined with streptomycin against *E. coli, S. aureus* and *P. aeruginosa*. Cinnamon essential oil when combined with colistin demonstrates a synergistic action against multidrug resistant *P. aeruginosa*; and with piperacillin demonstrates a synergistic interaction against *E. coli*. Synergistic combinations could decrease the minimum effective dose of the drugs, thus reducing their possible adverse effects and the costs of treatment.

Several combinations of spice essential oils were tested by García-Díez and coworkers (2017) using the checkerboard assay, for possible synergism among different essential oils in order to control food borne pathogens *Salmonella* spp., *Listeria monocytogenes, Escherichia coli* and *Staphylococcus aureus*. Thyme-cumin combination achieved 30-fold decrease in the MIC of cumin essential oil against *Salmonella* spp., about 60-fold against *E. coli* and 12-fold against *S. aureus*. Similarly thyme/cinnamon, thyme/garlic cinnamon/cumin, cinnamon/ parsley combinations showed a decrease in MIC values from 2 to 60-fold depending on bacterial species. The agar well diffusion method has been used by some workers (Chowdhury et al. 2019) to determine potential synergism among spice extracts against *E. coli, Vibrio cholerae, S. aureus, Ba. cereus, S. typhi* and *Pseudomonas* sp. The combination of garlic-turmeric, garlic-ginger, garlic-cinnamon and also turmeric-cinnamon showed synergistic properties against all of the tested bacteria. Denture Stomatitis (DS) or chronic erythematous candidiasis is associated with *Candida albicans* and *S. aureus*. Nystatin is recommended for oral and chronic erythematous candidiasis. To justify any synergic action, Zainal et al. 2021 identified synergistic action of nystatin with allicin for *C. albicans* and allicin-Chlorhexidine combination for *S. aureus*.

Very often it appears that *in vitro* assays fail to match *in vivo* results. This occurs when there is a discrepancy between laboratory determinations predicting drug efficacy, based on Minimum Inhibitory Concentrations (MICs) according to standardized protocols and the failure of drug treatment under clinical conditions.

Conclusions

Secondary metabolites of spices exert inhibitory effects on microbial cells, but detailed mechanisms underlying antimicrobial actions still need to be intensified. *In vitro* antimicrobial effects should match *in vivo* studies without any side effects. Hence there is a greater need for studying *in vivo* effects in order to apply the spice component for therapeutic application. In order to develop a specific treatment product on a commercial scale, the supply of particular variety of spice plant should be considered. The bioactive principle usually lies in a particular component, and gives more reproducible results than using whole plant material. Secondly the purified component shows much better antimicrobial activities, and this factor need more attention before developing a process on a commercial scale. Some spice essential oils and their components have potential toxic effects in the human body and this issue should be correctly addressed in future. Potent synergism is involved in antimicrobial efficacies and hence indepth studies in combinations of different spices and spices with known antibiotics can be an excellent approach to achieve the targets.

References

Adamczak, A., Ożarowski, M., Karpiński, T.M. (2020). Curcumin, a natural antimicrobial agent with strain-specific activity. Pharmaceuticals 13(7): 153.

Ahmad, I., Mehmood, Z., Mohammad, F. (1998). Screening of some Indian medicinal plants for their antimicrobial properties. J Ethnopharmacol 62(2): 183–193.

Aktuğ, Ş.E., Karapinar, M. (1986). Sensitivity of some common food-poisoning bacteria to thyme, mint and bay leaves. Int J Food Microbiol 3(6): 349–354.

Ankri, S., Mirelman, D. (1999). Antimicrobial properties of allicin from garlic. Microbes Infect 1(2): 125–129.

Arora, D.S., Kaur, J. (1999). Antimicrobial activity of spices. Int J Antimicrob Agents 12(3): 257–262.

Azzouz, M.A., Bullerman, L.B. (1982). Comparative antimycotic effects of selected herbs, spices, plant components and commercial antifungal agents. J Food Prot 45(14): 1298–1301.

Bakhru, H.K. (2004). Indian Spices and Condiments as Natural Healers. 3rd Edn. Jaico Publishing House, Mumbai.

Bassam, A.S. (2004). Antibacterial activities of some plant extracts utilized in popular medicine in Palestine. Turk J Biol 28(2-4): 99–102.

Beuchat, L.R. (1994). Antimicrobial properties of spices and their essential oils. Nat Antimicrob Syst Food Preserv 12: 257–262.

Biharee, A., Sharma, A., Kumar, A., Jaitak, V. (2020). Antimicrobial flavonoids as a potential substitute for overcoming antimicrobial resistance. Fitoterapia 104720.

Burt, S.A., Reinders, R.D. (2003). Antibacterial activity of selected plant essential oils against *Escherichia coli* O157: H7. Lett Appl Microbiol 36(3): 162–167.

Campbell, B.C., Chan, K.L., Kim, J.H. (2012). Chemosensitization as a means to augment commercial antifungal agents. Front Microbiol 3: 1–20.

Chao, S.C., Young, D.G., Oberg, C.J. (1998). Effect of a diffused essential oil blend on bacterial bioaerosols. J Essent Oil Res 10(5): 517–523.

Chowdhury, S.P., Banik, A., Aurin, S.A., Saima, S. (2019). Comparison of individual and synergistic antimicrobial activity of common spices against certain infectious pathogen in Bangladesh. Am J Plant Sci 10(9): 1599–1611.

Clemente, I., Aznar, M., Silva, F., Nerín, C. (2016). Antimicrobial properties and mode of action of mustard and cinnamon essential oils and their combination against foodborne bacteria. Innov Food Sci Emerg Technol 36: 26–33.

Conner, D.E., Beuchat, L.R. (1984). Effects of essential oils from plants on growth of food spoilage yeasts. J Food Sci 49(2): 429–434.

Cowan, M.M. (1999). Plant products as antimicrobial agents. Clin Microbiol Rev 12(4): 564–582.

de Oliveira Pereira, F., Mendes, J.M., de Oliveira Lima, E. (2013). Investigation on mechanism of antifungal activity of eugenol against *Trichophyton rubrum*. Med Mycol 51(5): 507–513.

De, M., Krishna De, A., Banerjee, A.B. (1999). Antimicrobial screening of some Indian spices. Phytotherapy Research: An International Journal Devoted to Pharmacological and Toxicological Evaluation of Natural Product Derivatives 13(7): 616–618.

Deans, S.G., Ritchie, G. (1987). Antibacterial properties of plant essential oils. Int J Food Microbiol 5(2): 165–180.

Dhara, L., Tripathi, A. (2013). Antimicrobial activity of eugenol and cinnamaldehyde against extended spectrum beta lactamase producing enterobacteriaceae by *in vitro* and molecular docking analysis. Eur J Integr Med 5(6): 527–536.

Dorman, H.D., Deans, S.G. (2000). Antimicrobial agents from plants: antibacterial activity of plant volatile oils. J Appl Microbiol 88(2): 308–316.

Duke, J.A., Ayensu, E.S. (1985). Medicinal Plants of China. Reference Publ. Inc., Algonac, MI.

Elgayyar, M., Draughon, F.A., Golden, D.A., Mount, J.R. (2001). Antimicrobial activity of essential oils from plants against selected pathogenic and saprophytic microorganisms. J Food Prot 64(7): 1019–1024.

Elizabeth, K.M. (2001). Antimicrobial activity of *Allium sativum* on some pathogenic bacteria. Indian J Microbiol 41(4): 321–323.

Farag, R.S., Daw, Z.Y., Hewedi, F.M., El-Baroty, G.S.A. (1989). Antimicrobial activity of some Egyptian spice essential oils. J Food Prot 52(9): 665–667.

Faria, N.C.G., Kim, J.H., Gonçalves, L.A.P., Martins, M.D.L., Chan, K.L., Campbell, B.C. (2011). Enhanced activity of antifungal drugs using natural phenolics against yeast strains of *Candida* and *Cryptococcus*. Lett Appl Microbiol 52(5): 506–513.

Friedman, M. (2017). Chemistry, antimicrobial mechanisms, and antibiotic activities of cinnamaldehyde against pathogenic bacteria in animal feeds and human foods. J Agric Food Chem 65(48): 10406–10423.

García-Díez, J., Alheiro, J., Pinto, A.L., Falco, V., Fraqueza, M.J., Patarata, L. (2017). Synergistic activity of essential oils from herbs and spices used on meat products against food borne pathogens. Nat Prod Commun 12(2): 1934578X1701200236.

Gootz, T.D. (2010). The global problem of antibiotic resistance. Crit Rev Immunol 30(1).

Hassan, H.A., Geniady, M.M., Abdelwahab, S.F., Abd-Elghany, M.I., Sarhan, H.A., Abdelghany, A.A., Alio, J.L. (2018). Topical eugenol successfully treats experimental *Candida albicans*-induced keratitis. Ophthalmic Res 60(2): 69–79.

He, T.F., Zhang, Z.H., Zeng, X.A., Wang, L.H., Brennan, C.S. (2018). Determination of membrane disruption and genomic DNA binding of cinnamaldehyde to *Escherichia coli* by use of microbiological and spectroscopic techniques. J Photochem Photobiol B: Biol 178: 623–630.

Hitokoto, H., Morozumi, S., Wauke, T., Sakai, S., Kurata, H. (1980). Inhibitory effects of spices on growth and toxin production of toxigenic fungi. Appl Environ Microbiol 39(4): 818–822.

Huhtanen, C.N. (1980). Inhibition of *Clostridium botulinum* by spice extracts and aliphatic alcohols. J Food Prot 43(3): 195–196.

Jeyakumar, G.E., Lawrence, R. (2021). Mechanisms of bactericidal action of Eugenol against *Escherichia coli*. J Herb Med 26: 100406.

Kamble, V.A., Patil, S.D. (2007). Antimicrobial effects of certain spices and condiments used in an Indian indigenous system of medicine. Zeitschrift fur Arznei-& Gewurzpflanzen 12(4): 188–193.

Karapinar, M., Aktuğ, Ş.E. (1987). Inhibition of foodborne pathogens by thymol, eugenol, menthol and anethole. Int J Food Microbiol 4(2): 161–166.

Kerekes, E.B., Deák, É., Takó, M., Tserennadmid, R., Petkovits, T., Vágvölgyi, C., Krisch, J. (2013). Anti-biofilm forming and anti-quorum sensing activity of selected essential oils and their main components on food-related micro-organisms. J Appl Microbiol 115(4): 933–942.

Kowalczyk, A., Przychodna, M., Sopata, S., Bodalska, A., Fecka, I. (2020). Thymol and thyme essential oil—new insights into selected therapeutic applications. Molecules 25(18): 4125.

Lambert, R.J.W., Skandamis, P.N., Coote, P.J., Nychas, G.J. (2001). A study of the minimum inhibitory concentration and mode of action of oregano essential oil, thymol and carvacrol. J Appl Microbiol 91(3): 453–462.

Lange, D. (2002, August). Medicinal and aromatic plants: trade, production, and management of botanical resources. pp. 177–197. *In*: XXVI International Horticultural Congress: The Future for Medicinal and Aromatic Plants, 629.

Leja, K.B., Czaczyk, K. (2016). The industrial potential of herbs and spices? A mini review. Acta Sci Pol Technol Aliment 15(4): 353–365.

Lu, K., Wang, Q., Huo, X., Gao, Y., Feng, F. (2019). Composition analysis of acetone extract of Cinnamomum cassia and its inhibition on 5 plant pathogens. Southwest China J Agric Sci 32(4): 798–802.

Mason, T.L. (1987). Inactivation of red beet β-glucan synthase by native and oxidized phenolic compounds. Phytochemistry 26(8): 2197–2202.

Meena, M.F. (1994). Antimicrobial activity of essential oils from spices. J Food Sci Technol 31: 68–70.

Mostafa, A.A., Al-Askar, A.A., Almaary, K.S., Dawoud, T.M., Sholkamy, E.N., Bakri, M.M. (2018). Antimicrobial activity of some plant extracts against bacterial strains causing food poisoning diseases. Saudi J Biol Sci 25(2): 361–366.

Nair, K.K.S., Narasimha Rao, D., Nair, R.B. (1996). Effect of spices and organic acids used in traditional meat products on some bacteria. Ind Food Pack 50(1): 5–10.

Nanasombat, S., Lohasupthawee, P. (2005). Antibacterial activity of crude ethanolic extracts and essential oils of spices against salmonellae and other enterobacteria. Curr Appl Sci Technol 5(3): 527–538.

Nazzaro, F., Fratianni, F., De Martino, L., Coppola, R., De Feo, V. (2013). Effect of essential oils on pathogenic bacteria. Pharmaceuticals 6(12): 1451–1474.

Nguefack, J., Budde, B.B., Jakobsen, M. (2004). Five essential oils from aromatic plants of Cameroon: their antibacterial activity and ability to permeabilize the cytoplasmic membrane of *Listeria innocua* examined by flow cytometry. Lett Appl Microbiol 39(5): 395–400.

Oliveira Ribeiro, S., Fontaine, V., Mathieu, V., Zhiri, A., Baudoux, D., Stévigny, C., Souard, F. (2020). Antibacterial and cytotoxic activities of ten commercially available essential oils. Antibiotics 9(10): 717.

Pruthi, J.S. (1976). In Spices and Condiments. 1st Edition. National Book Trust, India, Green Park, New Delhi. pp. 1–269.

Rosengarten Jr., F. (1969). The Book of Spices. Jove Publ., Inc., New York. pp. 23–96.

Russell, A.D. (1991). Mechanisms of bacterial resistance to non-antibiotics: Food additives and food and pharmaceutical preservatives. J Appl Bacteriol 71(3): 191–201.

Sagdic, O., Karahan, A.G., Ozcan, M., Ozkan, G. (2003). Note: effect of some spice extracts on bacterial inhibition. Food Sci Technol Int 9(5): 353–358.

Shapiro, S., Meier, A., Guggenheim, B. (1994). The antimicrobial activity of essential oils and essential oil components towards oral bacteria. Oral Microbiol Immunol 9(4): 202–208.

Shaskolskiy, B., Dementieva, E., Leinsoo, A., Runina, A., Vorobyev, D., Plakhova, X., Gryadunov, D. (2016). Drug resistance mechanisms in bacteria causing sexually transmitted diseases and associated with vaginosis. Front Microbiol 7: 747.

Shukla, H.S., Tripathi, S.G. (1987). Antifungal substance in the essential oil of anise (*Pimpinella anisum* L.). Agric Biol Chem 51(7): 1991–1993.

Singh, G., Kiran, S., Marimuthu, P., Isidorov, V., Vinogorova, V. (2008). Antioxidant and antimicrobial activities of essential oil and various oleoresins of *Elettaria cardamomum* (seeds and pods). J Sci Food Agric 88(2): 280–289.

Souissi, M., Azelmat, J., Chaieb, K., Grenier, D. (2020). Antibacterial and anti-inflammatory activities of cardamom (*Elettaria cardamomum*) extracts: Potential therapeutic benefits for periodontal infections. Anaerobe 61: 102089.

Spellberg, B., Guidos, R., Gilbert, D., Bradley, J., Boucher, H.W., Scheld, W.M., Infectious Diseases Society of America. (2008). The epidemic of antibiotic-resistant infections: a call to action for the medical community from the Infectious Diseases Society of America. Clin Infect Dis 46(2): 155–164.

Subbulakshmi, G., Naik, M. (2002). Nutritive value and technology of spices: current status and future perspectives. J Food Sci Technol (Mysore) 39(4): 319–344.

Syed, M., Hanif, M., Chaudhry, F.M., Bhatty, M.K. (1986). Antimicrobial activity of the essential oils of the Umbelliferae family Part 1. *Cuminum cyminum, Coriandrum sativum, Foeniculum vulgare* and *Bunium persicum* oils. Pak J Sci Ind Res 29: 183–188.

Talbot, G.H., Bradley, J., Edwards Jr, J.E., Gilbert, D., Scheld, M., Bartlett, J.G. (2006). Bad bugs need drugs: an update on the development pipeline from the Antimicrobial Availability Task Force of the Infectious Diseases Society of America. Clin Infect Dis 42(5): 657–668.

Thyagaraja, N., Hosono, A. (1996). Effect of spice extract on fungal inhibition. LWT-Food Sc Technol 29(3): 286–288.

Ultee, A., Kets, E.P.W., Smid, E.J. (1999). Mechanisms of action of carvacrol on the food-borne pathogen *Bacillus cereus*. Appl Environ Microbiol 65(10): 4606–4610.

Unemo, M., Jensen, J.S. (2017). Antimicrobial-resistant sexually transmitted infections: gonorrhoea and *Mycoplasma genitalium*. Nat Rev Urol 14(3): 139.

Vasconcelos, N.G., Croda, J., Simionatto, S. (2018). Antibacterial mechanisms of cinnamon and its constituents: A review. Microb Pathog 120: 198–203.

Wakdikar, S. (2004). Global health care challenge: Indian experiences and new prescriptions. Electron J Biotechnol 7(3): 02–03.

WHO. (2002). WHO Traditional Medicine Strategy 2002–2005, Geneva WHO/EDM/TRM 2002.

Witkowska, A.M., Hickey, D.K., Alonso-Gomez, M. and Wilkinson, M. (2013). Evaluation of antimicrobial activities of commercial herb and spice extracts against selected food-borne bacteria. J Food Res 2(4): 37.

Yamada, Y., Azuma, K. (1977). Evaluation of the *in vitro* antifungal activity of allicin. Antimicrob Agents Chemother 11(4): 743–749.

Yap, P.S.X., Lim, S.H.E., Hu, C.P., Yiap, B.C. (2013). Combination of essential oils and antibiotics reduce antibiotic resistance in plasmid-conferred multidrug resistant bacteria. Phytomedicine 20(8-9): 710–713.

Zacchino, S.A., Yunes, R.A., Filho, V.C., Enriz, R.D., Kouznetsov, V., Ribas, J.C. (2003). The need new antifungal drugs: Screening for antifungal compounds with a selective mode of action with emphasis on the inhibitors of the fungal cell wall. pp. 1–47. *In*: Mahendra Rai, Donatella Mares (eds.). Plant-derived Antimycotics Current Trends and Future Prospects. Food Product Press®, New York.

Zainal, M., Zain, N.M., Amin, I.M., Ahmad, V.N. (2021). The antimicrobial and antibiofilm properties of allicin against *Candida albicans* and *Staphylococcus aureus*—A therapeutic potential for denture stomatitis. Saudi Dent J 33: 105–111.

6

Biobased Products for Viral Diseases

Gleice Ribeiro Orasmo,[1,*] *Giovanna Morghanna Barbosa do Nascimento,*[2] *Maria Gabrielly de Alcântara Oliveira*[1] and *Jéssica Missilany da Costa*[1]

Introduction

For long, medicinal plants have been used to treat many infectious diseases, in most cases without a scientific basis to support their use (Khan et al. 2005). Crude plant extracts or plant-derived bioactives have been effective against different viral diseases, or at least have minimized their adverse effects (Ali et al. 2021).

Herbaceous plants, in particular, have been used for centuries against different diseases, including viral diseases and have become an important source of new compounds for the treatment of bacterial and viral diseases. Currently, there is great emphasis on establishing scientific evidence and justifying the use of preparation from medicinal plants (Khan et al. 2005). According to Vermani and Garg (2002), several research efforts are underway to identify plants and their active components that have an activity against sexually transmitted pathogens, such as the Human Immunodeficiency Virus (HIV), in order to provide an effective approach to preventing HIV transmission and treating AIDS.

The use of natural herbal products by folk medicine has been described for many years, and their properties are being exploited for the development of new drugs. Several bioactives have been described and their therapeutic properties have been proven as antiviral, anti-inflammatory and antioxidant (Lago et al. 2014; Santana et al. 2016). Medicinal plants used in traditional medicines are an excellent source of bioactive proteins and peptides, especially those traditionally prepared by extracting water for use as tea or food supplement (Panya et al. 2019).

Medicinal plants and functional foods have a beneficial influence on human and animal health. These benefits are due to the bioactive compounds present in them, which are able to reduce or prevent the action of pathogens, preventing the risk of disease (Szajdek and Borowska 2008). Many studies provide scientific evidence for the traditional use of plant extracts, this shows the sustainable use of the natural resource, and highlights the importance of conservation of medicinal plants as an origin for the discovery of new drugs and/or medicines (Wee et al. 2020). Several studies have been employed in order to

[1] Department of Biology, Natural Sciences Center, Federal University of Piaui, Teresina, Piauí, Brazil.
[2] Programa de Pós-Graduação em Desenvolvimento e Meio Ambiente, Teresina, Piauí, Brazil.
* Corresponding author: gleice@ufpi.edu.br

identify compounds and molecules with bioactive potential and possible drug candidates, with applications in large-scale industry (Atanasov et al. 2015).

For many years, medicinal plants have been used in different communities in the interior, backlands and indigenous peoples, developing traditional medicines for use in the treatment of various diseases. Herbal medicine has occurred naturally and provides a wide variety of natural products, which can help unravel the many mysteries behind human disease (Ahmad et al. 2020). Herbal compounds are effective in reducing infectious conditions, and these are the only treatments available before antibiotics are introduced. Thus, herbal medicines provide a rich tool for the production of new antivirals.

A variety of herbs have been investigated and their effects against viral infections have been identified (Lin et al. 2014). Ali et al. (2021) provided details of new studies of natural products and nutrients against different viral diseases, including recent findings for the treatment and prevention of SARS-CoV-2. According to the authors, plants or herbal products have been effective against different viral diseases or, at least, have minimized their effects.

In this chapter, bioproducts obtained from extracts or molecules isolated from plants, or even from other organisms, are addressed. These have been listed as antivirals, presenting findings in literature regarding their effects and potential use in treatments for various viral diseases, as well as, in the discovery of future new drugs.

Medicinal plants and their respective bioproducts as antiviral sources

Medicinal plants have been documented to treat a variety of diseases in humans for thousands of years, and offer a valuable resource for new therapeutics (Wee et al. 2020). Below, some plants and their respective active ingredients, found in literature, as well as their use for various viral diseases are reported.

1. Palm, Indian Fig or Cactus Pear (*Opuntia* spp.)

The palm (*Opuntia ficus-indica* (L.) Mill) also known as prickly pear, missionary cactus or smooth prickly pear, belongs to the Cactaceae family. This plant originates from Mexico, being widely distributed throughout Central America, South America, Australia, South Africa and the Mediterranean (Leo et al. 2010). The palm is a cactus widely used as forage, being an alternative for animal feed in the semi arid region of Brazil during periods of water scarcity (Castro et al. 2020).

According to Andrade et al. (2006) the palm has multiple therapeutic uses for communities in the interior of northeastern Brazil, the backlands of Bahia pointed out its use in combating fever, vaginal inflammation, flu, stomach pain, dryness, rheumatism and urinary infection. The cladodes (modified leaves) and the root were the parts most cited by the community as useful for the preparation of medicines, from teas to baths.

Recent studies have shown that the palm fruit, or pear cactus and cladodes are excellent candidates for healthy food development (Feugang et al. 2006), but they are hardly considered as such, although Native Americans and ancient medicine have traditionally known the pharmacological properties of *Opuntia* species with antidiabetic and anti-inflammatory action.

As reported by Feugang et al. (2006) the study by Ahmad et al. (1996) demonstrated that the administration of a cactus stem extract *Opuntia streptacantha* L. to mice, horses and humans, inhibits the intracellular replication of a number of DNA and RNA viruses, such as herpes simplex virus type 2, herpes equine virus, pseudorabies virus, influenza virus, respiratory syncytial disease virus and human immunodeficiency virus type 1. The

Figure 6.1. The palm, *Opuntia ficus-indica* (L.) Mill. Source: https://commons.wikimedia.org/wiki/File:Opuntia_streptacantha_kz1.JPG.

authors, however, did not investigate the active inhibitory components of the palm extract used in this study (Fig. 6.1).

2. Licorice (*Glycyrrhiza glabra* L.)

Licorice (*Glycyrrhiza glabra* L.) is a species of a flowering plant that belongs to the Fabaceae family. It has sweet roots, rich in glycyrrhizin, from which a syrup used in confectionery, cough medicine and in the production of some types of beer that is extracted. Glycyrrhizin is the main component isolated from licorice roots, which inhibited 50% of Hepatitis C Virus (HCV) (Ashfaq et al. 2011), also protected from influenza A virus, causing a reduction in lung cells infected with influenza (Wolkerstorfer et al. 2009). In addition to being effective against hepatitis and influenza viruses, it has also shown its effect in the treatment of Covid-19, against SARS-CoV-2 (Jovic et al. 2020).

Glycyrrhizin can be considered a potent antagonistic agent of the chikungunya and semliki forest viruses (Briolant et al. 2004). Cinatl et al. (2003) showed that glycyrrhizin effectively inhibited SARS-CoV virus replication at an early stage. In order to identify the active licorice molecule against different protein targets of COVID-19, using an *in silico* approach, Sinha et al. (2020) showed that the SARS-CoV-2 replication process was affected by two components of licorice, glycyrrhizic acid, which prohibited the virus from entering the host cell, and glicasperin A, which inhibited the virus replication process.

3. Garlic (*Allium sativum* L.)

Garlic (*Allium sativum* L.) is used in different cuisines from South America and Central America. Since ancient times, garlic has been used in folk remedies to treat fevers, intestinal worms, headaches and dysentery (Iciek et al. 2009). Garlic extracts exhibit antibacterial, antifungal, antiviral, immunostimulant, antioxidant and cholesterol-lowering properties (Stajner et al. 2006; Pittler and Ernst 2007). Bulb extracts of Italian *A. sativum* L. ecotypes were analyzed and their phenolic constituents determined. The results showed that all extracts exhibited antioxidant, antimicrobial and antiproliferative activity (Fratianni et al. 2016).

A. sativum L. extract has an inhibitory effect on avian infectious bronchitis virus, which is a single-stranded RNA and belongs to coronaviruses (Ali et al. 2021). In addition, it has demonstrated its effect on other viruses, including Influenza A and B (Fenwick and Hanley 1985), Cytomegalo Virus (Meng et al. 1993), Herpes Simplex Virus-1 (HSV-1) (Tsai et al. 1985) and Herpes Simplex Virus-2 (HSV-2) (Weber et al. 1992). Molecular docking studies revealed that essential oil or garlic extract can be used as an antiviral source against SARS-CoV-2 virus (Thuy et al. 2020).

Figure 6.2. Garlic (*Allium sativum* L.). Source: the authors.

Carvalho and co-workers (2020) reviewed studies that evidenced the antiviral and immunostimulating potential of garlic (*Allium sativum* L.), demonstrating its effectiveness against several pathogens (Fig. 6.2).

4. Elderberry (*Sambucus* spp.)

Sambucus is a botanical genus belonging to the Adoxaceae family, grouping the trees and shrubs known by the common name of elderberries. Several studies have shown the antiviral potential of elderberry (*Sambucus* spp.) species for different diseases, demonstrating their efficiency, including against coronaviruses (Carvalho et al. 2020). The genus originates from temperate and subtropical regions of both hemispheres, but mainly in the Northern Hemisphere. Some species are of interest as ornamental plants, others in the flower market, and the berries are used to make drinks and sweets, and some species are considered medicinal.

Sambucus formosana Nakai is a species of elderberry, a traditional Chinese medicinal herb known to have anti-inflammatory and antiviral potential. The chemical components of *S. formosana* N. include sambuculin A, oleanolic acid, α-amyrin and β-amyrin, β-sitosterol, ursolic acid, pomolic acid, lupeol palmitate, glycyrrhetic acid, phenolic acid constituents and flavonoids (Zhang et al. 2010; Weng et al. 2019).

The results of the study by Weng et al. (2019) revealed that the ethanol extract from the stem of *S. formosana* N. exhibited a strong anti-HCoV-NL63 effect, one of the common human coronaviruses (Li and Lin 2013; Huang et al. 2017). The study also evaluated some constituents of phenolic acid (caffeic acid, chlorogenic acid and gallic acid), among them, caffeic acid significantly inhibited the replication of HCoV-NL63 and specifically blocked virus binding, which is reported as a vital component in the activity anti-HCoV-NL63 (Weng et al. 2019).

Other *Sambucus* species have active components similar to *S. formosana* N. species, such as *S. nigra* L. and *S. lanceolate* R., with reports of antioxidant, anti-radical, antiviral, antimicrobial and anti-inflammatory activities (Krawitz et al. 2011; Mandrone et al. 2014). In addition, the *S. nigra* L. species showed antiviral activity against influenza A and B viruses, human immunodeficiency virus (HIV) and herpes simplex virus type 1 (HSV-1) (Krawitz et al. 2011; Mandrone et al. 2014). Likewise, the study by Fink et al. (2009) reported that compounds present in elderberry, cinnamon and green tea extracts bind to and block HIV type 1 (HIV-1) infection in target cells.

The black elderberry (*Sambucus nigra* L.) (Fig. 6.3) has a long ethnobotanical history in many cultures as a treatment for viral infection and, according to Porter and Bode (2017), it is currently one of the most used medicinal plants worldwide, however, according to the authors, substantial research to confirm its antiviral properties is lacking so far. Phenolic

Figure 6.3. The black elderberry, *Sambucus nigra* L. Source: https://pt.wikipedia.org/wiki/Sambucus_nigra#/media/Ficheiro:Sambucus_nigra-fruit001.jpg.

acids from *S. nigra* L., such as caffeic acid, showed a strong inhibitory effect on the *in vitro* replication of the influenza A virus (Utsunomiya et al. 2014; Porter and Bode 2017).

Pre-treatment with *Sambucus nigra* L. extract showed a great inhibition of the avian infectious bronchitis virus, the decrease was attributed to the polyphenols present, however, the authors do not rule out that the present lecithins can bind to viral proteins and inhibit the infection (Chen et al. 2014).

5. Alder (*Alnus japônica* (Thunb.) Steud.)

Alnus japônica T. also known as East Asian alder is a deciduous with a very fast growth rate, the leaf and bark extract has been used to increase immunity against influenza (Ra et al. 2010). Studies have shown that methanol extract of *A. japônica* T. bark has a strong antiviral activity against avian influenza virus subtype H9N2 (Tung et al. 2010), as well as isolated diarylheptanoids showed inhibitory activity on SARS-CoV replication (Park et al. 2012).

In addition, a study conducted by Kwon et al. (2010) described that the extract of *A. japônica* T. has high antiviral activity, being useful for the prevention of viral diseases, such as flu in humans and other species of mammals and birds, and against the influenza virus.

6. Fish mint (*Houttuynia cordata* T.)

Houttuynia cordata T. also known as fish mint, fish leaf, fish grass, Chinese lizard tail, among other names. It is an aromatic medicinal plant native to Southeast Asia mainly confined to a humid environment. Studies have shown that *H. cordata* T. ethyl acetate extract is effective in inhibiting dengue virus infectivity (Xie et al. 2013; Chiow et al. 2016). The *H. cordata* T. extract also inhibited avian infectious bronchitis virus (Yin et al. 2011) and influenza A (Choi et al. 2009) in the early stages of infection.

The distillate from the stem of *H. cordata* T. containing the components: methyl n-nonyl ketone, lauryl aldehyde and capryl aldehyde, showed inhibitory activity against herpes simplex virus type 1 (HSV-1), influenza virus and human immunodeficiency virus type 1 (HIV-1), without any cytotoxicity (Hayashi et al. 1995). The *H. cordata* T. extract successfully inhibited SARS-CoV3C as well as RNA-dependent protease and RNA polymerase (Lau et al. 2008).

7. Ginger (*Zingiber officinale* Rosc.)

Zingiber officinalle R. (Fig. 6.4), popularly known as ginger, originates from Tropical Asia and the Malay Archipelago (Nicacio et al. 2018) and belongs to the Zingiberacae family. It is a common spice used in cooking in many parts of the world, with teas made from fresh herb, flavorful and spicy.

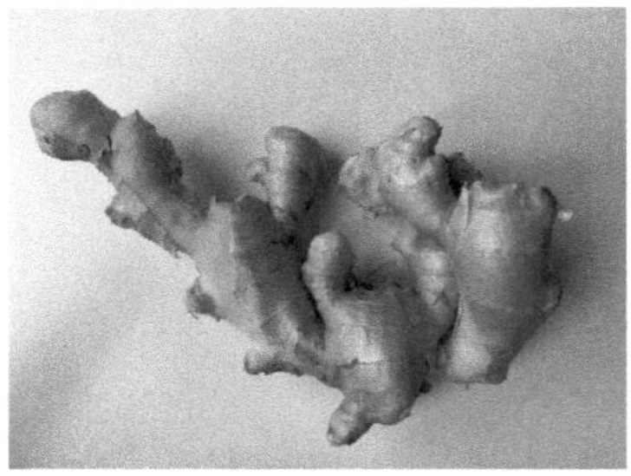

Figure 6.4. Ginger (*Zingiber officinale* R.). Source: the authors.

Ginger is considered a medicinal plant, being widely used in ancient China (Chang et al. 2013). In folk medicine, ginger is used to treat flu, colds and is also thermogenic (Nicacio et al. 2018). It has been shown to have several pharmacological activities, such as anti-nausea action, relieve joint pain and exhibit antimicrobial activities against various bacteria, fungi and nematodes (Ali et al. 2008), in addition, it has been proven to be effective for various viruses (Chrubasik et al. 2005).

Carvalho et al. (2020) reviewed studies that evidenced the antiviral and immunostimulating potential of ginger species; the bioactive products of these plants proved to be able to reinforce the protection against pathogens. Gingerol, a molecule present in ginger, showed significant results in the treatment of Tulane Virus (TV) and Hepatitis A Virus (HAV) (Patwardhan et al. 2020). Rasool et al. (2017) also showed the effectiveness of ginger in the treatment of avian influenza (H9N2).

A study by Zhang et al. (2020) evaluated the interaction of theaflavin 3,3′-digalate, a phenolic compound present in ginger, with ACE2 and M protease receptors. The host cell ECA2 protein is the same receptor as the SRAS-Cov-2 virus. The results suggest that theaflavin 3,3′-digalate is capable of inhibiting the ACE2 protein and, consequently, preventing the binding of SARS-CoV-2.

Ginger has been shown to have antiviral activity against the Human Respiratory Syncytial Virus (HRSV), responsible for causing respiratory tract and lung infections in newborns (Chang et al. 2013). The authors concluded that fresh but not dry ginger is effective against HRSV-induced plaque formation in the airways of the epithelium by blocking viral attachment and internalization. The study showed that fresh ginger was able to inhibit 70% of the infection against HRSV.

8. Red Spider Lily (*Lycoris radiata* (L'Her.))

Lycoris radiata (L'Her.) Herb, commonly known as red spider lily, belongs to the Amaryllidaceae family, is originally from China and was introduced in Japan and, from there, to the United States and other countries. The plant blooms in late summer or autumn, often in response to heavy rains. The active component of *Lycoris radiata* (L'Her.) Herb, licorine, successfully inhibited the virus responsible for the avian influenza infection (H5N1) (He et al. 2012; Yang et al. 2019). It has also been shown that licoricidin derivatives, obtained from the bulb of *L. radiata* (L'Her.) Herb, showed antiviral activity of hepatitis C and have inhibitory activity against tobacco mosaic virus (Chen et al. 2013; Yang et al. 2018).

Lycorin obtained from the ethanol extract of the stem cortex of *Lycoris radiata* (L'Her.) Herb successfully inhibited SARSCoV (Li et al. 2005). Although the mechanism of action of licorine on SARS-CoV is not yet clear, it has shown promising results as an antiviral.

9. Guduchi or Giloy (*Tinospora cordifolia* (Willd.) Miers)

Guduchi or Giloy (*Tinospora cordifolia* (Willd.) Miers) is an herbaceous vine with heart-shaped leaves and a climbing shrub of the Menispermaceae family found in India, which has been used for centuries in traditional medicine to treat various disorders. The stem extract of *Tinospora cordifolia* (Willd.) Miers has antibacterial, antifungal and antiviral properties (Batista et al. 2007). Guduchi stem extract inhibited herpes simplex virus (Pruthvish and Gopinatha 2018) and human immunodeficiency virus (HIV) (Kalikar et al. 2008), in addition, showed a decrease in the incidence of various symptoms associated with HIV.

The antiviral activity of natural compounds from *T. cordifolia* (Willd.) Miers was evaluated for their effectiveness against four main targets of SARS-CoV-2: the surface glycoprotein, receptor binding domain, RNA-dependent RNA polymerase and protease involved in attachment and replication of the virus (Sagar and Kumar 2020). The compounds berberine, isocolumbin, magnoflor and tinocordiside showed high binding efficacy against all four targets, validating the use of *T. cordifolia* (Willd.) Miers in the clinical management of infection caused by SARS-CoV-2.

10. Mushrooms

There are approximately 12,000 known species of mushrooms worldwide, of which at least 2,000 species are edible (Cheung 2008). *Lentinula, Auricularia, Hericium, Grifola, Flammulina, Pleurotus, Lactarius, Pisolithus, Tremella, Russula, Agaricus* and *Cordyceps* are known genus of edible mushrooms, consumed in their natural form or as food supplements (Cheung 2008; Rahi and Malik 2016). Many edible mushrooms have been described in traditional Asian culture as medicinal remedies for a variety of disorders and ailments (Santoyo et al. 2012).

Several bioactive compounds beneficial to human health are derived from mushrooms, including polysaccharides, carbohydrate-binding proteins, peptides, proteins, enzymes, polyphenols, triterpenes, triterpenoids (Rahi and Malik 2016), as well as some compounds exert antiviral activity against DNA and RNA viruses. According to Seo and Choi (2021) the targets of antiviral compounds aim mainly at the entry of the virus into the cell, the replication of the viral genome, viral and cell proteins and immune modulation, being evaluated through *in vitro* studies and *in vivo*. The authors report that some of these mushroom bioactive compounds have shown antiviral effects comparable to those of commonly used antiviral drugs against HIV, influenza A virus and hepatitis C virus. Therefore, mushroom bioactive compounds may be candidates for the treatment of viral infections (Seo and Choi 2021).

Compounds from mushrooms have been reported to fight diseases caused by influenza A virus (El Mekkawy et al. 1998), hepatitis C virus (Zajac et al. 2019), human immunodeficiency virus type 1 (HIV-1) (El Mekkawy et al. 1998; Tsai et al. 2017) and herpes simplex virus type 1 (HSV-1) (Santoyo et al. 2012).

The antiviral properties of aqueous and methanolic extracts of edible mushrooms: *Lentinula edodes* (shiitake), *Boletus edulis* B. (porcini) and *Pleurotus ostreatus* (Jacq.) P. Kumm. (black shimeji or oyster mushroom) were evaluated against herpes simplex virus type 1 (HSV-1) (Santoyo et al. 2012). The authors concluded that the aqueous extracts of mushrooms were able to significantly inhibit virus replication *in vitro,* more than methanol extracts. The β-glucan compounds present in the polysaccharide fractions showed greater antiviral activity.

Several triterpenes from *Ganodema lucidum* C. have been reported as active agents against human immunodeficiency virus type 1 (HIV-1) (El Mekkawy et al. 1998). Ganodermadiol, lucidadiol and aplanoxidic acid G., isolated from *G. pfeifferi* B. and other strains of *Ganodema*, showed antiviral activity *in vitro* against the influenza A virus (El Mekkawy et al. 1998), in addition, ganodermadiol was shown to be active against the herpes simplex virus type 1 (HSV-1) (Mothana et al. 2003).

11. Three-leaf chaste tree (*Vitex trifolia* L.)

Vitex trifolia L. is a medicinal plant commonly known as Panikisanbhalu (Hindi), three-leaf chaste tree (English), belonging to the Verbenaceae family, which grows in countries such as China, India, Australia and Singapore (Chan et al. 2016). Its leaves are used in traditional Chinese medicine to treat inflammation-related illnesses such as rheumatic pain, and its dried ripe fruits are used to treat illnesses such as rhinitis and dizziness (Wee et al. 2020).

The leaves and bark of the plant contain essential oils such as flavones and artemetin (Chan et al. 2016), as well as monoterpenes, diterpenes, beta-sitosterol, are present in these plants. Essential oils from the leaf of *V. trifolia* L. have been used as an anti-inflammatory and against headache, cold, cough, liver disease and antagonist to human immunodeficiency virus (anti-HIV) (Devi and Singh 2014).

Due to the high content of phenolic compounds, such as phenolic acid, flavones and flavonols, it was found that the methanol extract of *V. trifolia* L. has antioxidant and anticancer properties (Yao et al. 2016). Just as methanol extracts of *Sphaeranthus indicus* L. and its active ingredient 7-hydroxy frullanoid (7-HF) suppressed cytokine production (Srivastava et al. 2015), so *Vitex trifolia* L. blocks the pathways that reduce cytokines. Thus, a similar mechanism is being used in respiratory distress in SARS-CoV (Alam et al. 2002).

12. *Strobilanthes cusia* (Nees) Kuntze

Strobilanthes cusia (Nees) Kuntze is a member of the family Acanthaceae, is a perennial medicinal herb widely distributed from the Himalayas to Indochina. According to Zhou et al. (2017) *S. cusia* (Nees) Kuntze root (called "Nan-Ban-Lan-Gen" in Chinese) has been commonly used to treat infections with respiratory viruses such as influenza virus, mumps virus and Severe Acute Respiratory Syndrome (SARS) coronavirus.

Several bioactive components from the *S. cusia* (Nees) Kuntze root, including strobilanthes A, 3H-benzoxazolinone and aurantiamide acetate, have exhibited antiviral activity against influenza A and hepatitis B virus infections (Zhou et al. 2017).

S. cusia (Nees) Kuntze leaf methanol extract exhibited anti-HCoV-NL63 activity (Tsai et al. 2020), among the six components identified, tryptanthrin and indigodole B (5aR-ethyltryptanthrin) exhibited potent antiviral activity in reducing the cytopathic effect and progeny virus production. The authors identified triptantrin as the key active component of *S. cusia* (Nees) Kuntze extract, against HCoV-NL63. The results specify that tryptanthin has antiviral potential against HCoV-NL63 infection.

13. Turmeric (*Curcuma longa* L.)

Curcuma longa L. belongs to the Zingiberaceae family, it is a perennial herb distributed in all tropical and subtropical regions of the world, being widely cultivated in Asian countries, mainly in India and China. In India it is popularly known as 'Haldi' (Araújo and Leon 2001). The powder, called saffron, is widely used as a seasoning in Asian food preparations. In traditional Indian medicine it is used in powder form against various diseases and functional disorders.

Curcumin is the main curcuminoid obtained from the rhizomes of turmeric (*Curcuma longa* L.) (Patwardhan et al. 2020). In order to determine and compare the values for heat sensitization of hepatitis A virus and tulane virus, Patwardhan et al. (2020) used grape seed extract, gingerol (from ginger) and curcumin, which are foods 'generally recognized as safe' (GRA), which have anti-inflammatory, immunomodulatory and antiviral properties. The results suggest that GRA substances may help to inactivate hepatitis A virus and tulane virus.

Turmeric (*Curcuma longa* L.) and its polyphenolic compound curcumin are traditionally identified with antimicrobial properties (Araújo and Leon 2001). Curcumin also restricts virus delivery by acting on specific viral proteins such as integrase and protease on human immunodeficiency virus type 1 (HIV-1) (Mazumder et al. 1995; Vajragupta et al. 2005).

Kutluay et al. (2008) also reported that curcumin can inhibit herpes simplex virus (HSV) infection. In addition to antiviral activities, Sordillo and Helson (2015), hypothesized that curcumin could help fight the cytokine storm in Ebola fever. Curcumin analogos have also been identified with anti-influenza activity. In a silica approach, curcumin and its natural derivatives inhibited Ebola viral proteins (Baikerikar 2017). In addition to curcumin, other turmeric metabolites, bisdemethoxycurcumin, desmethoxycurcumin and tetrahydrocurcumin showed inhibitory activity against the Zairean Ebola viral proteins.

14. Basil (*Ocimum basilicum* L.)

Basil (*Ocimum basilicum* L.) is a well-known herb in traditional Chinese medicine. The different components of *O. basilicum* L. are used as remedies for the treatment of disorders such as eye, respiratory and liver viral infections (Lin and Kan 1990; De Feo and Senatore 1993).

Selected and purified aqueous extracts and active components of basil were used to identify possible antiviral activities against DNA viruses: herpes virus, adenovirus and hepatitis B virus, as well as were tested against RNA viruses: coxsackievirus B1 (CVB1) and enterovirus 71 (EV71) (Chiang et al. 2005). The authors reported that the crude aqueous and ethanol extracts of *O. basilicum* L. and the active components apigenin, linalool and ursolic acid, exhibited a broad spectrum of antiviral activity. However, no activity was observed for the components carvone, cineole, β-caryophyllene, farnesol, fenchone, geraniol, β-myrcene and α-thujone. It was found that the action of ursolic acid against CVB1 and EV71 occurs during the infection process and the replication phase. The authors reported that the potential use of ursolic acid for the treatment of CVB1 and EV71 infection deserves further investigation.

Below, some species of plants, or other organisms, as well as their respective active compounds, that have shown antiviral activity and have been used to combat the action of several viral pathogens are listed (Table 6.1).

Viral diseases and natural products with antiviral activities

Lin et al. (2014) reviewed the antiviral effects reported for various natural and herbal products, combating some viral pathogens, including: coronavirus (CoV), coxsackievirus (CV), dengue virus (DENV), enterovirus 71 (EV71), hepatitis B virus (HBV), hepatitis C virus (HCV), herpes simplex virus (HSV), human immunodeficiency virus (HIV), influenza virus, measles virus and respiratory syncytial virus (RSV). The authors concluded that many natural products and herbal ingredients have robust antiviral activity and their findings may further help to develop drug derivatives and alternative therapeutic pathways.

A variety of herbs have been investigated, and their effects against viral infections have been identified (Ahmad et al. 2020). The discovery of chebulagic acid and punicalagin, that are able to inhibit the entry of several viruses due to their concurrent properties to cell surface glycosaminoglycan, may inhibit the entry of herpes simplex virus type 1 (HSV-1)

Table 6.1. Species, their respective active compounds against different viral diseases.

Plant/Species	Bioproduct	Virus
1. Palm, Indian Fig or Cactus Pear *Opuntia ficus-indica* (L.) Mill	stem extract	Herpes Simplex Virus-2 (HSV-2) Influenza Virus Human Immunodeficiency Virus (HIV) Respiratory Syncytial Disease Virus (SRV)
2. Licorice *Glycyrrhiza glabra* L.	glycyrrhizin, glycyrrhizic acid, glycasperin A	Hepatitis C Virus Influenza-A Virus SARS-CoV SARS-CoV-2 Chikungunya Virus Semliki Forest Virus
3. Garlic *Allium sativum* L.	ethanol extracts from bulbs	Influenza A and B Virus Cytomegalovirus Herpes Simplex Virus-1 (HSV-1) Herpes Simplex Virus-2 (HSV-2) SARS CoV
4. Elderberry *Sambucus formosana* N. *Sambucus nigra* L.	caffeic acid chlorogenic acid gallic acid	Influenza A and B Virus Human Immunodeficiency Virus (HIV) Herpes Simplex Virus - 1 (HSV-1) HCoV-NL63
5. Alder *Alnus japônica* T.	diarylheptanoids	Avian Influenza Virus (H9N2) SARS-CoV Common Flu Influenza Virus
6. Fish mint *Houttuynia cordata* T.	methyl n-nonyl ketone, lauryl aldehyde and capryl aldehyde	Dengue Virus Herpes Simplex Virus-1 (HSV-1) Influenza Virus Human Immunodeficiency Virus (HIV-1) SARS-CoV3C
7. Ginger *Zingiber officinale* R.	gingerol, theaflavin 3,3′-digalate	Tulane Virus (TV) Hepatitis A Virus (HAV) Avian Influenza Virus (H9N2) SRAS-Cov-2 Human Respiratory Syncytial Virus (HRSV)
8. Red Spider Lily *Lycoris radiata* (L'Her.) Herb	liquorine licoricidin	Avian influenza (H5N1) Hepatitis C Virus (HCV) SARSCoV
9. Guduchi or Giloy *Tinospora cordifolia* (Willd.) Miers	berberine, isocolumbine, magnoflor and tinocordiside	Herpes Simplex Virus (HSV) Human Immunodeficiency Virus (HIV) SARS-CoV-2
10. Mushrooms *Lentinula edodes, Boletus edulis* B., *Pleurotus ostreatus* (Jacq.) P. Kumm., *Ganodema lucidum* C., *Ganodema pfeifferi* B.	ganodermadiol, lucidadiol, aplanoxidic acid G	Influenza A Virus Hepatitis C Virus (HCV) Human Immunodeficiency Virus (HIV-1), Herpes Simplex Virus (HSV-1)
11. Three-leaf chaste tree *Vitex trifolia* L.	flavones, artemetin, monoterpenes, diterpenes, beta-sitosterol	SARS-CoV Human Immunodeficiency Virus (HIV) Hepatitis Virus

Table 6.1 contd. ...

...Table 6.1 contd.

Plant/Species	**Bioproduct**	**Virus**
12. *Strobilanthes cusia* (Nees) Kuntze	strobilanthes A, 3H-benzoxazolinone, aurantiamide acetate, tryptanthrin, indigodole B (5aR-ethyltryptanthrin)	Influenza A and B Virus Mumps Virus SARS-CoV HCoV-NL63
13. Turmeric *Curcuma longa* L.	curcumin, bisdemethoxycurcumin, desmethoxycurcumin, tetrahydrocurcumin	Influenza Virus Human Immunodeficiency Virus (HIV) Ebola Virus Herpes Simplex Virus (HSV)
14. Basil *Ocimum basilicum* L.	raw extracts, apigenin, linalool, ursolic acid	Herpes Simplex Virus (HSV) Adenovirus Hepatitis B Virus (HBV) Coxsackievirus B1 Virus (CVB1) Enterovirus 71 (EV71)

and its cell-to-cell propagation (Lin et al. 2014). This can help to develop a broad spectrum of antivirals for the prevention and control of these viral pathogens.

It is important to promote studies aimed at the characterization of bioactive ingredients, define the mechanism of action, as well as assess the effectiveness and potential for application *in vivo*, which could help to develop antiviral treatments (Lin et al. 2014).

Listed below are some viral diseases, as well as some of the natural products that are antagonistic to them, already described in literature.

1. COVID-19/SARS

The severe acute respiratory syndrome caused by coronavirus 2 (SARS-CoV-2) is a new virus that was discovered in 2019, not previously identified in humans, responsible for COVID-19, having originated in the city of Wuhan in China, has spread rapidly worldwide and became a pandemic in 2020 (Ali et al. 2021).

CoV are single-stranded RNA viruses (ss-RNA) belonging to the Coronaviridae family. The CoV family consists of several species that cause upper respiratory and gastrointestinal tract infections in mammals and birds. In humans, it mainly causes common colds, but complications can occur, including pneumonia and SARS (severe acute respiratory syndrome). The already known SARS-CoV coronavirus, which causes the best known severe acute respiratory syndrome, caused a global threat with high human mortality in 2003 (Lin et al. 2014) and in 2019 the SARS-CoV-2 coronavirus, responsible for COVID-19, spread rapidly around the world and became a pandemic in 2020 (Ali et al. 2021). Hosts infected with severe acute respiratory syndrome coronavirus 2 (SRAS-CoV-2), present symptoms of acute respiratory infection, rapidly developing into acute respiratory failure and other severe complications (Zhang et al. 2020).

No specific treatment for COVID-19 has been approved so far, and repositioned drugs have been used to combat the disease, however, some compounds from natural herbs have shown encouraging antiviral properties (Ahmad et al. 2020). In China, traditional herbal medicine has often been used in conjunction with conventional medicine to treat SARS, identifying the antiviral mechanisms of these natural products has been elucidated about where and how they interact or the mechanisms that interrupt the viral cycle, including virus entry, replication, assembly and release (Li et al. 2005).

In a recent study Santana et al. (2021) reviewed the effects of natural flavonoids, such as aglycone and glycosides, in controlling COVID-19 and other coronavirus-related diseases. The study also sought to establish a relationship between biological potential and chemical properties of flavonoids, understand the mechanisms of action of these bioactives and their effects on the virus. The authors reported the potential use of natural flavonoids, as they showed antiviral effects, however, clinical studies are still needed.

New studies and findings of natural products and nutrients against different viruses that cause disease in humans, including SARS-CoV-2, have been reported, pointing to new possibilities in the prevention and treatment of COVID-19 (Ali et al. 2021). Cheng et al. (2006) reported that triterpene glycosides isolated from medicinal plants such as *Heteromorpha* spp., *Bupleurum* spp. and *Scrophularia scorodonia* L. exert antiviral activity against HCoV-22E9; these compounds effectively inhibited the early stage of the virus, preventing its penetration and fixation.

Traditional Chinese medicine has recommended the use of licorice (*Glycyrrhiza glabra* L.) in the treatment of infections caused by SARS-CoV-2, since the active agent, glycyrrhizin, induced effective antiviral activity against SARS coronavirus, inhibiting virus replication (Cinatl et al. 2003). More recently, a study showed that the virus ACE-2 receptor was disturbed by glycyrrhizic acid and glicasperin A, which showed high affinity for the endoribonuclease Nsp15, inhibiting virus replication. The authors concluded that glycosperin A and glycyrrhizic acid can be considered the best licorice molecules, useful against COVID-19 (Sinha et al. 2020).

Bailly and Vergoten (2020) also analyzed the anticoronavirus potential of the natural product glycyrrhizic acid, a drug used to treat liver diseases (including viral hepatitis) and specific skin inflammation (such as atopic dermatitis) in some countries. Glycyrrhizic acid has shown antiviral activities against different viruses, including human and animal coronaviruses associated with SARS.

Park and co-workers (2012) isolated the diarylheptanoid compounds, from the bark of *Alnus japônica* L., which showed inhibitory activity against protease similar to papain (necessary for the replication of SARS-CoV), among them, hirsutenone had more potent inhibitory activity, in addition to catechol and the α,β-carbonyl unsaturated molecule were present in the molecule, which are the main requirements for inhibition of the SARS-CoV cysteine protease.

More than 200 extracts of Chinese medicinal herbs were analyzed for coronavirus, of which, four extracts obtained from the species *Artemisia annua* L., *Lycoris radiata* (L'Her.) Herb, *Pyrrosia lingua* (Thunb.) Farw and *Lindera aggregate* T., showed moderate to potent antiviral activity, exhibiting anti-SAR-CoV effect (Li et al. 2005). *Lycoris radiata* (L'Her.) Herb extract showed the greatest potential, and lycorin was pointed out as the anti-SARS-CoV component.

Phenolic compounds extracted from the root of *Isatis indigotica* L. (Lin et al. 2005) and eight diterpenoids and four biflavonoids isolated from the ethanol extract of *Torreya nucifera* (L.) Siebold & Zucc. leaves (Ryu et al. 2010) showed activity against the SARS-CoV 3CL (pro) enzyme. Eighteen active substances, including 17 organosulfur compounds found in garlic essential oil or garlic extract (*Allium sativum* L.), inhibited the SARS-CoV-2 virus (Thuy et al. 2020). The results show that the 17 compounds, responsible for 99.4% of the contents of the garlic essential oil, have strong interactions with the amino acids of the ACE2 protein and the main protease of SARS-CoV-2.

Houttuynia cordata T. showed significant inhibition of SARS-CoV, the aqueous extract exhibited different antiviral mechanisms against SARS-CoV, including inhibition of viral 3CL protease and blocking of viral RNA-dependent RNA polymerase activity (Lau et al.

2008). More than 200 Chinese medicinal herb extracts have been tested for antiviral activities against SARS-CoV. Among the species evaluated, four extracts of different species *Lycoris radiate* (L'Her.) Herb, *Artemisia annua* L., *Pyrrosia lingua* (Thunb.) Farw and *Lindera aggregate* S., exhibited significant inhibition effects in the assay for virus-induced cytopathic effect. The results suggest that the four herbal extracts, and the compound licorine, are potential candidates for the development of new anti-SARS-CoV drugs in the treatment of SARS (Li et al. 2005).

In an *in silico* approach against SARS-CoV-2, Sagar and Kumar (2020) tested natural compounds from *Tinospora cordifolia* (Willd.) Miers., which showed high binding efficacy against the main targets: surface glycoprotein, receptor binding domain, dependent RNA polymerase of RNA and major protease. The compounds berberine, isocolumbin, magnoflor and tinocordiside from *T. cordifolia* (Willd.) Miers. showed high binding efficacy against all four main targets of SARS-CoV-2, involved in virus binding and replication, thus validating the use of *T. cordifolia* (Willd.) Miers. in the clinical management of infection caused by SARS-CoV-2.

In order to circumvent the respiratory difficulty in SARS-CoV, through pathways that reduce inflammatory cytokines, as has been shown with the use of methanol extract of *Sphaeranthus indicus* L., whose active ingredient 7-hydroxy frullanoid (7-HF), suppressed production of cytokines (Srivastava et al. 2015) and the *Vitex trefolia* L. extract blocked pathways that reduce inflammatory cytokines (Alam et al. 2002).

Strobilanthes cusia (Nees) Kuntze. is a Chinese herbal medicine used to treat respiratory viral infections (Tsai et al. 2020). The genus *Strobilanthes* belongs to the Acanthaceae family and is widely distributed in Northeastern India, Bangladesh, Southern China, the Himalayas, Myanmar and Taiwan, the plant is harvested from the wild for local use as a medicine and dye (Lin et al. 2018). The methanol extract of *S. cusia* (Nees) Kuntze. leaf, and its main chemical components sitosterol, indirubin, triptantrin, betulin, indigodol A and indigodol B, were evaluated for their anti-HCoV-NL63 activity (Tsai et al. 2020). According to the authors, the antiviral action against the human coronavirus still needs to be elucidated, however, the extract effectively inhibited the infection by HCoV-NL63, highlighting the triptantrin and indigodol B components, which exhibited strong viricidal activity.

Strobilanthes cusia (Nees) Kuntze. root has been commonly used to treat respiratory virus infections such as influenza virus, mumps virus, as well as for coronavirus of severe acute respiratory syndrome (SARS) (Zhou et al. 2017).

The SRAS-CoV-2 has the ability to bind the host's ACE (Angiotensin-Converting Enzyme), which predisposes the host to myocarditis and respiratory failure. Zhang et al. (2020) evaluated the interaction of theaflavin 3,3'-digalate, a phenolic compound present in ginger (*Zingiber officinale* R.), with ACE2 and M protease receptors. The host cell ACE2 protein is the same receptor as the SRAS-Cov-2 virus, thus, the results suggest that theaflavin 3,3'-digalate is able to inhibit the ACE2 protein and, consequently, prevent the binding of the SARS-COV-2 .

Babich et al. (2020) reviewed medicinal plants that grow in Northeast Asia and whose antioxidant, antiviral, anti-inflammatory and immunomodulatory characteristics are already known. The authors consider that the development of new effective anti-coronavirus drugs and therapies is important, but prevention and epidemiological protection measures cannot be dispensed with. Many foods and herbs have antiviral, immunomodulatory and anti-inflammatory properties, thus the authors emphasize that the combined use of available drugs with medicinal plants, which have such properties, can significantly increase protection and result in beneficial effects, significantly reducing the load virus of SARS-CoV-2, and also reducing the recovery period in patients with COVID-19.

2. Herpes

Viral infections, particularly infections caused by the herpes simplex virus (HSV), represent one of the most serious public health problems worldwide due to their devastating impact (Bag et al. 2012). The search for new antiviral agents, especially those with different mechanisms of action, is a crucial goal and there is an unmet need for alternative and complementary therapy against HSV infection.

Herpes simplex viruses (HSV-1 and -2) are important pathogens for humans, especially for highly susceptible adults. In addition, HSV-2 has been reported to be a high risk factor for HIV infection (Khan et al. 2005). Therefore, the discovery of new anti-HSV drugs deserves great efforts. Herpes simplex virus (HSV) is differentiated into two antigenic types of type 1 (HSV-1) and type 2 (HSV-2), their genome sequences are highly related, for the most part, and closely collinear.

A study by Bag and co-workers (2012) aimed to evaluate the antiviral potential of the crude methanolic extract of *Mallotus peltatus* (Geiseler) Muell.Arg., its active fraction and pure compound, against HSV-1 F and HSV-2 G. *Mallotus* is a botanical genus belonging to the Euphorbiaceae family, with some species found in tropical Africa and Madagascar, but most are found in East and Southeast Asia, as well as North and East Australia. The study, characterizing the ethnomedicine of the Andaman Islands, revealed that the bioactive fraction A and the ursolic acid isolated from *M. peltatus* (Geiseler) Muell.Arg. has good anti-HSV activity, probably inhibiting the early stage of multiplication, suggesting its potential use as anti-HSV agents.

Herpes simplex Virus antagonistic activity by bioactive extracts of *Antrodia camphorata* (M. Zang & C.H. Su) mycelia was evaluated (He et al. 2016). *A. camphorata* (M. Zang & C.H. Su) is a popular medicinal mushroom commonly used in Taiwan and Asia to treat many inflammatory disorders, as well as to treat liver diseases and cancers. Even at a very low concentration, the crude ethanol extract and isolated constituents showed inhibition of HSV replication. Fraction A and the isolated component anthrodin A showed a viral inhibitory effect, with a reduction in virus dissemination. The authors concluded that extracts from *Antrodia camphorata* (M. Zang & C.H. Su) mycelia and the compound anthrodin A may have potential use as anti-HSV agents and are promising candidates for future antiviral drugs for Herpes Simplex.

Acyclovir-resistant clinical isolates of herpes simplex virus type 1 (HSV-1) were analyzed *in vitro* for their susceptibility to essential oils of ginger (*Zingiber officinale* R.), thyme (*Thymus vulgaris* L.), hyssop (*Hyssopus officinalis* L.) and sandalwood (*Santalum album* L.). All essential oils exhibited high levels of virucidal activity against acyclovir-sensitive KOS strain and acyclovir-resistant HSV-1 clinical isolates and significantly reduced plaque formation (Schnitzler et al. 2007). Li et al. (2018) reviewed natural products with potential anti-HSV-1 activity, providing not only a reference for the safe dose of these agents, but also presenting the main pharmacodynamic mechanisms of traditional Chinese medicine. The potential of these active components in the development of anti-HSV-1 drugs has been described.

The antiviral properties of aqueous and methanolic extracts of edible mushroom species: *Lentinula edodes* Berk. (shiitake), *Boletus edulis* Bull. (porcini) and *Pleurotus ostreatus* (Jacq.) P. Kumm. (black shimeji or oyster mushroom) were evaluated against herpes simplex virus type 1 (HSV-1) (Santoyo et al. 2012). The aqueous extracts of mushrooms were able to significantly inhibit the replication of the virus *in vitro*, whereas the methanolic extracts exhibited a lower antiviral activity for all components evaluated. The β-glucan compounds present in the polysaccharide fractions showed greater antiviral activity than the complete aqueous extracts, except for *B. edulis* Berk. in which other compounds (probably chitin-binding lectins) may have contributed to the observed activity (Santoyo et al. 2012).

Genital herpes is one of the most prevalent sexually transmitted diseases worldwide, and is the most common cause of genital ulcers. The anti-herpes activity of several essential oils from different plant sources, as well as various essential oil constituents, have been demonstrated (Koch et al. 2008).

The essential oils of anise (*Illicium verum* H.), hyssop (*Hyssopus officinalis* L.), thyme (*Thymus vulgaris* L.), ginger (*Zingiber officinale* R.), chamomile (*Matricaria recutita* L.) and sandalwood (*Santalum album* L.) were investigated against herpes simplex virus type 2 (Koch et al. 2008). Essential oils were added at different stages during the viral infection cycle, and dose-dependent anti-HSV-2 activity was demonstrated for all tested essential oils. However, no inhibitory effect was observed when essential oils were added to the cells before infection with HSV-2 or after the adsorption period. Among the tested oils, chamomile oil exhibited a high selectivity index and proved to be a potent candidate as a viricidal agent in the treatment of genital herpes.

Illicium verum H. (Fig. 6.5), popularly known as star anise, Chinese anise, Siberian anise, badian or Chinese fennel, is a plant native to China and Vietnam. It is considered a spice for medicinal and culinary use. It has an aroma identical to anise because it contains the same oil, although it is stronger.

Kutluay and co-workers (2008) reported that curcumin, the active substance in turmeric, also known as turmeric (*Curcuma longa* L.), can inhibit herpes simplex virus infection. The entry of viral DNA into the host cell nucleus and the binding of viral protein VP16 to Immediate Early (IE) gene promoters was not affected by curcumin, but the recruitment of RNA polymerase II to these promoters was significantly decreased.

The compound ganodermadiol, isolated from the mushroom *Ganodema pfeifferi* B., showed antiviral activity against herpes simplex virus type 1 (HSV-1) (Mothana et al. 2003). Extracts from several medicinal plants and herbs, such as *Sambucus nigra* L. (Mandrone et al. 2014), *Tinospora cordifolia* (Willd.) Miers. (Pruthvish and Gopinatha 2018) and *Houttuynia cordata* T. (Hayashi et al. 1995), showed inhibitory activity against HSV-1.

Khan et al. (2005) reviewed the extracts and molecules of anti-HSV medicinal plants of herbal medicines, reported in studies from several laboratories. Traditional medicines such as 'Ayurvedic' from traditional Indian medicine and 'Chakma' herbal medicines used by the Chakma tribe of Bangladesh have been found to be a promising source for new anti-HSV drugs. Regarding the structure-activity relationship of several anti-HSV compounds, a large number of small molecules, such as phenolic compounds, polyphenols, terpenes, flavonoids, sugar-containing compounds, were considered promising antiherpetic agents.

Figure 6.5. Star anise (*Illicium verum* H.). Source: the authors.

The authors concluded that natural medicinal plant extracts are very important sources of anti-HSV agents.

3. Ebola

Ebola viral disease is a serious and mostly fatal disease in humans, caused by the ebola virus, which belongs to the Filoviridae family, and is a single-stranded RNA virus (Baikerikar 2017). The 2014 ebola virus (EBOV) epidemic in West Africa was the biggest outbreak since the disease was first reported in 1976. The disease is associated with a high risk of death, killing an average of about 50% of infected people, with the highest mortality rate of up to 90% (Formenty 2014).

There is no single treatment for this disease, which raises the need to identify a new therapy to control and treat this fatal condition. Sordillo and Helson (2015) hypothesized that curcumin, isolated from turmeric (*Curcuma longa* L.), can help fight the cytokine storm in Ebola fever. In fact, curcumin, one of the bioactives in turmeric, had the proven antiviral property for ebola (Baikerikar 2017). Turmeric is a plant widely used in Indian cuisine, being famous for its golden-yellow color and its spicy taste. Turmeric is also known as saffron, yellow ginger, among others. Curcumin is the active substance in turmeric, a mild aromatic stimulant used in the production of curry powder, which gives turmeric its yellow color.

In an *in silica* approach, curcumin and its natural derivatives inhibited ebola viral proteins (Baikerikar 2017). In addition to curcumin, the study evaluated the inhibitory activity of bisdemethoxycurcumin, desmethoxycurcumin and tetrahydrocurcumin against the viral proteins of ebola from Zaire. Bisdemethoxycurcumin showed better binding affinity than curcumin for most viral proteins. The tetrahydrocurcumin metabolite also showed comparable binding affinity to curcumin. These results indicate that curcumin, curcuminoids and the metabolite tetrahydrocurcumin may be potential lead compounds for the development of a new therapy for the viral disease ebola.

A study sought to identify compounds from African flora with the potential to inhibit the activity of the ebola virus protein (VP24), using pharmacoinformatics and molecular docking (Kwofie et al. 2019). An integrated library of 7675 natural products originating in Africa obtained from the AfroDB and NANPDB databases, as well as known inhibitors, were tested against VP24. Four natural products, ZINC000095486070, ZINC000003594643, ZINC000095486008 and sarcophin, were considered to be potential inhibitory molecules of the ebola viral proteins. The authors also considered that these natural compounds are worthy of being incorporated into the treatment for ebola virus.

In order to discover a new drug to fight ebola, Nasution et al. (2018) examined about 190,084 natural product composites from the ZINC15 database in virtual screening and soft-fit simulation. Two best ligands have progressed to molecular dynamics simulation for the ebola virus nucleoprotein (EBOV NP), one of the seven proteins that ebola virus encodes, which plays a key role in the EBOV proliferation cycle. The two proposed ligands, identified as: ZINC56874155 and ZINC85628951, showed promising results to be developed as a main compound in the treatment of ebola. An experimental study is necessary to validate the inhibition activities against EBOV NP.

4. Dengue

Dengue virus infection (DENV) is a global health problem that is associated with high morbidity and mortality in humans. DENV infection causes a spectrum of clinical manifestations ranging from mild Dengue Fever (DF) to severe Dengue Hemorrhagic Fever (DHF) to Dengue Shock Syndrome (DSS), which can be fatal (Martina et al. 2009; Jarver et al. 2012). It is estimated that approximately 50 million people are infected with DENV annually, with an estimated 20,000 deaths per year. These data consolidate the status of DENV infection as one of the main infectious diseases worldwide (Panya et al. 2019).

New bioactive peptides, demonstrating anti-viral activity for dengue, isolated from 33 Thai medicinal plants, were identified (Panya et al. 2019). Among the tested extracts, *Acacia catechu* L. peptides showed the most pronounced anti-DENV activity. *A. catechu* L. is a legume species of the genus *Acacia*, belonging to the Fabaceae family, an Asian medicinal plant, whose seeds are a good source of protein. Two bioactive peptides isolated from *A. catechu* L. inhibited the formation of DENV foci by more than 90%. The authors reported that the two peptides are recommended for further investigation as antivirals against DENV infection. The extract of the Asian medicinal plant *Houttuynia cordata* T., known as fish mint, prevented dengue virus infectivity (Xie et al. 2013; Chiow et al. 2016).

In order to examine antiviral properties of four types of bioflavonoids quercetin, hesperetin, naringin and daidzein, from an extract of citrus fruit against dengue virus type 2 (DENV-2), was verified (Zandi et al. 2011). Hesperetin is a flavonone, in which its glycoside form, hesperidin, is water soluble and can be found in many citrus fruits. The results revealed that quercetin exhibited the most significant antiviral activity against DENV-2 among the bioflavonoids tested. Daidzein showed very weak anti-dengue activity, while naringin and hesperetin did not show any anti-dengue activity. These findings suggest that a select group of bioflavonoids, including quercetin and fisetin, exhibit significant inhibitory activities against the dengue virus. According to the authors, it is interesting that the group of flavonoids, flavonol, was investigated to elucidate the mechanisms of inhibition of dengue virus replication.

5. Zika

Zika virus (ZIKV) is a flavivirus transmitted by *Aedes* mosquitoes, including *A. aegypti*, which is widespread worldwide, for which there is no vaccine or specific therapies. ZIKV has been spreading rapidly across South America, Central America and the Caribbean, and epidemiologists predict that the virus will continue to spread widely throughout the Western Hemisphere, including the United States (Rausch et al. 2017). The World Health Organization has declared that the current ZIKV outbreak is a global emergency, due to the rapid spread of the virus and the association of viral infection with microcephaly (Oliveira-Melo et al. 2016; Tetro 2016).

Nanchangmycin was identified as a potent inhibitor of zika virus entry into all cell types tested. Nanchangmycin is a natural product of *Streptomyces nanchangensis* O. that has been shown to have insecticidal activity against silkworms and antibacterial activity *in vitro* (Ouyang et al. 1993; Sun et al. 2002). However, little is known about this drug and its potential mechanism of action against ZIKV.

Rausch et al. (2017) tested 2000 compounds in three cell types against zika virus, of which 38 compounds showed antiviral activity in at least one cell type. Nacangmycin showed antiviral activity in all types of cells tested, also being active against several arboviruses. The authors concluded that bioactive screening revealed nacanngmycin as a broad-spectrum antiviral active against zika virus. In addition, nacangmycin was also active against other clinically relevant viruses, including West Nile virus, dengue virus and chikungunya virus, which use a similar entry route, and thus the study provides a small-molecule resource for study the pathogenesis of the zika virus.

6. Chikungunya

Chikungunya virus (CHIKV) and Semliki Forest Virus (SFV) are alphaviruses transmitted by arthropods, members of the Family Togaviridae (Briolant et al. 2004). Alphaviruses are enveloped viruses whose genome is made up of a single strand of RNA. semliki forest virus (SFV) is widely spread in Africa and infection in humans is relatively common (Lundstrom 1999). Chikungunya virus (CHIKV) is responsible for an acute disease in man, characterized by a triad of fever, arthralgia, maculopapular and rash (Brighton et al. 1983; McGill 1995), and there is no treatment for this viral disease as yet.

Eight compounds were tested *in vitro* for replication of chikungunya and semliki forest viruses, namely: ribavirin, 6-azauridine, glycyrrhizin, iota-carrageenan, fucoidan and dextran sulfate, IFN-2a and IFN-2b (Briolant et al. 2004). Among the compounds evaluated, some of them are bioproducts that have shown antiviral activity for several diseases, such as glycyrrhizin, the main component isolated from the roots of licorice (*Glycyrrhiza glabra* L.), which has been used against the hepatitis C virus (Ashfaq et al. 2011), SARS-CoV-2 (Cinatl et al. 2003; Jovic et al. 2020) and the influenza virus (Wolkerstorfer et al. 2009), and fucoidan, a sulfated polysaccharide, whose main component is a fucose, extracted from brown algae. The results indicate that several compounds, including ribavirin, recombinant human IFN-α, 6-azauridine and glycyrrhizin, can be considered as selective and potent anti-CHIKV and anti-SFV agents.

7. AIDS

The human immunodeficiency syndrome (AIDS) is a disease caused by the human immunodeficiency virus (HIV), belonging to the Retroviridae family, *Lentivirus* genus. AIDS is a disease of the human body's T-cell or lymphocyte-mediated immune system in which the body's natural defense system is destroyed by HIV infection (Kaur et al. 2020). According to the World Health Organization (WHO), it is estimated that about 75 million people have been infected with the human immunodeficiency virus (HIV), and about 37 million people are still struggling.

Although several advances have been made in the field of antiretroviral combination therapy, HIV is still responsible for a considerable number of deaths in Africa. Currently, the biodiversity of the plant kingdom is being explored by several researchers for the discovery of powerful anti-HIV drugs with different mechanisms of action. There is a great demand for herbal medicines as well as their derivatives to evaluate. Several studies show plants that exhibited significant anti-HIV activity.

According to Salehi et al. (2018), since the beginning of the HIV epidemic, there have been many promising treatments with antiretroviral drug cocktails, however, many of those infected have adverse reactions or do not have immediate access to currently available chemotherapies. Thus, there is a need to discover new anti-HIV agents to complement the current arsenal and provide therapeutic options for populations with limited resources.

Kaur et al. (2020) reviewed the most potent plants and natural products with anti-HIV activity along with their mechanisms of action. In addition to plants, fungi are also investigated for anti-HIV activities, thus, a new cadinano sesquiterpene was isolated from cultures of the basidiomycete, *Tyromyces chioneus* (Fr.) P.Karst. (Liu et al. 2007). Its structure was established based on spectral measurements and showed significant anti-HIV-1 activity.

Baicalin is a flavonoid purified from the medicinal plant *Scutellaria baicalensis* Georgi and has been shown to have anti-inflammatory and anti-HIV-1 activities (Li et al. 2000). The flavonoid baicalin inhibits HIV-1 infection on entry of the virus. The authors proposed that the flavonoid baicalin can be used as a basis for the development of new anti-HIV-1 agents.

Salehi and co-workers (2018) carried out a survey of plants with anti-HIV activity. A total of 717 plant species were listed, which have inhibitory activity in all phases of viral replication. The authors also observed that many compounds tested inhibited almost all stages of virus replication, including: alkaloids, sulfated polysaccharides, polyphenols, flavonoids, coumarins, phenolics, tannins, triterpenes, lectins, among others.

According to Aiken and Chen (2005), betulinic acid derivatives are organic compounds of low molecular weight synthesized from a natural product of plant origin, and several of them are potent and highly selective HIV-1 inhibitors. In recent decades, it has been observed that the most favorable therapeutic management for this infection consists mainly of the use of specific inhibitors of the viral replication cycle, which, administered together, reduce the viral load to undetectable levels. Betulinic acid derivatives have potential as new therapies for HIV-1. Aiken and Chen (2005) reviewed recent studies and discussed the new mechanisms of action of this interesting class of antiviral compounds.

The activity of the constituents of *Juglans mandshurica* M. against human immunodeficiency virus-type 1 was evaluated (Min et al. 2002). *J. mandshurica* M. (Juglandaceae) also known as the Manchurian nut, is a deciduous tree, native to the eastern Asian region. Three naphthalene glycosides, four flavonoids and two galloyl glycosides were isolated from the stem bark of *J. mandshurica*. Among the isolated compounds, taxifolin had the most potent HIV-induced cytopetic activity. However, naphthalene glycosides, flavonoids and galloyl tannins were inactive against anti-HIV-1 activity.

Tan et al. (1991) evaluated 156 pure natural products known to be reverse transcriptase (HIV-1) inhibitors, which is a useful approach in the prophylaxis and intervention of AIDS. A total of 100 plant extracts were also evaluated and 15 of these extracts showed significant inhibitory activity. The ethanolic and methanolic extracts of the plants *Arctium lappa* L., *Viola yedoensis japonica* (Nakai) Hashim. ex E.Hama, *Epidemium grandiflorum* C. Morren and *Prunella vulgaris* L. present a significant reverse transcriptase inhibition rate, thus, it is possible that they contain high concentrations of TR-HIV-1 inhibitory bioproducts in these compounds (Tan et al. 1991). The authors reported that potent HIV-1 reverse transcriptase inhibitors from natural products discovered using the present protocol include compounds from the alkaloid classes benzophenanthridine, protoberberine alkaloid, isoquinoline alkaloid and iridoid.

Several plant extracts showed antiviral activity against the human immunodeficiency virus (HIV), such as: elderberry *Sambucus nigra* L. (Fink et al. 2009; Krawitz et al. 2011), turmeric (*Curcuma longa* L.) (Mazumder et al. 1995; Vajragupta et al. 2005), the Chinese herbs: guduchi (*Tinospora cordifolia* (Willd.) Miers.) (Kalikar et al. 2008), and various by-products of mint-fish stem distillate (*Houttuynia cordata* T.) (Hayashi et al. 1995). In addition, several triterpenes from the mushroom *Ganodema lucidum* C. have been found to be active agents against human immunodeficiency virus type 1 (HIV-1) (El Mekkawy et al. 1998).

8. Hepatitis

Hepatitis C virus (HCV) is a major cause of chronic liver disease that can lead to permanent liver damage, hepatocellular carcinoma and death. The currently available treatment with interferon plus ribavirin has limited benefits due to adverse side effects such as anemia, depression, fatigue and flu-like symptoms. Although there is an effective vaccine against the hepatitis B virus (HBV), chronic infection poses an enormous health burden to the world (Hoofnagle 1990).

Glycyrrhizin is the main component isolated from licorice roots, which inhibited 50% of the hepatitis C virus (HCV) (Ashfaq et al. 2011). The study aimed to study the antiviral effect of glycyrrhizin against HCV and liver cells infected with HCV were treated with glycyrrhizin in non-toxic doses. The results demonstrated that glycyrrhizin inhibited HCV in a dose-dependent manner and resulted in a 50% reduction in HCV. Comparative studies were done with interferon alpha to investigate the synergistic effects between the antiviral compound and interferon. The results suggest that glycyrrhizin inhibits HCV viral particles and exhibited a synergistic effect when combined with interferon. The authors concluded that it is likely that, in the future, glycyrrhizin together with interferon will be the best option to treat HCV infection.

HCV infection is considered one of the main causes of human hepatitis. HCV belongs to the Flaviviridae family, as does the Bovine Viral Diarrhea Virus (BVDV). The discovery of diphenylmethane analogues as viral anti-bovine diarrhea agents is important because HCV does not replicate efficiently in cell cultures or animals. Based on the antiviral screening of our diphenylmethane derivatives prepared as steroid substitutes, one 1,1-diphenylcyclobutane analogue and two diethyldiphenylsilane analogues were identified as lead compounds with potent activity against bovine viral diarrhea virus (BVDV), having 50% of effectiveness (Hosoda et al. 2009). Thus, BVDV is considered a good model for human HCV.

Laggera alata D. is a traditional Chinese medicinal herb that belongs to the Asteraceae family. The genus *Laggera* is distributed mainly in Africa and Southeast Asia, with *L. alata* D. being one of the only species of the genus found in China, used in folk medicine for 300 years (Malik and Lee 2000). Hao and co-workers (2012) examined the hepatoprotective and antiviral properties of isochlorogenic acid A, isolated from *Laggera alata* D., against hepatitis B virus infection. The isochlorogenic acid A markedly declined the content of hepatitis B virus. The authors concluded that isochlorogenic acid A has potential for anti-hepatitis B activity. The anti-HBV target of isochlorogenic acid A is likely associated with blocking HBV replication, reducing the stability of the main HBV protein. Furthermore, the hepatoprotective effect of isochlorogenic acid A could be achieved by its antioxidant property and HO-1 induction.

Some aqueous extracts and active components of different plants showed antiviral activity against hepatitis A virus (HAV) and tulane virus (TV), such as gingerol, present in ginger (*Zingiber officinale* R.), curcumin, from the rhizomes of turmeric (*Curcuma longa* L.) and grape seed extract (Patwardhan et al. 2020).

Several compounds derived from plants as well as other organisms have shown antiviral activities for hepatitis viruses. Licoricidin, obtained from the bulb of *L. radiata* (L'Her.) Herb, proved to be antagonistic to the hepatitis C virus (Chen et al. 2013), active components of the aqueous extract of basil (*Ocimum basilicum* L.) were identified as antivials for the hepatitis B virus (Chiang et al. 2005), in addition, bioactive compounds from mushrooms have shown antiviral effects against the hepatitis C virus (Seo and Choi 2021).

9. Severe Respiratory Infection

Severe respiratory infection can be caused by several viruses, with the human respiratory syncytial virus (HRSV) being one of the most common airway viruses. HRSV is a non-segmented enveloped RNA negative virus of the Paramyxoviridae family. The stem extract of the *Opuntia streptacantha* L. cactus inhibited the intracellular replication of the human respiratory syncytial virus (Ahmad et al. 1996).

Ginger (*Zingiber officinalle* R.) has been shown to have anti-HRSV activity (Chang et al. 2013). The authors tested the ability of ginger to stimulate antiviral cytokines through an enzyme-linked immunosorbent assay for human respiratory syncytial virus (HRSV). Dry ginger did not show any dose-dependent inhibition, however, fresh ginger was effective when administered before virus inoculation, which in high concentration, contributed to neutralize the viral infection. Thus, it is noteworthy that fresh ginger was effective against the human respiratory syncytial virus (HRSV), blocking viral fixation and internalization. It is recommended for people using ginger as soon as cold symptoms are present.

10. Influenza or Flu

Influenza or flu, is caused by the influenza virus, is a seasonal illness that affects millions of people worldwide each year. Anti-influenza compounds derived from herbs have many advantages, such as low cost and low toxicity, in addition to being an extensive source. In addition, herbal medicines often have multi-targeted effects, which not only act as antiviral agents, but also stimulate the immune system. Therefore, in recent years medicinal plant extracts and phytochemicals have attracted increasing attention as new potential sources for the development of new antiviral drugs emerge (Ping et al. 2016).

Thus, new therapeutic agents that control the Influenza A Virus (IAV), the propagation and attenuation of excessive pro-inflammatory responses are needed. Polysaccharide extract from *Radix isatidis* L., a traditional Chinese herbal medicine, exerted potent *in vitro* anti-IAV activity against human seasonal influenza viruses (H1N1 and H3N2) and avian influenza viruses (H6N2 and H9N2) (Li et al. 2017). *R. isatidis* L. root polysaccharide extract therefore has the potential to be used as an adjunct to antiviral therapy for the treatment of IAV infection.

Bioactive fractions and representative compounds extracted from banlangen (*Radix isatidis* L.) were tested for antiviral activities against influenza virus (Ping et al. 2016). *R. isatidis* dried root has been used to treat colds, sore throats and headaches for hundreds of years in China. Ping and co-workers (2016) evaluated the antiviral activities of the compounds clemastanin B, epigoitrine, phenylpropanoid portion and the mixture of phenylpropanoids, alkaloids and organic acid fractions from *Radix isatidis* L. All compounds exerted anti-influenza activity by inhibiting virus multiplication and virus blocking. The main mode of action of phenylpropanoid and the mixture of phenylpropanoids, alkaloids and organic acid fractions was the inhibition of virus replication. Inhibitory effects on virus attachment and multiplication are the main modes for epigoitrin.

Many medicinal plants have been reported as antivirals for influenza virus. The *Sambucus nigra* L. species showed antiviral activity against influenza A and B viruses (Krawitz et al. 2011) and the caffeic acid compound from this elderberry showed a strong inhibitory effect on influenza A virus replication (Porter and Bode 2017). The methanol extract of alder (*Alnus japonica* T.) has strong antiviral activity against the avian influenza virus subtype H9N2 (Tung et al. 2010). *Houttuynia cordata* T. extract inhibited influenza A virus (Choi et al. 2009), as well as its compounds methyl n-nonyl ketone, lauryl aldehyde and capryl aldehyde, showed anti-Influenza activity (Hayashi et al. 1995). Rasool et al. (2017) found the effectiveness of ginger (*Zingiber officinale* R.) in the treatment of avian influenza (H9N2). The active compound licorine, obtained from *Lycoris radiata* (L'Her.) Herb, inhibited the avian influenza virus (H5N1) (He et al. 2012; Yang et al. 2019).

Mushroom bioactive compounds showed antiviral effects for the influenza A virus, such as ganodermadiol, lucidadiol and aplanoxidic acid G., isolated from the mushroom *Ganodema pfeifferi* B., which showed *in vitro* antiviral activity against the influenza A virus (El Mekkawy et al. 1998).

Below some viral diseases, the antagonistic bioproducts to them, as well as their respective species from which they are obtained are listed (Table 6.2).

Table 6.2. Viral diseases, antagonistic bioproducts and their respective origins.

Viral disease	**Bioproducts and their respective origins**
1. COVID-19/SARS	- triterpene glycosides (*Heteromorpha* spp., *Bupleurum* spp., *Scrophularia scorodonia* L.); - glycyrrhizin, glycyrrhizic acid, glicasperin A (*Glycyrrhiza glabra* L.); - diarylheptanoid (*Alnus japônica* L.); - lycorin (*Lycoris radiata* (L'Her.) Herb); - diterpenoids and biflavonoids (*Torreya nucifera* (L.) Siebold & Zucc.); - phenolic compounds (*Isatis indigotica* L.); - organosulfur compounds (*Allium sativum* L.); - aqueous extract (*Houttuynia cordata* T.); - herb extracts (*Lycoris radiata* (L'Her.) Herb, *Artemisia annua* L., *Pyrrosia lingua* (Thunb.) Farw, *Lindera aggregate* S.); - berberine, isocolumbin, magnoflor, tinocordiside (*Tinospora cordifolia* (Willd.) Miers.); - 7-hydroxy frullanoid (7-HF) (*Sphaeranthus indicus* L.); - sitosterol, indirubin, triptantrin, betulin, indigodol A, indigodol B (*Strobilanthes cusia* (Nees) Kuntze.); - theaflavin 3,3'-digalate (*Zingiber officinale* R.);
2. Herpes	- bioactive fraction A, ursolic acid (*Mallotus peltatus* (Geiseler) Muell.Arg.); - anthrodin A (*Antrodia camphorata* (M. Zang & C.H. Su); - essential oils (*Zingiber officinale* R., *Thymus vulgaris* L., *Hyssopus officinalis* L., *Santalum album* L.); - aqueous extracts, β-glucan (*Lentinula edodes* Berk., *Boletus edulis* Bull. and *Pleurotus ostreatus* (Jacq.) P. Kumm.); - essential oils (*Illicium verum* H., *Hyssopus officinalis* L., *Thymus vulgaris* L., *Zingiber officinale* R., *Matricaria recutita* L. and *Santalum album* L.); - curcumin (*Curcuma longa* L.); - ganodermadiol (*Ganodema pfeifferi* B.); - plants and herbs extracts (*Sambucus nigra* L., *Tinospora cordifolia* (Willd.) Miers. and *Houttuynia cordata* T.);
3. Ebola	- curcumin, bisdemethoxycurcumin, desmethoxycurcumin and tetrahydrocurcumin (*Curcuma longa* L.);
4. Dengue	- bioactive peptides (*Acacia catechu* L.); - quercetin and daidzein (citrus fruit extract);
5. Zika	- nancangmycin (*Streptomyces nanchangensis* O.);
6. Chikungunya	- glycyrrhizin (*Glycyrrhiza glabra* L.); - fucoidan (extracted from brown algae);
7. AIDS	- cadinano sesquiterpene (*Tyromyces chioneus* (Fr.) P.Karst.); - baicalin (*Scutellaria baicalensis* Georgi); - betulinic acid (organic compounds of plant origin); - taxifolin (*Juglans mandshurica* M.); - plant extracts (*Sambucus nigra* L., *Curcuma longa* L., *Tinospora cordifolia* (Willd.) Miers. and *Houttuynia cordata* T.); - triterpenes (*Ganodema lucidum* C.);
8. Hepatitis	- glycyrrhizin (*Glycyrrhiza glabra* L.); - isochlorogenic acid A (*Laggera alata* D.); - gingerol (*Zingiber officinale* R.); - curcumin (*Curcuma longa* L.); - grape seed extract (*Vitis* spp.); - licoricidin (*Lycoris radiata* (L'Her.) Herb); - aqueous extract (*Ocimum basilicum* L.);
9. Severe Respiratory Infection	- stem extract (*Opuntia streptacantha* L.); - fresh ginger extract (*Zingiber officinale* R.);

Table 6.2 contd. ...

...Table 6.2 contd.

Viral disease	Bioproducts and their respective origins
10. Influenza or Flu	- clemastanin B, epigoitrine, phenylpropanoid (*Radix isatidis* L.); - caffeic acid (*Sambucus nigra* L.); - methanol extract (*Alnus japonica* T.); - methanol extract, methyl n-nonyl ketone, lauryl aldehyde and capryl aldehyde (*Houttuynia cordata* T.); - ginger extract (*Zingiber officinale* R.); - licorine (*Lycoris radiata* (L'Her.) Herb); - ganodermadiol, lucidadiol and aplanoxidic acid G. (*Ganodema pfeifferi* B.);

Conclusions

A range of medicinal plants, extracts and compounds isolated from plants or other organisms have been investigated, which have been shown to have antiviral activity for various diseases. Studies about natural products and their pharmacological properties have been the target of growing interest by researchers around the world. Studies have, in most cases, confirmed the potential of medicinal plants or promising antiviral molecules. The efficiency of these bioproducts has been demonstrated acting at different stages of viral infection, adsorption, virus entry into the host cell, virus replication, assembly and dispersion. Bioactives discovered and positively evaluated are potential candidates in the development of new antiviral drugs and this has been the ultimate goal of research.

References

Ahmad, A., Davies, J., Randall, S., Skinner, G.R.B. (1996). Antiviral properties of extract of *Opuntia streptacantha*. Antivir Res 30: 75–85.

Ahmad, A., Rehman, M.U., Alkharfy, K.M. (2020). An alternative approach to minimize the risk of coronavirus (Covid-19) and similar infections. Eur Rev Med Pharmacol Sci 24(1): 4030–4034.

Aiken, C., Chen, C.H. (2005). Betulinic acid derivatives as HIV-1 antivirals. Trends Mol Med 11(1): 31–6.

Alam, G., Wahyuono, S., Ganjar, I.G., Hakim, L., Timmerman, H., Verpoorte, R. (2002). Tracheospasmolytic activity of viteosin-A and vitexicarpin isolated from *Vitex trifolia*. Planta Med 68: 1047–1049.

Ali, B.H., Blunden, G., Tanira, M.O., Nemmar, A. (2008). Some phytochemical, pharmacological and toxicological properties of ginger (*Zingiber officinale* Roscoe): a review of recent research. Food Chem Toxicol 46: 409–420.

Ali, S.G., Ansari, M.A., Alzohairy, M.A., Almatroudi, A., Alomary, M.N., Alghamdi, S., Rehman, S., Khan, H.M. (2021). Natural products and nutrients against different viral diseases: prospects in prevention and treatment of SARS CoV-2. Med 57(2): 169.

Andrade, C.T.S., Marques, J.G.W., Zappi, D.C. (2006). The use of medicinal cacti by locals at the semi-arid in Bahia State, Brazil. Rev Bras Pl Med 8(3): 36–42.

Araújo, C.C., Leon, L.L. (2001). Biological activities of *Curcuma longa* L. Mem Inst Oswaldo Cruz 96: 723–8.

Ashfaq, U.A., Masoud, M.S., Nawaz, Z., Riazuddin, S. (2011). Glycyrrhizin as antiviral agent against Hepatitis C Virus. J Transl Med 9: 112.

Atanasov, A.G., Waltenberger, B., Wenzig, E.M.P., Linder, T., Wawrosch, C., Uhrin, P., Stuppne, H. (2015). Discovery and resupply of pharmacologically active plant-derived natural products: A review. Biotechnol Adv 33(8): 1582–1614.

Babich, O., Sukhikh, S., Prosekov, A., Asyakina, L., Ivanova, S. (2020). Medicinal plants to strengthen immunity during a pandemic. Pharmaceuticals 13(10): 313.

Bag, P., Chattopadhyay, D., Mukherjee, H., Ojha, D., Mandal, N., Sarkar, M.C. Chatterjee, T., Das, G., Chakraborti, S. (2012). Anti-herpes virus activities of bioactive fraction and isolated pure constituent of Mallotus peltatus: an ethnomedicine from Andaman Islands. Virol J 9: 98.

Baikerikar, S. (2017). Curcumin and natural derivatives inhibit ebola viral proteins: an *in silico* approach. Pharmacognosy Res 9(Suppl 1): S15–S22.

Bailly, C., Vergoten, G. (2020). Glycyrrhizin: An alternative drug for the treatment of COVID-19 infection and the associated respiratory syndrome? Pharmacol Ther 214: 107618.

Batista, A.C.M., Ribeiro, P.A.M., Arantes, M.C.B., Oliveira, L.M.G., Freitas, M.R.F., Silva, L.N.M., Nogueira, J.C.M., Paula, J.R., Bara, M.T.F. (2007). Estudo Farmacognóstico de Tinospora cordifolia (Wild) Miers Ex Hook. F. & Thoms, Menispermaceae. Rev Eletr Farm 2(2).

Brighton, S.W., Prozesky, O.W., de la Harpe, A.L. (1983). Chikungunya virus infection. A retrospective study of 107 cases. S Afr Med J 63: 313–315.

Briolant, S., Garin, D., Scaramozzino, N., Jouan, A., Crance, J.M. (2004). *In vitro* inhibition of Chikungunya and Semliki Forest viroses replication by antiviral compounds: synergistic effect of interferon-α and ribavirin combination. Antivir Res 61: 111–117.

Carvalho, M.das G.de S., Bellini, J.D.S., Amaral, J.L.C., Moraes, L.C. de, Alves, L.R. de C., Carvalho Filho, C.G. de, Pereira, F.H., Oliveira, F.L. de, Garcia, E.K.I., Garcia, J.A.D. (2020). Elderberry, garlic and ginger: antivirals and/or immunostimulants? Res Soc Dev v. 9, n. 9.

Castro, I.N. de, Santos, S.R. dos, Kondo, M.K., Maia, V.M., Santos, C.C.R. dos. (2020). Agronomic performance and water use efficiency of irrigated cactus pear cultivars. Rev Caatinga, Mossoró 33(2): 529–540.

Chan, E.W.C., Baba, S., Chan, H.T., Kainuma, M., Tangah, J. (2016). Medicinal plants of sandy shores a short review on *V. trifolia* L. and Iomoea pes caprae. Indian J Nat Prod Resour 7: 107–115.

Chang, J.S., Wang, K.C., Yeh, C.F., Shieh, D.E., Chiang, L.C. (2013). Fresh ginger (*Zingiber officinale*) has anti-viral activity against human respiratory syncytial virus in human respiratory tract cell lines. J Ethnopharmacol 145(1): 146–151.

Chen, C., Zuckerman, D.M., Brantley, S., Sharpe, M., Childress, K., Hoiczyk, E., Pendlenton, A.R. (2014). *Sambucus nigra* extracts inhibit infectious bronchitis virus at an early point during replication. BMC Vet Res 10(24): 2–12.

Chen, D.Z., Jiang, J.D., Zhang, K.Q., He, H.P., Di, Y.T., Zhang, Y., Cai, J.-Y., Wang, L., Li, S.-L., Yi, P., Peng, Z.G., Hao, X.J. (2013). Evaluation of anti-HCV activity and SAR study of (+)-lycoricidine through targeting of host heat-stress cognate 70 (Hsc70). Bioorg Med Chem Lett 23: 2679–2682.

Cheng, P.W., Ng, L.T., Chiang, L.C., Lin, C.C. (2006). Antiviral effects of saikosaponins on human coronavirus 229E *in vitro*. Clin Exp Pharmacol Physiol 33: 612–616.

Cheung, P.C. (2008). Mushrooms as Functional Foods. Cheung, P.C.K., Ed.; John Wiley & Sons: Hoboken, NJ, USA.

Chiang, L., Ng, L., Cheng, P., Chiang, W., Lin, C. (2005). Antiviral activities of extracts and selected pure constituents of *Ocimum basilicum*. Clin Exp Pharmacol Physiol 32(10): 811–816.

Chiow, K.H., Phoon, M.C., Putti, T., Tan, B.K., Chow, V.T. (2016). Evaluation of antiviral activities of houttuynia cordatathunb, extract, quercetin, quercetrin and cinanserin on murine coronavirus and dengue virus infection. Asian Pac J Trop Med 9: 1–7.

Choi, H.J., Song, J.H., Park, K.S., Kwon, D.H. (2009). Inhibitory effects of quercetin 3-rhamnoside on influenza A virus replication. Eur J Pharm Sci 37: 329–333.

Chrubasik, S., Pittler, M.H., Roufogalis, B.D. (2005). Zingiberis rhizoma: A comprehensive review on the ginger effect and efficacy profiles. Phytomedicine 12(9): 684–701.

Cinatl, J., Morgenstern, B., Bauer, G., Chandra, P., Rabenau, H., Doerr, H.W. (2003). Glycyrrhizin, an active component of liquorice roots, and replication of SARS-associated coronavirus. Lancet 361: 2045–2046.

De Feo, V., Senatore, F. (1993). Medicinal plants and phytotherapy in the Amalfitan coast, Salerno province, Campania, Southern Italy. J Ethnopharmacol 39(1): 39–51.

Devi, W.R., Singh, C.B. (2014). Chemical composition, anti-dermatophytic activity, antioxidant and total phenolic content within the leaves essential oil of Vitex trifolia. Int J Phytocosmetics Nat Ingred 1: 5–10.

El Mekkawy, S., Meselby, M.R., Nakamura, N., Tekuza, Y., Hattori, M., Kakiuchi, N., Shimotohno, K., Kawahata, T., Otake, T. (1998). Anti-HIV-1 and anti HIV-1-protease substances from *Ganoderma lucidum*. Phytochemistry 49: 1651–1657.

Fenwick, G.R., Hanley, A.B. (1985). Allium species poisoning. Vet Rec 116: 28.

Feugang, J.M., Konarski, P., Zou, D., Stintzing, F.C., Zou, C. (2006). Nutritional and medicinal use of Cactus pear (*Opuntia* spp.) cladodes and fruits. *In*: Front Biosci 11: 2574–2589, Sep 1.

Fink, R.C., Roschek Jr, B., Alberte, R.S. (2009). HIV type-1 entry inhibitors with a new mode of action. Antivir Chem Chemother 19: 243–255.

Formenty, P. (2014). Ebola virus disease. Emerg Infect Dis 328: 121–134.

Fratianni, F., Ombra, M.N., Cozzolino, A., Riccardi, R., Spigno, P., Tremonte, P., Nazzaro, F. (2016). Phenolic constituents, antioxidant, antimicrobial and anti-proliferative activities of different endemic Italian varieties of garlic (*Allium sativum* L.). J Funct Foods 21: 240–248.

Hao, B., Wu, Y., Wang, J., Hu, S., Keil, D.J., Hu, H. (2012). Hepatoprotective and antiviral properties of isochlorogenic acid A from Laggera alata against hepatitis B virus infection. J Ethnopharmacol 144: 190–194.

Hayashi, K., Kamiya, M., Hayashi, T. (1995). Virucidal effects of the steam distillate from Houttuynia cordata and its components on Hsv-1, Influenza-Virus, and HIV. Planta Med 61: 237–241.

He, J., Qi, W.B., Wang, L., Tian, J., Jiao, P.R., Liu, G.Q., Ye, W.-C., Liao, M. (2012). Amaryllidaceae alkaloids inhibit nuclear-to-cytoplasmic export of ribonucleoprotein (RNP) complex of highly pathogenic avian influenza virus H5N1. Influenza Other Respir Viruses 7: 922–931.

He, Y., Lu, Z., Shi, P., Hao, J., Zhao, Z., Xie, H., Mao, P. Chen, S.J. (2016). Anti-herpes simplex virus activities of bioactive extratos from Antrodia camphorata mycelia. Antivir Ther 21(5): 377–383.

Hoofnagle, J.H. (1990). Alpha-interferon therapy of chronic hepatitis B. Current status and recommendations. J Hepatol 11: S100–S107.

Hosoda, S., Aoyama, H., Goto, Y., Salim, M.T.A., Okamoto, M., Hashimoto, M., Baba, M., Hashimoto, Y. (2009). Discovery of diphenylmethane analogs as anti-bovine diarrhea viral agentes. Bioorg Med Chem Lett 19: 3157–3161.

Huang, S.H., Su, M.C., Tien, N., Huang, C.J., Lan, Y.C., Lin, C.S., Chen, C.H., Lin, C.W. (2017). Epidemiology of human coronavirus NL63 infection among hospitalized patients with pneumonia in Taiwan. J Microbiol Immunol Infect 50(6): 763–770.

Iciek, M., Kwiecieri, I., Wlodek, L. (2009). Biological properties of garlic and garlic-derived organosulfur compounds. Environ Mol Mutat 50: 247–265.

Jarver, P., Coursindel, T., Andaloussi, S.E., Godfrey, C., Wood, M.J., Gait, M.J. (2012). Peptide-mediated cell and *in vivo* delivery of antisense oligonucleotides and siRNA. Mol Ther Nucleic Acids 1: e 27.

Jovic, T.H., Ali, S.R., Ibrahim, N., Jessop, Z.M., Tarassoli, S.P., Dobbs, T.D., Holford, P., Thornton, C.A., Whitaker, I.S. (2020). Could vitamins help in the fight against COVID-19? Nutrients 12: 2550.

Kalikar, M.V., Thawani, V.R., Varadpande, U.K., Sontakke, S.D., Singh, R.P., Khiyani, R.K. (2008). Immunomodulatory effect of *Tinospora cordifolia* extract in human immuno-deficiency virus positive patients. Indian J Pharmacol 40: 107–110.

Kaur, R., Sharma, P., Gupta, G.K., Ntie-Kang, F., Kumar, D. (2020). Structure-activity-relationship and mechanistic insights for Anti-HIV natural products. Molecules 25(9): 2070.

Khan, M.T.H., Ather, A., Thompson, K.T., Gambari, R. (2005). Extracts and molecules from medicinal plants against herpes simplex viroses. Antiviral Res 67(3): 107–119.

Koch, C., Reichling, J., Schneele, J., Schnitzler, P. (2008). Inhibitory effect of essential oils against herpes simplex virus type 2. Phytomedicine 15: 71–78.

Krawitz, C., Mraheil, M.A., Stein, M., Imirzalioglu, C., Domann, E., Pleschka, S., Hain, T. (2011). Inhibitory activity of a standardized elderberry liquid extract against clinically relevant human respiratory bacterial pathogens and influenza A and B viruses. BMC Complement Altern Med 11: 16.

Kutluay, S.B., Doroghazil, J., Roemerd, M.E., Triezenberg, S.J. (2008). Curcumin inhibits herpes simplex virus immediate-early gene expression by a mechanism independent of p300/CBP histone acetyltransferase activity. Virology 10; 373(2): 239–247.

Kwofie, S.K., Broni, E., Teye, J., Quansah, E., Issah, I., Wilson, M.D., Miller III, W.A., Tiburu, E.K., Bonney, J.H.K. (2019). Pharmacoinformatics-based identification of potential bioactive compounds against Ebola virus protein VP24. Comput Biol Med 113: 103414.

Kwon, J.H., Cho, S.H., Kim, S.J., Ahn, Y.J., Ra, J.C. (2010). Antiviral Composition Comprising Alnus Japonica Extracts. US Patent US 2010/0189827 A1.

Lago, J.H.G., Toledo-Arruda, A.C., Mernak, M., Barrosa, K.H., Martins, M.A., Tibério, I.F.L.C., Prago, C.M. (2014). Structure-activity association of flavonoids in lung diseases. Molecules 19: 3570–3595.

Lau, K.M., Lee, K.M., Koon, C.M., Cheung, C.S., Lau, C.P., Ho, H.M., Lee, M.Y., Au, S.W., Cheng, C.H., Lau, C.B., Tsui, S.K., Wan, D.C., Waye, M.M., Wong, K.B., Wong, C.K., Lam, C.W., Leung, P.C., Fung, K.P. (2008). Immunomodulatory and anti-SARS activities of Houttuynia cordata. J Ethnopharmacol 118: 79–85.

Leo, M., Bruzual De Abreu, M., Pawlowska, A.M., Cioni, P.L., Braca, A. (2010). Profiling the chemical contente of opuntia fícus indica flowers by HPLC-P-DA-ESI-MS and GCEIMS analyses. Phytochem Lett 3(1): 48–52.

Li, B.Q., Fu, T., Dongyan, Y., Mikovits, J.A., Ruscetti, F.W., Wang, J.M. (2000). Flavonoid baicalin inhibits HIV-1 infection at the level of viral entry. Biochem Biophys Res Commun 276(2): 534–538.

Li, S.W., Lin, C.W. (2013). Human coronaviruses: clinical features and phylogenetic analysis. Biomed 3: 43–50.

Li, S.Y., Chen, C., Zhang, H.Q., Guo, H.Y., Wang, H., Wang, L. (2005). Identification of natural compounds with antiviral activities against SARS-associated coronavirus. Antivir Res 67: 18–23.

Li, W., Wang, X., Luo, Z., Liu, L., Yan, C., Yan, C.Y. Chen, G.D., Gao, H., Duan, W.J., Kurihara, H., Li, Y.F., He, R.R. (2018). Traditional chinese medicine as a potential source for HSV-1 therapy by acting on virus or the susceptibility of host. Int J Mol Sci 19: 3266.

Li, Z., Li, L., Zhou, H., Zeng, L., Chen, T., Chen, Q., Zhou, B. Yutao Wang, Y., Chen, Q., Hu, P., Yang, Z. (2017). *Radix isatidis* polysaccharides inhibit influenza a virus and influenza a virus-induced inflammation via suppression of host TLR3 signaling *in vitro*. Molecules 22(1): 116.

Lin, C.W., Tsai, F.J., Tsai, C.H., Lai, C.C., Wan, L., Ho, T.Y., Hsieh, C.C., Chao, P.D. (2005). Anti-SARS coronavirus 3C-like protease effects of *Isatis indigotica* root and plant-derived phenolic compounds. Antiviral Res 68: 36–42.

Lin, C.C., Kan, W.S. (1990). Medicinal plants used for the treatment of hepatitis in Taiwan. Am J Chin Med 18: 35–43.

Lin, L., Hsu, W., Lin, C. (2014). Antiviral natural products and herbal medicines. J Tradit Complement Med 4(1): 24–35.

Lin, W., Huang, W., Ning, S., Wang, X., Ye, Q., Wei, D. (2018). *De novo* characterization of the *Baphicacanthus cusia* (Nees) Bremek transcriptome and analysis of candidate genes involved in indican biosynthesis and metabolism. PLoS ONE 13: e0199788.

Liu, D., Wang, F., Yang, L., Zheng, Y., Liu, J. (2007). A new cadinane sesquiterpene with significant anti-HIV-1 activity from the cultures of the basidiomycete *Tyromyces chioneus*. J Antibiot 60(5): 332–334.

Lundstrom, J.O. (1999). Mosquito-borne viruses in western Europe: a review. J Vector Ecol 24: 1–39.

Malik, A.H., Lee, W.M. (2000). Chronic hepatitis B virus infection: treatment strategies for the next millennium. Ann Intern Med 132: 723–731.

Mandrone, M., Lorenzi, B., Maggio, A., La Mantia, T., Scordino, M., Bruno, M., Poli, F. (2014). Polyphenols pattern and correlation with antioxidant activities of berries extracts from four different populations of sicilian *Sambucus nigra* L. Nat Prod Res 28(16): 1246–1253.

Martina, B.E., Koraka, P., Osterhaus, A.D. (2009). Dengue virus pathogenesis: an integrated view. Clin Microbiol Rev 22: 564.

Mazumder, A., Raghavan, K., Weinstein, J., Kohn, K.W., Pommier, Y. (1995). Inhibition of human immunodeficiency virus type-1 integrase by curcumin. Biochem Pharmacol 18; 49(8): 1165–70.

McGill, P.E. (1995). Viral infections: alpha-viral arthropathy. Baillieres Clin Rheumatol 9: 145–150.

Meng, Y., Lu, D., Guo, N., Zhang, L., Zhou, G. (1993). Anti-HCMV effect of garlic components. Virol Sin 8: 147–150.

Min, B.S., Lee, H.K., Lee, S.M., Kim, Y.H., Bae, K.H., Otake, T., Nakamura, N., Hattori, M. (2002). Anti-human immunodeficiency virus-type 1 activity of constituents from *Juglans mandshurica*. Arch Phma Res 25(4): 441–445.

Mothana, R.A.A., Awadh, N.A.A., Jansen, R., Wegner, U., Mentel, R., Lindequist, U. (2003). Antiviral lanostanoid triterpenes from the fungus *Ganoderma pfeifferi* BRES. Fitoterapia 74: 177–180.

Nasution, M.A.F., Toepak, E.P., Alkaff, A.H., Tambunan, U.S.F. (2018). Flexible docking-based molecular dynamics simulation of natural product compounds and Ebola virus Nucleocapsid (EBOV NP): a computational approach to discover new drug for combating Ebola. BMC Bioinformatics 19(Suppl 14): 419.

Nicacio, G.L.S., Moura, S.C., Costa, J.V.J., Sena, C.R., Cruz, T.B.F., Lopes, G.N.M., Cecílio, A.B. (2018). Breve revisão sobre as propriedades fitoterápicas do *Zingibe rofficinale* Roscoe o gengibre. Sinapse Múlt 7(2): 74–80.

Oliveira Melo, A.S., Malinger, G., Ximenes, R., Szejnfeld, P.O., Alves Sampaio, S., Bispo de Filippis, A.M. (2016). Zika virus intrauterine infection causes fetal brain abnormality and microcephaly: tip of the iceberg? Ultrasound Obstet Gynecol 47: 6–7.

Ouyang, L.T.G., Gao, Y., Zhang, P., Xie, X. (1993). Two insecticidal antibiotics produced by Streptomyces nanchangensis. J Jiangxi Agric Univ 15: 148–153.

Panya, A., Yongpitakwattana, P., Budchart, P., Sawasdee, N., Krobthong, S., Paemanee, A., Roytrakul, S., Rattanabunyong, S., Choowongkomon, K., Yenchitsomanus, P.T. (2019). Novel bioactive peptides demonstrating anti-dengue virus activity isolated from the Asian medicinal plant Acacia Catechu. Chem Biol Drug Des 93: 100–109.

Park, J.Y., Jeong, H.J., Kim, J.H., Kim, Y.M., Park, S.J., Kim, D., Park, K.H., Lee, W.S., Ryu, Y.B. (2012). Diarylheptanoids from Alnus japonica inhibit papain-like protease of severe acute respiratory syndrome coronavirus. Biol Pharm Bull 35: 2036–2042.

Patwardhan, M., Morgam, M.T., Dia, V., D'Souza, D.H. (2020). Heat sensitization of hepatitis A virus and Tulane virus using grapeseed extract, gingerol and curcumin. Food Microbiol 90: N.103461.

Ping, X., Weiyang, Y., Jianwei, C., Xiang, L. (2016). Antiviral activities against influenza virus (FM1) of bioactive fractions and representative compounds extracted from Banlangen (*Radix Isatidis*). J Tradit Chin Med 36(3): 369–376.

Pittler, M.H., Ernst, E. (2007). Clinical effectiveness of garlic (*Allium sativum*). Mol Nutr Food Res 51: 1382–1385.

Porter, R.S., Bode, R.F. (2017). A review of the antiviral properties of black Elder (*Sambucus nigra* L.) products. Phytother Res 31(4): 533–554.

Pruthvish, R., Gopinatha, S.M. (2018). Antiviral prospective of *Tinospora cordifolia* on HSV-1. Int J Curr Microbiol Appl Sci 7: 3617–3624.

Ra, J.C., Kwon, H.J., Kim, B.G., You, S.H. (2010). Anti-Influenza Viral Composition Containing Bark or Stem Extract of Alnus japonica. US Patent US 2010/0285160 A1, 11.

Rahi, D.K., Malik, D. (2016). Diversity of mushrooms and their metabolites of nutraceutical and therapeutic significance. J Mycol 2016: 1–18.

Rasool, A., Khan, M., Ali, M., Anjum, A., Amhed, I., Aslam, A., Nawaz, M. (2017). Anti-Avian influenza virus H9N2 activity of aqueous extracts of *Zingiber officinalis* (Ginger) & *Allium sativum* (Garlic) in chick embryos. Pak J Pharm Sci 30: 1341–1344.

Rausch, K., Hackett, B.A., Weinbren, N.L., Reeder, S.M., Sadovsky, Y., Hunter, C.A., Schultz, D.C., Coyne, C.B., Cherry, S. (2017). Screening bioactives reveals nanchangmycin as a broad spectrum antiviral active against zika virus. Cell Rep 18: 804–815.

Ryu, Y.B., Jeong, H.J., Kim, J.H., Kim, Y.M., Park, J.Y., Kim, D., Nguyen, T.T., Park, S.J., Chang, J.S., Park, K.H., Rho, M.C., Lee, W.S. (2010). Biflavonoids from *Torreya nucifera* displaying SARS-CoV 3CL(pro) inhibition. Bioorg Med Chem 18: 7940–7947.

Sagar, V., Kumar, A.H.S. (2020). Efficacy of natural compounds from *Tinospora cordifolia* against SARS-CoV-2 protease, surface glycoprotein and RNA polymerase. BEMS Rep 6: 6–8.

Salehi, B., Kumar, N.V.A., Sener, B., Sharifi-Rad, M., Mahady, M.K.G.B., Vlaisavljevic, S., Iriti, M., Kobarfard, F., Setzer, W.N., Ayatollahi, S.A., Ata, A., Sharifi-Rad, J. (2018). Medicinal plants used in the treatment of human immunodeficiency virus. J Mol Sci 19: 1459.

Santana, F.P.R., Pinheiro, N.M., Mernak, M.I.B., Righetti, R.F., Martins, M.A., Lago, J.H.G., Prado, C.M. (2016). Evidences of herbal medicine-derived natural products effects in inflammatory lung diseases. Mediators Inflamm 2348968–2348981.

Santana, F.P.R., Thevenard, F., Gomes, K.S., Taguchi, L., Câmara, N.O.S., Stilhano, R.S., Ureshino, R.P., Prado, C.M., Ghilardi, Lago, J.H.G. (2021). New perspectives on natural flavonoids on COVID-19-induced lung injuries. Phytother Res 1–19.

Santoyo, S., Ramírez-Anguiano, A.C., Aldars-García, L., Reglero, G., Soler-Rivas, C. (2012). Antiviral activities of *Boletus edulis, Pleurotus ostreatus* and *Lentinus edodes* extracts and polysaccharide fractions against Herpes simplex virus type 1. J Food Nutr Res 51(4): 225–235.

Schnitzler, P., Koch, C., Reichling, J. (2007). Susceptibility of drug-resistant clinical herpes simplex virus type 1 strains to essential oils of ginger, thyme, hyssop, and sandalwood. Antimicrob Agents Chemother 51(5): 1859–1862.

Seo, D.J., Choi, C. (2021). Antiviral bioactive compounds of mushrooms and their antiviral mechanisms: a review. Viruses 13(1): 350.

Sinha, S.K., Prasad, S.K., Islam, M.A., Gaurav, S.S., Patil, R.B., Al-Faris, N.A., Aldayel, T.S., AlKehayez, N.M., Wabaidur, S.M., Shakya, A. (2020). Identification of bioactive compounds from Glycyrrhiza glabra as possible inhibitor of SARS-CoV-2 spike glycoprotein and non-structural protein-15: A pharmacoinformatics study. J Biomol Struct Dyn 1–15.

Sordillo, P.P., Helson, L. (2015). Curcumin suppression of cytokine release and cytokine storm. A potential therapy for patients with Ebola and other severe viral infections. *In Vivo* 29(1): 1–4.

Srivastava, R.A.K., Mistry, S., Sharma, S. (2015). A novel anti-inflammatory natural product from Sphaeranthus indicus inhibits expression of VCAM1 and ICAM1, and slows atherosclerosis progression independent of lipid changes. Nutr Metab 12: 20.

Stajner, D., Milic-DeMarino, N., Canadanovic-Brunet, J., Kapor, A., Stajner, M., Popovic, B.M. (2006). Exploring *Allium* species as a source of potential medicinal agents. Phytother Res 20: 581–584.

Sun, Y., Zhou, X., Liu, J., Bao, K., Zhang, G., Tu, G., Kieser, T., Deng, Z. (2002). '*Streptomyces nanchangensis*', a producer of the insecticidal polyether antibiotic nanchangmycin and the antiparasitic macrolide meilingmycin, contains multiple polyketide gene clusters. Microbiology 148: 361–371.

Szajdek, A., Borowska, E.J. (2008). Bioactive compounds and health-promoting properties of berry fruits: a review. Plant Foods Hum Nutr 63(4): 147–56.

Tan, G.T., Pezzuto, J.M., Kinghorn, A.D., Hughes, S.H. (1991). Evaluation of natural products as inhibitors of human immunodeficiency vírus type 1 (HIV-1) reverse transcriptase. J Nat Prod 54(1): 143–154.

Tetro, J.A. (2016). Zika and microcephaly: causation, correlation, or coincidence? Microbes Infect 18: 167–168.

Thuy, B.T.P., My, T.T.A., Hai, N.T.T., Hieu, L.T., Hoa, T.T., Loan, H.T.T.P., Triet, N.T., Anh, T.T.V., Quy, P.T., Tat, P.V., Hue, N.V., Quang, D.T., Trung, N.T., Tung, V.T., Huynh, L.K. (2020). Investigation into SARS-CoV-2 resistance of compounds in garlic essential oil. ACS Omega 5: 8312–8320.

Tsai, H.C., Chen, I.T., Wu, K.S., Tseng, Y.T., Sy, C.L., Chen, J.K., Lee, S.S.J., Chen, Y.S. (2017). High rate of HIV-1 drug resistance in treatment failure patients in Taiwan, 2009–2014. Infect Drug Resist 10: 343.

Tsai, Y.C., Lee, C.L., Yen, H.R., Chang, Y.S., Lin, Y.P., Huang, S.H., Lin, C.W. (2020). Antiviral action of tryptanthrin isolated from *Strobilanthes cusia* leaf against human coronavirus NL63. Biomolecules 10(3): 366.

Tsai, Y., Cole, L.L., Davis, L.E., Lockwood, S.J., Simmons, V., Wild, G.C. Antiviral properties of garlic. (1985). *In vitro* effects on influenza B, herpes simplex and coxsackie viruses. Planta Med 5: 460–461.

Tung, N.H., Kwon, H.J., Kim, J.H., Ra, J.C., Kim, J.A., Kim, Y.H. (2010). An anti-influenza component of the bark of Alnus japonica. Arch Pharm Res 33: 363–367.

Utsunomiya, H., Ichinose, M., Ikeda, K., Uozaki, M., Morishita, J., Kuwahara, T., Koyama, A.H., Yamasaki, H. (2014). Inhibition by caffeic acid of the influenza A virus multiplication *in vitro*. Int J Mol Med 34(4): 1020–1024.

Vajragupta, O., Boonchoong, P., Morris, G.M., Oslon, A.J. (2005). Active site binding modes of curcumin in HIV-1 protease and integrasse. Bioorganic Med Chem Lett 15(14): 3364–8.

Vermani, K., Garg, S. (2002). Herbal medicines for sexually transmitted diseases and AIDS. J Ethnopharmacol 80(1): 49–66.

Weber, N.D., Andersen, D.O., North, J.A., Murray, B.K., Lawson, L.D., Hughes, B.G. (1992). *In vitro* virucidal effects of Allium sativum (garlic) extract and compounds. Planta Med 58: 417–423.

Wee, H.N., Neo, S.Y., Singh, D., Yew, H.C., Qiu, Z.Y., Tsai, X.R.C., How, S.Y., Yip, K.-Y.C., Tan, C.-H., Koh, H.-L. (2020). Effects of *Vitex trifolia* L. leaf extracts and phytoconstituents on cytokine production in human U937 macrophages. BMC Complement Med Ther 20: 91.

Weng, J.R., Lin, C.S., Lai, H.C., Lin, Y.P., Wang, C.Y., Tsai, Y.C., Wu, K.C., Huang, S.H., Lin, C.W. (2019). Antiviral activity of *Sambucus formosana* Nakai ethanol extract and related phenolic acid constituents against human coronavirus NL63. Virus Res 273: 197767.

Wolkerstorfer, A., Kurz, H., Bachhofner, N., Szolar, O.H. (2009). Glycyrrhizin inhibits influenza A virus uptake into the cell. Antivir Res 83: 171–178.

Xie, M.L., Phoon, M.C., Dong, S.X., Tan, B.K.H., Chow, V.T. (2013). Houttuynia cordata extracts and constituents inhibit the infectivity of dengue virus type 2 *in vitro*. Int J Integr Biol 14: 78–85.

Yang, D.Q., Chen, Z.R., Chen, D.Z., Hao, X.J., Li, S.L. (2018). Anti-TMV effects of Amaryllidaceae alkaloids isolated from the bulbs of Lycoris radiata and Lycoricidine derivatives. Nat Prod Bioprospect 8: 189–197.

Yang, L., Zhang, J.H., Zhang, X.L., Lao, G.J., Su, G.M., Wang, L., Li, Y.L., Ye, W.C., He, J. (2019). Tandem mass tag-based quantitative proteomic analysis of lycorine treatment in highly pathogenic avian influenza H5N1 virus infection. Peer J 7: e7697.

Yao, J.L., Fang, S.M., Liu, R., Oppong, M.B., Liu, E.W., Fan, G.W., Zhang, H. (2016). A review on the terpenes from Genus Vitex. Molecules 21: 1179.

Yin, J., Li, G., Li, J., Yang, Q., Ren, X. (2011). *In vitro* and *in vivo* effects of Houttuynia cordata on infectious bronchitis virus. Avian Pathol 40: 491–498.

Zajac, M., Muszalska, I., Sobczak, A., Dadej, A., Tomczak, S., Jelinska, A. (2019). Hepatitis C—new drugs and treatment prospects. Eur J Med Chem 165: 225–249.

Zandi, K., Teoh, B., Sam, S., Wong, P., Mustafa, M.R., AbuBakar, S. (2011). Antiviral activity of four types of bioflavonoid against dengue virus type-2. Virol J 8: 560.

Zhang, T., Yang, D., Wang, Y., Chen, X., Bi, K. (2010). Simultaneous determination of four components in the herbs of *Sumbucus chinensis* Lindl. by high performance liquid chromatography. Asian J Tradit Med 5(2): 70–74.

Zhang, J., Shen, X., Yan, Y., WANG, Y., Cheng, Y. (2020, March 2). Discovery of anti-SARS-CoV-2 agents from commercially available flavor via docking screening. https://doi.org/10.31219/osf.io/vjch2.

Zhou, B., Yang, Z., Feng, Q., Liang, X., Li, J., Zanin, M., Jiang, Z., Zhong, N. (2017). Aurantiamide acetate from *Baphicacanthus cusia* root exhibits anti-inflammatory and anti-viral effects via inhibition of the NF-kappaB signaling pathway in Influenza A virus-infected cells. J Ethnopharmacol 199: 60–67.

7

Brazilian Medicinal Plant Extracts with Antimicrobial Action Against Microorganisms that Cause Foodborne Diseases

Luiza Helena da Silva Martins,[1,*] *Sabrina Baleixo da Silva,*[2] *Carissa Michelle Goltara Bichara,*[1] *Johnnat Rocha Allan de Oliveira,*[3] *Adilson Ferreira Santos Filho,*[2] *Rafaela Cristina Barata Alves,*[1] *Andrea Komesu*[4] and *Mahendra Rai*[5]

Introduction

Foodborne diseases represent a public health hazard worldwide. World Health Organization (WHO) data show approximately 600 million cases of foodborne infections in its database in 2010 and for children aged 5 years, the impact is global, reaching 217 million episodes of contamination for children in this age (Kirk et al. 2015; Aik et al. 2020). In Brazil, the biggest cause of foodborne disease is undoubtedly the inappropriate handling of food, which is directly associated with outbreaks that occur in households and food services in the country according to data from Brasil (2016). Over the years, with modernization and technology, these outbreaks continue to occur, requiring studies and research with strategies that aim to improve the aspect of food security in the country (Andrade et al. 2020).

Microorganisms are highly versatile and can enter the food chain at all stages before they are ready for final consumption. By adaptations to environments, they are able to multiply, produce toxic compounds in contaminated food and colonize new hosts. It is important to note that the infective dose of food pathogens is generally less than the amount of microorganisms needed to alter the organoleptic characteristics visible in food;

[1] Institute of Animal Health and Production (ISPA), Federal Rural University of Amazonia (UFRA), Av Presidente Tancredo Neves N° 2501; Terra Firme, 66.077-830, Belem, Pará, Brazil.
[2] Graduate Program in Food Science and Technology, Federal University of Pará, Augusto Corrêa Avenue, Guamá, 66.075-900, Belém, Pará, Brazil.
[3] Institute of Health Sciences, Faculty of Nutrition (FANUT), Federal University of Pará (UFPA), R. Augusto Corrêa, 01 - Guamá, 66075-110 Belém, PA, Brazil.
[4] Department of Marine Sciences (DCMar), Federal University of São Paulo (UNIFESP), R. Silva Jardim, n° 136, Vila Matias, 11015-020, Santos, SP, Brazil.
[5] Department of Microbiology, Nicolaus Copernicus University, Torun, Poland.
* Corresponding author: luiza.martins@ufra.edu.br

this explains why most food outbreaks have been related to eating attractive foods, normal taste and odor (Greig and Ravel 2009; Oliveira et al. 2010).

Pathogenic microorganisms are responsible for infectious diseases. They are one of the main causes of premature death in humans, which is due to the strong resistance of pathogenic microorganisms to antibiotics. Such facts suggest the use of natural compounds that are capable of inhibiting the development of microorganisms causing infectious diseases (Bezerra et al. 2019). In this context, the use of roots, leaves and even seeds of medicinal plants is encouraged, as their composition contains bioactive compounds with antimicrobial potential including Essential Oils (EOs), terpenes, alkaloids, flavonoids, phenolic compounds, among others (Bhavaniramya et al. 2019).

Several plants have been widely used in folk medicine for long by communities in Brazil and worldwide. These uses include anti-inflammatory, antioxidant, antibacterial, antifungal properties, as well as the treatment of respiratory problems, rheumatism, antitumoral, among others. Several parts of the plants have been mentioned in research, such as leaves, stems, roots, bark, flowers and fruits.

The compounds responsible for these activities are known as phytochemicals and to some extent EOs (Prakash et al. 2015; Makhuvele et al. 2020). The phytochemicals are well known as antimicrobial bioactive, which can include several major groups as phenolic compounds, flavonoids, phytosterols, carotenoids, tocols, saponins, tannins EOs, organic acids, among others (Abdel-Fattah et al. 2018; Makhuvele et al. 2020).

The evaluation of medicinal plants extracts is of great importance to produce new products with food preservatives and disinfectant properties and food processing environments, as well as for the development of innovative food packaging (Borges et al. 2013; Carraro et al. 2014; Du et al. 2020). Furthermore, it is possible that after mutagenicity, cytotoxicity and genotoxicity tests, to prove their safety, these extracts can be associated with existing products, and thus act synergistically in the conservation of the target product.

This chapter intends to address the potential extracts of Brazilian medicinal plants to combat pathogenic and deteriorating microorganisms that can cause food poisoning and compromise food quality, emphasizing the properties of phytochemicals, mechanisms of action and technologies for extraction.

Foodborne diseases and their main microorganisms

The growth of the international food trade and displacement facilities has accelerated the spread of multiregional and even multinational Outbreaks of Foodborne Diseases (OFD) or Foodborne Diseases (FD), which have increased the population's vulnerability to pathogens and foodborne contaminants. These agents represent an emerging risk and are responsible for serious global public health problems, such as morbidity and mortality, causing medical care costs, in addition to significant economic losses by governments and industries, becoming a potential threat in developed and developing countries (Oliveira et al. 2010; Valeriano et al. 2012).

Outbreaks of foodborne infections or OFD are called when two or more people develop the same symptoms after eating foods of the same origin, which can occur in the form of infection, intoxication and/or intoxication (Mendonça et al. 2020). Some diseases that affect millions of people annually are associated with symptoms that are reflected in disorders of the digestive system of the host and can last from a few hours to several days, varying according to the microorganism or toxin found and the amount of contaminated food eaten. In most cases, the most common symptoms include stomach pain, nausea, vomiting, diarrhea, and sometimes fever. In susceptible individuals, such as children, the elderly, pregnant women and the immunosuppressed, the clinical picture can be more severe and prolonged, with consequences such as dehydration, bloody diarrhea, acute renal failure and even lead to death, depending on the patient's physical condition (Oliveira et al. 2010; Câmara Flores and Melo 2015).

In recent years, in many countries there have been several outbreaks of emerging and reemerging foodborne diseases in the world. The World Health Organization (WHO) has expressed that approximately an average of 420,000 losses of healthy lives are reported each year after eating contaminated food. In addition, children under the age of five die every year about 125,000. In 2011, in the United States they estimated that OFDs were responsible for approximately 48 million illnesses, 128,000 hospitalizations and 3,000 deaths. In China, about 748 million cases of acute gastrointestinal disease and 420 million medical consultations occur each year (Li et al. 2018). In Europe in 2010, 5,262 foodborne outbreaks occurred, involving about 43,500 people, of which about 4,600 were hospitalized and there were 25 deaths. In the UK, there are more than 1.7 million cases of foodborne illnesses each year, resulting in almost 22,000 hospitalized and 700 deaths. In France, foodborne pathogens have been estimated to cause between 10,000 and 20,000 hospital admissions per year. Moreover, in 2011, there was a serious outbreak in Germany, in which 3,186 cases (22% of cases) developed a severe complication, the Hemolytic-Uremic Syndrome (HUS) (O'Brien 2014).

These facts from around the world have put the health authorities in countries on alert about the need to revise their standards and launch foodborne disease surveillance programs to improve safety strategies aimed at ensuring food security in food supply and preventing these diseases due to proliferation of pathogens (Li et al. 2018).

Foods are vehicles for many pathogenic microbial agents that result from the association of factors that can influence their contamination, allowing the proliferation or survival of these agents in food. The main reasons for the outbreaks caused by pathogenic agents are related to faults in handling, mainly in the preparation of contaminated raw or partially cooked food, due to insufficient cooking time and temperature during cooking. In addition to the prolonged exposure time of foods to room temperature, which are related to improper storage, preparation before consumption and food without prior inspection, reasons that favor microbial proliferation (Hughes et al. 2007; Câmara Flores and Melo 2015).

The population of pathogens that cause food outbreaks of greatest importance for food safety are bacteria, parasites, viruses and fungi. More than 200 types of disease are known, and many studies point to bacteria as the main agents involved (Oliveira et al. 2010). It is important to point out that, despite little notification, viral food outbreaks often occupy a prominent place in the numbers of deaths and financial expenditures for governments. As they cause disease in the world, are easy to spread and resistant to environments. The main pathogens that cause viral DOA are rotavirus, norovirus, hepatitis A virus and polio (Silva and Lavinas 2010).

The national and international outbreaks recorded and investigated are usually those that involve a large number of people or that present more severe symptoms, which leads patients seeking health services. The reports of these eating disorders can be investigated about their pathogenic agents, mechanism, place of occurrence and other pertinent characteristics through laboratory etiological identification, bromatological exams or by epidemiological criteria. However, because most food pathogens cause mild symptoms and have rapid treatment, many cases of DOA are underreported in the health surveillance or incomplete records related to the laboratory diagnosis of the pathogens involved. Often the diseases cannot be diagnosed, showing a worldwide problem of official systems of health and epidemiological authorities. This hinders the traceability of DOA-causing foods and the action of these authorities (Coutinho et al. 2009; Oliveira et al. 2010; Sirtoli and Comarella 2018).

Among the foods most frequently implicated in the reported outbreaks, are mainly animal foods as a source for contamination, such as meat, milk and dairy products, eggs and products based on eggs, fish and crustaceans. And less often foods like salads, vegetables and fruits (Hughes et al. 2007; Greig and Ravel 2009; Oliveira et al. 2010).

Among the most reported foodborne pathogens, according to Mendonça et al. (2020), in a recent survey on the main emerging diseases and their respective etiological agents, include 69% mentioned salmonellosis, 38% staphylococcal intoxication, 23% *Escherichia coli* gastroenteritis, 15% botulism, *Bacillus cereus* gastroenteritis and *Clostridium perfringens* and toxinfection 8% for listeriosis were reported in publications in English and Portuguese between 1988 and 2020. These data corroborate with data reported from Brazil. According to the data released by the federal government, from 2009 to 2018, 6,093 outbreaks were recorded, with 672,873 exposed people and 99 deaths, from diseases caused by the same etiological agents of high occurrence in outbreaks, with emphasis on *Salmonella* spp., *Escherichia coli* and *Staphylococcus aureus* (Brasil 2019). Moreover, in Europe, in addition to the report *Salmonella* was the pathogen with the highest frequency of outbreaks (30.5%), viruses (15%) and *Campylobacter jejuni* (8.9%) appeared (O'Brien 2014).

In view of these data, it is important to disclose the characteristics of the main etiological agents responsible for these eating disorders reported worldwide. In this case, salmonellosis is a zoonosis caused by *Salmonella* spp. and is considered the most important cause of DOA in the world, with great significance for public health and high morbidity. This bacterium is Gram-negative, non-spore-forming and facultative anaerobic and its primary habitat is the gastrointestinal tract of man (Kottwitz et al. 2010; Mendonça et al. 2020). According to Appiah et al. (2020), who evaluated an outbreak of salmonellosis from 1989 to 2018 in 47 publications found reports of 45,215 cases in outbreaks that occurred in 25 countries. Similarly, in another study, between 1999 and 2008 in a state in Brazil, the authors reported that 286 outbreaks occurred, where 5,641 people were affected and 881 were hospitalized (Kottwitz et al. 2010).

Escherichia coli is a bacterium found naturally in the intestines of humans and animals and causes gastroenteritis. In addition, it has Gram-negative, non-sporogenic, facultative anaerobic characteristics and some strains can cause serious foodborne illnesses with more serious problems of hemorrhagic cases or that reach the brain and cause something even irreversible (Millezi et al. 2014; Mendonça et al. 2020).

Another important cause of gastroenteritis is staphylococcal enterotoxicosis, caused by poisoning by *Staphylococcus aureus*, a pathogenic bacterium of great importance and responsible for 45% of the toxinfections in the world, due to the ingestion of the enterotoxin produced by this species. This pathogenic bacterium is Gram-positive, has the shape of coconuts and its habitat that facilitates its presence in food is the skin, armpit, and especially in humans and animals, the nasopharyngeal mucosa. Another characteristic of this microorganism is its resistance to heat, osmotic pressure and environmental stresses (Millezi et al. 2014; Mendonça et al. 2020).

Botulism is a serious food poisoning caused by a toxin produced by the bacterium *Clostridium botulinum*, which affects the nervous system and can lead to death. Its characteristics are bacillus, Gram-positive, strict anaerobic and spore forming. The toxins produced are exotoxin in food botulism and endotoxin in the intestine. Listeirosis is an infectious disease caused by the bacterium *Listeria monocytogenes*. This species is rod-shaped, Gram-positive, non-sporulated and facultative anaerobic. Listeriosis can cause septicemia, meningoencephalitis, skin lesions and intrauterine infection in pregnant women that can lead to abortion in the second or third trimester or stillbirths (Câmara Flores and Melo 2015; Mendonça et al. 2020).

Thus, the dissemination of the main related aspects of OFD and its etiological agents is necessary in a scientific approach as a way to understand the factors involved in these diseases and how to treat them, however, there are still a considerable number of new challenges with investigations of emerging or re-emerging microorganisms that require a detailed report of outbreaks transmitted by food and water, which are essential for disease control, prevention and information. In this context, the discovery of new substances with potential inhibition on these OFD-causing microorganisms becomes essential, such as medicinal plants that have attracted the attention of researchers, as it is a promising source of substances that can be used in the control of pathogens of these diseases (Diaz et al. 2010; Valeriano et al. 2012; Millezi et al. 2014).

Bio-based antimicrobial agents from Brazilian plant extracts—an overview

The extraction technique plays an important role for raw material quality maintaining, which has led to improve conventional techniques in favor of eco-friendly technologies that allow to preserve better target bioactive compounds, operating at lower temperatures and avoiding as much as possible the use of solvents, with more sustainable processing and reduced energy use and environmental pollution (Pateiro et al. 2020).

In addition, a promising future is the use of these bioactive agents as food preservatives as an alternative to synthetic substances. A possibility to this use would be the application of antimicrobial plants extracts in food packaging development, that have already been proved to be an effective and reliable alternative replacing synthetic additives (Orsuwan and Sothornvit 2018; Karam et al. 2019).

Many species native to Brazil have already proven antimicrobial effects against foodborne pathogens and spoiling bacteria, whether in the presence of EOs, spices, herbs or crude extracts. Many antimicrobial properties of these foods are linked to bioactive compounds that play a certain influence in causing damage to the microorganism's cytoplasmic membrane (Tsuchiya and Iinuma 2000), the possibility of interaction with macromolecules such as proteins (Rice-Evans et al. 1997; Martins et al. 2016), interferences in the inhibition of nucleic acid synthesis (Plaper et al. 2003) and energy metabolism (Uzel et al. 2005), enabling an antimicrobial action.

Ponzilacqua et al. (2019) evaluated the aqueous extracts of sweet passion fruit (*Passiflora alata*), araçá (*Psidium cattleianum*), rosemary (*Rosmarinus officinalis*) and oregano (*Origanum vulgare*) for their ability to degrade aflatoxin B1 (AFB1) *in vitro*, with interesting results: a time-dependent decrease in AFB1 levels was observed in the aqueous leave extracts of all plants. Rosemary extract had the highest AFB1 reduction, with mean percentages of 49.0 to 60.3% at 24–48 h, followed by oregano after 48 hours of incubation (38.3%) and araçá at 36 h (30.7%).

Naves et al. (2019) evaluated the biological activities of ethanolic and aqueous extract of leaf, bark and seed of *Garcinia brasiliensis*. The leaf aqueous extracts displayed significant antimicrobial activity against *Sthapylococcus aureus* (concentrations that inhibit 50% - IC50 - of 1 μg/mL), an important Gram-positive strain responsible for serious problems to human health.

Silva et al. (2020) obtained ethanolic extracts of different plants of the *Mimosa* genus, constituted by phenolic compounds, including flavonoids and tannins. The plants presented satisfactory activity against fungal strains and *S. aureus*.

EOs are already well established as antimicrobial agents against a range of pathogens. There are many valuable compounds that are part of their composition, such as terpenoids, phenylpropanoids, alcohols, aldehydes, esters, ethers, ketones, phenols, fatty acids, oxides and sulfur derivatives (Burt 2004; Radaelli et al. 2016; Pateiro et al. 2020).

The efficient antimicrobial and antioxidant action of EOs has led to several researches on the use of these compounds in technological applications, among them, food preservatives with a great potential to control and inhibit the growth of microorganisms in meat and meat products (Bajpai et al. 2012; Aminzare et al. 2016; Pateiro et al. 2020).

Although all the EOs have shown to exert antimicrobial efficacy, the concentration and the quality of these compounds depend on several factors, which mainly depend on the species, the harvest conditions, and the place where they have grown (for example, the abundance and/or lack of mineral components), environmental conditions and the portion of the plant used for distillation (Salgueiro et al. 1997; Bajpai et al. 2012).

Duarte et al. (2007) obtained EOs from leaves of 29 medicinal plants commonly used in Brazil and screened against 13 different *Escherichia coli* serotypes. EOs from *Cymbopogon martinii* exhibited a broad inhibition spectrum, presenting strong activity against 10 out of 13 *Escherichia coli* serotypes (three enterotoxigenic, two enteropathogenic, three enteroinvasive and two shiga-toxin producers). *C. winterianus* inhibited strongly two enterotoxigenic, one enteropathogenic, one enteroinvasive and one shiga-toxin producer serotypes. *Aloysia triphylla* also shows good potential to kill *E. coli* with moderate to strong inhibition.

Marques et al. (2015) assessed the activity of the EOs from *Origanum vulgare* L. (oregano) and *Origanum majorana* L. (marjoram) against *S. aureus* isolated from poultry meat marketed. The authors obtained nine isolates showing genes for enterotoxin A and three for enterotoxin B, and four isolates were antibiotic-resistant. However, the results were interesting, once all isolates were sensitive to the oils tested.

Orlanda and Nascimento (2015) investigated the antibacterial activities of oil isolated from the fresh leaves of *Ruta graveolens* L., with activity against Gram-positive and Gram-negative bacteria (with MIC values of 0.75–1.40 $\mu g{\cdot}mL^{-1}$), such as *Bacillus cereus* and *S. aureus* which are the most susceptible bacteria.

In another study, Radaelli et al. (2016) evaluated the antimicrobial activities of EOs of six condiments commonly used in Brazil: *Ocimum basilicum* L. (basil), *Rosmarinus officinalis* L. (rosemary), *Origanum majorana* L. (marjoram), *Mentha* × *piperita* L. var. Piperita (peppermint), *Thymus vulgaris* L. (thyme) and *Pimpinella anisum* L. (anise) against *Clostridium perfringens* strain A. All oils showed bactericidal activity, except anise oil, which was only bacteriostatic. The authors highlight that the oxygenated compound (oxygenated monoterpenes and phenylpropanoids) might be responsible for the antimicrobial activity against *C. perfringens*, but the synergistic effects of these chemicals with other minor constituents should also be considered.

Perigo et al. (2016) tested the EOs from 11 species of *Piper*. The majority of the species exhibited inhibitory activity against pathogenic Gram negative and positive bacteria, reaching up to 30% of the inhibition levels of commercial antibiotics. The contents of bicyclogermacre and γ-muurolene were positively associated to *E. coli* inhibition; germacrene D and trans-caryophyllene were associated to the inhibitory activity against all tested bacteria. Limonene and cis-β-ocymene were associated with *Staphylococci* inhibition.

Sandri et al. (2007) obtained the EOs from aerial parts of six Brazilian species of the genus *Cunila* Mill. (Lamiaceae) currently used in beverages and food preparation, and in folk medicine and screened for antibacterial activity against 15 bacterial species. The oil of *C. galioides* citral efficiently controlled the growth of *Bacillus* sp., *Listeria monocytogenes*,

S. aureus, *Aeromonas hydrophila*, and *Enterocoocus faecalis*. Although less efficient, the other EOs studied were effective against *Bacillus* species and *S. aureus*. In general, the antimicrobial activity of *Cunila* EOs was more pronounced against Gram-positive than against Gram-negative bacteria.

Baldim et al. (2018) reported the antimicrobial activity of *Ocimum basilicum* and observed efficiency against *B. cereus*. The authors found that the concentration of the major compound (linalool - from 24.7 to 60.2%), was not the main feature for the activity against the microorganisms. The antimicrobial action depends on a sum of factors, such as synergism with important minor constituents that can increase the activity.

Silva et al. (2018) also tested the EOs of another species *Ocimum canum* and found the activity against both Gram-positive (*S. aureus*) and Gram-negative bacteria (*E. coli* and *Pseudomonas aeruginosa*), showing greater halo against *E. coli* strain. The MIC assay showed that oil showed interesting antimicrobial activity that can also be applied against *S. aureus*, being even more effective against *P. aeruginosa* when compared to standard eugenol.

Nobre et al. (2018) assessed EOs of *Orbignia speciosa* (babaçu) and *Mauritia flexuosa* (buriti) against Gram negative bacteria, where the major inhibitory activities were *E. coli* (MIC 32 μg/mL), *S. aureus* (MIC 256 μg/mL)—two multiresistant strains—and *E. coli* ATCC 25922 (MIC 512 μg/mL) for fixed oil of *O. speciosa*. For *M. flexuosa*, the greatest inhibitory activity was for *S. aureus* (MIC 256 μg/mL).

Mechanisms of antimicrobial action of plant extracts against pathogenic microorganisms of food origin

When microorganisms reach food products, they use the nutrients present in the food to grow and produce metabolites that will cause food to deteriorate. According to Kirk et al. (2017), foodborne illnesses can be considered a widespread problem of food security, which is caused by the consumption of contaminated food products, becoming a significant safety concern for public health.

Food loss due to microorganisms affects a wide variety of foods and turns into waste, even in developed countries. According to Gustavsson et al. (2011), it is possible to estimate annual food losses of up to 40% due to several factors, including deterioration by microorganisms. Most of the deteriorations observed in food occur due to the action of bacteria, yeasts and molds, which are responsible for the loss of a considerable number of foods and food products (Lianou et al. 2016). Figure 7.1 shows the action mechanisms of bioactive compounds to avoid microorganisms.

There is a considerable concern among consumers regarding the risk of using synthetic additives for human health, which has led to a decrease in the use of these chemicals in food preservation. Thus, the use of antimicrobial agents obtained from plant extracts against foodborne microorganisms is an excellent strategy for food preservation without the use of synthetic preservatives, many of which can already be related to the development of diseases such as cancer. However, the exact mechanism of action of these antimicrobial products is not well documented. There are countless studies focused on obtaining plant extracts with antimicrobial properties, which can be used to combat the emergence of diseases appearing from the consumption of food contaminated by pathogenic microorganisms (Suppakul et al. 2016; Clarke et al. 2017).

According to Ceruso et al. (2020), the number of cases of listeriosis in the United States and the United Kingdom is large, which can be minimized by the inhibitory effect of plant extracts on growth of the foodborne pathogen such as *Listeria monocytogenes*. A large part

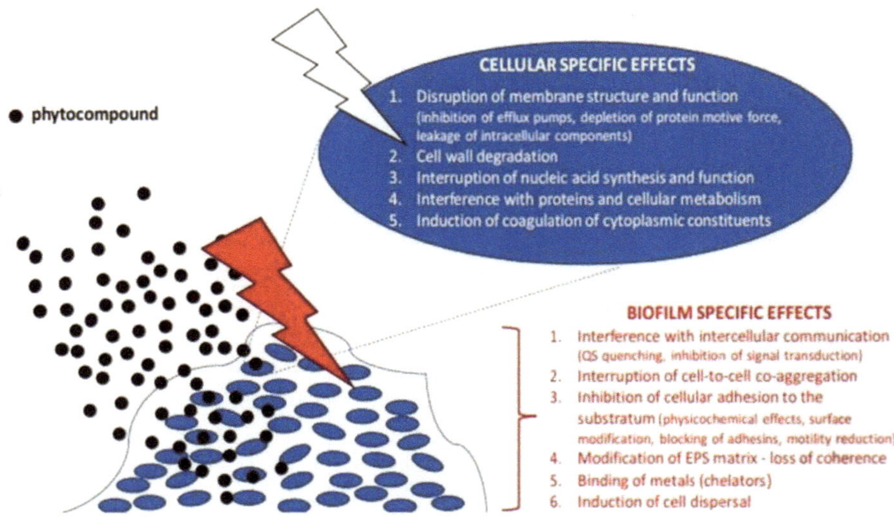

Figure 7.1. Mechanisms of action of phytocompounds in microorganism (biofilm) adapted from Sakarikou et al. (2020) with the permission of food research international.

of plant extracts shows in their composition the presence of phenolic compounds, which have plant protective activity against bacteria, viruses and fungi (Górniak et al. 2018; Burlacu et al. 2020). According to Ceruso et al. (2020), the antimicrobial activity against *Listeria monocytogenes* occurs due to the high content of phenolic compounds present in the evaluated plant extracts.

The use of such extracts in addition to having the inhibitory effect of foodborne pathogens is capable of promoting food preservation, thus being a natural food preservative, which has become a consumer preference when compared to synthetic preservatives. In the case of listeriosis outbreaks due to *L. monocytogenes*, foods such as meat, dairy products, and fresh products contain it in which this bacterium grows even in adverse environmental conditions, such as high salt concentration, low pH and low temperature (Allen et al. 2015; Rothrock et al. 2019; Bahrami et al. 2020).

According to Field and Lettinga (1992), the inhibition of bacteria by the action of phenolic compounds is considered complex, and may occur due to the chelating effect that these compounds have on iron, which is important for the survival of many bacteria. Phenolic compounds also act in conjunction with nonspecific forces, causing an effect on microbial membranes, adhesins, enzymes and transport proteins of the cell envelope, and polyphenols interact with proteins and/or phospholipids in the lipid bilayer, increasing membrane permeability, modifying ion transport processes and damaging cell membranes (Borges et al. 2013; Carraro et al. 2014; Liu et al. 2017; Ceruso et al. 2020).

In addition to *Listeria monocytogenes*, other microorganisms such as *E. coli*, *Pseudomonas aeruginosa* and *S. aureus* are inhibited by the activity of phenolic compounds, which can undergo irreversible changes in their membranes, such as ruptures and formation of intracellular pores (Borges et al. 2013). It is possible to verify that phenolic compounds can inhibit bacterial motility through the loss of flagella and inhibit the formation of biofilms, an effect observed by Carraro et al. (2014) for olive leaf extracts.

In another interesting study, Gonelimali et al. (2018) reported the importance of evaluating the internal pH (pHint) and the membrane potential when it is exposed to plant extracts, as this evaluation can provide a better understanding of the mechanisms of action of the antimicrobial activity of these extracts. According to the authors, the results indicated that the plant extracts significantly affected microorganism Gram-positive and Gram-negative cells, as there was a decline in pHint and hyperpolarization of the cell membrane. These plant extracts are of great value as natural antimicrobials and can be safely used as food preservatives against *Staphylococcus aureus* (SA) and *Escherichia coli* (EC).

Regarding existing studies, it is possible to verify from several researches a wide variety of studies based on vegetables known to popular consumption, for example, the raw extracts of cinnamon, garlic, basil, curry, ginger, sage, and mustard exhibit antimicrobial properties against a wide range of Gram-positive and Gram-negative bacteria (Alzoreky and Nakahara 2003; Castro et al. 2008).

Gonelimali et al. (2018) observed a decrease in cytoplasmic pH (pHint) and cell wall disruption in cells that were evaluated using plant extracts, this observation shows a possible mechanism of antibacterial action, indicating that plant extracts tested by the author could be used as natural preservatives in foods to eliminate or control the growth of pathogenic and spoilage of microorganisms.

It is possible to verify that the phenolic compounds can still have an antimicrobial effect through mechanisms such as modification in the normal functioning of the membrane, with alteration in the active transport and in the metabolic activities of fundamental enzymes, increase in the concentration of protons, with consequent direct reduction of the substrate pH and interruption of substrate transport by altering the permeability of the cell membrane (Davidson 1997). According to Ricke (2003) short-chain organic acids interfere with energy metabolism, altering the structure of the cytoplasmic membrane through interaction with membrane proteins.

According to Goni et al. (2009), EOs extracted from plant matrices demonstrate effective antimicrobial activity, as they produce structural and functional damage to the bacterial cell membrane. Furthermore, the authors found that antimicrobial compounds in plants can attack the cell membrane in the lipid bilayer part. Other compounds such as flavonoids have the ability to penetrate the cell membrane of bacteria (Davidson and Naidu 2000).

Plant extracts can act as an alternative when one wants to find new drugs against various infectious diseases caused by pathogenic microorganism. In recent years, there has been an alarming increase in the occurrence of dangerous and stronger antibiotics (Saleem et al. 2010). Most of the contamination can come through microorganisms grouped in the form of biofilms, which constitute a significant amount of all microbial and chronic infections in animals and humans, as well as in food spoilage (Jafri 2019).

Phytochemical compounds have been shown to influence the establishment and growth of bacterial biofilm, as well as bacterial adhesion, motility and quorum detection (QS). According to Borges et al. (2013) plant extracts, rich in flavonoids do not demonstrate the same problem as conventional antibiotics concerning bacterial resistance.

According to Gorniak et al. (2019), several research teams have reported that the flavonoids inhibit biofilms. Awolola et al. (2014) showed a significant antibiofilm activity of isovitexin (apigenin-6-C-glycoside 14), Epicatechin, and 5,7,4′-trihydroxiflavanol against *S. aureus* ATCC 29213 and El-Adawi (2012) observed a decrease of 55–66% in the biofilm formation of *S. mutans* after exposure to 2–15% Epicatechin. It is reported in research that flavonoids are significant topoisomerase inhibitors, which contribute to their antimicrobial activity (Gorniak et al. 2019).

In literature, studies that demonstrate the activity of flavonoids against pathogens of the plant itself are reported, which could be effectively applied against human pathogens. In addition, the antibacterial activities of many plant-derived flavonoids use different mechanisms than conventional drugs and, therefore, may be important in improving antibacterial therapy (Pandey and Kumar 2013). There is a wide variety of plant matrices that could be used to obtain plant extracts for antimicrobial purposes. In the Amazon region, this variety is even greater, due to the enormous variety of plants still unknown, such as the murucizeiro, an Amazonian fruit tree that shows high concentrations of phenolic compounds (Oliveira et al. 2020).

Studies have shown the significant action of flavonoids as topoisomerase inhibitors, contributing to their antimicrobial activity. According to Plaper et al. (2003), DNA gyrase is defined as an essential enzyme so that DNA replication can be performed, being exclusive to prokaryotes, and an attractive target for drugs with antimicrobial action. Quercetin, apigenin and sakuranet are flavonoids capable of inhibiting *Helicobacter pylori* 3-hydroxyacyl ACP dehydrase (Zhang et al. 2008). Eleven flavanones with different configurations of hydroxyl groups were evaluated to verify the antimicrobial activity on *E. faecalis* (Jeong et al. 2009) and many were shown to be efficient, mainly being naringenin and taxifolin. Mori et al. (1987) reported that some flavonoids can affect the DNA of microorganisms, and act in the inhibition of bacterial nucleic acid synthesis. They noted that the incubation with epigallocatechin gallate, myricetin and robinetin resulted in the reduction of DNA, RNA and protein synthesis by *Proteus vulgaris* and *S. aureus.*

Technologies for the extraction of antimicrobial compounds from medicinal plants

Medicinal plants are well known worldwide, for their powerful pharmacological action. In addition, among the functions of medicinal plants, the antimicrobial potential is widely investigated in the scientific community. Although some medicinal plants have a certain cytotoxicity, this can be avoided, if consumed correctly. Medicinal plants also play an important role in the treatment of diseases, and can serve as an alternative to the use of antibiotics, especially from traditional communities, such as indigenous, riverside, quilombola, settlements and even families in the urban area. Medicinal plants can be grown in backyards, or bought at street markets, in which the ways of using and acting on the plant are shared by family members or vendors (Ricardo et al. 2018).

Secondary metabolites from plants with antimicrobial potential are extracted both conventionally for example, solid-liquid extraction, maceration, pressing, infusion, steam distillation and soxhlet extraction, as well as by new extraction technologies. The difference is that conventional methods help in obtaining less sustainable volatile extracts and may have the aromatic components degraded at the end of the extraction, due to the high temperatures required in the process, whereas new extraction technologies, work exceptionally in obtaining quality extracts, while causing less damage to the environment. Therefore, they are an alternative to the industrial sector for obtaining secondary compounds with high added value.

The chemical and pharmaceutical industry is interested in new extraction technologies, which present current and innovative processes with a focus on obtaining high quality plant extracts and EOs containing bioactive ingredients. This highlights interesting technologies applied to the extraction of antimicrobial compounds from plants, such as super critical fluid technology, ultrasound-assisted technology and enzyme-assisted technology. Table 7.1 highlights the kinds of extraction, the main compounds extracted and some relevant data on the potential of medicinal plant antimicrobials.

The supercritical fluid technology is a unitary operation largely used in the extraction of secondary metabolites from the plants, including those that have antimicrobial action. It is a very interesting technology, mainly due to the selectivity of the extracts obtained.

Table 7.1. Brazilian medicinal plants tested against pathogenic microorganisms that cause foodborne diseases.

Extraction form	Plant	Extracted compounds	Microorganism	Minimum and maximum inhibitory concentration	Inhibition zone	References
Maceration and solid-liquid extraction	*Moringa stenopetala*	Oleic acid, tetradecanoic acid, octadecanoic acid, n-hexadecanoic acid, phenol, 2,4-bis (1,1-dimethyl ethyl), beta.-1-rhamnopyranoside, 5-o-acetyl-thio-octyl, di-(2-ethylhexyl) phthalate and phytol	*Staphylococcus aureus*	125 ~ 250 µg/ml	-	Manilal et al. 2020
Liquid-solid extraction	*Cnidoscolus quercifolius*	Lupeol-3β-O-cinnamate and lupeol β-O-dihydrocinnamate and phylacantone	*E. coli* *K. pneumoniae* *S. marcescens*	0,5 e 0,25 mg.ml	-	De Oliveira-Júnior et al. 2018
Solid-liquid extraction and agitation	*Brassica carinata A. Braun*	Alkaloids, flavonoids, terpenoids, estra-1,3,5 (10)-trien-17á-ol (4.09%), 9,12.15 octadecatrienoic acid, ethyl ester (2.95%), 9.12 acid, 15-octadecatrienoic, 2,3-dihydroxypropyl ester, (Z, Z, Z) (2.59%) 2-pentadecanone, 6,10,14-trimethyl (2.41%), 2,3-dihydroxypropyl ester (2.19%)	*Staphylococcus* *E. coli* spp.	-	15 mm 12 mm	Murugesan et al. 2020
Hydrodistillation	*Dracocephalum kotschyi*	Cyclohexylalene (52.63%) and limonene (35.88%), α-pinene (13.66%), (E)-citral (12.89%), neral (11.25%), methyl geranate (8.66%), limonene (8.33%), campholenal (6.22%) and geraniol (5.69%)	*Aspergillus brasiliensis*	31,25 µg/ml	-	Ghavam et al. 2020
Hydrodistillation	*Piper betle* L.	eugenol (46.02%) estragole (15.12%), linalool (11.62%) α-copaene (6.54%) and anethole (2.62% and 2.65%)	*Mycobacterium smegmatis*	0,5 e 1 mL	-	Madhumita et al. 2020
Super critical fluid	*Campomanesia xanthocarpa*	β-Caryophyllene (11,67%), α-Cadinol (7,17%), β-Selinene (5,36%), Linalool (5,09%),	*Pseudomonas aeruginosa*	5,0 mg mL^{-1}	-	Capeletto et al. 2020
Microwave	*Satureja montana*	Carvacrol (825–950 µg/mg)	*Escherichia coli ATCC 25922,* *Salmonella enterica sv Anatum SF2* *Staphylococcus aureus ATCC 6538*	25 µg/mL 250 µg/mL 150 µg/mL	-	Santos et al. 2020

To obtain extracts on both the laboratory and industrial scale, sample preparation and characterization is necessary. Initially, the leaves are dried at temperatures up to 40°C and ground to improve the penetration power of the solvent in the matrix, then the sample is characterized in terms of its particle size and actual density. After deciphering the characteristics of the raw materials, the extraction bed is set up, with conditions that can vary in terms of temperature, pressure, granulometry, extraction time and types of solvents, this earlier method will influence the yields and characteristics of the compound obtained, in addition to greater selectivity of extraction and shorter processing time, making this technology a promising alternative in the recovery of antimicrobial compounds present in medicinal plants.

In a research carried out by Silva et al. (2020), the leaves of *Lippia thymoides*, popularly known as bush rosemary, were used for supercritical extraction of EOs, the optimized condition showed a yield of 1.63 ± 0.01%, at a temperature of 50°C and a pressure of 300 bar, an EOs identified by gas chromatography, mass spectrometry, showed 88.56% of thymol, a substance belonging to the group of terpenes. Thymol has anti-inflammatory and antiseptic properties reported in literature, in addition to being a potential compound with antimicrobial action against the strains of *C. albicans* and *C. tropicalis*, with inhibition halos of 10 mm and 8 mm, respectively.

Another very interesting technology applied to the extraction of antimicrobial compounds from medicinal plants is the extraction by microwaves. Microwaves emit radiations from the electromagnetic spectrum that vary in frequency from 300 MHz (radio radiation) to 300 GHz, an optimized condition of microwave extraction depends mainly on the frequency emitted and the ability of the plant matrix to absorb energy and dissipate heat, this is a basic and necessary characteristic for the material to be subjected to microwave extraction. Plants, in turn, are good absorbent materials; consequently, they are able to heat and extract compounds efficiently.

Wei et al. (2018), extracted *Salvia officinalis* EOs using the solventless microwave extraction system and with double cooling. The *S. officinalis* leaves were initially immersed in 71% water for 1 hour followed by microwave extraction, at a power of 650 watt for 35 minutes, after being subjected to different humidity conditions, extraction time and irradiation power. The authors used a conventional collector and a cooled collector as a way to decrease the temperature of the EOs obtained in the collector during the extraction process. This extraction system resulted in yield of 1.630% of EOs consisting of 68.00% of oxygenated monoterpenes, 6.75% of monoterpenes and 6.78% of sesquiterpene hydrocarbons. It showed antimicrobial activity against *B. subtillis* and *S. epidermidis* strains, with 0.16 mg/mL and 0.31 mg/mL of concentrations that were considered as minimal, respectively. The authors concluded that the combination of the cooling system coupled with microwave extraction technology enhanced the extraction, allowing better yield and quality in the composition of EOs extracts.

Current literature data highlights enzyme-assisted technology, and its potential for extracting antimicrobial compounds from medicinal plants (Chotphruethipong et al. 2019; Qadir et al. 2019; Shimotori et al. 2020). Enzymes can also be used in the pre-treatment stage of extraction for both lipophilic and hydrophilic compounds. In addition, they have advantages such as reduced time, ability to extract at low temperatures, have high specificity, because even the molecules located inside the cytoplasmic matrix, organelles and cell wall of the plant can be extracted Fig. 7.2. In addition, one can observe a decrease in the time of the filtration and purification steps. Extraction assisted by enzymes in conjunction with other extraction technologies is capable of reaching maximum levels of yield; the extraction process does not require numerous stages of installations compared

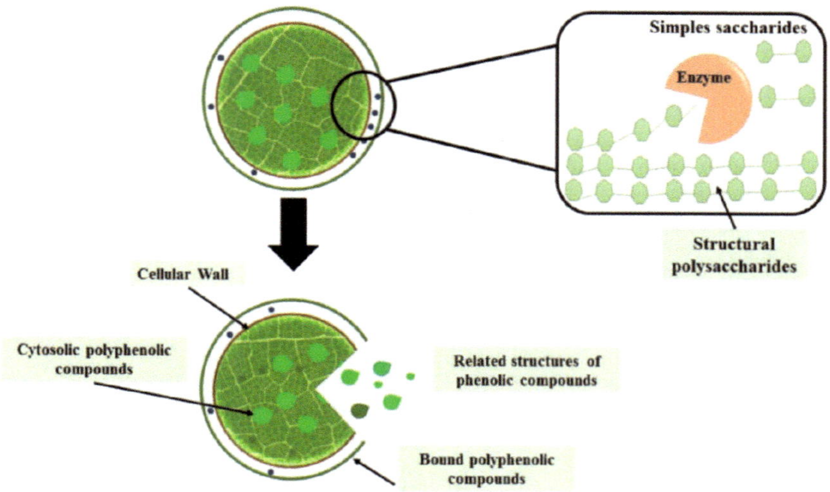

Figure 7.2. Process of enzymatic degradation of plant cell walls and the release of phenolic compounds. Adapted from Gligor et al. (2019) with the permission of Trends in Food Science and Technology.

to other industrial plants. However, to obtain a satisfactory result, one should consider the parameters of moisture and granulometry of the plant matrix, and also observe the specificity and quantity of enzyme (Gligor et al. 2019).

Conclusion

The Amazon is very rich in its biodiversity, with products of plant origin that have high antioxidant and antimicrobial potential. Moreover, one of the applications of these extracts proved to be effective in their antimicrobial action against pathogenic microorganisms that cause foodborne disease. However, more research needs to be carried out to consolidate these results. Since, the use of food additives of natural origin would be viable as they do not cause any harm to health. They are able to conserve food for a longer time and avoid the growth of foodborne microorganisms in these matrices. Therefore, the natural plant extracts could be an alternative for this use, keeping in mind that good handling practices must be maintained in all stages of food processing.

Acknowledgment

The authors acknowledge the institutions, Federal Rural University of Amazonia, Federal University of Pará, Federal University of Sao Paulo, and SGB Amravati University for supporting research and for the access to didactic materials and scientific papers.

MR thankfully acknowledges the Polish National Agency for Exchange (NAWA) for financial support to visit the Department of Microbiology, Nicolaus Copernicus University, Toruń, Poland.

References

Abdel-Fattah, S.M., Badr, A.N., Ali, S.M., Hassan, R.A. (2018). Antifungal and antimycotoxigenic impact of eco-friendly extracts of wild stevia. J Biol Sci 18: 488–499.

Aik, J., Turner, R.M., Kirk, M.D., Heywood, A.E., Newall, A.T. (2020). Evaluating food safety management systems in Singapore: A controlled interrupted time-series analysis of foodborne disease outbreak reports. Food Control 117: 107324.

Allen, K., Wałecka-Zacharska, E., Kosek-Paszkowska, K., Devlieghere, F., Van Meervenne, E., Osek, J., Wieczorek, K., Bania, J. (2015). *Listeria monocytogenes*—An examination of food chain factors potentially contributing to antimicrobial resistance. Food Microbiol 54: 178–189.

Alzoreky, N.S., Nakahara, K. (2003). Antibacterial activity of extracts from some edible plants commonly consumed in Asia. Int J Food Microbiol 80: 223–230.

Aminzare, M., Hashemi, M., Hassanzad Azar, H., Hejazi, J. (2016). The use of herbal extracts and essential oils as a potential antimicrobial in meat and meat products: a review. J Hum Environ Health Promot 2: 63–74.

Andrade, M.L., Stedefeldt, E., Zanin, L.M., da Cunha, D.T. (2020). Food safety culture in food services with different degrees of risk for foodborne diseases in Brazil. Food Control 112: 107152.

Appiah, G.D., Chung, A., Bentsi-Enchill, A.D., Kim, S., Crump, J.A., Mogasale, V., Slayton, R.B., Mintz, E.D. (2020). Typhoid outbreaks, 1989–2018: implications for prevention and control. Am J Trop Med Hyg 102: 1296.

Awolola, G.V., Koorbanally, N.A., Chenia, H., Shode, F.O., Baijnath, H. (2014). Antibacterial and anti-biofilm activity of flavonoids and triterpenes isolated from the extracts of *Ficus Sansibarica warb.* Subsp. *Sansibarica* (Moraceae) extracts. Afr J Tradit Complement Altern Med 11: 124–131.

Bahrami, A., Baboli, Z.M., Schimmel, K., Jafari, S.M., Williams, L. (2020). Efficiency of novel processing technologies for the control of *Listeria monocytogenes* in food products. Trends Food Sci Technol 96: 61–78.

Bajpai, V.K., Baek, K.H., Kang, S.C. (2012). Control of Salmonella in foods by using essential oils: A review. Food Res Int 45: 722–734.

Baldim, J.L., Silveira, J.G.F., Almeida, A.P., Carvalho, P.L.N., Rosa, W., Schripsema, J., Chagas-Paula, D.A., Soares, M.G., Luiz, J.H.H. (2018). The synergistic effects of volatile constituents of *Ocimum basilicum* against foodborne pathogens. Ind Crops Prod 112: 821–829.

Bezerra, J.W.A., Costa, A.R., de Freitas, M.A., Rodrigues, F.C., de Souza, M.A., da Silva, A.R.P., Lima, J.R.S., Morais-Braga, M.F.B. (2019). Chemical composition, antimicrobial, modulator and antioxidant activity of essential oil of *Dysphania ambrosioides* (L.) Mosyakin & Clemants. Comp Immunol Microbiol Infect Dis 65: 58–64.

Bhavaniramya, S., Vishnupriya, S., Al-Aboody, M.S., Vijayakumar, R., Baskaran, D. (2019). Role of essential oils in food safety: Antimicrobial and antioxidant applications. Grain Oil Sci Technol 2: 49–55.

Borges, A., Ferreira, C., Saavedra, M.J., Simões, M. (2013). Antibacterial activity and mode of action of ferulic and gallic acids against pathogenic bacteria. Microb Drug Resist 19: 256–265.

Brasil. (2016). Doenças Transmitidas por Alimentos. 2016. Retrieved June 29, 2016. http://portalsaude.saude.gov.br/images/pdf/2016/marco/10/Apresenta——oM.L.dadosgerais-DTA-2016.pdf > site.

Brasil. (2019). Surtos de Doenças Transmitidas por Alimentos no Brasil Informe 2018. Retrieved June 04, 2021. https://www.paho.org/pt/brasil>site.

Burlacu, E., Nisca, A., Tanase, C. (2020). A comprehensive review of phytochemistry and biological activities of *Quercus* species. Forests 11(9): 904.

Burt, S. (2004). Essential oils: their antibacterial properties and potential applications in foods—a review. Int J Food Microbiol 94: 223–253.

Câmara Flores, A.M.P., de Melo, C.B. (2015). Principais bactérias causadoras de doenças de origem alimentar. Ver Bras Med Vet 37(1): 65–72.

Capeletto, C., Conterato, G., Scapinello, J., Rodrigues, F.S., Copini, M.S., Kuhn, F., Dal, M.J., Oliveira, J.V. (2016). Chemical composition, antioxidant and antimicrobial activity of guavirova (*Campomanesia xanthocarpa* Berg) seed extracts obtained by supercritical CO_2 and compressed n-butane. J Supercrit Fluids 110: 32–38.

Carraro, L., Fasolato, L., Montemurro, F., Martino, M.E., Balzan, S., Servili, M., Novelli, E., Cardazzo, B. (2014). Polyphenols from olive mill waste affect biofilm formation and motility in *Escherichia coli* K-12. Microb Biotechnol 7: 265–275.

Castro, S.B.R., Leal, C.A., Freire, F.R., Carvalho, D.A., Oliveira, D.F., Figueiredo, H.C.P. (2008). Antibacterial activity of plant extracts from Brazil against fish pathogenic bacteria. Braz J Microbiol 39: 756–760.

Ceruso, M., Clement, J.A., Todd, M.J., Zhang, F., Huang, Z., Anastasio, A., Pepe, T., Liu, Y. (2020). The inhibitory effect of plant extracts on growth of the foodborne pathogen, *Listeria monocytogenes*. Antibiot 9: 319.

Chotphruethipong, L., Benjakul, S., Kijroongrojana, K. (2019). Extração assistida por ultrassom de fenólicos antioxidantes das folhas de cajueiro (*Anacardium occidentale* L.). J Food Sci Technol 56: 1785–1792.

Clarke, D., Tyuftin, A.A., Cruz-Romero, M.C., Bolton, D., Fanning, S., Pankaj, S.K., Bueno-Ferrerd, C., Cullen, J.P. (2017). Surface attachment of active antimicrobial coatings onto conventional plastic-based laminates and performance assessment of these materials on the storage life of vacuum packaged beef sub-primals. Food Microbiol 62: 196–201.

Coutinho, J.G., Cardoso, A.J.C., Toral, N., Silva, A.C.F.D., Ubarana, J.A., Aquino, K.K.N.D.C.D., Fagundes, A., Vasconcellos, A.B. (2009). A organização da Vigilância Alimentar e Nutricional no Sistema Único de Saúde: histórico e desafios atuais. Rev Bras Epidemiol 12: 688–699.

Davidson, P.M. (1997). Chemical preservatives and natural antimicrobial compounds. *In*: Doyle, M.P., Beuchat, L.R., Montville, T.J. (eds.). Food Microbiology: Fundamentals and Frontiers. ASM Press, Washington DC. Intake. 1997.

Davidson, P.M., Naidu, A.S. (2000). Phyto-phenols. *In*: Naidu, A.S. (ed.). Natural Food Antimicrobial Systems. Boca Raton, FL: CRC Press. Intake. 2000.

de Oliveira-Júnior, R.G., Ferraz, C.A.A., Pontes, M.C., Cavalcante, N.B., da Cruz Araujo, E.C., de Oliveira, A.P., Picot, L., Rolim, L.A., da Silva Almeida, J.R.G. (2018). Antibacterial activity of terpenoids isolated from *Cnidoscolus quercifolius* Pohl (Euphorbiaceae), a Brazilian medicinal plant from Caatinga biome. Eur J Integr Med 24: 30–34.

Diaz, M.A., Rossi, C.C., Mendonça, V.R., Silva, D.M., Ribon, A.D.O., Aguilar, A.P., Muñoz, G.D. (2010). Screening of medicinal plants for antibacterial activities on *Staphylococcus aureus* strains isolated from bovine mastitis. Rev Bras Farmacogn 20: 724–728.

Du, R., Qu, Y., Qi, P.X., Sun, X., Liu, Y., Zhao, M. (2020). Natural flagella-templated Au nanowires as a novel adjuvant against *Listeria monocytogenes*. Nanoscale 12: 5627–5635.

Duarte, M.C.T., Leme, E.E., Delarmelina, C., Soares, A.A., Figueira, G.M., Sartoratto, A. (2007). Activity of essential oils from Brazilian medicinal plants on *Escherichia coli*. J Ethnopharmacol 111: 197–201.

El-Adawi, H. (2012). Inhibitory effect of grape seed extract (GSE) on cariogenic bacteria. J Med Plants Res 6: 4883–4891.

Field, J.A., Lettinga, G. (1992). Toxicity of tannic compounds to microorganisms. *In*: Hemingway, R.W., Laks, P.E. (eds.). Plant Polyphenols. Springer US: London, UK. Intake. 1992.

Ghavam, M., Manconi, M., Manca, M.L., Bacchetta, G. (2021). Extraction of essential oil from Dracocephalum kotschyi Boiss. (Lamiaceae), identification of two active compounds and evaluation of the antimicrobial properties. Ethnopharmacol 267: 113513.

Gligor, O., Mocan, A., Moldovan, C., Locatelli, M., Crişan, G., Ferreira, I.C. (2019). Enzyme-assisted extractions of polyphenols—A comprehensive review. Trends in Food Science & Technology 88: 302–315.

Gonelimali, F.D., Lin, J., Miao, W., Xuan, J., Charles, F., Chen, M., Hatab, S.R. (2018). Antimicrobial properties and mechanism of action of some plant extracts against food pathogens and spoilage microorganisms. Front Microbiol 9: 1639.

Goni, P., Lopez, P., Sanchez, C., Gomez-Lus, R., Becerril, R., Nerín, C. (2009). Antimicrobial activity in the vapour phase of a combination of cinnamon and clove essential oils. Food Chem 116: 982–989.

Górniak, I., Bartoszewski, R., Króliczewski, J. (2019). Comprehensive review of antimicrobial activities of plant flavonoids. Phytochem Rev 18(1): 241–272.

Greig, J.D. and Ravel, A. (2009). Analysis of foodborne outbreak data reported internationally for source attribution. Int J Food Microbiol 130(2): 77–87.

Gustavsson, J., Cederberg, C., Sonesson, U., Otterdijk, R., Maybeck, A. (2011). Global Food Losses and Food Waste: Extent, Causes, and Prevention. Düsseldorf: FAO.

Hughes, C., Gillespie, I.A., O'Brien, S.J. (2007). Foodborne transmission of infectious intestinal disease in England and Wales, 1992–2003. Food Control 18: 766–772.

Jafri, H. (2019). Mixed species biofilms of *Candida albicans* and bacterial pathogens and interference of biofilms by selected anti-infective plant extracts/compounds (Doctoral dissertation, Aligarh Muslim University).

Jeong, K.W., Lee, J.Y., Kang, D.I., Lee, J.U., Shin, S.Y., Kim, Y. (2009). Screening of flavonoids as candidate antibiotics against *Enterococcus faecalis*. J Nat Prod 72: 719–724.

Karam, L., Roustom, R., Abiad, M.G., El-Obeid, T., Savvaidis, I.N. (2019). Combined effects of thymol, carvacrol and packaging on the shelf-life of marinated chicken. Int J Food Microbiol 291: 42–47.

Kirk, M.D., Pires, S.M., Black, R.E., Caipo, M., Crump, J.A., Devleesschauwer, B., Hall, A.J. (2015). World Health Organization estimates of the global and regional disease burden of 22 foodborne bacterial, protozoal, and viral diseases, 2010: a data synthesis. PLoS Med 12: e1001921.

Kirk, M.D., Angulo, F.J., Havelaar, A.H., Black, R.E. (2017). Diarrhoeal disease in children due to contaminated food. Bull World Health Organ 95: 233–234.

Kottwitz, L.B.M., de Oliveira, T.C.R.M., Alcocer, I., Farah, S.M.D.S.S., Abrahão, W.S.M., dos Prazeres Rodrigues, D. (2010). Avaliação epidemiológica de surtos de salmonelose ocorridos no período de 1999 a 2008 no Estado do Paraná, Brasil. Acta Sci Health Sci 32(1): 9–15.

Li, W., Wu, S., Fu, P., Liu, J., Han, H., Bai, L., Pei, X., Li, N., Liu, X., Guo, Y. (2018). National molecular tracing network for foodborne disease surveillance in China. Food Control 88: 28–32.

Lianou, A., Panagou, E.Z., Nychas, G.J.E. (2016). Microbiological spoilage of foods and beverages. The Stability and Shelf Life of Food. 3–42. doi: 10.1016/B978-0-08-100435-7.00001-0.

Liu, Y., McKeever, C.L., Malik, N.S.A. (2017). Assessment of the antimicrobial activity of olive leaf extract against foodborne bacterial pathogens. Front Microbiol 8: 113.

Madhumita, M., Guha, P., Nag, A. (2019). Extraction of betel leaves (*Piper betle* L.) essential oil and its bio-actives identification: Process optimization, GC-MS analysis and anti-microbial activity. Ind Crops Prod 138: 111578.

Makhuvele, R., Naidu, K., Gbashi, S., Thipe, V.C., Adebo, O.A., Njobeh, P.B. (2020). The use of plant extracts and their phytochemicals for control of toxigenic fungi and mycotoxins. Heliyon 6(10): e05291.

Manilal, A., Sabu, K.R., Shewangizaw, M., Aklilu, A., Seid, M., Merdekios, B., Tsegaye, B. (2020). *In vitro* antibacterial activity of medicinal plants against biofilm-forming methicillin-resistant *Staphylococcus aureus*: efficacy of Moringa stenopetala and Rosmarinus officinalis extracts. Heliyon 6(1): e03303.

Marques, J.L., Volcão, L.M., Funck, G.D., Kroning, I.S., Silva, W.P., Fiorentini, A.M., Ribeiro, G.A. (2015). Antimicrobial activity of essential oils of *Origanum vulgare* L. and *Origanum majorana* L. against *Staphylococcus aureus* isolated from poultry meat. Ind Crops Prod 77: 444–450.

Martins, N., Barros, L., Ferreira, I.C.F.R. (2016). *In vivo* antioxidant activity of phenolic compounds: Facts and gaps. Trends Food Sci Technol 48: 1–12.

Mendonça, L.P., Costa, M.G.A., Macedo, R.C.B.S., Júnior, R.A.S., Santana, F.E.O., Silva, L.R.C. (2020). Emerging foodborne diseases: an integrative review. Rev Bra Edu Saúde 10(3): 1–6.

Millezi, A.F., Baptista, N.N., Caixeta, D.S., Rossoni, D.F., Cardoso, M.G., Piccoli, R.H. (2014). Caracterização química e atividade antibacteriana de óleos essenciais de plantas condimentares e medicinais contra *Staphylococcus aureus* e *Escherichia coli*. Rev Bras de Plantas Medicinais 16(1): 18–24.

Mori, A., Nishino, C., Enoki, N., Tawata, S. (1987). Antibacterial activity and mode of action of plant flavonoids against *Proteus vulgaris* and *Staphylococcus aureus*. Phytochem 26: 2231–2234.

Murugesan, K., Mulugeta, K., Hailu, E., Tamene, W., Yadav, S.A. (2020). Insights for integrative medicinal potentials of Ethiopian Kale (*Brassica carinata*): Investigation of antibacterial, antioxidant potential and phytocompounds composition of its leaves. Chin Herb Med 13(2): 250–254.

Naves, V.M.L., dos Santos, M.H., Ribeiro, I.S., da Silva, C.A., Silva, N.C., Silva, M.A., da Silva, G.A., Dias, A.L.T., Ionta, M., Dias, D.F. (2019). Antimicrobial and antioxidant activity of *Garcinia brasiliensis* extracts. S Afr J Bot 124: 244–250.

Nobre, C.B., de Sousa, E.O., de Lima Silva, J.M.F., Melo Coutinho, H.D., Martins da Costa, J.G. (2018). Chemical composition and antibacterial activity of fixed oils Mauritia flexuosa and Orbignia speciosa associated with aminoglycosides. Eur J Integr Med 23: 84–89.

O'Brien, S.J. (2014). Foodborne diseases: prevalence of foodborne diseases in Europe. Encyclopedia of Food Safety, 302–311. https://doi.org/10.1016/B978-0-12-378612-8.00077-9.

Oliveira, A.B.A., Paula, C.M.D., Capalonga, R., Cardoso, M.R.I.C., Tondo, E.C. (2010). Foodborne diseases, main etiologic agents and general aspects: a review. Rev HCPA 30(3): 279–85.

Oliveira, J.A.R., Komesu, A., Martins, L.H.S., Rogez, H., Pena, R.S. (2020). Enzyme-assisted extraction of phenolic compounds from murucizeiro leaves (*Byrsonima crassifolia*). Sci Plena 16(5): 2–9.

Orlanda, J.F.F., Nascimento, A.R. (2015). Chemical composition and antibacterial activity of *Ruta graveolens* L. (Rutaceae) volatile oils, from São Luís, Maranhão, Brazil. S Afr J Bot 99: 103–106.

Orsuwan, A., Sothornvit, R. (2018). Active banana flour nanocomposite films incorporated with garlic essential oil as multifunctional packaging material for food application. Food Bioprocess Technol 11(6): 1199–1210.

Pandey, A.K., Kumar, S. (2013). Perspective on plant products as antimicrobials agents: a review. Pharmacol 4: 469–480.

Pateiro, M., Munekata, P.E.S., Sant'Ana, A.S., Domínguez, R., Rodríguez-Lázaro, D., Lorenzo, J.M. (2020). Application of essential oils as antimicrobial agents against spoilage and pathogenic microorganisms in meat products. Int J Food Microbiol 337: 108966.

Perigo, C.V., Torres, R.B., Bernacci, L.C., Guimarães, E.F., Haber, L.L., Facanali, R., Vieira, M.A.R., Quecini, V., Marques, M.O.M. (2016). The chemical composition and antibacterial activity of eleven *Piper* species from distinct rainforest areas in Southeastern Brazil. Ind Crops Prod 94: 528–539.

Plaper, A., Golob, M., Hafner, I., Oblak, M., Solmajer, T., Jerala, R. (2003). Characterization of quercetin binding site on DNA gyrase. Biochem Biophys Res Commun 306: 530–536.

Ponzilacqua, B., Rottinghaus, G.E., Landers, B.R., Oliveira, C.A.F. (2019). Effects of medicinal herb and Brazilian traditional plant extracts on *in vitro* mycotoxin decontamination. Food Contr 100: 24–27.

Prakash, B., Kedia, A., Mishra, P.K., Dubey, N.K. (2015). Plant essential oils as food preservatives to control moulds, mycotoxin contamination and oxidative deterioration of agri-food commodities—Potentials and challenges. Food Contr 47: 381–391.

Qadir, R., Anwar, F., Gilani, M.A., Yaqoob, M.N., Ahmad, M. (2019). Enzyme-assisted extraction for optimized recovery of phenolic bioactives from *Peganum hermala* leaves using response surface methodology. Curr Top Nutraceutical Res 17(4): 349–354.

Radaelli, M., Silva, B.P., Weidlich, L., Hoehne, L., Flach, A., Costa, L.A.M.A., Ethur, E.M. (2016). Antimicrobial activities of six essential oils commonly used as condiments in Brazil against *Clostridium perfringens*. Braz J Microbiol 46: 424–430.

Ricardo, L.M., Dias, B.M., Mügge, F.L., Leite, V.V., Brandão, M.G. (2018). Evidence of traditionality of Brazilian medicinal plants: The case studies of *Stryphnodendron adstringens* (Mart.) Coville (barbatimão) barks and *Copaifera* spp. (copaíba) oleoresin in wound healing. J Ethnopharmacol 219: 319–336.

Rice-Evans, C., Miller, N., Paganga, G. (1997). Antioxidant properties of phenolic compounds. Trend Plant Sci 2: 152–159.

Ricke, S.C. (2003). Perspectives on the use of organic acids and short chain fatty acids as antimicrobials. Poult Sci 82(4): 632–639.

Rothrock, M.J., Micciche, A.C., Bodie, A.R., Ricke, S.C. (2019). *Listeria* occurrence and potential control strategies in alternative and conventional poultry processing and retail. Front Sustain Food Syst 3: 33.

Sakarik, O., Mocan, A., Moldovan, C., Locatelli, M., Crişan, G., Ferreira, I.C. (2019). Enzyme-assisted extractions of polyphenols—A comprehensive review. Trends Food Sci Technol 88: 302–315.

Sakarikou, C., Kostoglou, D., Simões, M., Giaouris, E. (2020). Exploitation of plant extracts and phytochemicals against resistant Salmonella spp. in biofilms. Food Res Int 128: 108806.

Saleem, M., Nazir, M., Ali, M.S., Hussain, H., Lee, Y.S., Riaz, N., Jabbar, A. (2010). Antimicrobial natural products: an update on future antibiotic drug candidates. Nat Prod Rep 27: 238–25.

Salgueiro, L.R., Vila, R., Tomi, F., Figueiredo, A.C., Barroso, J.G., Canigueral, S., Casanova, J., Proença, A.C., Adzet, T. (1997). Variability of essential oils of *Thymus caespititius* from Portugal. Phytochem 45: 307–311.

Sandri, I.G., Zacaria, J., Fracaro, F., Delamare, A.P.L., Echeverrigaray, S. (2007). Antimicrobial activity of the essential oils of Brazilian species of the genus Cunila against foodborne pathogens and spoiling bacteria. Food Chem 103: 823–828

Santos, J.D., Coelho, E., Silva, R., Passos, C.P., Teixeira, P., Henriques, I., Coimbra, M.A. (2019). Chemical composition and antimicrobial activity of *Satureja montana* byproducts essential oils. Ind Crops Prod 137: 541–548.

Shimotori, Y., Watanabe, T., Kohari, Y., Chiou, T.C., Ohtsu, N., Nagata, Y., Murata, M. (2020). Extração assistida por enzima de fitoquímicos bioativos da hortelã-pimenta japonesa (*Mentha arvensis* L. cv.'Hokuto '). J Oleo Sci 56: ess19181.

Silva, M.P., Lavinas, F.C. (2010). Virose alimentar: microbiologia das principais doenças de origem alimentar transmitidas por vírus. Saúde & Ambiente em Revista 5(1): 33–43.

Silva, S.A.N.M., Barros, A.B., Souza, J.M.T., Moura, A.F., Araújo, A.R., Mendes, M.G.A., Daboit, T.C., Silva, D.A., Araújo, A.J., Filho, J.D.B.M. (2020). Phytochemical and biological prospection of *Mimosa* genus plants extracts from Brazilian northeast. Phytochem Lett 39: 173–181.

Silva, V.D., Almeida-Souza, F., Teles, A.M., Neto, P.A., Mondego-Oliveira, R., Filho, N.E.M., Taniwaki, N.N., Abreu-Silva, A.L., Calabrese, K.S., Filho, V.E.M. (2018). Chemical composition of *Ocimum canum* Sims. essential oil and the antimicrobial, antiprotozoal and ultrastructural alterations it induces in Leishmania amazonensis promastigotes. Ind Crops Prod 119: 201–208.

Sirtoli, D.B., Comarella, L. (2018). O papel da vigilância sanitária na prevenção das doenças transmitidas por alimentos (DTA). Rev Saúd e Desenv 12(10): 197–209.

Suppakul, P., Thanathammathorn, T., Samerasut, O., Khankaew, S. (2016). Shelf life extension of "fios de ovos", an intermediate-moisture egg-based dessert, by active and modified atmosphere packaging. Food Contr 70: 58–63.

Tsuchiya, H., Iinuma, M. (2000). Reduction of membrane fluidity by antibacterial sophoraflavanone G isolated from *Sophora exigua*. Phytomed 7: 161–165.

Uzel, A., Sorkun, K., Önçağ, Ö., Çoğulu, D., Gençay, Ö., Salih, B. (2005). Chemical compositions and antimicrobial activities of four different Anatolian propolis samples. Microbiol Res 160: 189–195.

Valeriano, C., Piccoli, R.H., Cardoso, M.G., Alves, E. (2012). Antimicrobial activity of essential oils against sessile and planktonic pathogens of food source. Rev Bras Plantas Medicinais 14(1): 57–67.

Wei, Z.F., Zhao, R.N., Dong, L.J., Zhao, X.Y., Su, J.X., Zhao, M., Li, L., Bian, J.B., Zhang, L.J. (2018). Dual-cooled solvent-free microwave extraction of *Salvia officinalis* L. essential oil and evaluation of its antimicrobial activity. Ind Crops Prod 120: 71–76.

Zhang, L., Kong, Y., Wu, D., Zhang, H., Wu, J., Chen, J., Ding, J., Hu, L., Jiang, H., Shen, X. (2008). Three flavonoids targeting the beta-hydroxyacyl-acyl carrier protein dehydratase from *Helicobacter pylori*: crystal structure characterization with enzymatic inhibition assay. Protein Sci 17: 1971–1978.

8

Recent Discoveries of Natural Products as Antimicrobial Alternatives for Bovine Mastitis Treatment

Pâmella B. A. Domingues,[1] *João Paulo L. Morgado,*[2]
Maria Aparecida S. Moreira,[3] *Valdir F. Veiga-Júnior*[2,*] and *Fábio A. Pieri*[1,*]

Introduction

In the context of veterinary medicine, mastitis is an inflammation that affects the parenchyma of the mammary gland, harming the animal's health and altering milk production (Barros et al. 2017). This disease is complex and multifactorial, mainly caused by bacteria, fungi and other pathogens that infect mammary epithelial cells (Gong et al. 2018). The bacterium, *Staphylococcus aureus*, causes subclinical and clinical mastitis and has the potential for contagious spread to other cows in the herd. This bacterium has high virulence and great invasion power in cells of the mammary gland, it is persistent and can replicate within the host cell, as well as produce toxins and hydrolytic enzymes that suppress host defences and damage the cells and tissues of the host (Mordmuang et al. 2019). *Escherichia coli* is a Gram-negative pathogenic bacterium that causes environmental mastitis. With the presence of lipopolysaccharides (LPS) on its surface, it can induce an exacerbated inflammatory response through pro-inflammatory mediators such as Tumour Necrosis Factor alpha (TNF-α), interleukin-6 (IL-6), and interleukin-1β (IL-1β) (Gong et al. 2018). A summary representation of the pathophysiology of mastitis can be seen in Fig. 8.1.

According to clinical manifestations, mastitis can be classified into clinical or subclinical mastitis. Clinical mastitis in cattle affects the udder, causing inflammation with signs such as heat, swelling, discolouration and abnormal discharge. The infected animal may experience fever, loss of appetite and even death. In most cases of subclinical mastitis, there are no signs of inflammation or infection, but laboratory tests on milk samples enable the detection of high somatic cell counts, predominantly neutrophils, due to the host's immune response to the infection, or positive gelation in the samples caused by the DNA

[1] Life Sciences Institute, Federal University of Juiz de Fora-Campus Governador Valadares, Rua São Paulo, 745, Centro, Governador Valadares, MG, CEP 35010180, Brazil.
[2] Chemical Engineering Section, Military Institute of Engineering (IME), Praça General Tibúrcio, 80, Praia Vermelha, Urca, CEP: 22290-270, Rio de Janeiro, RJ, Brazil.
[3] Veterinary Department, Universidade Federal de Viçosa (UFV), Peter Henry Rolfs Avenue, Viçosa - MG, Zip Code 36570-900, Brazil.
* Corresponding authors: valdir.veiga@gmail.com; pierifabio@hotmail.com

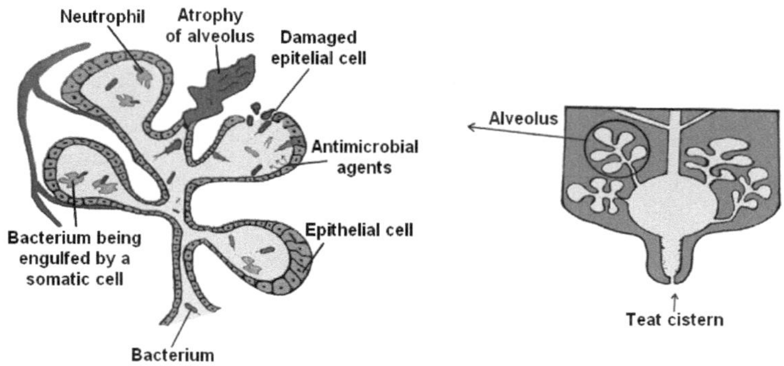

Figure 8.1. Mastitis physiopathology: the sphincter and the keratinized epithelium of the roof are part of the first line of defence against invading bacteria; if the bacteria are not destroyed by the immune system, they continue to multiply and invade the smaller ducts and alveolar areas of the breast tissue. Milk-secreting cells damaged by bacteria toxins and other irritants release substances that lead to increased permeability of blood tissues. Leukocytes move to the site of the infection, entering the alveolar tissue in large numbers. Fluids, minerals and clotting factors also penetrate the affected area. Milk clots can close the ducts, isolating the infected regions. As the infection persists and the ducts remain clogged, the trapped milk causes the secretory cells to revert to a state of rest (non-producing) and the alveolus begins to shrink.

of infiltrating somatic cells in the test referred to as the California Mastitis Test (CMT) (Pumipuntu et al. 2019).

Mastitis has a negative effect on the economy. To dairy farms, in addition to a reduction in milk production and quality, the disease also affects the fertility of cows due to irregular or extended oestrus, resulting in delivery problems. A herd affected by mastitis requires treatment, intervention and infection control, and labour, all of which come at a high cost to the farm; furthermore, there may even be loss of the animals (Pumipuntu et al. 2019).

Although subclinical mastitis is more prevalent, the economic impact is more difficult to quantify due to the variability and intensity of the disease, in addition to going unnoticed in some cases. The incidence of the disease in the dry season is lower, as due to the absence of symptoms, there are no daily observations of breast secretions, so there is a lack of diagnosis. At the beginning of lactation, the identification of clinical mastitis is greater as it also includes the diagnosis of infections contracted during the dry period (Rollin et al. 2015).

The economic impact of clinical mastitis is greater than any other medical illnesses in dairy cattle. An analysis of the average cost of clinical mastitis was performed by Rollin et al. (2015), indicating a total cost of US$ 444 per case (direct and indirect costs). Indirect costs include the milk production loss, premature disposal and loss of replacement of the cattle, and the future reproductive loss, representing 71% of the total cost per case of mastitis. The sooner cows with mastitis are diagnosed, the greater the chances of the producer making decisions related to prevention and management on the farm.

The major concern regarding the use of antibiotics to control mastitis is the development of drug resistance, reduced efficacy and the appearance of super bacteria (Barros et al. 2017; Gong et al. 2018), mainly because of off-label use, with longer duration and higher frequency (Lardé et al. 2020). In addition to the losses in milk production and animal health, there may also be implications to human health, through the presence of resistant strains in the diet and allergic reactions due to the presence of antimicrobials in milk (Barros et al. 2017). In the last decade, the "One Health" approach has received considerable attention in the international scientific community, with topics related to the interaction of human medicine, veterinary medicine and the environment, in addition to the introduction of good practices. The comprehension of this approach allows the improvement in dairy

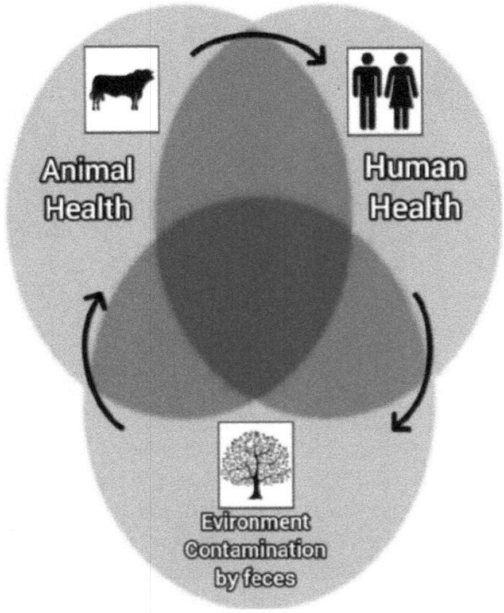

Figure 8.2. "One Health" approach related to mastitis.

production when introduced during the management of animals and milk production to prevent the spread of pathogens and ensure the quality of products. Daily producers must also be aware of the necessary measures so that waste is disposed of correctly and does not contaminate the environment (Garcia et al. 2019). Mastitis is an ideal target for the "One Health" approach as it is a multi-species, painful disease that has consequences to cattle and human health, along with economic, environmental and public health effects (Fig. 8.2).

To decrease the number of microorganisms capable of causing infection, a disinfection of the cows' roof must be carried out prior to milking, termed pre-dipping, and to prevent the transmission of microorganisms from the environment to the mammary gland after milking, called post-dipping (Campos et al. 2016; Picoli et al. 2016). In addition to the search for alternatives to antibiotics due to bacterial resistance, there is an incentive to apply organic and agroecological animal husbandry practices that use sustainable and renewable sanitary resources, in order to replace or complement conventional products. The use of plant extracts is on the rise due to the variety of bioactive compounds and biological actions within them (Campos et al. 2016). Some of them, such as the Chinese medicinal herb *Scutellaria baicalensis* and its main compounds baicalein and baicalin are used to improve the production performance of dairy cows and act therapeutically on diseases caused by pathogenic microorganisms, such as mastitis (Yin et al. 2021).

In view of the concern with resistance to conventional antimicrobials, interest in the use of medicinal plants for therapeutic purposes has increased considerably. Plants have been used in traditional medicine for centuries, and still used to this day, and are rich sources of bioactive substances able to act on microorganisms (Shin and Park 2018). Data from the World Health Organization (WHO) show that medicinal plants are great sources of herbal medicines, so it is of great importance to know their characteristics and effects to take full advantage of their benefits (Shin and Park 2018; Okmen and Balpınar 2017). Natural extracts generally have fewer side effects and good biological activity when compared with conventional antimicrobials (Gong et al. 2018). The objective of this study was to evaluate the original research published between 2015 and 2020 on natural products, extracted from plants, used in the prevention or treatment of mastitis.

Natural extracts and compounds with antimicrobial activity

A literature search was carried out in the Scopus, Web of Science, PubMed, and SciELO databases, using the descriptors: "natural products", "natural extract", "mastitis", and "antimicrobial", alone or in combination. The publications found were filtered for those published in the period from 2015 to 2021, and a selection was made by analysing the titles and abstracts, of the remaining publications, the entire text was analysed. A total of 159 articles were initially identified, and after excluding duplicates and some papers not related to natural products (content verification), 25 articles, published between 2015 and 2021, were entirely analysed in for use in the present study. About 17 of them address only antimicrobial activities of extracts, essential oils or compounds isolated from plants. Of these, 12 articles address therapeutic action alone, two prophylactics actions alone, and three both therapeutic and prophylactic, as shown in Table 8.1.

To minimise the effects caused by resistance to conventional antibiotics (treatment failure, dissemination of resistance genes, etc.), safe and effective alternatives are being investigated. In the context of mastitis, there is an urgent demand for both treatment and prevention options, since this is the main infectious disease of dairy herds that causes so many complications to the dairy industry (Gomes et al. 2018).

Considering the parameters adopted by Ríos and Recio (2005) and Gibbons (2008), a plant-based extract is considered viable to be used in treatment if it presents a Minimum Inhibitory Concentration (MIC) less than 100 µg/mL. This precept was not met in the studies by Freitas et al. (2020), Fiordalisi et al. (2016), and Okmen and Balpınar (2017), where several plant extracts showed activities with the MIC values ranged from 750 µg/mL to 75 mg/mL (Table 8.1).

Essential Oils (EOs) are produced by plants as a defence against infections and herbivores, in addition to attracting pollinating insects. EOs consist of a mixture that ranges from 1 to 60 compounds, mainly terpenoids, but also phenylpropanoids and fatty ester derivatives. Due to the rich source of chemical compounds with potential medicinal applicability and control of microorganisms, interest in this natural product has grown (de Jesus et al. 2020). The EOs of *Chromolaena squalida, Campomanesia sessiliflora,* and *Myrsine guianensis* showed antibacterial activity against a multi-resistant strain of *S. aureus*, NEWP0023, isolated from cow's milk with mastitis. The *C. squalida* EO presented the best result with an MIC of 7.80 µg/mL against *Staphylococcus* sp. 841 multiresistant strain, while the EOs of the other two plant species had the same MIC value of 31.25 µg/mL against the same *Staphylococcus* strain. As for the Gram-negative bacteria, *P. aeruginosa* and *E. coli,* which have an external membrane giving it additional protection and some resistance to antimicrobial agents, the EOs showed moderate activity (MIC values of 500 µg/mL), except that of *C. squalida.* These results suggest the viability of using EOs in the development of alternative antimicrobials (de Jesus et al. 2020).

Most studies that investigate natural products as antimicrobial agents use plants from a specific geographic region. Despite satisfactory results, according to Paşca et al. (2017 and 2020), studies with acclimated plants are necessary to enable greater access. Thus, the authors tested extracts from 11 indigenous species or acclimated plants against 10 species of microorganisms, including *Staphylococcus* spp. and *E. coli.* The best MIC results were obtained for the following species: *Evernia prunastri, Artemisia absinthium* and *Lavandula angustifolia*, between 0.9 µg/mL and 6.25 µg/mL, similar to the antimicrobial activity of the known antimicrobials, florfenicol and enrofloxacin.

Rhodomyrtus tomentosa is a plant of the *Myrtaceae* family, and its ethanolic extract, obtained from the leaves, has shown good activity against Gram-positive bacteria, especially

Table 8.1. Natural antimicrobial products to mastitis.

Product feature	Natural product/ isolated compound	Target microorganism	Effective dose	Notes	Reference
Essential oils	Leaves of *Chromolaena squalida*	*Staphylococcus aureus* (NEWP0023)	MIC 7.8 µg/mL		de Jesus et al. 2020
	Campomanesia sessiliflora	*Escherichia coli* (NEWP0022)	MIC 31.25 µg/mL		
	Myrsine guianensis	*Escherichia coli* (NEWP0018, producer of β-lactamase)	MIC 31.25 µg/mL		
	Matayba guianensis	*Pseudomonas aeruginosa* (NEWP0027)			
	Siparuna guianensis	*Staphylococcus* sp. 841			
	Ocotea minarum	*Staphylococcus* sp. 873			
	Endlicheria paniculata	*Salmonella typhi* 905			
Ethanolic extract	Peel of *Punica granatum*	*Escherichia coli*	75 mg/mL		Freitas et al. 2020
		Staphylococcus aureus			
		Streptococcus agalactiae			
Aqueous extract	*Mangifera indica* leaf	*Staphylococcus* spp.	MIC 1.8–7.5 mg/mL	MIC 1.8 mg/mL to *S. aureus* ATCC; 3.7–7.5 mg/mL *S. haemolyticus* and *S. aureus*	Manzur et al. 2020
Ethanolic extract		*Staphylococcus aureus* (ATCC 25923)	45.3 mg/mL was capable of reducing mature biofilms in rubber and stainless steel		
Hydro-organic (HO) L extract	White grape marc *Vitis vinifera*	*Staphylococcus aureus* (CECT 59)	IC50 HO L (0.809); IC50 HO P (<< 0.625)		Rama et al. 2020
Hydro-organic (HO) P extract		*Streptococcus uberis* (CECT 994)	IC50 HOL (<< 0.625); IC50 HO P (1349)		

Table 8.1 contd. ...

...Table 8.1 contd.

Product feature	Natural product/ isolated compound	Target microorganism	Effective dose	Notes	Reference
Methanolic extract	*Persicaria pensylvanica* flower	*Staphylococcus aureus* (Newbould 305)	MIC 62.5 μg/mL		Abdi and Kerro Dego 2019
	Persicaria pensylvanica leaf		MIC 31.25 μg/mL		
	Persicaria pensylvanica stem		MIC 7.8 μg/mL		
	Persicaria pensylvanica root		MIC ≤ 46.9 μg/mL		
Methanolic extract	*Aloe barbadensis* Miller	*Escherichia coli* (ATCC 25922)		Percentage of bacteria with cell membrane disruption after 24-hour incubation	Forno-Bell et al. 2019
		Staphylococcus aureus (ATCC 29213)			
		Streptococcus uberis (ATCC 9927)	16 mg/mL		
		Staphylococcus aureus (ATCC 43300) (MRSA)	10 mg/mL		
Essential oils	*Tagetes minuta* L.	*Staphylococcus aureus* (ATCC 25923)	MIC 1 mg/mL		Sperandio et al. 2019
		Staphylococcus aureus (ATCC 33592) (MRSA)	MIC 1 mg/mL		
		Escherichia coli (ATCC 8739)	MIC 1 mg/mL		
		Escherichia coli	MIC 3 mg/mL		
Methanolic extract	*Juglans regia*	*Staphylococcus aureus* (ATCC 25923)	MIC 0.78–1.56 mg/mL		Gomes et al. 2018
	Eucalyptus globulus	6 clinical strains	MIC 0.19–0.39 mg/mL		
	Foeniculum vulgare		MIC 3.125–6.25 mg/mL		
Hydro-alcoholic extract	Brown propolis	*Staphylococcus aureus*	25 mg/mL		Picoli et al. 2016
		Staphylococcus epidermidis			
		Staphylococcus intermedius			
		Streptococcus agalactiae			
		Streptococcus dysgalactiae			
		Streptococcus uberis			
		Escherichia coli			

Table 8.1 contd. ...

...Table 8.1 contd.

Product feature	Natural product/ isolated compound	Target microorganism	Effective dose	Notes	Reference
Ethanolic extract	Brazilian propolis from Urupema	*Staphylococcus aureus* (ATCC 25923)	750 μg/mL		Fiordalisi et al. 2016
	Green propolis from Minas Gerais	*Staphylococcus aureus* (isolated from mastitis milk)			
		Escherichia coli (ATCC 8740)			
		Escherichia coli (isolated from mastitis milk)			
Hydro-alcoholic extract	*Achyrocline satureioides* flower	*Candida albicans*	0.05 g/mL		Campos et al. 2016
		Candida krusei	0.06 g/mL		
		Candida rugosa	0.07 g/mL		
		Candida albicans (ATCC 14053)	0.025 g/mL–5 g/mL		
Ethanolic extract	*Rhodomyrtus tomentosa* leaf	*Staphylococcus aureus* (ATCC 29213)	-		Mordmuang and Voravuthikunchai 2015
		Coagulase-positive staphylococci	64 μg/mL		
		Coagulase-negative staphylococci	64 μg/mL		
Active substances isolated from the fruits of *Garcinia brasiliensis*	Guttiferone A	*Streptococcus agalactiae*	MIC Gut-A 7.81 μg/mL 7-epi 0.98 μg/mL		Maia et al. 2018
	7-epiclusianone	*Streptococcus uberis*	MIC Gut-A 15.62 μg/mL 7-epi 3.90 μg/mL		
Ethanolic extract	*Rhodomyrtus tomentosa* leaves	*Staphylococcus aureus* Newbould (ATCC 29740)	MIC 8–16 μg/mL		Mordmuang et al. 2019
Pure compound isolated from *Rhodomyrtus tomentosa*	Rhodomyrtone		MIC 0.25–0.5 μg/mL		
Compound extracted from *Rheedia brasiliensis*	7-epiclusianone	*Streptococcus agalactiae* (Isolated from mastitis)	MIC 7.8 μg/mL		Barros et al. 2017
	Complex 7-epiclusianone-Cu	*Streptococcus uberis* (Isolated from mastitis)	MIC 7.8 μg/mL		

Table 8.1 contd. ...

...Table 8.1 contd.

Product feature	Natural product/ isolated compound	Target microorganism	Effective dose	Notes	Reference
Alcoholic extract	*Evernia prunastri*	*Staphylococcus* spp.	MIC 0.09–6.25 µg/mL		Paşca et al. 2017
	Artemisia absinthium	*Vibrio fluvialis*	MIC 0.09–3.12 µg/mL		
	Lavandula angustifolia	*Serratia liquefaciens*	MIC 0.09–6.25 µg/mL		
	Agastache foeniculum	*Escherichia coli*			
	Althaea officinalis	*Lactococcus lactis*			
	Melissa officinalis	*Enterobacter intermedius*			
	Mentha pulegium	*Bacillus cereus*			
	Nepeta cataria	*Yersinia ruckeri*			
	Origanum vulgare	*Aeromonas* spp.			
	Plantago lanceolata	*Kytococcus sedentarius*			
	Populus nigra	*Kytococcus sedentarius*			
Methanolic extract	*Hypericum perforatum*	*Staphylococcus aureus*	MIC 1625–812.5 µg/mL		Okmen and Balpınar 2017
		Coagulase-negative Staphylococci	MIC 1625–3250 µg/mL		

MIC: Minimum Inhibitory Concentration

S. aureus, coagulase-negative *Staphylococci* and *Streptococcus* spp. (Mordmuang and Voravuthikunchai 2015; Mordmuang et al. 2019). Besides the extract, a compound isolated from this plant, rhodomyrtone (Fig. 8.3), also showed satisfactory antibacterial activity with MIC values of 8 mg/mL, 16 mg/mL and 0.25 mg/mL and 0.5 mg/mL, against strains of *S. aureus* including its persistent phenotype (SCV: Small Colony Variant) and Methicillin Resistant *S. aureus* (MRSA), respectively. The isolated compound had similar results to vancomycin, a conventional antibiotic important in human medicine. In addition to their own antibacterial activity, the ethanolic extract and the isolated rhodomyrtone enhanced the activity of some traditional antibiotics when used together, although the mechanism of action was not fully elucidated and required further investigation (Mordmuang et al. 2019). Furthermore, the ethanolic extract from the leaves of *R. tomentosa* tested against *S. aureus* ATCC 29213 strain and a strain isolated from cows with mastitis had MIC values that ranged from 6 mg/mL to 64 mg/mL (Mordmuang and Voravuthikunchai 2015).

Plants of the *Garcinia* genus (Guttiferae family) are rich sources of phytochemicals. *G. brasiliensis*, for example, a species native to the Amazon region, is rich in gutiferone A (Gut-A) and 7-epiclusianone (7-epi) (Fig. 8.3). Gut-A has several reports of bioactivities including as an antimicrobial against *S. aureus* and HIV, trypanocidal agent, antispasmodic, antioxidant, antitumour and protease inhibitor (Gustafson et al. 1992; Ngouela et al. 2006; Coelho et al. 2008; Martins et al. 2009a; Martins et al. 2009b). The 7-epi compound also presents several biological activities, such as toxicity to *Trypanosoma cruzi* trypomastigotes

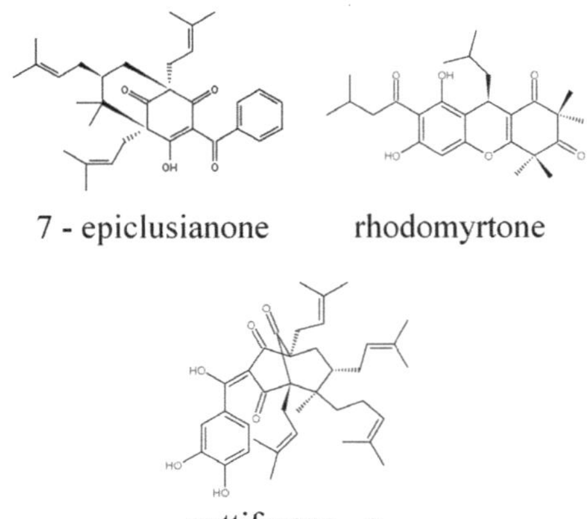

Figure 8.3. Rhodomyrtone, guttiferone a (Gut-A), and 7-epiclusianone (7-epi).

(Alves et al. 1999), anti-HIV (Piccinelli et al. 2005), vasodilation (Cruz et al. 2006), antianaphylaxis (Neves et al. 2007), antinociceptive and anti-inflammatory (Santa-Cecilia et al. 2011), antimicrobial activity against *Streptococcus* spp. (Almeida et al. 2008; Barros et al. 2017), inhibition of cell reproduction in Madin-Darby canine kidney cells (Vieira et al. 2009) and antiproliferative activity against cancer cells (Almeida et al. 2008).

Maia et al. (2018) evaluated the antimicrobial activity of compounds isolated from *G. brasiliensis* against clinical isolates of *S. agalactiae* and *Streptococcus uberis* obtained from cows affected by mastitis. The authors observed low MIC values for both Gut-A and 7-epi, the latter with values even lower than 0.98 µg/mL against *S. agalactiae* and 3.90 µg/mL against *S. uberis*. Barros et al. (2017) also evaluated compounds isolated from *G. brasiliensis* against the same species of mastitis-derived isolates. When tested against *S. agalactiae* and *S. uberis*, 7-epidemonstrated an MIC greater than the previous study, at 7.8 µg/mL against the two bacteria. When 7-epi was associated with copper (7-epi-Cu), the MIC value was maintained; however, the Minimum Bactericidal Concentration (MBC) decreased from 15.5 µg/mL to 7.3 µg/mL. The authors expected this behaviour with the addition of copper, since metals have the ability to induce microbial toxicity through protein dysfunction, production of Reactive Oxygen Species (ROS), and disrupting membrane function. According to Ríos and Recio (2005) and Gibbons (2008), the isolated compounds of *G. brasiliensis* have relevant antimicrobial activities, since the MIC values are below 10 µg/mL.

The methanolic extracts of 14 plant species were evaluated for their antimicrobial activity against two *S. aureus* strains: ATCC and a clinical isolate (Gomes et al. 2018). *Eucalyptus globulus* and *Juglans regia* extracts were found to be the most effective, with the inhibition halo in the diffusion disc test ranging from 8 mm to 15 mm and 12 to 16 mm, respectively. When the antimicrobial actions were further evaluated, *E. globulus* and *J. regia* extracts showed total inhibition of bacterial growth with MIC values ranging from 0.19 mg/mL to 0.39 mg/mL and 0.78 mg/mL to 1.56 mg/mL, respectively. Of the extracts analysed, *E. globulus* was the most efficient against the tested strains, with the lowest concentration necessary to inhibit bacterial growth. The phenolic compounds present in the sample of *E. globulus,* responsible for the activity were gallotannins, ellagic acid glycosides and quercetin derivatives. In view of the results, the extracts of *E. globulus* and *J. regia* were highlighted as promising agents for use in mastitis control (Gomes et al. 2018). Other flavonoid commonly observed in natural extracts, myricetin were described in *Mimosa tenuiflora* and *Commiphora leptophloeos* ethanolic extracts from the leaves (43.2 and 294.9 mg in 10 g, respectively). These extracts were evaluated against 33 *Staphylococcus* spp. with MBC inhibition range between 195.30 and 3125.00 µg mL^{-1} and between 781.20 and 6250.00 µg mL^{-1} for the extracts of *M. tenuiflora* and *C. leptophloeos*, respectively (Souza et al. 2020).

Tagetes minuta L. is a native species from South America, the extracts and essential oils of which present biological activities such as an insecticide, repellent, nematicide, antioxidant, anti-inflammatory and antimicrobial. Such activities are related to the presence of substances like cis-tagetone, dihydrotagetone, trans-ocimenone and cis-ocimenone in their chemical composition (Fig. 8.4). The EO of *T. minuta*, at the concentration of 1 mg/mL, was able to reduce approximately 80% of the growth of *S. aureus* strains including a standard ATCC 25923, an MRSA strain and field isolates. Whilst against *E. coli* strains, *T. minuta* EO had MIC values ranging from 1 mg/mL to 3 mg/mL against standard strains and field isolates. The field strains are subject to selective pressures that usually results in higher genetic diversity, generating different levels of resistance. However, this EO showed cytotoxicity to mammary gland epithelial cells at a concentration above 10 µg/mL, which requires attention in its use. The study suggests that this EO would be effective in external use at concentrations between 1 mg/mL and 3 mg/mL, being able to prevent the entry of bacteria in the bovine mammary gland or even to treat already established infections. Its activities in low concentration on bacteria resistant to methicillin are of great importance, considering the morbidity and mortality it can cause. However, due to its toxicity more studies are needed to guarantee its safety (Sperandio et al. 2019).

The chemical evaluation of aqueous and ethanolic extracts obtained from mango leaves (*Mangifera indica*) showed different compositions. The aqueous extract had more polar substances, such as tannins, condensed tannins, flavones and flavonoids. The ethanolic extract also had condensed tannins, but together with gallic acid and fatty acids. Comparing the antimicrobial action of the two extracts, the ethanolic extract demonstrated better results, with the MIC ranging from 1.8 mg/mL to 7.5 mg/mL for *S. aureus* ATCC 25923 and clinical isolates. The ethanolic extract, but not the aqueous extract, also showed the ability

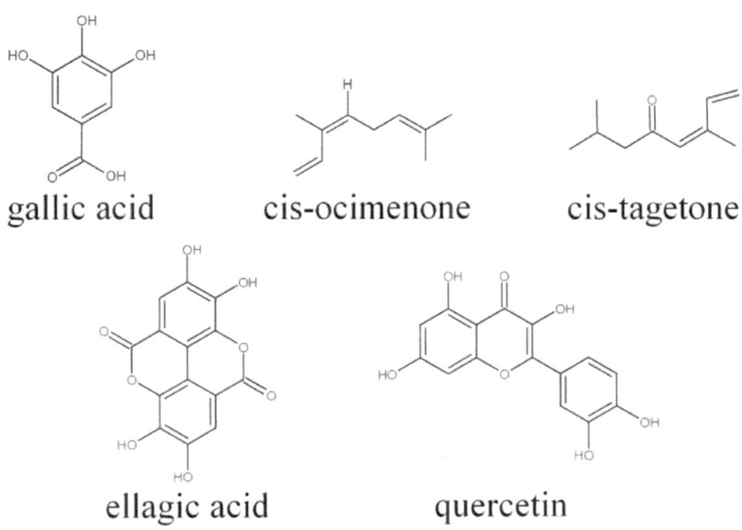

Figure 8.4. Phenolic (gallic acid, ellagic acid, quercetin) and volatile compounds (cis-tagetone, cis-ocimenone) present in ethanolic extracts and essential oils obtained from *Tagetes minuta* L.

to remove mature biofilms from steel or rubber at a concentration of 45.3 mg/mL within 5 minutes of incubation (Manzur et al. 2020).

Achyrocline satureioides, a plant native to southern Brazil (known as "macela"), is popularly used due to its anti-inflammatory, antispasmodic, analgesic, sedative, antioxidant and immunostimulant activities. The hydroalcoholic extract obtained from the flowers of *A. satureioides* at a concentration of 5 g/100 mL showed fungicidal action, with logarithmic reduction of population density, in addition to total inactivation of *Candida* strains and clinical isolates depending on the contact time. Thus, the flower extract of *A. satureioides* could be used in the disinfection, if the time criteria are based on the sensitivity of the microorganisms that cause mastitis. Since no fungi was observed after the first hour of treatment, the authors concluded that the use of the extract for a shorter time will not present an effective action against the disease (Campos et al. 2016). In another study performed with this same species by Machado et al. (2020), described four substances, three very common (luteolin, quercetin and 3-O-methylquercetin) and one more specific, the chalcone achyrobichalcone at the extracts. The authors observed the action of nanoemulsions of *Linum usitatissimun* oil loaded with this "macela" extract and observed the *in vitro* antimicrobial activity against MRSA with MIC values ranging from 1.2 to 10% (v/v).

Propolis has over 300 chemical compounds, which together confer biological activities such as an antiviral, antifungal, antiprotozoal, antibacterial, anti-inflammatory, healing, antioxidant, immunomodulatory, among others. The chemical composition of the hydroalcoholic extract of brown propolis presented caffeic, ferulic and chlorogenic acids, together with rutin and in particular, p-coumaric acid, the latter largely responsible for its antibacterial action. Studying the time necessary for the antibacterial action of the brown propolis extract (at a concentration of 25 mg/mL), it was detected that within 15 minutes, an effect was observed on *Staphylococcus* sp., *Streptococcus* sp., and *E. coli*. The total elimination of the bacteria occurred between 2 and 4 hours. From these results, the brown propolis extract shows the potential to be used in hygiene practices before milking, as a preventive agent for bovine mastitis (Picoli et al. 2016).

In vitro cytotoxicity tests on natural products should be performed to assess cytotoxic behaviour, predict possible risks, and thus, select the most suitable ones for *in vivo* evaluations. Of the studies with promising antimicrobial activity, only Maia et al. (2018), Barros et al. (2017), Mordmuang et al. (2019), and Sperandio et al. (2019) performed cytotoxicity tests. Cytotoxicity was assessed by the lactate dehydrogenase assay (Mordmuang et al. 2019) and in bovine mammary gland epithelial cells by the MTT method (3-[4,5-dimethyl-thiazol-2-yl]-2 bromide), 5-diphenyl tetrazolium. The *T. minuta* EO, which showed *S. aureus* and *E. coli* inhibitory activities at low concentrations, also showed cytotoxicity from 30 μg/mL, requiring attention for its use (Sperandio et al. 2019). The compound, 7-epi did not show cytotoxicity in the studies performed by Barros et al. (2017) and Maia et al. (2018). However, the presented cytotoxic of 0.5 MIC was above the MIC needed to kill bacteria. Synergistic effects with other antimicrobial agents also allow the use of Gut-A in lower concentrations. The extract of *R. tomentosa* showed cytotoxicity of 10 to 12% at the MIC value, while the isolated compound of this extract, rhodomyrtone, showed cytotoxicity at lower concentrations of 1 μg/mL to 32 μg/mL.

Natural extracts and compounds with anti-inflammatory activity

Since bovine mastitis is an inflammatory disease, knowledge about alternative herbal anti-inflammatories is also of great importance, as they can act in more than one way and

Table 8.2. Natural anti-inflammatory drugs used for mastitis.

Natural product/Isolated compound	Product feature	Effective dose	Reference
Moringa oleifera	Methanol extract	200 μg/mL ^{T}	Cheng et al. 2019
Minthostachys verticillata	Essential oil	100 μg/mL ^{P}	Montironi et al. 2019
Rhodomyrtus tomentosa leaf	Ethanolic extract	8–256 μg/mL ^{T}	Mordmuang and Voravuthikunchai 2015
Peiminine	Alkaloid extracted from *Fritillaria*	5 mg/kg ^{T}	Gong et al. 2018
Ricinus communis, Ocimum sanctum, Eucalyptus globulus and *Mentha piperita*	Combination of essential oils	*registered spray ^{T}	Selvam et al. 2015
Gambogic acid	Isolated compound	50 nM ^{T,P}	Tang et al. 2020

^{T}: Therapeutical use; ^{P}: Prophylactic use

through more than one mechanism of action. In the present study, six articles address the use of products from natural sources to potentially control the inflammation of mastitis, as shown in Table 8.2.

The ethanolic extract obtained from the leaves of *R. tomentosa*, in addition to the antimicrobial action described earlier, also revealed an anti-inflammatory action (Mordmuang and Voravuthikunchai 2015). The authors described that the extract demonstrated the ability to prevent protein denaturation and stabilise red cell membranes in a concentration-dependent manner, ranging from 8 μg/mL to 256 μg/mL. This activity is related to the presence of phenolic compounds, flavonoids, triterpenoids, steroids and tannins.

One of the inflammation pathways that occurs in mastitis is through the release of LPS present in the outer membrane of Gram-negative bacteria. LPS activates nuclear factor kappa B (NF-κB), pro-inflammatory cytokines such as: TNF-α, IL-1β and IL-6, as well as other mediators such as cyclooxygenase-2 (COX-2) (Gong et al. 2018; Cheng et al. 2019). Gong et al. (2018) evaluated the anti-inflammatory activity of peiminine (Fig. 8.5), an alkaloid extracted from *Fritillaria* plants, which are commonly used in Chinese medicine and have a variety of biological actions. This alkaloid (at a concentration of 5 mg/kg) did not present toxic side effects in an LPS-induced mastitis mouse model, in addition to suppressing pathological changes in the mammary gland, inhibiting myeloperoxidase activity and the production of pro-inflammatory mediators (TNF-α, IL-6, IL-1β, COX-2, and iNOS). Cheng et al. (2019) also assessed the activity of a plant extract from *Moringa oleifera* in the case of LPS-induced mastitis, but in bovine mammary epithelial cells. The leaf extract of this plant is rich in antioxidants such as phenolic acids, flavonoids, vitamin C, and also tannins, saponins, phytate, oxalate and alkaloids. The methanolic extract of *M. oleifera* at 200 μg/mL attenuated the expression of COX-2 and inhibited the activation of NF-κB and other pro-inflammatory cytokines. The effect of gambogic acid on LPS-induced mastitis in mouse mammary epithelial cells was evaluated by Tang et al. (2020). Gambogic acid (Fig. 8.5) showed anti-inflammatory, antimicrobial and neuroregulatory effects. At a concentration of 50 nM, the compound presented an effect remarkably similar to that of dexamethasone (steroidal anti-inflammatory). Gambogic acid was able to decrease the production of IL-1β, IL-6, and TNF-α, protect cells from apoptosis, maintain a stable mitochondrial membrane, and inhibit the overproduction of ROS.

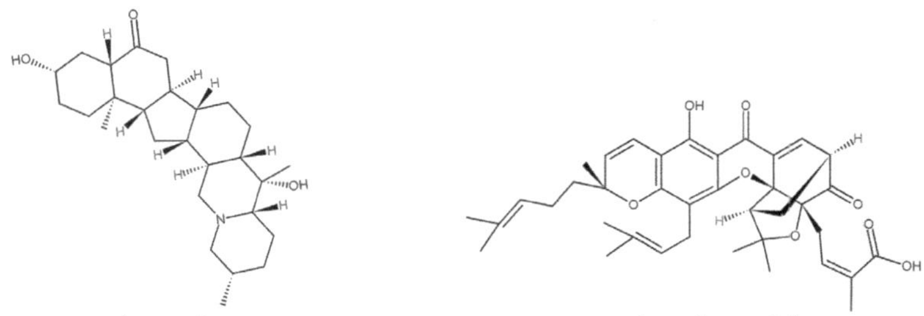

Figure 8.5. Natural substances as anti-inflammatories for mastitis treatment.

The EO of *Minthostachys verticillata* was evaluated for its ability to induce macrophage phagocytosis and modulate the inflammatory response (Montironi et al. 2019). The authors observed that the EO was able to modulate the mechanisms of adhesion (such as adhesins, membrane receptors, extracellular matrix proteins) and the phagocytic capacity of macrophages, as well as inhibiting interferon-gamma (IFN-γ) indicating anti-inflammatory activity. Another study evaluated Wisprec® Advanced (M/S Natural Remedies Private Limited, Bangalore, Karnataka, India) a spray based on a combination of EOs (*Ricinus communis, Ocimum sanctum, Eucalyptus globulus*, and menthol isolated from *Mentha piperita*) in lactating cows with clinical mastitis (Selvam et al. 2015). The spray caused a significant decrease in fever, pain and swelling of the udder, showing peripheral analgesic and anti-inflammatory effects through the inhibition of cyclooxygenase, lipo-oxygenase and other inflammatory mediators. The authors reported the advantage of topical application compared to injections, in terms of the absence of pain on application, generation of medical waste, and risk of cross contamination through the reuse of materials.

Natural extracts and compounds with other actions

The high milk production and intense cellular metabolism observed in a dairy herd promote a predisposition to oxidative stress by the accumulation of ROS. The chronic accumulation of ROS can trigger tissue damage and inflammatory responses, causing inflammation in the udders of dairy cows (Cheng et al. 2019). The ability of some plants, such as *Moringa oleifera* and *Hypericum perforatum* to eliminate free radicals favours the cure of the disease.

The *M. oleifera* leaf extract presents antioxidant compounds such as phenolic acids, flavonoids and vitamin C, as well as also other biologically active substances, such as alkaloids, cardenolides and cardiac glycosides. In addition to the antioxidant effect, this extract of *M. oleifera* has antifibrotic, anti-inflammatory, antimicrobial, anti-glycaemic, anti-tumour and anti-cancer actions (Cheng et al. 2019). The antioxidant activity of the methanol extract of *M. oleifera* was evaluated by Cheng et al. (2019), who reported significant results. In the concentration range of 6.25 μg/mL and 200 μg/mL, the *Moringa* extract was able to inhibit the ABTS (2,2'-Azino-bis(3-ethylbenzothiazoline-6-sulphonic acid) diammonium salt) free radical by 5 to 90%, varying according to the dose. The radical scavenging was also satisfactory in the DPPH (2,2-diphenyl-1-picryl-hydrazyl) assay, with 81% of the sequestering activity at a concentration of 6.25 μg/mL and approximately 90% at concentrations of 12.5 μg/mL to 200 μg/mL. This result, according to the authors, may be due to the difference in solubility between ABTS and DPPH. In addition to these, the *Moringa* extract was capable of eliminating intracellular ROS produced by LPS, at concentrations between 50 μg/mL and 200 μg/mL.

St. John's *wort, Hypericum perforatum* L., has therapeutic effects reported in literature on burns, bruises, swelling, anxiety and depression, in addition to antiviral, wound healing, analgesic, hepatoprotective, antioxidant and antimicrobial activities. The effects occur by the presence of compounds such as naftodiantrons, flavonol glycosides, biflavones, proanthocyanidins and phenylpropans (Okmen and Balpınar 2017). The antioxidant activity of the methanolic extract obtained from the flowers of *H. perforatum* was evaluated by Okmen and Balpınar (2017) using the DPPH test. The authors observed that 32% inhibition of free radicals was obtained with the extract at a concentration of 100 mg/mL, equivalent to 0.83 mM/g of Trolox, the control used in the study. Comparing the results of the DPPH tests carried out with methanolic extracts of *M. oleifera* leaves and flowers of *H. perforatum* L., the *Moringa* extract appears to have better antioxidant activities, as it eliminates a higher percentage of free radicals at a lower concentration (Okmen and Balpınar 2017).

As a way of circumventing the adverse effects of drugs used to treat mastitis and the resistance of bacteria using antibiotics, researchers are evaluating ways to modulate the host immune response to an infection. Immunomodulatory substances or modifiers of the immune response act in a non-specific way in the host's immune system. Plant-derived compounds showing this action were discovered, such as the EO of *Minthostachys verticillata.* The main oil components are the monoterpenes pulegone and menthone, isomenthone, limonene, menthol, α-pinene, carvone, β-pinene, piperitenone, sabinene, myrcene and β-ocimene. *M. verticillata* EO has the ability to stimulate human $CD4^+$ and $CD8^+$ T cells, and induce IFN-γ synthesis, in addition to inhibiting pro-inflammatory mediators (Montironi et al. 2019).

Conclusion

The use of herbal medicines in cattle herds is encouraged in organic and agroecological farming, using sustainable and renewable health supplies and resources to replace or complement conventional products. Several medicinal plants, from genera *Mangifera Achyrocline, Minthostachys, Moringa, Hypericum, Fritillaria, Tagetes* and *Rhodomyrtus* showed to be potential sources of herbal medicines, presenting less adverse effects, low toxicity, good biological activities, and an alternative to conventional antibiotics. Whether extracts, essential oils or their isolated compounds, they present antimicrobial, immunomodulatory, anti-inflammatory and antioxidant actions and could be used both in the prevention and treatment of the disease.

Acknowledgments

This study was financed in part by the Foundation of the State of Minas Gerais (FAPEMIG), the Carlos Chagas Foundation for Research Support in the State of Rio de Janeiro (FAPERJ), and the National Council for Scientific and Technological Development (CNPq).

Abbreviations

7-epi	7-epiclusianone
ABTS	2,2′-Azino-bis(3-ethylbenzothiazoline-6-sulfonic acid) diammonium salt
CMT	California Mastitis Test
COX-2	Cyclooxygenase-2
DNA	Desoxyribonucleic Acid

DPPH	2,2-diphenyl-1-picryl-hydrazyl
EO	Essential Oil
Gut-A	Gutiferone a
HIV	Human Immunodeficiency Virus
IFN-γ	Interferon-gamma
IL-1β	Interleukin-1β
IL-6	Interleukin-6
LPS	Lipopolysaccharides
MBC	Minimum Bactericidal Concentration
MIC	Minimal Inhibitory Concentration
MRSA	Methicillin-Resistant *Staphylococcus aureus*
MTT	3-[4,5-dimethyl-thiazol-2-yl]-2 bromide), 5-diphenyl tetrazolium
NF-κB	Nuclear Factor kappa B
ROS	Reactive Oxygen Species
SCV	Small Colony Variant
TNF-α	Tumour Necrosis Factor Alpha

References

Abdi, R.D., Kerro-Dego, O. (2019). Antimicrobial activity of *Persicaria pensylvanica* extract against *Staphylococcus aureus*. Eur J Integr Med 29: 100921.

Ali, M., Ikram, M., Ijaz, M. (2020). Green synthesis and evaluation of n-type ZnO nanoparticles doped with plant extract for use as alternative antibacterials. Appl Nanosci 10: 3787–3803.

Almeida, L.S.B., Murata, R.M., Yatsuda, R., Dos Santos, M.H., Nagem, T.J., Alencar, S.M., Koo, H., Rosalen, P.L. (2008). Antimicrobial activity of *Rheedia brasiliensis* and 7-epiclusianone against *Streptococcus mutans*. Phytomedicine 15: 886–891.

Alves, T.M.A., Alves, R.O., Romanha, A.J., Santos, M.H., Nagem, T.J., Zani, C.L. (1999). Biological activities of 7-epiclusianone. J Nat Prod 62: 369–371.

Barros, M., Perciano, P.G., Dos Santos, M.H., Oliveira, L.L., Costa, E.D., Moreira, M.A.S. (2017). Antibacterial activity of 7-epiclusianone and its novel copper metal complex on *Streptococcus* spp. isolated from bovine mastitis and their cytotoxicity in MAC-T cells. Molecules 22(5): 823.

Campos, F.L., Valente, P., Ethur, E.M., Avancini, C.A.M. (2016). Disinfectant activity of the crude hydroalcoholic extract of *Achyrocline satureioides* (Asteraceae) on *Candida* spp. isolated in problem-situations of bovine mastites. Acta Vet Bras 10(4): 327–333.

Cheng, W.N., Jeong, C.H., Seo, H.G., Han, S.G. (2019). Moringa extract attenuates inflammatory responses and increases gene expression of casein in bovine mammary epithelial cells. Animals (7): 391.

Coelho, L.P., Serra, M.F., Pires, A.L., Cordeiro, R.S., Silva, P.M.R., Dos Santos, M.H., Martins, M.A. (2008). 7-Epiclusianone, a tetraprenylated benzophenone, relaxes airway smooth muscle through activation of the nitric oxide-cGMP pathway. J Pharmacol Exp Ther 327(1): 206–214.

Cruz, A.J., Lemos, V.S., Dos Santos, M.H., Nagem, T.J., Cortes, S.F. (2006). Vascular effects of 7-epiclusianone, a prenylated benzophenone from *Rheedia gardneriana*, on the rat aorta. Phytomedicine 13: 442–445.

De Jesus, G.S., Micheletti, A.C., Padilha, R.G., De Paula, J.S., Alves, F.M., Leal, C.R.B., Garcez, F.R., Garcez, W.S., Yoshida, N.C. (2020). Antimicrobial potential of essential oils from cerrado plants against multidrug-resistant foodborne microorganisms. Molecules 25(14): 3296.

Fiordalisi, S.A.L., Honorato, L.A., Loiko, M.R., Avancini, C.A.M., Veleirinho, M.B.R., Machado-Filho, L.C.P., Kuhnen, S. (2016). The effects of Brazilian propolis on etiological agents of mastitis and the viability of bovine mammary gland explants. J Dairy Sci 99(3): 2308–2318.

Forno-Bell, N., Bucarey, S.A., Garcia, D., Iragüen, D., Chacón, O., San Martin, B. (2019). Antimicrobial effects caused by *Aloe barbadensis* Miller on bacteria associated with mastitis in dairy cattle. Nat Prod Comm 14(12): 1–5.

Freitas, T.F., Marques, F.M., Kuster, R.M., Salles, J.B., Victório, C.P., Assis, M.C. (2020). *Punica granatum* L. inhibits the growth of microorganisms associated with bovine mastitis. Nat Prod J 10(5): 2020.

Garcia, S.N., Osburn, B.I., Cullora, J.C.A. (2019). One health perspective on dairy production and dairy food safety. One Health 7(2019): 100086.

Gibbons, S. (2008). Phytochemicals for bacterial resistance—strengths, weaknesses and opportunities. Planta Med 74(6): 594–602.

Gomes, F., Martins, N., Barros, L., Rodriguesa, M.E., Oliveira, M.B.P.P., Henriques, M., Ferreira, I.C.R.F. (2018). Plant phenolic extracts as an effective strategy to control *Staphylococcus aureus*, the dairy industry pathogen. Ind Crops Prod 112: 515–520.

Gong, Q., Li, Y., Ma, H., Guo, W., Kan, X., Xu, D., Liu, J., Fu, S. (2018). Peiminine protects against lipopolysaccharide-induced mastitis by inhibiting the AKT/NF-κB, ERK1/2 and p38 signaling pathways. Int J Mol Sci 19(9): 2637.

Gustafson, K.R., Blunt, J.W., Munro, M.H.G., Fuller, R.W., McKee, T.C., Cardelina II, J.H., McMahon, J.B., Cragg, G.M., Boyd, M.R. (1992). The guttiferones, HIV-inhibtory benzophenones from *Symphonia globulifera, Garcinia livingstonei, Garcinia ovalifolia* and *Clusia rosea*. Tetrahedron 48(46): 10093–10102.

Hughes, K., Watson, C.J. (2018). The mammary microenvironment in mastitis in humans, dairy ruminants, rabbits and rodents: a one health focus. J Mamm Gland Biol Neopl 23(1): 27–41.

Lardé, H., Dufour, S., Archambault, M., Léger, D., Loest, D., Roy, J.P., Francoz, D. (2020). Assignment of Canadian defined daily doses and Canadian defined course doses for quantification of antimicrobial usage in cattle. Front Vet Sci 7: 10.

Machado, G.T.P., Veleirinho, M.B., Honorato, L.A., Kuhnen, S. (2020). Formulation and evaluation of anti-MRSA nanoemulsion loaded with Achyroclinesatureioides: a new sustainable strategy for the bovine mastitis. Nano Express 1, 3: 030004.

Maia, N.L., Barros, M., Oliveira, L.L., Cardoso, S.A., Dos Santos, M.H., Pieri, F.A., Tamalho, T.C., Da Cunha, E.F.F., Moreira, M.A.S. (2018). Synergism of plant compound with traditional antimicrobials against *Streptococcus* spp. isolated from bovine mastitis. Front Microbiol 9: 1203.

Manzur, A.G.B., Junior, V.S.M., Morais-Costa, F., Mariano, E.G.A., Careli, R.T., Da Silva, L.M.V., Coelho, S.G., De Almeida, A.C., Duarte, E.R. (2020). Extract of *Mangifera indica* L. leaves may reduce biofilms of *Staphylococcus* spp. in stainless steel and teatcup rubbers. Food Sci Tech Int 26(1): 11–20.

Martins, F.T., Assis, D.M., Dos Santos, M.H., Camps, I., Veloso, M.P., Juliano, M.A., Alves, L.C., Doriguetto, A.C. (2009a). Natural polyprenylated benzophenones inhibiting cysteine and serine proteases. Eur J Med Chem 44(3): 1230–1239.

Martins, M.D., Marques, M.M., Bussadori, S.K., Mesquita-Ferrari, R.A., Pavesi, V.C.S., Wadt, N.S., Fernandes, K.P. (2009b). Citotoxicidade *in vitro* de extratos de arnica brasileira (*Solidago microglossa*) e arnica paulista (*Porophyllum ruderale*). ConScientiaeSaúde 8(1): 99–104.

Mordmuang, A., Voravuthikunchai, S.P. (2015). *Rhodomyrtus tomentosa* (Aiton) Hassk. leaf extract: an alternative approach for the treatment of staphylococcal bovine mastitis. Res Vet Sci 102: 242–246.

Mordmuang, A., Brouillette, E., Voravuthikunchai, S.P., Malouin, F. (2019). Evaluation of a *Rhodomyrtus tomentosa* ethanolic extract for its therapeutic potential on *Staphylococcus aureus* infections using *in vitro* and *in vivo* models of mastitis. Vet Res 50: 49.

Montironi, I.D., Reinoso, E.B., Paullier, V.C., Siri, M.I., Pianzzola, M.J., Moliva, M., Campra, N., Bagnis, G., LaRocque-de-Freitas, I.F., Decote-Ricardo, D., Freire-de-Lima, C.G., Raviolo, J.M., Cariddi, L.N. (2019). *Minthostachys verticillata* essential oil activates macrophage phagocytosis and modulates the innate immune response in a murine model of *Enterococcus faecium* mastitis. Res Vet Sci 125: 333–344.

Neves, J.S., Coelho, L.P., Cordeiro, R.S.B., Veloso, M.P., Silva, P.M.R., Dos Santos, M.H., Martins, M.A. (2007). Antianaphylactic properties of 7-epiclusianone, a tetraprenylated benzophenone isolated from *Garcinia brasiliensis*. Planta Med 73(7): 644–649.

Ngouela, S., Lenta, B.N., Noungoue, D.T., Ngoupayo, J., Boyom, F.F., Tsamo, E., Gut, J., Rosenthal, P.J., Connolly, J.D. (2006). Anti-plasmodial and antioxidant activities of constituents of the seed shells of *Symphonia globulifera* Linn F. Phytochemistry 67(3): 302–306.

Okmen, G., Balpınar, N. (2017). The biological activities of *Hypericum perforatum* L. African J Trad Compl Altern Med 14(1): 213–218.

Paşca, C., Mărghitaş, L., Dezmirean, D., Bobiş, O., Bonta, V., Chirilă, F., Matei, I., Fiţ, N. (2017). Medicinal plants based products tested on pathogens isolated from mastitis milk. Molecules 22(9): 1473.

Paşca, C., Mărghitaş, L.A., Dezmirean, D.S., Matei, I.A., Bonta, V., Paşca, I., Chirilă, F., Cîmpean, A., Iosif Fiţ, N. (2020). Efficacy of natural formulations in bovine mastitis pathology: alternative solution to antibiotic treatment. J Vet Res 64(4): 523–529.

Piccinelli, A.L., Cuesta-Rubio, O., Chica, M.B., Mahmood, N., Pagano, B., Pavone, M., Barone, V., Rastraelli, L. (2005). Structural revision of clusianone and 7-epi-clusianone and anti-HIV activity of polyisoprenylated benzophenones. Tetrahedron 61(34): 8206–8211.

Picoli, T., Peter, C.M., Hoffmann, J.F., Latosinski, G.S., Zani, J.L., Vargas, G.A., Hubner, S., Fischer, G. (2016). Chemical characterization and antibacterial action of brown propolis extract from Southern Brazil. Rev Bras Med Vet 38(4): 365–371.

Pumipuntu, N., Tunyong, W., Chantratita, N., Diraphat, P., Pumirat, P., Sookrung, N., Chaicumpa, W., Indrawattana, N. (2019). *Staphylococcus* spp. associadas à mastite bovina subclínica nas províncias central e nordeste da Tailândia. Peer J 7: e6587.

Rama, J.L.R., Mallo, N., Biddau, M., Fernandes, F., De Miguel, T., Sheiner, L., Choupina, A., Lores, M. (2020). Exploring the powerful phytoarsenal of white grape marc against bacteria and parasites causing significant diseases. Environ Sci Pol Res. DOI: 10.1007/s11356-019-07472-1.

Ríos, J.L., Récio, M.C. (2005). Medicinal plants and antimicrobial activity. J Ethnopharmacol 100(1-2): 80–84.

Rollin, E., Dhuyvetter, K.C., Overton, M.W. (2015). The cost of clinical mastitis in the first 30 days of lactation: An economic modeling tool. Prev Vet Med 122(3): 257–264.

Santa-Cecília, F.V., Freitas, L.A.S., Vilela, F.C., Veloso, C.D.C., Da Rocha, C.Q., Moreira, M.E.C., Dias, D.F., Giusti-Paiva, A., Santos, M.H. (2011). Antinociceptive and anti-inflammatory properties of 7-epiclusianone, a prenylated benzophenone from *Garcinia brasiliensis*. Eur J Pharmacol 670(1): 280–285.

Selvam, R., Sureshbabu, G., Saravanakumar, M., Prashanth, S. (2015). Evaluation of a polyherbal topical aerosol spray as a supportive therapy for clinical mastitis in dairy cows. J Adv Vet Anim Res 2(3): 285–290.

Shin, B., Park, W. (2018). Zoonotic diseases and phytochemical medicines for microbial infections in veterinary science: current state and future perspective. Front Vet Sci 5: 166.

Sousa, M.M., Araújo, R.M.P., Libório, R.C., Costa, E.C., Araújo, B.N., Lima, M.S., Gouveia, G.V., Costa, M.M., Peixoto, L.J.S., Peixoto, R.M. (2020). Antimicrobial potential of Jurema preta and umburana, native species of the Caatinga biome, on Staphylococcus isolated from small ruminants with mastitis. Semina 41(5) supp 1: 2231–2241.

Sperandio, J., Veleirinho, B., Honorato, L.A., Campestrini, L.H., Kuhnen, S. (2019). *In vitro* antibacterial and cytotoxicity activities of *Tagetes minuta* L. essential oil towards bovine mastitis treatment. Arq Bras Med Vet Zootec 71(4): 1251–1259.

Tang, X., Chang, L., Li, T., Lin, C., Hao, Z., Zhang, H., Zhao, G., Chen, Y., Guo, A., Hu, C. (2020). Gambogic acid alleviates inflammation and apoptosis and protects the blood-milk barrier in mastitis induced by LPS. Int Immunopharmacol 86: 106697.

Tadee, P., Chukiatsiri, K., Amornlerdpisan, D., Paserakung, A., Kittiwan, N. (2020). Antimicrobial effect of Japanese pumpkin (Cucurbita maxima) extract on local mastitis pathogen. Vet Integr Sci 18(3): 141–152.

Vieira, F.T., Maia, J.R.S., Vilela, M.J., Ardisson, J.D., Santos, M.H., Oliveira, T.T., Nagem, T.J. (2009). Spectroscopic investigation of organotin (IV) derivatives of 7-epiclusianone: a preliminary *in vitro* antitumor evaluation of HN-5 human carcinoma cell. Main Group Metal Chem 32(5): 235–246.

Yin, B., Li, W., Qin, H., Yun, J., Sun, X. (2021). The use of Chinese Skullcap (*Scutellaria baicalensis*) and its extracts for sustainable animal production. Animals 11: 1039.

9

Plant-based Nanomaterials and their Antimicrobial Activity

Mayuri Napagoda,[1,*] *Priyalatha Madhushanthi*,[1] *Dharani Wanigasekara*[1] and *Sanjeeva Witharana*[2]

Introduction

Antimicrobial agents could be considered as one of the most successful types of chemotherapeutic agents in the history of medicine with respect to their role in the control of infectious diseases that save millions of lives (Aminov 2010). Despite the availability of several reports on the use of various kinds of antibiotic agents even before the advent of modern medicine, the discovery of arsphenamine by Paul Ehrlich in 1909 was one significant achievement. Arsphenamine was an arsenic derivative that showed activity against *Treponema pallidum*, the causative agent of syphilis (Durand et al. 2019). The accidental discovery of penicillin in 1928 marked the dawn of a new era in this branch of medicine. After returning from a holiday, Alexander Fleming, who was a bacteriologist at St. Mary's hospital in London, observed that one of his petri dishes inoculated with *Staphylococci* and left on a laboratory bench was contaminated with a fungus (Swann 1983). He noticed that the fungal contaminant in the petri dish was suppressing the growth of nearby bacteria. The contaminant fungus was identified as *Penicillium notatum*, and the antibacterial compound that was produced by this fungus was named "penicillin". Fleming published his findings in "The British Journal of Experimental Pathology" in 1929 to disclose that penicillin was capable of inhibiting bacterial growth *in vitro* (Fleming 1929). However, he did not manage to purify or characterize the activity of penicillin. Instead, he thought that it could be useful as a local antiseptic (Aminov 2010).

After Fleming's great discovery, Howard Florey together with Ernst Chain decided to design a methodology to culture *Penicillium* and to produce it in adequate quantities to be used in further testing of its antimicrobial potential. In the mid-1940s, this research group evaluated the *in vitro* efficacy of penicillin while the effectiveness of penicillin *in vivo* was also determined using a mouse model (Cranston and Sidebottom 2016). The first human

[1] Department of Biochemistry, Faculty of Medicine, University of Ruhuna, Galle 80 000, Sri Lanka.
[2] Department of Mechanical & Manufacturing Engineering, Faculty of Engineering, University of Ruhuna, Galle 80 000, Sri Lanka.
* Corresponding author: mayurinapagoda@yahoo.com

to receive penicillin as part of its toxicity test was a woman with terminal cancer. After receiving a penicillin injection, she developed fever and rigor caused by the impurities of pyrogenic origin that were present in the penicillin mixture (Cranston and Sidebottom 2016; Lobanovska and Pilla 2017). Thereafter, Edward Abraham and Chain worked to remove the residual pyrogens before injecting it into patients. A patient with a severe *Staphylococcal* and *Streptococcal* infection became the second person to receive the purified penicillin. The repeated intravenous injections of penicillin for 5 days had a profound effect on his recovery (Fletcher 1984; Lobanovska and Pilla 2017). Fleming, Florey, and Chain were awarded the Nobel Prize in 1945 for the discovery of penicillin and its curative effect in various infectious diseases (Lobanovska and Pilla 2017).

The discovery of penicillin dawned the "golden era" of antibiotic research (1940–1962) leading to a new pathway in treating infectious diseases (Singh and Barrett 2006). Research on the development of other antimicrobials soon started and as a result, widely used antibiotics such as streptomycin, chloramphenicol and tetracycline were discovered. Further, the first-discovered antibiotics played a significant role in the treatment of infections during World War II (Lerner 2004). In that period, antibiotics were so successful and considered as "miracle drugs" that could treat any infectious disease. As a result of the phenomenal success of antibiotics, infectious bacterial diseases were naively thought to be permanently defeated (Penesyan et al. 2015).

This chapter aims to present a state of the art review of the plant-based nanomaterials and their antimicrobial activity. The discussion is arranged as follows: first the emergence of resistance to antimicrobial (antibiotic and anti-fungal) agents since their discovery are investigated. From this, is an acute need to switch to alternative approaches are witnessed. These alternative approaches are described next. One such approach is plant-based antimicrobials. The traditional knowledge on plant-based antimicrobials that has been accumulated over thousands of years is discussed. Scientific interrogations conducted on some of these medicinal plants are presented hereafter. Nanotechnology is a branch in science that has the potential to uncover the hitherto unseen properties of materials. The nanotechnological approaches employed to enhance the antimicrobial activity of phytoconstituents are described next. Biogenic metal nanoparticles and their antimicrobial activity are discussed after this. The toxicity concerns involving aforesaid nanoparticles are now described. The chapter ends with concluding remarks.

The emergence of antimicrobial resistance

The first incident of antibiotic resistance was reported soon after the discovery of penicillin. It was an *E. coli* strain capable of producing penicillinase enzyme to inactivate penicillin (Abraham and Chain 1988). In 1942, four penicillin resistant *Staphylococcus aureus* strains were found in hospitalized patients (Rammelkamp and Maxon 1942). There was a rapid rise in the reported infections caused by penicillin-resistant *S. aureus* in the next few years. In the late 1960s, more than 80% of both hospital and community-acquired strains of *S. aureus* were found to be penicillin-resistant (Lowy 2003). The rapid rise of penicillin resistance temporarily came to a halt with the introduction of second-generation, semi-synthetic methicillin in the 1960s. Nevertheless, methicillin-resistant bacterial strains soon emerged (Hartman and Tomasz 1981).

The *Enterobacteriaceae* group, particularly *Escherichia coli* species were identified with high rates of penicillin resistance (Nordmann 1998). Between 1950–2001, nearly two-thirds of human disease-causing *E. coli* strains were found to be ampicillin-resistant (Bouza and Cercenado 2002).

The development of antibiotic resistance came together with the introduction of new generations of penicillin into clinical practice. Over 150 antibiotics have been introduced since the discovery of penicillin, and for the majority of existing antibiotics, resistance has emerged. Further, the recent increase of multi/pan-drug resistant strains has occurred with enhanced mortality and morbidity. Overall, failure of the antibiotic treatments to superbug infections has been responsible for the persistence and spread of multi-drug resistant species worldwide which represents a global public health threat (Bergman et al. 2009). The World Health Organization (WHO) in its "Report of global antimicrobial resistance-2014" revealed high levels of antibiotic resistance in the common infection-causing bacteria (Brown 2015). During the past few decades, infections caused by MRSA (Methicillin-Resistant *Staphylococcus aureus*), fluoroquinolone-resistant *Pseudomonas aeruginosa*, and VRE (Vancomycin-Resistant *Enterococcus faecium*) had been rapidly increasing, while, the incidence of pan-antibiotic resistant infections due to Acinetobacter species, multi drug-resistant *P. aeruginosa* and carbapenem-resistant *Klebsiella* species (Souli et al. 2008) had significantly inclined. Moreover, the occurrence of multidrug-resistant 'ESKAPE' (*Enterococcus* species, *Staphylococcus aureus*, *Klebsiella* species, *Acinetobacter baumannii*, *Pseudomonas aeruginosa*, and *Enterobacter* species) has turned out to be a public health threat (Boucher et al. 2009). There are strains of some Gram-negative bacteria such as *A. baumannii* that have acquired resistance to all currently available antibiotics and have been widely spread in clinics also due to their strong viability and colonization occurrence rate (Higgins et al. 2010; Zhi-Wen et al. 2015).

Normally, antibiotics are capable of shutting down or subverting the essential bacterial cell functions, while bacterial antibiotic resistance mechanisms appear to exploit each possible strategy of preventing a drug from attacking its target. Some of the well-understood resistance mechanisms include

i) target modification; for instance, a mutation in the 30S ribosomal protein which confers resistance to streptomycin, efflux of the drug and restricted penetration
ii) destruction of the antibiotic; e.g., Distortion of the β-lactam ring by β-lactamases target modification (Walsh 2003; Alekshun and Levy 2007).

Antibiotic tolerance is a different phenomenon where bacterial cells tolerate the antimicrobial drug. The main culprit responsible for antibiotic tolerance of pathogens is a specialized survivor called a "persister" (Lewis 2007; Lewis 2010), and the role of these persisters in the recalcitrance and relapse of bacterial infections has gained recognition over the recent past (Fisher et al. 2017). Persisters are phenotypic variants of actively dividing cells of a population produced stochastically, but are not mutants and the relative abundance of persisters reaches 1% at the late exponential phase of growth (Keren et al. 2004). Thus, persisters are described as slow-growing or growth-arrested cells with the ability to resume growth after a lethal stress. Their tolerance to bactericidal antimicrobial drugs depends on the active targets for killing the cell (Keren et al. 2004; Dorr et al. 2010; Fisher et al. 2017).

Persisters also play a significant role in the development of conventional antibiotic-resistant mutants. These are killed very slowly, if at all, and resume growth when antibiotic concentrations fall. It results in relapsing infection with a large population size that facilitates the development of antibiotic resistance (Levin and Rozen 2006). The significance of persisters in the recalcitrance of incurable infectious diseases raises the need of developing promising therapeutic approaches that effectively kill both dormant and actively dividing pathogenic bacteria.

Once considered a rare incidence, the development of resistance to antifungal therapy is on the rise in many clinical settings nowadays. Some fungal genera possess intrinsic resistance to antifungal therapy while others acquire the resistance during therapy. Fungal infections are considered as one of the major causes of morbidity and mortality in immunocompromised patients (Perea and Patterson 2002; Cowen et al. 2014; Srinivasan et al. 2014). Factors like the expansion of the immunocompromised and severely ill patient population including HIV infected patients, organ transplant recipients receiving immune-suppressants and cancer patients suffering from chemotherapy-induced neutropenia have been identified as major causes for the occurrence of life-threatening systemic fungal infections. In addition, treatment with broad-spectrum antifungals and glucocorticoids, frequent use of invasive medical procedures, receiving hemodialysis or peritoneal dialysis also contributed to the above situation (Lortholary and Dupont 1997; Perea and Patterson 2002).

The major causative agent for nosocomial fungal infections is *Candida* species, especially *C. albicans* being the most abundant etiological agent of systemic fungal infections in the bloodstream. *Aspergillus* infections together with *Candida* infections, which commonly affect granulocytopenic and other immunocompromised patients comprise 90% of all prevalent nosocomial fungal infections (Fridkin and Jarvis 1996). The major opportunistic pathogenic fungus that is responsible for the fungal infections in immunocompromised individuals such as HIV patients, cancer patients receiving immunosuppressive treatments, individuals undergoing organ transplantation and patients with serious medical conditions is *Cryptococcus neoformans* (Mitchell and Perfect 1995). Moreover, rarely found deep fungal infections are caused by Zygomycetes members such as *Rhizomucor pusillus*, *Rhizopus arrhizus* and *Absidia corymbifera* and Ascomycota members like *Fusarium* spp., *Blastoschizomyces capitatus*; Basidiomycetes like *Trichosporon beigelii* are intrinsically becoming resistant to the available antifungal medications (Fridkin and Jarvis 1996).

According to Perea and Patterson (2002), currently available antifungal drugs for the treatment of invasive mycoses can be categorized into four different groups based on their mechanism of action. They are;

1) Membrane alternation function (e.g., amphotericin B)
2) Inhibition of the synthesis of DNA/RNA (e.g., flucytosine)
3) Inhibition of ergosterol biosynthesis (e.g., azoles such as fluconazole, miconazole, ketoconazole, itraconazole, voriconazole, posaconazole and ravuconazole)
4) Inhibition of glucan synthesis (e.g., echinocandins such as micafungin, caspofungin and anidulafungin) (Perea and Patterson 2002)

The antifungal resistance depends on a combination of factors such as the host, antifungal agent or the fungal pathogen. However clinical resistance is not always able to predict, it highlights the significance of individualizing antifungal treatment strategies based on the clinical situation (Kanafani and Perfect 2008).

Before the introduction of antiretroviral therapy for AIDS patients, there was a rise in the prevalence of the fluconazole-resistant *C. albicans* with oropharyngeal or esophageal candidiasis. Heavy use of fluconazole is believed as the major cause of azole resistance (Goldman et al. 2000). A study conducted by Law et al. (1994) revealed that one-third of AIDS patients have fluconazole-resistant *C. albicans* in their oral cavities (Law et al. 1994). Krcmery and Barnes (2002) reported the fluconazole resistance in Non-Albicans *Candida* (NAC) species such as *C. tropicalis*, *C. glabrata*, *C. krusei* and *C. parapsilosis* that are responsible for bloodstream and nosocomial infections (Krcmery and Barnes 2002). The decreased drug concentration at the site of action due to the development of active efflux pumps, target site alteration, development of bypass pathways and up-regulation of target enzyme are the possible mechanisms for azole resistance in these fungal strains (Kanafani and Perfect 2008).

Despite the fact that the resistance to amphotericin-B among *Candida* strains is uncommon, the increasing Minimum Inhibitory Concentrations (MICs) of amphotericin B have been frequently reported for *C. krusei* and *C. glabrata* isolates (Pfaller et al. 2004). Moreover, intrinsic polyene resistance is commonly reported in *Candida lusitaniae* and *Trichosporon beigelii* (Walsh et al. 1990; Pfaller et al. 1994). Filamentous fungi exhibit more reduced susceptibility than yeasts to polyenes. Among *Aspergillus* species, *A. terreus* is resistant to amphotericin-B (Sabatelli et al. 2006). In addition, it is reported that the polyene resistance is rising among other *Aspergillus* species, such as *A. flavus* and *A. fumigatus*, which traditionally show the highest susceptibility towards amphotericin-B (Messer et al. 2006; Sabatelli et al. 2006).

The mutations in the gene responsible for ergosterol biosynthesis lead to the accumulation of other sterols in the fungal membrane. Hence, polyene resistant *Candida* and *Cryptococcus* isolates possess lower ergosterol content compared to polyene susceptible isolates (Dick et al. 1980). Sokol-Anderson et al. (1986) stated that amphotericin-B resistance of *C. albicans* may be mediated by enhanced catalase activity together with the decreasing susceptibility to oxidative damage (Sokol-Anderson et al. 1986).

On the other hand, echinocandins have a high potential against *Candida* and *Aspergillus* species. However, these agents are not effective against *Cryptococcus, Fusarium, Trichosporon* and *Scedosporium* species and other zygomecytes (Espinel-Ingroff 2003). *Candida parapsilosis* and *Candida guilliermondii* have higher MIC values than *Candida albicans* for echinocandins (Mora-Duarte et al. 2002). In addition, breakthrough *Candida* infections caused by *C. parapsilosis* have been reported in individuals receiving echinocandins for other indications (Cheung et al. 2006). The mechanism for echinocandin resistance in *Candida* spp. involves mutations in the genes that encode the subunits of the glucan synthase enzyme (Perlin et al. 2017).

The prevalence of flucytosine resistance rates is reported as 1–2% among *C. albicans* and 2% among *C. neoformans* (Price et al. 1994; Brandt et al. 2001). However, the rate of the resistance development of yeasts to flucytosine has lead clinicians to use flucytosine in combination with other antifungal drugs mainly amphotericin-B (Hospenthal and Bennett 1998).

Thus the above observations underline the need of an effective antifungal stewardship program to control the antifungal drug resistance while introducing novel antifungal agents to the currently available narrow arsenal.

The need to switch to alternative approaches

Although antibiotic resistance against both Gram-positive and Gram-negative pathogenic bacteria continues to increase in the community, new classes of antibiotics have not been discovered since the introduction of daptomycin and linezolid in the 1980s (Durand et al. 2019). In recent years, only the optimization or combination of already known compounds was performed resulting in some commercialized products and as a result, the amount of newly approved antimicrobial agents has been decreasing (Singh and Barrett 2006; Donadio et al. 2010).

Similarly, the limitations in currently available antifungal therapies such as toxicity, interactions, restricted routes of administration and drug resistance, demands the more potent drugs with novel mechanisms of actions to overcome the antifungal resistance as well as new formulations with enhanced bioavailability and safety profiles (Rauseo et al. 2020). Therefore, alternative strategies such as plant-based bioactive components, phage therapy and antimicrobial peptides should be investigated, as these approaches could offer alternative routes to prevent antibiotic resistance (Lewis 2013).

Plant-based antimicrobials: Revisiting the traditional knowledge

Traditional herbal medicines have been in use for thousands of years in various parts of the world for the treatment of infectious diseases (Anyanwu and Okoye 2017). For example, in ancient civilizations, plants like frankincense and myrrh were employed as mouthwashes (Cowan 1999) whereas *Tagetes minuta* had been used in Argentinian folk medicine to treat multiple infectious diseases (Tereschuk et al. 1997). Moreover, propolis was prescribed by Hippocrates in ancient Greece to treat sores and ulcers while *Scutellaria baicalensis* had been applied topically in Chinese traditional medicine to treat periodontal abscesses and infected oral wounds (Cushnie and Lamb 2005). Although literature reveals the bioactivities of medicinal plants for the treatment of infections of the skin, lungs and gastrointestinal tract, etc. (Carlson 2002), only a few comprehensive ethnobotanical studies have been conducted so far on herbal remedies used against infective diseases. Some significant findings of these ethnobotanical studies are given below.

Amuka et al. (2014) studied the use of plant materials for infective diseases by the Ogiek community in the Mau forests in the central Rift Valley of Kenya. Their study revealed the use of 50 plant species belonging to 32 plant families for various microbial ailments like stomachache and dysentery, pneumonia, tuberculosis, gonorrhea, leprosy and syphilis. *Hypoestes verticillaris, Gomphocarpus semilunatus, Lagenaria sphaerica, Momordica rostrata, Rhamnus prinoides* and *Zanthoxylum chalybeum* are some of the plant species mentioned by the traditional healers who participated in the study. The majority of the plant remedies were formulated in water, while the others were either formulated in honey or cooked with meat soup. In addition, several plant raw materials were soaked in water for some time and an aliquot was used after decanting the solid material (Amuka et al. 2014).

The ethnobotanical survey conducted in Nkonkobe Municipality of the Eastern Cape Province, South Africa recorded 27 plant species used for the treatment of ear, nose and throat infections. These plants belonged to 21 families and out of those Asteraceae was identified as the most prominent plant family followed by the Lamiaceae, Alliaceae and Rutaceae. *Artemisia afra, Lippia javanica* and *Allium sativum* were frequently mentioned against throat infections while *Sansevieria hyacinthoides* and *Cotyledon orbiculata* were widely used to treat ear infections. *Asclepias fruticosa* was reported as the most popular plant remedy against nose infections. The herbal materials were administered orally as a drink to heal throat infections, powdered leaf materials were sniffed for nose infections, and the leaf sap was squeezed directly into the ear (Dyubeni and Buwa 2012).

In an ethnobotanical study carried out in the Bunda District, Mara Region of Tanzania, 52 plants belonging to 29 Families were reportedly used for various infectious diseases. Out of these families, Papilionaceae was the most frequent, followed by Compositae, Anacardiaceae, Labiatae and Mimosaceae. *Chenopodium ambrosioides, Hoslundia opposita, Euphorbia tirucalli, Combretum adenogonium, Dichrostachys cinerea, Cardiospermum halicacabum* were some of the commonly used plant species among the communities in this region. A large number of plant species were used in the treatment of venereal diseases, particularly syphilis and gonorrhea (Maregesi et al. 2007). Moreover, 52 plant species in 25 families were identified in a more recent study conducted for the documentation of traditional knowledge on medicinal plants in the management of sexually transmitted infections in

Sesheke District of the Western Province, Zambia. Diseases like gonorrhea, syphilis and genital herpes were frequently controlled with the use of plants such as *Terminalia sericea, Strychnos cocculoides, Cassia occidentalis, Momordica balsamina* and *Peltophorum africanum* (Chinsembu 2016).

In remote areas of the Haripur District, Khyber Pakhtunkhwa, Pakistan, ethnobotanical surveys were conducted to gather knowledge on plant species used by the inhabitants against different infectious diseases. Subsequently, 10 plant species were evaluated in detail for their antimicrobial activity against several Gram-positive and Gram-negative bacterial species. The results had revealed a potentially promising activity in ethanolic extract of *Azadirachta indica*, ethanolic extract of *Punica granatum* and aqueous and ethanolic extracts of *Eucalyptus globulus* and *Bergenia ciliata* (Khan et al. 2014).

Omwenga et al. (2009) conducted an ethnobotanical survey for the purpose of documenting the plant species used by the Samburu Community in Kenya for the treatment of diarrhea. They performed *in vitro* antimicrobial activity studies on the most commonly mentioned plant species against several Gram-positive and Gram-negative bacterial species. Sixteen medicinal plants including six *Acacia* species and two *Cissus* species were identified in this study. The antibacterial activity of these 16 plant species was evaluated by the agar disk diffusion method against *Staphylococcus aureus, Bacillus subtillis, Salmonella typhi, Escherichia coli* and *Pseudomonas aeruginosa*. Most of the plants exhibited a prominent antimicrobial activity against *E. coli* and *S. typhi*, the causative organisms of diarrhea (Omwenga et al. 2009).

Numerous attempts have been made to collect information on the utilization of herbal remedies for the treatment of infectious diseases by local communities in the Indian subcontinent. Fifty-four species of medicinal plants were in use against diarrhea and dysentery by the people inhabiting in Pachalur hills of Dindigul district in Tamil Nadu of South India. These plants belonged to 31 plant families. Of these, species like *Acorus calamus, Murraya koenigii, Sesbania grandiflora* and *Shorea robusta* were used in the treatment of both dysentery and diarrhea (Shanmugam et al. 2011). A similar study was conducted in Mayurbhanj district, Odisha, India to identify the plants used by tribes of Similipal forest. This study revealed the use of 47 plant species belonging to 33 families by the local community to treat diarrhea and dysentery, of which *Acorus calamus, Aegle marmelos, Centella asiatica, Curculigo orchioides, Emblica officinalis, Oroxylum indicum, Syzygium cumini* and *Terminalia bellirica* were identified as most cited species (Rout and Panda 2017). Another study conducted among the tribal communities in Meghalaya, Northeast India had revealed the utilization of 58 plant species to treat and cure diarrhea and dysentery (Laloo and Hemalatha 2011) while 53 plant species belonging to 34 families were frequently used by the Lushai community from Bandarban district in Bangladesh owing to the antidysentery and antidiarrheal activities (Uddin et al. 2015).

In another study, 35 indigenous medicinal plants were identified in Urmia, Iran that were effective against infectious diseases of urinary, reproductive, digestive, respiratory tracts and the skin. Those plant species belonged to 17 plant families and the Lamiaceae had the highest frequency of plants for the treatment of infectious diseases. *Althaea hirsuta, Mentha longifolia* were some examples of plant species that had been used against pulmonary infections. *Alhagi camelorum* and *Glycyrrhiza glabra* were commonly used against intestinal infections while *Dipsacus laciniatus* and *Equisetum arvense* were used against urinary tract infections. *Lamium album* was employed as a remedy for both urinary tract infections and vaginitis. Some examples of plants that have been used against skin infections included *Sanguisorba minor, Verbascum agrimonifolium* and *Ixiolirion tataricum*. Plant leaves were the most widely used part in preparation of herbal remedies and most of the medicinal herbs were prescribed in the boiled forms (Bahmani et al. 2015).

Recently, Phumthum and Balslev (2020) carried out a meta-analysis to identify the plants used by the Karen community in northern and western Thailand for treating infectious diseases caused by protozoa, bacteria, fungi, helminth worms and viruses. This revealed that *Tinospora crispa, Aesculus assamica, Cleidion javanicum* and *Tadehagi triquetrum* as plant species with the high Cultural Index (CI) against bacterial infections while *Senna alata* and *Tithonia diversifolia* were identified as powerful plants in the treatment of fungal infections (Phumthum and Balslev 2020). In 2018, Bizuayehu and Garedew reviewed 48 published and unpublished ethnobotanical studies on medicinal plants used for the treatment of gonorrhea in Ethiopia which enabled them to compile a total of 100 anti-gonorrhea plants belonged 46 families. The highest number of anti-gonorrhea medicinal plant species were reported in Euphorbiaceae (13 species), Cucurbitaceae (9 species) and Solanaceae (8 species) families. Roots were found to be the most commonly used plant parts while most of the preparations were taken orally (Bizuayehu and Garedew 2018).

An ethnobotanical survey carried out in Guatemala for the purpose of collecting information about the commonly used species for treating vaginitis recorded over 100 plant species and the *in vitro* anticandidal activity studies were performed using 71 of the reported plant species. Out of the tested plant species, the most potent activity was observed in *Solanum nigrescens*. The clinical trials conducted on non-pregnant women with confirmed *Candida albicans* vaginitis who had been treated with intra-vaginal suppositories containing *S. nigrescens* maceration revealed that this herbal preparation was as efficacious as the reference drug nystatin (Giron et al. 1988).

Thus, it is clear that the ethnobotanical observations are inevitable not only in the selection of plant species for pharmacological screening but also to preserve the local knowledge that has been transferred over generations without any written pieces of evidence. Some of the scientific investigations conducted on the antimicrobial activity of popular medicinal plants are described next.

Phytochemicals with antimicrobial activity: A brief outlook

A large number of plant extracts have demonstrated *in vitro* antimicrobial activities and some examples are summarized below.

Prasannabalaji et al. (2012) evaluated the *in vitro* antimicrobial activity of ethanol, methanol and acetone extracts of South Indian traditional medicinal plants *Ocimum sanctum, O. gratissimum, Aegle marmelos, Adhatoda vasica* against *Staphylococcus aureus, Salmonella typhi, Salmonella paratyphi, Escherichia coli* and *Klebsiella pneumoniae*. They reported a MIC Value of 0.039 mg/mL for the methanolic extract of *O. gratissimum* against *S. typhi* and *S. paratyphi*, 0.0625 mg/mL against *E. coli* and 0.078 mg/mL against *S. aureus*. Similarly, the methanolic extract of *A. marmelos* also showed prominent inhibitory action against the above organisms with MIC values in the range of 0.156–0.625 mg/mL (Prasannabalaji et al. 2012).

In another study, antibacterial activity was observed in ethanolic extracts of *Psidium guajava, Phyllanthus niruri, Ehretia microphylla* and *Piper betle* leaves against the Gram-positive methicillin-resistant *S. aureus* and vancomycin-resistant *Enterococcus*. The highest antimicrobial activity was found in *P. betle* extract with MIC value of 19–156 µg/mL (Valle et al. 2015). The antimicrobial activity of 20 plant species used by an aborigine tribe of Kalahandi district, in India was determined by Dubey et al. (2012) against 10 multidrug-resistant pathogenic bacteria. The ethanolic extracts of *Anthocephalus cadamba* and *Pterocarpus santalinus* displayed conspicuous activity against all the bacterial species (Dubey et al. 2012). Similarly, another research group in India investigated the effect of *Acacia nilotica, Syzygium aromaticum, Cinnamomum zeylanicum, Terminalia arjuna* and *Eucalyptus globulus* on multidrug-resistant (MDR) and control strains of Gram-positive and Gram-negative bacterial species. The most potent activity was observed in the ethanolic extract of *A. nilotica* with a MIC range of 9.75–31.3 µg/mL (Khan et al. 2009).

Several studies conducted in Brazil also revealed the promising antimicrobial activities of traditional medicinal plants. The extracts prepared from 10 plants collected from the Amazon region in the Amapa State were subjected to antibacterial activity studies against MDR bacterial strains. This study revealed that *Geissospermum argenteum, Uncaria guianensis, Brosimum acutifolium, Copaifera reticulata, Licania macrophylla, Ptychopetalum olacoides* and *Dalbergia subcymosa* extracts were effective against MDR *S. aureus* and *P. aeruginosa* (Correia et al. 2008). In another study, Violante et al. (2012) screened the extracts prepared from six plant species endemic to the Cerrado of the Central-West region of Brazil, i.e., *Erythroxylum suberosum, Hyptis crenata, Roupala brasiliensis, Simarouba versicolor, Guazuma ulmifolia* and *Protium heptaphyllum* against several bacterial and fungal pathogens. All the plants showed antimicrobial properties to a certain extent, while *R. brasiliensis* displayed the broadest spectrum of action by inhibiting all the fungal strains tested (*Candida albicans, C. glabrata, C. krusei, C. parapsilosis, C. tropicalis* and *Cryptococcus neoformans*) with the strongest activities against *C. glabrata* (MIC = 15.6 µg/mL) and *C. krusei* (MIC = 31.3 µg/mL). Furthermore, the results obtained from this study supported the indigenous use of the investigated plants in the treatment of respiratory, gastrointestinal and skin diseases (Violante et al. 2012).

The family Clusiaceae is comprised of many plant species with strong antimicrobial potencies. For example, the hexane and DCM fractions of *Calophyllum rubiginosum* exhibited MIC values of 100 and 12.5 µg/mL-against *S. aureus* and *Bacillus cereus* respectively (ALkhamaiseh et al. 2011). The resins of *Calophyllum antillanum* displayed MIC of 31.8 µg/mL against *S. aureus* and 23.8 µg/mL against *Trichophyton rubrum* while the resins of *C. inophyllum* also inhabited *S. aureus* and *T. rubrum* with MIC of 1.1 and 3.3 µg/mL respectively. The ethanolic extracts prepared from resin, leaves and fruits of *Garcinia mangostana* showed antimicrobial activity against *S. aureus* and *E. coli* with a MIC of 0.1 mg/mL (Cunha et al. 2014). Recently, Napagoda et al. (2020) examined the antimicrobial activity of the extracts prepared from *Garcinia cambogia* against Gram-positive *S. aureus, Staphylococcus saprophyticus, Enterococcus faecalis* as well as some clinical isolates of methicillin resistant *S. aureus* (MRSA). The hexane and dichloromethane extracts displayed a potent activity with MIC in the range of 31.25–250 µg/mL along with strong disinfectant potencies (Napagoda et al. 2020).

Apart from the crude plant extracts, the essential oils obtained from many plant species have shown potent antimicrobial properties. For example, essential oils derived from *Cymbopogon citratus* inhibited the growth of *E. coli, Salmonella typhimurium* and *S. aureus* with MIC values of 0.6, 2.5 and 0.6 µL/mL respectively (Burt 2004). Furthermore, the essential oils extracted from *Ocimum gratissimum* showed profuse antibacterial activity against *E. coli* (MIC = 6 µg/mL), *Pseudomonas aeruginosa* (MIC ≥ 24 µg/mL) and *S. aureus* (MIC = 0.75 µg/mL) (Nikolić et al. 2014). Similarly, the essential oils of *Lavandula angustifolia* displayed a MIC value of 0·25% (v/v) against *E. coli*, and 1·0% (v/v) against *S. aureus*. The essential oils of the same plant effectively inhibited the growth of *C. albicans* with a MIC of 0.25% (v/v). This study also revealed the strong antimicrobial potential of the essential oils extracted from *Thymus vulgaris* against *C. albicans* and *E. coli* [MIC of 0·12% (v/v)], and *Vetiveria zizanioides* against *C. albicans* and *E. faecalis* [MIC of 0.12% (v/v)]. Moreover, the essential oils of *V. zizanioides* displayed MIC of 0·06% (v/v) against *S. aureus* as well (Hammer et al. 1999).

Apart from the aforementioned *in vitro* studies, a few reports exist on clinical trials with respect to antimicrobial activity of herbal remedies. One of the notable studies was a randomized, double-blind, placebo-controlled clinical trial conducted to determine the effects of regular ingestion of cranberry juice on urinary tract infections. This study was performed using a group of elderly women who consumed 300 mL of a standard commercially available cranberry beverage per day. Their urine was collected monthly for six months and examined for the presence of bacteria and white blood cells. The findings revealed that the women who consumed cranberry juice had fewer bacteria in their urine than what was observed in untreated women (Avorn 1996).

The antimicrobial potential of a plant extract is determined by the presence of various phytochemicals in the given extract. The major groups of phytochemicals that are responsible for antimicrobial activity include alkaloids, saponins, terpenoids, and phenolic compounds (Gupta and Birdi 2017). Some examples for each of these classes are given in Table 9.1.

In addition to the aforementioned examples, different types of xanthones have been documented for their potent antimicrobial activities against MRSA strains (Fig. 9.1). For example, calozeyloxanthone isolated from *Calophyllum moonii* and *C. lankensis* inhibited MRSA strains with MIC = 4.1–8.1 µg/mL (Dharmaratne et al. 1999). Similarly, many xanthones like 1,7-dihydroxyxanthone, rubraxanthone, α-mangostin, β-mangostin, y-mangostin, isolated from *Garcinia mangostana* showed strong anti-MRSA activity. The

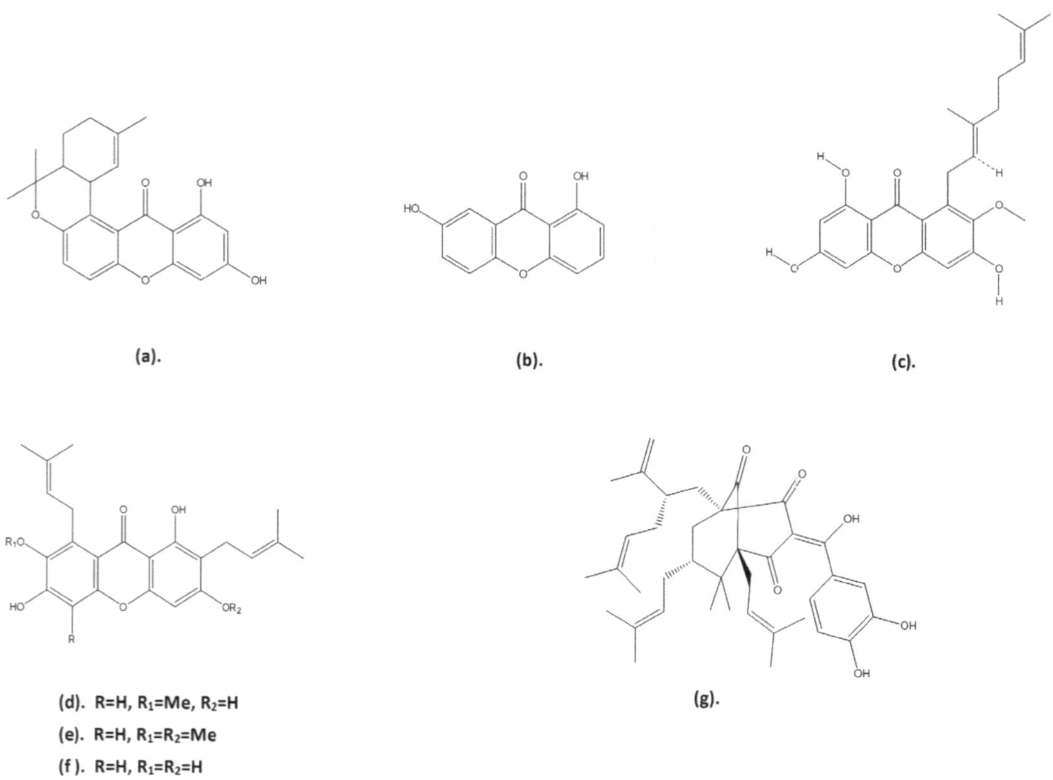

Figure 9.1. Phytochemicals possessing potent activity against MRSA. (a). calozeyloxanthone (b). 1,7-dihydroxyxanthone (c). rubraxanthone (d). α-mangostin (e). β-mangostin (f). γ-mangostin (g). garcinol.

Table 9.1. Some examples of phytochemicals with antimicrobial activities.

Class	Subclass	Example	Antimicrobial activity against	Reference
Phenolics	Phenolic acids	*p*-hydroxycinnamic acid	*Bacillus cereus, B. megaterium*	Fernandez et al., 1996
		Gallic acid	*Aeromonas hydrophila , A. sobria*	
	Quinones	Hypericin	*Staphylococcus aureus, S. epidermidis*	Feyzioğlu et al., 2013
	flavone	Chrysin	*Pseudomonas aeruginosa, Candida albicans*	Rojas et al., 1992
	flavonol	Galangin	*Aspergillus tamarii, A. flavus, Cladosporium sphaerospermum, Penicillium digitatum, P. italicum*	Afolayan and Meyer, 1997
	flavanol	Catechin	*Staphylococcus aureus, Salmonella typhimurium*	Peres et al., 1997
		Epicatechin	*Vibrio cholera*	Toda et al., 1992
	flavanone	Naringenin	MRSA (methicillin-resistant *S. aureus*)	Tsuchiya et al., 1996
	Tannins	Punicalagin (ellagitannins)	MRSA, MSSA (methicillin-sensitive *S. aureus*)	Machado et al., 2002
	Coumarin	Aegelinol	*Staphylococcus aureus, Salmonella typhi*	Basile et al., 2009
		Agasyllin	*Enterobacter cloacae, E. aerogenes*	
Terpenoids, Essential oils		Eugenol	*Candida albicans, C. neoformans, Escherichia coli, Enterobacter aerogenes, Proteus vulgaris, Salmonella typhimurium, Pseudomonas aeruginosa*	Marchese et al., 2017
		Thymol	*Candida lusitaniae, Bacillus subtilis, B.cereus*	Falcone et al., 2005
		Geraniol	*Staphylococcus aureus, Listeria monocytogenes, Escherichia coli, Salmonella enterica*	Zanetti et al., 2015
		Citral	*Escherichia coli, Salmonella typhimurium, Listeria monocytogenes, Staphylococcus aureus*	Fisher and Phillips, 2008
Alkaloid		Piperine	*Escherichia coli, Staphylococcus aureus, Proteus vulgaris, Pseudomonas aeruginosa Klebsiella pneumoniae, Candida albicans*	Aldaly, 2010
		Berberine	*Streptococcus agalactiae*	Peng et al., 2015
		Caffeine	*Staphylococcus aureus, Bacillus cereus, Escherichia coli, Proteus mirabilis, Klebsiella pneumoniae*	Nonthakaew et al., 2015
		Ephedrine	*Staphylococcus aureus, Acinetobacter*	Tulgar et al., 2018
Saponins		3-O-β-D-glucuronopyranosyl-oleanolic acid	*Escherichia coli, Shigella flexneri, Staphylococcus aureus*	Tagousop et al., 2018
		O-β-D-glucuronopyranosyloleanolic acid 28-O-β-D-glucopyranosyl ester	*Escherichia coli, Shigella flexneri, Staphylococcus aureus, Cryptococcus neoformans*	
		3-O-β-D-glucopyranosyl(1 → 2)-β-D-glucuronopyranosyl oleanolic acid	*Escherichia coli, Shigella flexneri Staphylococcus aureus, Candida albicans, C. parapsilosis, Cryptococcus neoformans*	

strongest inhibitory effect was observed with rubraxanthone with the MIC value of 0.313–1.25 µg/mL against both MRSA and MSSA isolates. The inhibitory strength was even higher than that of vancomycin (MIC = 3.13–6.25 µg/mL) which was used as the reference drug (Iinuma et al. 1996). In addition, garcinol, a polyisoprenylated benzophenone derivative isolated from *G. purpurea* also displayed potent activity against some MRSA strains with MIC = 6.25–25 µg/mL (Iinuma et al. 1996).

Enhancing the antimicrobial activity of phytoconstituents: Nanotechnological approaches

Although herbal extracts and plant-derived bioactive natural products exhibit potent activities with diverse mechanisms of action under *in vitro* conditions, these herbal drugs encounter many limitations at the stage of clinical applications. The low solubility of phytochemicals hinders the development of drug formulations resulting in inferior circulation in the bloodstream. Many of these drugs could exhibit low gastrointestinal absorption, while some other drugs get metabolized quickly limiting their effectiveness for only a short duration of time. In this respect, the use of nano-inspired or nano-enabled phytochemical delivery systems could circumvent the limitations of herbal drugs by improving the solubility, bioavailability and uptake, target specificity, and kinetics while lowering the toxicity (Wang et al. 2014).

Nanoemulsions (NE) are placed among the most popular classes of nanomaterials for topical and systemic delivery of active compounds. These nanomaterials are a heterogeneous mixture of two immiscible liquids either oil in water or water in oil, thus the incorporation of the active ingredients in the nucleus (hydrophilic or hydrophobic) protects these bioactive compounds against enzymatic degradation and prevents recognition by macrophages. As a result, circulation time increases and it allows for versatility of drug delivery. dos Santos Ramos et al. (2019) incorporated a fraction obtained from the crude extract of *Syngonanthus nitens* into a nanoemulsion composed of cholesterol as the oil phase, polyoxyethylene 20-cetyl ether and soy phosphatidylcholine as a surfactant and a phosphate buffer (pH 7.4) + chitosan polymer dispersion as the aqueous phase. This *S. nitens* extract loaded with nanoemulsions was tested *in vivo* using the vulvovaginal candidiasis (VVC) rodent model. Vulvovaginal candidiasis is one of the common disease conditions among women and is caused by the genus *Candida* thus requires effective antifungal therapies. The results of the *in vivo* VVC assay revealed that the inclusion of *S. nitens* fraction into nanoemulsion had promoted the antifungal activity, hence could be used in the treatment of acute VVC (dos Santos Ramos et al. 2019). In another study, Ghazy et al. (2021) formulated a nanoemulsion of anise seed extract and evaluated the bioactivity against some selected foodborne pathogenic bacteria. The anise seed extract was made using the cold solvent extraction method with ethanol. The HPLC analysis revealed the presence of anethole, naringenin and taxifolin as the major chemical compounds in the seed extract. The average droplet size of the nanoemulsion was determined as 440 nm and it displayed antimicrobial activity against *S. aureus*, *Bacillus cereus*, *E. coli*, *Yersinia enterocolitica* and *Listeria monocytogene*. Interestingly, higher antimicrobial activity was observed for the nanoemulsion in comparison to the bulk anise extract (Ghazy et al. 2021). Similarly, Chuesiang et al. (2021) demonstrated that the antimicrobial activity of cinnamon oil could be enhanced by nanoemulsion approach (Chuesiang et al. 2021).

Polymers with amphiphilic properties are capable of assembling into a wide variety of morphologies thus are ideal as drug delivery candidates (Thota and Jiang 2015). Among the various polymers that have been developed to formulate polymeric nanoparticles, Poly (Lactic-co-Glycolic Acid) (PLGA) has drawn immense attention over recent years. PLGA is a biodegradable co-polymer and is widely used in drug delivery systems (Danhier et al. 2012). In an earlier attempt, PLGA nanoparticles with entrapped eugenol and trans-cinnamaldehyde were synthesized by Gomes et al. (2011) using the emulsion evaporation method. Over 90% entrapment efficiency was observed for both types of nanoparticles that were found to be effective in inhibiting the growth of *Salmonella* spp. and *Listeria* spp. (Gomes et al. 2011). Thereafter, Hill et al. (2013) synthesized two types of PLGA nanoparticles with two different ratios of lactide to glycolide along with the cinnamon bark extract using an emulsion-solvent evaporation method. Both types of PLGA displayed a more gradual release of cinnamon bark extract from the polymer matrix and effectively inhibited *Salmonella* spp. and *Listeria* spp. than the free cinnamon bark extract (Hill et al. 2013). Later, PLGA nanoparticles of acerola (*Malpighia emarginata*), guava (*Psidium guajava*) and passion fruit (*Passiflora edulis*) by-product extracts were synthesized and evaluated for antimicrobial activity against *L. monocytogenes* Scott A and *E. coli*. Interestingly, all the fruit by-product encapsulated in PLGA nanoformulations were capable of inhibiting the growth of tested bacterial species at lower concentrations than corresponding unencapsulated fruit by-product. The acerola by-product extract was found to be superior over the other extracts for nanoencapsulation in terms of entrapment efficiency, size, controlled release profile and antibacterial potency (Silva et al. 2014). Recently, Pereira et al. (2018) synthesized PLGA nanoparticles by an adapted emulsion-evaporation method for the delivery of phenolic extract of *Campomanesia xanthocarpa* fruit with enhanced functional properties. Both free fruit extract and its loaded-PLGA nanoparticles inhibited *Listeria innocua*, but once the extract was nanoencapsulated with PLGA, the antibacterial activity had significantly increased (Pereira et al. 2018).

Chitosan is a biocompatible, biodegradable and non-toxic polymer having numerous applications in medicine and industry. Chen et al. (2009) prepared eugenol and carvacrol-grafted chitosan nanoparticles via the Schiff base reaction and the antibacterial activity of these nanomaterials was evaluated against *E. coli* and *S. aureus*. The essential oil component-grafted chitosan nanoparticles displayed less cytotoxicity towards the mouse fibroblasts than the free essential oils and higher antimicrobial activity than the unmodified chitosan nanoparticles (Chen et al. 2009).

Mesoporous silica nanoparticles have great potential as drug delivery vehicles and their biocompatibility as well as the dynamics of drug release could significantly enhance with a titania surface coating (Farooq et al. 2018). Brezoiu et al. (2020) investigated the nanoconfinement effect of plant extracts into mesopores of silica and titania nanomaterials on the radical scavenging and antimicrobial properties. In this study, polyphenolic extracts of *Salvia officinalis* and *Thymus serpyllum* were embedded into the above mesoporous inorganic matrices and the antibacterial, antifungal and radical scavenging activities have been evaluated. The stability of the phytochemicals was enhanced by embedding the extracts into mesoporous inorganic matrices, while the antimicrobial and radical scavenging potencies were higher in extract-loaded materials than the corresponding free extract (Brezoiu et al. 2020).

Besides the field of medicine, the antimicrobial activity of phytochemical-loaded nanomaterials are also useful in the food industry. For example, Ghosh et al. (2014) formulated an oil-in-water nanoemulsion using sesame oil, non-ionic surfactant and water for the delivery of eugenol. This eugenol-loaded nanoemulsion exhibited antibacterial activity against *S. aureus* by altering membrane permeability. This nanoemulsion was able to preserve orange juice against microbial spoilage (Ghosh et al. 2014). Similarly, nanoemulsions of anise oil were prepared and the effects of these nanomaterials on the survival of the food-borne pathogens, *L. monocytogenes* and *E. coli* were investigated. The results indicated that the nanoemulsions of anise oil possess better antimicrobial activity than the bulk anise oil (Topuz et al. 2016). A recent study that describes the antibacterial potential of the eugenol entrapped ethosome nanoparticles against fruit anthracnose revealed that the antibacterial activity of eugenol entrapped ethosome nanoparticles was significantly higher than that of the free eugenol (Jin et al. 2019).

Casein, the major milk protein of bovine milk is capable of self-assembling to the micellar structure in solution, with diameters in the range 10–400 nm. The amphiphilic nature of casein makes it an efficient nanocarrier that could encapsulate the poorly water-soluble compounds in the hydrophobic core of the micelle. Therefore, Xue et al. (2019) developed eugenol-casein nanoparticles with a satisfactory sustained-release effect and demonstrated that those nanoparticles could significantly potentiate the antifungal activity of eugenol reducing the anthracnose incidence. Thus eugenol-casein nanoparticles would be useful as environmental-friendly preservatives for postharvest fruits (Xue et al. 2019).

Nanofibers/nanocomposites have attracted tremendous attention over recent years as materials with diverse applications. Nanofibers can be prepared by chemical, electrochemical and electrospinning techniques, although electrospinning is recognized as the most versatile and viable method. Not only that, electrospun nanofibers are found to be superior over nanofibers fabricated by other methods with respect to the properties like high porosity, high surface area per unit mass, etc. A large number of natural and synthetic polymers were used for electrospinning and many researchers had focused on using nanofibers containing natural products for various practical applications (Xue et al. 2019).

Kegere et al. (2019) fabricated electrospun nanofibers from Poly(Vinyl Alcohol) (PVA), *Bidens pilosa* extract and chitosan and thereafter the antibacterial potential of this nanocomposite was investigated against *E. coli* and *S. aureus*. The nanofiber composites showed higher antibacterial activity than their parental forms (Kegere et al. 2019). In another study, PVA and inulin were used in the fabrication of composite nanofibers (CNFs). The prebiotic activity of the inulin/PVA CNFs was evaluated with *Lactobacillus* species while the antibacterial activity was determined against *E. coli* and *S. aureus*. A 33% increase in the *Lactobacillus* sp. growth as well as a higher antibacterial activity against *E. coli* and *S. aureus* were observed with inulin/PVA CNFs in comparison to the pure inulin solution (Wahbi et al. 2020).

Curcumin, a natural polyphenolic compound present in *Curcuma longa* was used to fabricate gelatin/poly(caprolactone)/Curcumin nanofibers by electrospinning. The antimicrobial activity of this nanomaterial was determined against Methicillin-Resistant *S. aureus* (MRSA) and Extended-Spectrum β-Lactamase (ESBL), the microorganisms that are responsible for nosocomial infections in hospitals. Interestingly, the fabricated nanofibers displayed potent antibacterial activity against the tested organisms, showing promise for antimicrobial applications (Fallah et al. 2016). A similar study was performed by Ramalingam et al. (2015) to produce curcumin loaded poly(2-hydroxy ethyl methacrylate) nanofiber using the electrospinning method. The drug release pattern and antibacterial potency against MRSA and ESBL were investigated and the controlled and sustained release pattern of curcumin was found to be highly effective in the treatment of infections caused by multidrug-resistant organisms (Ramalingam et al. 2015).

Electrospun nanofibers are widely applied as materials for wound dressing. The high oxygen porosity and different pore sizes, high surface to volume ratio as well as the comparable texture of these materials to the natural extracellular matrix in the skin could facilitate the wound healing process. In an early attempt, polycaprolactone (PCL) and polyvinyl pyrrolidone nanofiber mat containing *Tecomella undulata* extract was fabricated and the antibacterial activity was evaluated against *P. aeruginosa*, *S. aureus* and *E. coli*. The *T. undulata* extract loaded PCL/PVP nanofiber mat inhibited the growth of the tested bacterial species and demonstrated the possible utility as a wound healing matrix (Suganya et al. 2011). More recently, Fayemi et al. (2018) fabricated polyacrylonitrile nanofibers containing moringa (MR) leaf extracts for wound dressing. These nanomaterials displayed a profound antibacterial activity and wound healing potency signifying their potential to develop into effective wound dressings (Fayemi et al. 2018). Zeyohanness et al. (2018) also prepared electrospun nanofibers containing a plant extract for wound dressing purposes. Ethyl acetate extract of *Rhodomyrtus tomentosa* and PVA solutions were used for the nanofiber production and the antimicrobial effect of the nanomaterial was determined against *E. coli, P. aeruginosa, B. subtilis* and *E. faecalis*. Interestingly, the *R. tomentosa* loaded PVA nanofiber mats were found to be active against all four bacterial strains tested (Zeyohanness et al. 2018).

Nanocomposites loaded with phytoconstituents also have various applications in the food industry. For example, Alboofetileh et al. (2018) developed a biodegradable nanocomposite films by incorporating montmorillonite nanoclay and marjoram essential oil in sodium alginate matrix. This nanocomposite film displayed antibacterial activity by inhibiting the growth of *E. coli, L. monocytogenes, B. cereus* and *S. aureus* and demonstrated its suitability to develop as biodegradable packaging materials capable of prolonging the shelf life of food products (Alboofetileh et al. 2018). Similarly, polyurethane nanofibers containing tea tree essential oils were fabricated by the electrospinning technique and their antimicrobial and cytotoxicity properties have been evaluated. The findings confirmed the potential applicability of these nanofibers as fruit packaging materials (Lee et al. 2020).

In another study, the essential oils of *Laurus nobilis* and *Rosmarinus officinalis* were incorporated within a zein protein matrix by an electrospinning technique to develop novel active edible films. Significant antimicrobial activity was observed against *L. monocytogenes, S. aureus* and aerobic mesophilic bacteria in cheese slices coated with the essential oil loaded zein nanofibers films (Göksen et al. 2020).

The discussion above suggests that the phytoengineered nanomaterials with antimicrobial potential would offer a broad spectrum of applications in medicine, agriculture industry and others in the near future.

Biogenic metal nanoparticles and their antimicrobial activity

Nanoparticles exhibit hitherto unseen yet favorable properties due to their size and shape. Metal nanoparticles can be synthesized by physical, chemical and biological methods. While keeping in mind some drawbacks related to the physical and chemical synthesis methods, biogenic synthesis of metallic nanoparticles has received significant attention in the recent past. Biological synthesis of metal nanoparticles is an eco-friendly approach that utilizes biological organisms like bacteria, fungi, plants, viruses, actinomycetes, algae and yeasts. This approach does not require any toxic chemicals for the synthesis or releases hazardous by-products to the environment (Dhillon et al. 2012; Sable et al. 2012; Ingle et al. 2021). Hence, it is inexpensive, energy-saving, safe and sustainable (Prasad et al. 2011). Metal nanoparticles that are produced using plant extracts and their antimicrobial activity are focused here.

In this approach, biomolecules that are secreted from leaves, flowers, seeds, bark, root, etc. of the plant are responsible for the reduction of the metal ions and stabilization of the nanoparticles. Green synthesis of nanoparticles also follows the same mechanism as other methods where an aqueous solution of metal salts which are made up of metal ions are first reduced into metal atoms by using reducing substances and thereafter those metal atoms are aggregated together to form very small clusters. Then those clusters grow into small particles known as nanoparticles (Banker et al. 2010). This is a widely researched area. For example, Devadiga et al. (2015) synthesized silver nanoparticles (AgNP) using *Tectona grandis* (teak) leaf extract. Silver nanoparticles formed in this study were approximately 28 nm in size and spherical in shape. Their results showed a significant reduction of bioactive compounds such as flavonoids, polyphenols and antioxidants that acted as reducing agents and stabilizers during the green synthesis. These AgNPs showed antibacterial activity against *E. coli* and *S. aureus* with a minimum inhibitory concentration (MIC) of 25.6 µg/mL against both organisms (Devadiga et al. 2015). Murugan et al. (2014) produced AgNPs using bark extract of *Acacia leucophloea* and evaluated their antibacterial activity. These nanoparticles were spherical and had an average size of 17–19 nm. Fourier transform infrared spectroscopy suggested that primary amines, aldehyde/ketone, aromatic, azo and nitro compounds of the bark extract are responsible for the bioreduction and capping of the nanoparticles. These nanoparticles were found to be effective against *S. aureus*, *B. cereus*, *L. monocytogenes* and *S. flexneri* (Murugan et al. 2014). Arshad et al. (2021) synthesized AgNPs using an aqueous root extract of *Salvadora persica*. Fourier transform infrared spectroscopy (FTIR) suggested that OH–, C=N– and NH–functional groups of phenolic, proteins or nitrogenous compounds of the root extract are responsible for the reduction and stabilization of produced AgNPs. In addition to this, produced nanoparticles were in 37.5 nm size and demonstrated antibacterial activity against *E. coli* and *S. epidermis* with a MIC of 0.39 µg/mL and 0.19 µg/mL respectively (Arshad et al. 2021). Similarly, AgNPs synthesized using *Nymphae odorata* extract as well as the root extract of *Echinops kebericho* demonstrated significant antibacterial activity against *E. coli* and *S. aureus* (Gudimalla et al. 2021; Hairat et al. 2021).

Gold nanoparticles (AuNPs) have been synthesized from the leaf extract of *Nigella arvensis* which contained 212 µg/mL of flavonoids and 145 µg/mL of phenolic compounds. The resulted AuNPs were spherical and 3–37 nm in size. These nanoparticles exhibited strong antibacterial activity against *Staphylococcus epidermidis*, *Bacillus subtilis*, *S. aureus*, *E. coli*, *Serratia marcescens* and *P. aeruginosa* with MIC values of 250, 62.5, 125, 62.5, 125, and 250 µg/mL respectively. Based on these results, it was speculated that AuNPs could bind to the bacterial outer cell membrane and produce some structural changes causing degradation and finally cell death. In addition, inside the bacterial cell, AuNPs could interact with DNA and inhibit the transcription resulting in growth inhibition and cellular damage (Chahardoli et al. 2017). In another study, AuNPs were synthesized from *Euphorbia hirta* leaf extract and the antibacterial assay was conducted against *E. coli*, *P. aeruginosa* and *Klebsiella pneumoniae*. At a 200 µg/mL concentration, an inhibition of 88, 86 and 96% was found against *E. coli*, *P. aeroginosa* and *K. pneumoniae* respectively. It was concluded that the bioactive compounds of the plant had adhered to the AuNPs and the synergistic effect was responsible for the antibacterial activity (Annamalai et al. 2013).

Rajesh et al. (2018) produced copper nanoparticles (CuNPs) biogenically from the bud extract of *Syzygium aromaticum* and thereafter the antimicrobial activity was investigated. The SEM images revealed that the CuNPs were approximately 20 nm in size and spherical in shape. The small size of these CuNPs enabled these materials to interact easily with bacteria than larger particles. The resulted CuNPs exhibited antibacterial effects against *Staphylococcus* spp., *E. coli*, *Pseudomonas* spp. and *Bacillus* spp. This antibacterial activity of CuNPs was attributed to the release of copper ions (Cu^{+}) which could attach to the cell wall of the bacteria by electrostatic attraction. In addition, the metal ions could penetrate the membrane of the bacteria and exert toxic effects. The antifungal activity studies were also performed against *Aspergillus niger*, *A. flavus* and *Penicillium* spp. and the synthesized CuNPs exhibited strong fungicidal activity against *Penicillium* spp. (Rajesh et al. 2018). Shende et al. (2015) prepared biogenic CuNPs using *Citrus medica* juice. The synthesized nanoparticles were in the range of 10–60 nm. Antibacterial activity studies were performed against *E. coli*, *P. acne*, *K. pneumoniae*, *S. typhi* and *P. aeruginosa* while the antifungal activity was evaluated against *Fusarium oxysporum*, *F. graminearum* and *F. culmorum*. Among the tested microorganisms *E. coli* was the most sensitive bacteria and *F. culmorum* was the highest sensitive fungi. Considering earlier hypothesis on antimicrobial mechanisms, the investigators proposed a mechanism giving five possibilities; the accumulation of CuNPs on the cell surface would form pits resulting in a leakage, reduction of transmembrane electrochemical potential due to the interaction between CuNPs and the membrane which ultimately affects the membrane integrity, DNA damage due to the interactions between CuNPs and DNA, inactivation of proteins by interacting with sulfhydryl groups of the protein, development of oxidative stress with the entry of CuNPs and Cu^{2+} into the cell resulting in cell death (Shende et al. 2015).

Zinc oxide nanoparticles (ZnONPs) have also been synthesized using several biogenic materials. For example, *Nyctanthes arbor-tristis* flower extract was used for the synthesis of ZnONPs and their antifungal activity was examined. TEM images revealed that the produced nanoparticles were in 12–32 nm in size at optimum conditions. The antifungal assay was performed against *Alternaria alternata, A. niger, Botrytis cinerea. F. oxysporum, Penicillium expansum* which resulted in MIC values of 64, 16, 128, 64, and 128 µg/mL respectively (Jamdagni et al. 2018). Similarly, Senthilkumar and Sivakumar (2014) produced ZnONPs from green tea (*Camellia sinensis*). The average size was 16 nm when calculated using XRD data. A higher amount of phenolic compounds that are present in green tea extract provided the reducing action while proteins, aminoacids and lipids present in the extract helped to stabilize the nanoparticles. The antibacterial studies were performed against *K. pneumoniae, P. aeruginosa, E. coli* and *S. aureus*, however, only Gram-negative bacteria (*K. pneumoniae, P. aeruginosa, E. coli*) displayed a good sensitivity with a MIC value of 20 µg/mL. Similarly, antifungal activity of ZnONPs was also tested against *Aspergillus fumigatus, Penicillium* spp., *A. flavus* and *A. niger.* Inhibitory zones were found at the concentration of 20 µg/mL (Senthilkumar and Sivakumar 2014). Pachaiappan et al. (2021) synthesized ZnONPs by using *Justicia adhatoda* leaf extract and the prepared ZnONPs exhibited antibacterial activity against *S. aureus, E. coli, P. aeruginosa* and antifungal activity against *A. niger, A. flavus* and *A. fumigatus*. It was speculated that these nanoparticles could release Zn^{2+} and form reactive oxygen species which cause the destruction of microbes (Pachaiappan et al. 2021).

Thakur et al. (2019) synthesized titanium dioxide nanoparticles (TiO_2NPs) from *Azadirachta indica* (neem) leaf extract and evaluated their antimicrobial properties. The nanoparticles were spherical and 15–50 nm in size. FTIR data revealed the presence of terpenoids, flavonoids and proteins in this extract which were responsible for the stabilization and production of nanoparticles. Antibacterial assays were performed against *E. coli, Bacillus subtilis, S. typhi* and *K. pneumoniae* and the lowest MIC value of 10.42 µg/mL was seen against both *S. typhi* and *Escherichia coli* (Thakur et al. 2019). Subhapriya and Gothamipriya (2018) conducted experiments to synthesize TiO_2NPs using *Trigonella foenum-graecum* extract and thereafter to evaluate the antimicrobial activity. Resulting TiO_2NPs were spherical in shape and 20–90 nm in size. *S. aureus, S. faecalis, E. coli, P. vulgaris, E. faecalis, P. aeruginosa, Yersinia enterocolitica, B. subtilis* and fungus *C. albicans* were used in the antibacterial assay. Synthesized nanoparticles inhibited the growth of all above mentioned bacterial species (Subhapriya and Gothamipriya 2018).

Based on the aforementioned evidence, it is clear that some of the biogenetically synthesized metal nanoparticles possess strong antimicrobial activities, qualifying them to be effectively used to solve the antimicrobial resistance problem.

Possible toxicity issues

Despite nanotoxicity being a widely researched topic (Akçan et al. 2020), there are not many reports on the direct toxic effects of phytoengineered nanomaterials with antimicrobial activity. Some examples of nanotoxicity in general are highlighted below.

A toxicity study conducted on *Drosophila melanogaster* to examine the influence of particle size of AgNPs on the development of Drosophila's eggs revealed that nanoparticles that were < 100 nm in size were less toxic on eggs than nanoparticles > 100 nm in size. Furthermore, AgNPs reduced the ability of pupae to emerge as adults (Gorth et al. 2011). Cohen et al. (2013) studied the CuO NP toxicity on human skin and concluded that nano-sized particles have greater toxicity than micro-sized particles and are capable of inducing necrosis and cytokine secretion (Cohen et al. 2013). However, Mohammadyari et al. (2014) found that CuO nanoparticles are safe upto 400 ppm for biomedical applications based on a study that involved Wistar rats (Mohammadyari et al. 2014). Meanwhile, a study carried out on embryos of zebrafish revealed that the exposure of TiO_2 nanoparticles had increased the premature hatching of the zebrafish and decreased the time that was required for a normal mature (Samaee et al. 2015). In another study with TiO_2 nanoparticles of 4, 10, 30, and 134 nm sizes, Osborne et al. (2012) did not observe a toxic effect on zebrafish embryos (Osborne et al. 2012). On the other hand, several *in vitro* and *in vivo* studies carried out on animal and human skin revealed that nanoparticles were confined to the level of the stratum corneum after topical application, and the shedding and turnover of the stratum corneum would prevent the accumulation of nanoparticles (Jansen et al. 2013). These controversial views on nontoxicity demand more reliable and reproducible screening approaches to investigate bio-physicochemical interactions at the nano/bio interface and thereby to assess the hazardous effects of the nanomaterials.

Conclusion

Since the discovery of penicillin in the last century, antimicrobial agents have helped to control infectious diseases and saved millions of valuable lives. Their dominance however is diminishing fast due to the gradual and systematic development of microbial resistance. Mortality and morbidity rates are on the rise, stressing the need for a new class of antimicrobial substances with enhanced resistance and resilience. The new candidates may overcome the toxicity, interactions, restricted routes of administration and drug resistance, while ensuring the enhanced bioavailability and safety profiles. The best answer to many questions is often found in nature. For thousands of years, humans have been using wild plants to treat infectious diseases. Modern research has revealed a long list of plant extracts that have demonstrated *in vitro* antimicrobial activities of varying degree. The wide array of secondary metabolites such as phenolic compounds, alkaloids, terpenoids, etc., present in these plants could have elicited the pharmacological effects.

Although herbal extracts and plant-derived bioactive natural products exhibit potent activities with diverse mechanisms of action under *in vitro* conditions, these herbal drugs encounter many limitations at the stage of clinical applications. For instance, the low solubility and hydrophobicity, low gastrointestinal absorption, quick metabolizing, biodegradability and stability. Recent advances in nanotechnology have shown the potential to enhance the performance of these phytochemicals as antimicrobials. The use of nano-inspired or nano-enabled phytochemical delivery systems would circumvent the limitations of herbal drugs by improving the solubility, bioavailability and uptake, target specificity and kinetics while lowering the toxicity.

Nevertheless, the properties of nanomaterials is an area that is being rigorously interrogated by the researchers. Their less-known toxicity may have a direct appeal to human health, while their long-term stability would decide the efficacy and hence the commercial value of the product. These concerns need to be resolved before a nano-embedded antimicrobial agent is introduced in the market.

References

Abraham, E.P., Chain, E. (1988). An enzyme from bacteria able to destroy penicillin. 1940. Rev Infect Dis 10(4): 677–678.

Afolayan, A.J., Meyer, J.J. (1997). The antimicrobial activity of 3,5,7-trihydroxyflavone isolated from the shoots of *Helichrysum aureonitens*. J Ethnopharmacol 57: 177–181.

Akçan, R., Aydogan, H.C., Yildirim, M.Ş., Taştekin, B., Sağlam, N. (2020). Nanotoxicity; a challenge for future medicine. Turk J Med Sci 50(4): 1180–1196.

Alboofetileh, M., Rezaei, M., Hosseini, H., Abdollahi, M. (2018). Morphological, physico-mechanical, and antimicrobial properties of sodium alginate-montmorillonite nanocomposite films incorporated with marjoram essential oil. J Food Process Preserv 42(5): e13596. doi:10.1111/jfpp.13596.

Aldaly, Z.T.K. (2010). Antimicrobial activity of piperine purified from *Piper nigrum*. J Basrah Res (Sciences) 36 A(5).54.

Alekshun, M.N., Levy, S.B. (2007). Molecular mechanisms of antibacterial multidrug resistance. Cell 128(6): 1037–1050.

ALkhamaiseh, S.I., Taher, M., Ahmad, F. (2011). The phytochemical contents and antimicrobial activities of malaysian *Calophyllum rubiginosum*. Am J Appl Sci 8(3): 201–205.

Aminov, R.I. (2010). A brief history of the antibiotic era: lessons learned and challenges for the future. Front Microbiol 1: 134. doi: 10.3389/fmicb.2010.00134.

Amuka, O., Okemo, P.O., Machocho, A.K., Mbugua, P.K. (2014). Ethnobotanical survey of selected medicinal plants used by Ogiek communities in Kenya against microbial infections. Ethnobot Res Appl 12: 627–641.

Annamalai, A., Christina, V.L.P., Sudha, D., Kalpana, M., Lakshmi, P.T.V. (2013). Green synthesis, characterization and antimicrobial activity of Au NPs using *Euphorbia hirta* L. leaf extract. Colloids Surf B: Biointerfaces 108: 60–65.

Anyanwu, M.U., Okoye, R.C. (2017). Antimicrobial activity of Nigerian medicinal plants. J Intercult Ethnopharmacol 6(2): 240–259.

Arshad, H., Sami, M.A., Sadaf, S., Hassan, U. (2021). *Salvadora persica* mediated synthesis of silver nanoparticles and their antimicrobial efficacy. Sci Rep 11: 5996.

Avorn, J. (1996). The effect of cranberry juice on the presence of bacteria and white blood cells in the urine of elderly women. What is the role of bacterial adhesion? Adv Exp Med Biol 408: 185–186.

Bahmani, M., Saki, K., Shahsavari, S., Rafieian-Kopaei, M., Sepahvand, R., Adineh, A. (2015). Identification of medicinal plants effective in infectious diseases in Urmia, northwest of Iran. Asian Pac J Trop Biomed 5(10): 858–864.

Bankar, A., Joshi, B., Kumar, A.R., Zinjarde, S. (2010). Banana peel extract mediated synthesis of gold nanoparticles. Colloids Surf. B: Biointerfaces 80: 45–50.

Basile, A., Sorbo, S., Spadaro, V., Bruno, M., Maggio, A., Faraone, N., Rosselli, S. (2009). Antimicrobial and antioxidant activities of coumarins from the roots of *Ferulago campestris* (Apiaceae). Molecules 14(3): 939–952.

Bergman, M., Nyberg, S.T., Huovinen, P., Paakkari, P., Hakanen, A.J. (2009). Association between antimicrobial consumption and resistance in *Escherichia coli*. Antimicrob Agents Chemother 53(3): 912–917.

Bizuayehu, B., Garedew, B. (2018). A review on the ethnobotanical study of medicinal plants used for the treatment of gonorrhea disease in Ethiopia. Indian J Nat Prod Resour 9(3): 183–193.

Boucher, H.W., Talbot, G.H., Bradley, J.S., Edwards, J.E., Gilbert, D., Rice, L.B., Scheld, M., Spellberg, B., Bartlett, J. (2009). Bad bugs, no drugs: no ESKAPE! An update from the Infectious Diseases Society of America. Clin Infect Dis 48(1): 1–12.

Bouza, E., Cercenado, E. (2002). *Klebsiella* and *Enterobacter*: Antibiotic resistance and treatment implications. Semin Respir Infect 17(3): 215–230.

Brandt, M.E., Pfaller, M.A., Hajjeh, R.A., Hamill, R.J., Pappas, P.G., Reingold, A.L., Rimland, D., Warnock, D.W., Cryptococcal Disease Active Surveillance Group. (2001). Trends in antifungal drug susceptibility of *Cryptococcus neoformans* isolates in the United States: 1992 to 1994 and 1996 to 1998. Antimicrob Agents Chemother 45(11): 3065–3069.

Brezoiu, A.-M., Prundeanu, M., Berger, D., Deaconu, M., Matei, C., Oprea, O., Vasile, E., Negreanu-Pîrjol, T., Muntean, D., Danciu, C. (2020). Properties of *Salvia officinalis* L. and *Thymus serpyllum* L. extracts free and embedded into mesopores of silica and titania nanomaterials. Nanomaterials 10(5): 820. doi:10.3390/nano10050820.

Brown, D. (2015). Antibiotic resistance breakers: Can repurposed drugs fill the antibiotic discovery void? Nat Rev Drug Discov 14(12): 821–832.

Burt, S. (2004). Essential oils: their antibacterial properties and potential applications in foods—a review. Int J Food Microbiol 94(3): 223–253.

Carlson, T.J.S. (2002). Medical ethnobotanical research as a method to identify bioactive plants to treat infectious diseases. *In*: Iwu, M.M., Wootton, J.C. (eds.). Advances in Phytomedicine-Ethnomedicine and Drug Discovery 1: 45–53, Elsevier, NL.

Chahardoli, A., Karimi, N., Sadeghi, F., Fattahi, A. (2017). Green approach for synthesis of gold nanoparticles from Nigella arvensis leaf extract and evaluation of their antibacterial, antioxidant, cytotoxicity and catalytic activities. Artif Cells Nanomed Biotechnol 46(3): 579–588.

Chen, F., Shi, Z., Neoh, K.G., Kang, E.T. (2009). Antioxidant and antibacterial activities of eugenol and carvacrol-grafted chitosan nanoparticles. Biotechnol Bioeng 104(1): 30–39.

Cheung, C., Guo, Y., Gialanella, P., Feldmesser, M. (2006). Development of candidemia on caspofungin therapy: a case report. Infection 34(6): 345–348.

Chinsembu, K.C. (2016). Ethnobotanical study of medicinal flora utilised by traditional healers in the management of sexually transmitted infections in Sesheke District, Western Province, Zambia. Rev Bras Farmacogn 26(2): 268–274.

Chuesiang, P., Sanguandeekul, R., Siripatrawan, U. (2021). Enhancing effect of nanoemulsion on antimicrobial activity of cinnamon essential oil against foodborne pathogens in refrigerated Asian seabass (*Lates calcarifer*) fillets. Food Control 122: 107782.

Cohen, D., Soroka, Y., Ma'or, Z., Oron, M., Portugal-Cohen, M., Brégégère, F.M., Berhanu, D., Valsami-Jones, E., Hai, N., Milner, Y. (2013). Evaluation of topically applied copper(II) oxide nanoparticle cytotoxicity in human skin organ culture. Toxicol *In Vitro* 27(1): 292–298.

Correia, A.F., Segovia, J.F.O., Gonavalves, M.C.A., De Oliveria V.L., Silveria, D., Carvalho, J.C.T., Kanzaki, L.I.B. (2008). Amazonian plant crude extracts screening for activity against multidrug-resistant bacteria. Eur Rev Med Pharmacol Sci 12: 369–380.

Cowan, M.M. (1999). Plant products as antimicrobial agents. Clin Microbiol Rev 12(4): 564–582.

Cowen, L.E., Sanglard, D., Howard, S.J., Rogers, P.D., Perlin, D.S. (2014). Mechanisms of antifungal drug resistance. Cold Spring Harb Perspect Med 5(7): a019752. https://doi.org/10.1101/cshperspect.a019752.

Cranston, D., Sidebottom, E. (2016). Penicillin and the Legacy of Norman Heatley. 1st ed. Words By Design, Oxfordshire, UK.

Cunha, B.L.A., de França, J.P., Moraes, A.A.F.S., Chaves, A.L.F., Gaiba, S., Fontana, R., Kersul do Sacramento, C., Ferreira, L.M., Pereira de França, L. (2014). Evaluation of antimicrobial and antitumoral activity of *Garcinia mangostana* L. (mangosteen) grown in Southeast Brazil. Acta Cir Bras 29(Suppl. 2): 21–28.

Cushnie, T.P., Lamb, A.J. (2005). Antimicrobial activity of flavonoids. Int J Antimicrob Agents 26(5): 343–356.

Danhier, F., Ansorena, E., Silva, J.M., Coco, R., Le Breton, A., Préat, V. (2012). PLGA-based nanoparticles: an overview of biomedical applications. J Control Release 161(2): 505–522.

Devadiga, A., Shetty, K.V., Saidutta, M.B. (2015). Timber industry waste-teak (*Tectona grandis* Linn.) leaf extract mediated synthesis of antibacterial silver nanoparticles. Int Nano Lett 5: 205–214.

Dharmaratne, H.R.W., Wijesinghe, W.M.N.M., Thevanasem, V. (1999). Antimicrobial activity of xanthones from *Calophyllum* species, against methicillin-resistant *Staphylococcus aureus* (MRSA). J Ethnopharmacol 66(3): 339–342.

Dhillon, G.S., Brar, S.K., Kaur, S., Verma. M. (2012). Green approach for nanoparticle biosynthesis by fungi: current trends and applications. Crit Rev Biotechnol 32(1): 49–73.

Dick, J.D., Merz, W.G., Saral, R. (1980). Incidence of polyene-resistant yeasts recovered from clinical specimens. Antimicrob Agents Chemother 18(1): 158–163.

Donadio, S., Maffioli, S., Monciardini, P., Sosio, M., Jabes, D. (2010). Antibiotic discovery in the twenty-first century: Current trends and future perspectives. J Antibiot Res 63(8): 423–430.

Dörr, T., Vulić, M., Lewis, K. (2010). Ciprofloxacin causes persister formation by inducing the TisB toxin in *Escherichia coli*. PLoS Biol 8(2): e1000317. doi: 10.1371/journal.pbio.1000317.

dos Santos Ramos, M.A., da Silva, P.B., de Toledo, L.G., Oda, F.B., da Silva, I.C., Dos Santos, L.C., Dos Santos, A.G., de Almeida, M., Pavan, F.R., Chorilli, M., Bauab, T.M. (2019). Intravaginal delivery of *Syngonanthus nitens* (Bong.) Ruhland fraction based on a nanoemulsion system applied to vulvovaginal candidiasis treatment. J Biomed Nanotech 15(5): 1072–1089.

Dubey, D., Sahu, M.C., Rath, S., Paty, B.P., Debata, N.K., Padhy, R.N. (2012). Antimicrobial activity of medicinal plants used by aborigines of Kalahandi, Orissa, India against multidrug resistant bacteria. Asian Pac J Trop Biomed 2(2): S846–S854.

Durand, G.A., Raoult, D., Dubourg, G. (2019). Antibiotic discovery: History, methods and perspectives. Int J Antimicrob Agents 53(4): 371–382.

Dyubeni, L., Buwa, L.V. (2012). An ethnobotanical study of plants used for the treatment of ear, nose and throat (ENT) infections in Nkonkobe Municipality, South Africa. J Med Plant Res 6(14): 2721–2726.

Espinel-Ingroff, A. (2003). *In vitro* antifungal activities of anidulafungin and micafungin, licensed agents and the investigational triazole posaconazole as determined by NCCLS methods for 12,052 fungal isolates: review of the literature. Rev Iberoam Micol 20(4): 121–136.

Falcone, P., Speranza, B., Nobile, M.A.D., Corbo, M.R., Sinigaglia, M. (2005). A study on the antimicrobial activity of thymol intended as a natural preservative. J Food Prot 68(8): 1664–1670.

Fallah, M., Bahrami, S.H., Ranjbar-Mohammadi, M. (2016). Fabrication and characterization of PCL/gelatin/curcumin nanofibers and their antibacterial properties. J Ind Text 46(2): 562–577.

Farooq, A., Shukur, A., Astley, C., Tosheva, L., Kelly, P., Whitehead, D., Azzawi, M. (2018). Titania coating of mesoporous silica nanoparticles for improved biocompatibility and drug release within blood vessels. Acta Biomater 76: 208–216.

Fayemi, O.E., Ekennia, A.C., Katata-Seru, L., Ebokaiwe, A.P., Ijomone, O.M., Onwudiwe, D.C., Ebenso, E.E. (2018). Antimicrobial and wound healing properties of polyacrylonitrile-moringa extract nanofibers. ACS Omega 3(5): 4791–4797.

Fernandez, M.A., Garcia, M.D., Saenz, M.T. (1996). Antibacterial activity of the phenolic acids fraction of *Scrophularia frutescens* and *Scrophularia sambucifolia*. J Ethnopharmacol 53: 11–14.

Feyzioğlu, B., Demircili, M.E., Özdemir, M., Doğan, M., Baykan, M., Baysal, B. (2013). Antibacterial effect of hypericin. Afr J Microbiol Res 7(11): 979–982.

Fisher, K., Phillips, C. (2008). Potential antimicrobial uses of essential oils in food: Is citrus the answer? Trends Food Sci Technol 19(3): 156–164.

Fisher, R.A., Gollan, B., Helaine, S. (2017). Persistent bacterial infections and persister cells. Nat Rev Microbiol 15(8): 453–464.

Fleming, A. (1929). On the antibacterial action of cultures of a penicillium, with special reference to their use in the isolation of B. influenzæ. Br J Exp Pathol 10(3): 226–236.

Fletcher, C. (1984). First clinical use of penicillin. Br Med J (Clinical Research ed.), 289(6460): 1721–1723.

Fridkin, S.K., Jarvis, W.R. (1996). Epidemiology of nosocomial fungal infections. Clin Microbiol Rev 9(4): 499–511.

Ghazy, O.A., Fouad, M.T., Saleh, H.H., Kholif, A.E., Morsy, T.A. (2021). Ultrasound-assisted preparation of anise extract nanoemulsion and its bioactivity against different pathogenic bacteria. Food Chem 341(Pt 2): 128259.

Ghosh, V., Mukherjee, A., Chandrasekaran, N. (2014). Eugenol-loaded antimicrobial nanoemulsion preserves fruit juice against, microbial spoilage. Colloids Surf B: Biointerfaces 114: 392–397.

Girón, L.M., Aguilar, G.A., Cáceres, A., Arroyo, G.L. (1988). Anticandidal activity of plants used for the treatment of vaginitis in Guatemala and clinical trial of a *Solanum nigrescens* preparation. J Ethnopharmacol 22(3): 307–313.

Göksen, G., Fabra, M.J., Ekiz, H.I., López-Rubio, A. (2020). Phytochemical-loaded electrospun nanofibers as novel active edible films: Characterization and antibacterial efficiency in cheese slices. Food Control 112: 107133. doi:10.1016/j.foodcont.2020.107133.

Goldman, M., Cloud, G.A., Smedema, M., LeMonte, A., Connolly, P., McKinsey, D.S., Kauffman, C.A., Moskovitz, B., Wheat, L.J. (2000). Does long-term itraconazole prophylaxis result in *in vitro* azole resistance in mucosal *Candida albicans* isolates from persons with advanced human immunodeficiency virus infection? The National Institute of Allergy and Infectious Diseases Mycoses study group. Antimicrob Agents Chemother 44(6): 1585–1587.

Gomes, C., Moreira, R.G., Castell-Perez, E. (2011). Poly (DL-lactide-co-glycolide) (PLGA) nanoparticles with entrapped trans-cinnamaldehyde and eugenol for antimicrobial delivery applications. J Food Sci 76(2): N16–N24.

Gorth, D.J., Rand, D.M., Webster, T.J. (2011). Silver nanoparticle toxicity in *Drosophila*: size does matter. Int J Nanomedicine 6: 343–350.

Gudimalla, A., Jose, J., Varghese, R.J., Thomas, S. (2021). Green synthesis of silver nanoparticles using *Nymphae odorata* extract incorporated films and antimicrobial activity. J Polym Environ 29: 1412–1423.

Gupta, P.D., Birdi, T.J. (2017). Development of botanicals to combat antibiotic resistance. J Ayurveda Integr Med 8(4): 266–275.

Hairat, S., Abebe, H., Hairat, A., Esatu, H. (2021). Ecofriendly green synthesis and characterization of silver nanoparticles from root extract of *Echinops kebericho* and their antimicrobial activity against human pathogenic bacteria. Plant Cell Biotechnol Mol Biol 22(1-2): 121–135.

Hammer, K.A., Carson, C.F., Riley, T.V. (1999). Antimicrobial activity of essential oils and other plant extracts. J Appl Microbiol 86: 985–990.

Hartman, B., Tomasz, A. (1981). Altered penicillin-binding proteins in methicillin-resistant strains of *Staphylococcus aureus*. Antimicrob Agents Chemother 19(5): 726–735.

Higgins, P.G., Dammhayn, C., Hackel, M., Seifert, H. (2010). Global spread of carbapenem-resistant *Acinetobacter baumannii*. J Antimicrob Chemother 65(2): 233–238.

Hill, L.E., Taylor, T.M., Gomes, C. (2013). Antimicrobial efficacy of poly (DL-lactide-co-glycolide) (PLGA) nanoparticles with entrapped cinnamon bark extract against *Listeria monocytogenes* and *Salmonella typhimurium*. J Food Sci 78(4): N626–N632.

Hospenthal, D.R., Bennett, J.E. (1998). Flucytosine monotherapy for cryptococcosis. Clin Infect Dis 27: 260–264.

Iinuma, M., Tosa, H., Tanaka, T., Asai, F., Kobayashl, Y., Shimano, R., Miyauchi, K.-I. (1996). Antibacterial activity of xanthones from Guttiferaeous plants against Methicillin-resistant *Staphylococcus aureus*. J Pharm Pharmacol 48(8): 861–865.

Iinuma, M., Tosa, H., Tanaka, T., Kanamaru, S., Asai, F., Kobayashi, Y., Miyauchi, K., Shimano, R. (1996). Antibacterial activity of some *Garcinia* benzophenone derivatives against Methicillin-Resistant *Staphylococcus aureus*. Biol Pharm Bull 19: 311–314.

Ingle, A.P., Wagh, S., Biswas, J.K., Mondal, M., Feitosa, C.M., Rai, M. (2020). Phyto-fabrication of different nanoparticles and evaluation of their antibacterial and anti-biofilm efficacy. Curr Nanosci 17: 1–14.

Jamdagni, P., Khatri, P., Rana, J.S. (2018). Green synthesis of zinc oxide nanoparticles using flower extract of *Nyctanthes arbor-tristis* and their antifungal activity. J King Saud Univ Sci 30(2): 168–175.

Jansen, R., Osterwalder, U., Wang, S.Q., Burnett, M., Lim, H.W. (2013). Photoprotection: part II. Sunscreen: development, efficacy, and controversies. J Am Acad Dermatol 69(6): 867.e1–867.e14.

Jin, P., Yao, R., Qin, D., Chen, Q., Du, Q. (2019). Enhancement in antibacterial activities of eugenol-entrapped ethosome nanoparticles via strengthening its permeability and sustained release. J Agric Food Chem 67(5): 1371–1380.

Kanafani, Z.A., Perfect, J.R. (2008). Resistance to antifungal agents: mechanisms and clinical impact. Clin Infect Dis 46(1): 120–128.

Kegere, J., Ouf, A., Siam, R., Mamdouh, W. (2019). Fabrication of poly(vinyl alcohol)/chitosan/*Bidens pilosa* composite electrospun nanofibers with enhanced antibacterial activities. ACS Omega 4(5): 8778–8785.

Keren, I., Kaldalu, N., Spoering, A., Wang, Y., Lewis, K. (2004). Persister cells and tolerance to antimicrobials. FEMS Microbiol Lett 230(1): 13–18.

Keren, I., Shah, D., Spoering, A., Kaldalu, N., Lewis, K. (2004). Specialized persister cells and the mechanism of multidrug tolerance in *Escherichia coli*. J Bacteriol 186(24): 8172–8180.

Khan, N., Abbasi, A.M., Dastagir, G., Nazir, A., Shah, G.M., Shah, M.M., Shah, M.H. (2014). Ethnobotanical and antimicrobial study of some selected medicinal plants used in Khyber Pakhtunkhwa (KPK) as a potential source to cure infectious diseases. BMC Complement Altern Med 14: 122. doi: 10.1186/1472-6882-14-122.

Khan, R., Islam, B., Akram, M., Shakil, S., Ahmad, A., Ali, S.M., Siddiqui, M., Khan, A.U. (2009). Antimicrobial activity of five herbal extracts against multi drug resistant (MDR) strains of bacteria and fungus of clinical origin. Molecules 14(2): 586–597.

Krcmery, V., Barnes, A.J. (2002). Non-albicans *Candida* spp. causing fungaemia: Pathogenicity and antifungal resistance. J Hosp Infect 50(4): 243–260.

Laloo, D., Hemalatha, S. (2011). Ethnomedicinal plants used for diarrhea by tribals of Meghalaya, Northeast India. Pharmacogn Rev 5(10): 147–154.

Law, D., Moore, C.B., Wardle, H.M., Ganguli, L.A., Keaney, M.G.L., Denning, D.W. (1994). High prevalence of antifungal resistance in *Candida* spp. from patients with AIDS. J Antimicrob Chemother 34(5): 659–668.

Lee, J.Y., Lee, J., Ko, S.W., Son, B.C., Lee, J.H., Kim, C.S., Park, C.H. (2020). Fabrication of antibacterial nanofibrous membrane infused with essential oil extracted from tea tree for packaging applications. Polymers 12: 125. doi:10.3390/polym12010125.

Lerner, P.I. (2004). Producing penicillin. N Engl J Med 351(6): 524–524.

Levin, B.R., Rozen, D.E. (2006). Non-inherited antibiotic resistance. Nat Rev Microbiol 4(7): 556–562.

Lewis, K. (2007). Persister cells, dormancy and infectious disease. Nat Rev Microbiol 5(1): 48–56.

Lewis, K. (2010). Persister cells. Annu Rev Microbiol 64: 357–372.

Lewis, K. (2013). Platforms for antibiotic discovery. Nat Rev Drug Discov 12: 371–387.

Lobanovska, M., Pilla, G. (2017). Focus: drug development: penicillin's discovery and antibiotic resistance: lessons for the future? Yale J Biol Med 90(1): 135–145.

Lortholary, O., Dupont, B. (1997). Antifungal prophylaxis during neutropenia and immunodeficiency. Clin Microbiol Rev 10(3): 477–504.

Lowy, F.D. (2003). Antimicrobial resistance: the example of *Staphylococcus aureus*. J Clin Invest 111(9): 1265–1273.

Machado, T.de B., Leal, I.C.R., Amaral, A.C.F., antos, K.R.N. dos, Silva, M.G. da., Kuster, R.M. (2002). Antimicrobial ellagitannin of *Punica granatum* fruits. J Braz Chem Soc 13(5): 606–610.

Marchese, A., Barbieri, R., Coppo, E., Orhan, I.E., Daglia, M., Nabavi, S.F., Izadi, M., Abdollahi, M., Nabavi, S.M., Ajami, M. (2017). Antimicrobial activity of eugenol and essential oils containing eugenol: A mechanistic viewpoint. Crit Rev Microbiol 43(6): 668–689.

Maregesi, S.M., Ngassapa, O.D., Pieters, L., Vlietinck, A.J. (2007). Ethnopharmacological survey of the Bunda district, Tanzania: plants used to treat infectious diseases. J Ethnopharmacol 113(3): 457–470.

Messer, S.A., Jones, R.N., Fritsche, T.R. (2006). International surveillance of *Candida* spp. and *Aspergillus* spp.: Report from the SENTRY Antimicrobial Surveillance Program (2003). J Clin Microbiol 44(5): 1782–1787.

Mitchell, T.G., Perfect, J.R. (1995). Cryptococcosis in the era of AIDS-100 years after the discovery of *Cryptococcus neoformans*. Clin Microbiol Rev 8(4): 515–548.

Mohammadyari, A., Razavipour, S.T., Mohammadbeigi, M., Negahdary, M., Ajdary, M. (2014). Exploring vivo toxicity assessment of copper oxide nanoparticle in Wistar rats. J Biol Today's World 3(6): 124–128.

Mora-Duarte, J., Betts, R., Rotstein, C., Colombo, A.L., Thompson-Moya, L., Smietana, J., Lupinacci, R., Sable, C., Kartsonis, N., Perfect, J., Caspofungin Invasive Candidiasis Study Group. (2002). Comparison of caspofungin and amphotericin B for invasive candidiasis. N Engl J Med 347(25): 2020–2029.

Murugan, K., Senthilkumar, B., Senbagam, D., Al-Sohaibani, S. (2014). Biosynthesis of silver nanoparticles using *Acacia leucophloea* extract and their antibacterial activity. Int J Nanomedicine 9: 2431–2438.

Napagoda, M., Gerstmeier, J., Butschek, H., De Soyza, S., Pace, S., Lorenz, S., Qader, M., Witharana, S., Nagahawatte, A., Wijayaratne, G., Svatoš, A., Jayasinghe, L., Koeberle, A., Werz, O. (2020). The anti-inflammatory and antimicrobial potential of selected ethnomedicinal plants from Sri Lanka. Molecules 25(8): 1894.

Nikolić, M., Jovanović, K.K., Marković, T., Marković, D., Gligorijević, N., Radulović, S., Soković, M. (2014). Chemical composition, antimicrobial, and cytotoxic properties of five Lamiaceae essential oils. Ind Crops Prod 61: 225–232.

Nonthakaew, A., Matan, Na., Aewsiri, T., Matan, Ni. (2015). Caffeine in foods and its antimicrobial activity. Int Food Res J 22(1): 9–14.

Nordmann, P. (1998). Trends in β-lactam resistance among Enterobacteriaceae. Clin Infect Dis 27(Supplement_1): S100–S106.

Omwenga, E.O., Okemo, P.O., Mbugua, P.K., Ogol, C.K.P. (2009). Ethnobotanical survey and antimicrobial evaluation of medicinal plants used by the Samburu community (Kenya) for treatment of diarrhorea. Pharmacogn Mag 4(18): 165–176.

Osborne, O.J., Johnston, B.D., Moger, J., Balousha, M., Lead, J.R., Kudoh, T., Tyler, C.R. (2012). Effects of particle size and coating on nanoscale Ag and TiO_2 exposure in zebrafish (*Danio rerio*) embryos. Nanotoxicology 7(8): 1315–1324.

Pachaiappan, R., Rajendran, S., Ramalingam, G., Vo, D.N., Priya, P.M., Soto-Moscoso, M. (2021). Green synthesis of zinc oxide nanoparticles by *Justicia adhatoda* leaves and their antimicrobial activity. Chem Eng Technol 44(3): 551–558.

Peng, L., Kang, S., Yin, Z., Jia, R., Song, X., Li, L., Li, Z., Zou, Y., Liang, X., Li, L., He, C., Ye, G., Yin, L., Shi, F., Lv, C., Jing, B. (2015). Antibacterial activity and mechanism of berberine against *Streptococcus agalactiae*. Int J Clin Exp Pathol 8(5): 5217–5223.

Penesyan, A., Gillings, M., Paulsen, I.T. (2015). Antibiotic discovery: Combatting bacterial resistance in cells and in biofilm communities. Molecules 20(4): 5286–5298.

Perea, S., Patterson, T.F. (2002). Antifungal resistance in pathogenic fungi. Clin Infect Dis 35(9): 1073–1080.

Pereira, M.C., Oliveira, D.A., Hill, L.E., Zambiazi, R.C., Borges, C.D., Vizzotto, M., Mertens-Talcott, S., Talcott, S., Gomes, C.L. (2018). Effect of nanoencapsulation using PLGA on antioxidant and antimicrobial activities of guabiroba fruit phenolic extract. Food Chem 240: 396–404.

Peres, M.T.L.P., Monache, F.D., Cruz, A.B., Pizzolatti, M.G., Yunes, R.A. (1997). Chemical composition and antimicrobial activity of *Croton urucurana* Baillon (Euphorbiaceae). J Ethnopharmacol 56: 223–226.

Perlin, D.S., Rautemaa-Richardson, R., Alastruey-Izquierdo, A. (2017). The global problem of antifungal resistance: prevalence, mechanisms, and management. Lancet Infect Dis 17(12): e383–e392.

Pfaller, M.A., Messer, S.A., Hollis, R.J. (1994). Strain delineation and antifungal susceptibilities of epidemiologically related and unrelated isolates of *Candida lusitaniae*. Diagn Microbiol Infect Dis 20(3): 127–133.

Pfaller, M.A., Messer, S.A., Boyken, L., Tendolkar, S., Hollis, R.J., Diekema, D.J. (2004). Geographic variation in the susceptibilities of invasive isolates of *Candida glabrata* to seven systemically active antifungal agents: a global assessment from the ARTEMIS Antifungal Surveillance Program conducted in 2001 and 2002. J Clin Microbiol 42(7): 3142–3146.

Phumthum, M., Balslev, H. (2020). Anti-infectious plants of the Thai Karen: A meta-analysis. Antibiotics 9: 298. doi: 10.3390/antibiotics9060298.

Prasad, T.N.V.K.N., Kambala, V.S.R., Naidu, R. (2011). A critical review on biogenic silver nanoparticles and their antimicrobial activity. Curr Nanosci 7: 531–544.

Prasannabalaji, N., Muralitharan, G., Sivanandan, R., Kumaran, S., Pugazhvendan, S. (2012). Antibacterial activities of some Indian traditional plant extracts. Asian Pac J Trop Dis 2: S291–S295.

Price, M.F., LaRocco, M.T., Gentry, L.O. (1994). Fluconazole susceptibilities of *Candida* species and distribution of species recovered from blood cultures over a 5-year period. Antimicrob Agents Chemother 38(6): 1422–1424.

Rajesh, K.M., Ajitha, B., Reddy, Y.A.K., Suneetha, Y., Reddy, P.S. (2018). Assisted green synthesis of copper nanoparticles using *Syzygium aromaticum* bud extract: Physical, optical and antimicrobial properties. Optik (Stuttg.) 154: 593–600.

Ramalingam, N., Natarajan, T.S., Rajiv, S. (2015). Preparation and characterization of electrospun curcumin loaded poly(2-hydroxyethyl methacrylate) nanofiber—a biomaterial for multidrug resistant organisms. J Biomed Mater Res Part A 103(1): 16–24.

Rammelkamp, C.H., Maxon, T. (1942). Resistance of *Staphylococcus aureus* to the action of penicillin. Proc Soc Exp Biol Med 51(3): 386–389.

Rauseo, A.M., Coler-Reilly, A., Larson, L., Spec, A. (2020). Hope on the horizon: novel fungal treatments in development. Open Forum Infect Dis 7(2): ofaa016. doi:10.1093/ofid/ofaa016.

Rojas, A., Hernandez, L., Pereda-Miranda, R., Mata, R. (1992). Screening for antimicrobial activity of crude drug extracts and pure natural products from Mexican medicinal plants. J Ethnopharmacol 35: 275–283.

Rout, S.D., Panda, T. (2017). Ethnobotanical survey of medicinal plants used for the treatment of diarrhoea and dysentery by the tribals of Similipal forest, Mayurbhanj, Odisha, India. Appl Sci Rep 19(1): 9–18.

Sabatelli, F., Patel, R., Mann, P.A., Mendrick, C.A., Norris, C.C., Hare, R., Loebenberg, D., Black, T.A., McNicholas, P.M. (2006). *In vitro* activities of posaconazole, fluconazole, itraconazole, voriconazole, and amphotericin B against a large collection of clinically important molds and yeasts. Antimicrob Agents Chemother 50(6): 2009–2015.

Sable, N., Gaikwad, S., Bonde, S., Gade, A., Rai, M. (2012). Phytofabrication of silver nanoparticles by using aquatic plant *Hydrilla verticilata, Nusanta*. Bioscience 4(2): 45–49.

Samaee, S., Rabbani, S., Jovanovic, B., Mohajeri-Tehrani, M.R., Haghpanah, V. (2015). Efficacy of the hatching event in assessing the embryo toxicity of the nano-sized TiO_2 particles in zebrafish: A comparison between two different classes of hatching-derived variables. Ecotoxicol Environ Saf 116: 121–128.

Senthilkumar, S.R., Sivakumar, T. (2014). Green tea (*Camellia sinensis*) mediated synthesis of zinc oxide (ZnO) nanoparticles and studies on their antimicrobial activities. Int J Pharm Pharm Sci 6(6): 461–465.

Shanmugam, S., Kalaiselvan, M., Selvakumar, P., Suresh, K., Rajendran, K. (2011). Ethnomedicinal plants used to cure diarrhoea and dysentery in Sivagangai district of Tamil Nadu, India. Int J Res Ayurveda Pharm 2(5): 991–994.

Shende, S., Ingle, A.P., Gade, A., Rai, M. (2015). Green synthesis of copper nanoparticles by *Citrus medica* Linn. (Idilimbu) juice and its antimicrobial activity. World J Microbiol Biotechnol 31: 865–873.

Silva, L.M., Hill, L.E., Figueiredo, E., Gomes, C.L. (2014). Delivery of phytochemicals of tropical fruit by-products using poly (DL-lactide-co-glycolide) (PLGA) nanoparticles: synthesis, characterization, and antimicrobial activity. Food Chem 165: 362–370.

Singh, S.B., Barrett, J.F. (2006). Empirical antibacterial drug discovery—foundation in natural products. Biochem Pharmacol 71(7): 1006–1015.

Srinivasan, A., Lopez-Ribot, J.L., Ramasubramanian, A.K. (2014). Overcoming antifungal resistance. Drug Discov Today Technol 11: 65–71.

Sokol-Anderson, M.L., Brajtburg, J., Medoff, G. (1986). Amphotericin B-induced oxidative damage and killing of *Candida albicans*. J Infect Dis 154(1): 76–83.

Souli, M., Galani, I., Giamarellou, H. (2008). Emergence of extensively drug-resistant and pandrug-resistant Gram-negative bacilli in Europe. Eurosurveillance 13(47): 19045.

Subhapriya, S., Gomathipriya, P. (2018). Green synthesis of titanium dioxide (TiO_2) nanoparticles by *Trigonella foenum-graecum* extract and its antimicrobial properties. Microb Pathog 116: 215–220.

Suganya, S., Senthil Ram, T., Lakshmi, B.S., Giridev, V.R. (2011). Herbal drug incorporated antibacterial nanofibrous mat fabricated by electrospinning: An excellent matrix for wound dressings. J Appl Polym Sci 121(5): 2893–2899.

Swann, J.P. (1983). The search for synthetic penicillin during World War II. Br J Hist Sci 16(53 Pt 2): 154–190.

Tagousop, C.N., Tamokou, J-D., Kengne, I.C., Ngnokam, D., Voutquenne-Nazabadioko, L. (2018). Antimicrobial activities of saponins from *Melanthera elliptica* and their synergistic effects with antibiotics against pathogenic phenotypes. Chem Cent J 12: 97. doi: 10.1186/s13065-018-0466-6.

Tereschuk, M.L., Riera, M.V., Castro, G.R., Abdala, L.R. (1997). Antimicrobial activity of flavonoids from leaves of *Tagetes minuta*. J Ethnopharmacol 56: 227–232.

Thakur, B.K., Kumar, A., Kumar, D. (2019). Green synthesis of titanium dioxide nanoparticles using *Azadirachta indica* leaf extract and evaluation of their antibacterial activity. S Afr J Bot 124: 223–227.

Thota, N., Jiang, J. (2015). Computational amphiphilic materials for drug delivery. Front Mater 2: 64. doi:10.3389/fmats.2015.00064.

Toda, M., Ōkubo, S., Ikigai, H., Suzuki, T., Suzuki, Y., Hara, Y., Shimamura, T. (1992). The protective activity of tea catechins against experimental infection by *Vibrio cholerae* O1. Microbiol Immunol 36(9): 999–1001.

Topuz, O.K., Özvural, E.B., Zhao, Q., Huang, O., Chikindas, M., Gölükçü, M. (2016). Physical and antimicrobial properties of anise oil loaded nanoemulsions on the survival of foodborne pathogens. Food Chem 203: 117–123.

Tsuchiya, H., Sato, M., Miyazaki, T. (1996). Comparative study on the antibacterial activity of phytochemical flavanones against methicillin-resistant *Staphylococcus aureus*. J Ethnopharmacol 50: 27–34.

Tulgar, S., Alasehir, E.A., Selvi, O. (2018). A atividade antimicrobiana de efedrina e da combinação de efedrina e propofol: um estudo *in vitro* [The antimicrobial activity of ephedrine and admixture of ephedrine and propofol: an *in vitro* study]. Rev Bras Anestesiol 68(1): 69–74.

Uddin, S., Chowdhury, V., Uddin, S.B., Howlader, M.S.A. (2015). Ethnomedicinal plants used for the treatment of diarrhoea and dysentery by the Lushai community in Bandarban district, Bangladesh. J Adv Med Life Sci 2(4): 1–7.

Valle, D.L., Andrade, J.I., Puzon, J.J.M., Cabrera, E.C., Rivera, W.L. (2015). Antibacterial activities of ethanol extracts of Philippine medicinal plants against multidrug-resistant bacteria. Asian Pac J Trop Biomed 5(7): 532–540.

Violante, I.M., Hamerski, L., Garcez, W.S., Batista, A.L., Chang, M.R., Pott, V.J., Garcez, F.R. (2012). Antimicrobial activity of some medicinal plants from the Cerrado of the central western region of Brazil. Braz J Microbiol 43(4): 1302–1308.

Wahbi, W., Siam, R., Kegere, J., El-Mehalmey, W.A., Mamdouh, W. (2020). Novel inulin electrospun composite nanofibers: Prebiotic and antibacterial activities. ACS Omega 5(6): 3006–3015.

Walsh, T.J., Melcher, G.P., Rinaldi, M.G., Lecciones, J., McGough, D.A., Kelly, P., Lee, J., Callender, D., Rubin, M., Pisso, P.A. (1990). *Trichosporon beigelii*, an emerging pathogen resistant to amphotericin B. J Clin Microbiol 28(7): 1616–1622.

Walsh, C. (2003). Antibiotics: Actions, Origins, Resistance. 1st Ed. ASM Press, Washington, DC.

Wang, S., Su, R., Nie, S., Sun, M., Zhang, J., Wu, D., Moustaid-Moussa, N. (2014). Application of nanotechnology in improving bioavailability and bioactivity of diet-derived phytochemicals. J Nutr Biochem 25(4): 363–376.

Xue, J., Wu, T., Dai, Y., Xia, Y. (2019). Electrospinning and electrospun nanofibers: Methods, materials, and applications. Chem Rev 119(8): 5298–5415.

Xue, Y., Zhou, S., Fan, C., Du, Q., Jin, P. (2019). Enhanced antifungal activities of eugenol-entrapped casein nanoparticles against anthracnose in postharvest fruits. Nanomaterials (Basel, Switzerland) 9(12): 1777. doi: 10.3390/nano9121777.

Zanetti, M., Ternus, Z.R., Dalcanton, F., de Mello, M.M.J., de Oliveira, D., Araujo, P.H.H., Riella, H.G., Fiori, M.A. (2015). Microbiological characterization of pure geraniol and comparison with bactericidal activity of the cinnamic acid in gram-positive and gram-negative bacteria. J Microb Biochem Technol 7(4): 186–193.

Zeyohanness, S.S., Hamid, H.A., Zulkifli, F.H. (2018). Poly(vinyl alcohol) electrospun nanofibers containing antimicrobial *Rhodomyrtus tomentosa* extract. J Bioact Compat Polym 33(6): 1–12.

Zhi-Wen, Y., Yan-Li, Z., Man, Y., Wei-Jun, F. (2015). Clinical treatment of pandrug-resistant bacterial infection consulted by clinical pharmacist. Saudi Pharm J *23*(4): 377–380.

Section II
Biobased Metabolites Used in Microbial Diseases

10

Approaches for Identification and Validation of Antimicrobial Compounds of Plant Origin:

A Long Way from the Field to the Market

Lívia Maria Batista Vilela,[1] *Carlos André dos Santos-Silva,*[2] *Ricardo Salas Roldan-Filho,*[1] *Pollyanna Michelle da Silva,*[3] *Marx de Oliveira Lima,*[1] *José Rafael da Silva Araújo,*[1] *Wilson Dias de Oliveira,*[1] *Suyane de Deus e Melo,*[1] *Madson Allan de Luna Aragão,*[1] *Thiago Henrique Napoleão,*[3] *Patrícia Maria Guedes Paiva,*[3] *Ana Christina Brasileiro-Vidal*[1] and *Ana Maria Benko-Iseppon*[1,*]

Introduction

The relationship between human beings and plants is ancient, whether using plant parts as food, for obtaining raw material or for treating diseases. Especially in this last aspect, the number of new compounds of plant origin identified is increasing for phytotherapeutic use or as inspiration for later chemical synthesis.

In this scenario, identifying new antimicrobial compounds remains mandatory and urgent since available drugs are increasingly less efficient due to the emergence of increasingly resistant microorganisms. This hectic evolution of microorganisms, associated with a decrease in the launch of new drugs, has created a public health problem worldwide with an alarming number of deaths due to infections. In this context, bioactive molecules arising from the coevolutionary pressure between plant-pathogen appear as a promising option. Plants are regarded a rich source of new molecules with therapeutic properties, such as bactericidal, antifungal, antiviral activities, among others (Vilela et al. 2021; Santos-Silva et al. 2020; 2021; Benko-Iseppon et al. 2010; Benko-Iseppon and Crovella 2017).

Different biochemical methods can be applied to extract, purify and identify plant compounds, depending on the target molecules' chemical characteristics. Regarding the

[1] Departamento de Genética, Centro de Biociências, Universidade Federal de Pernambuco, Av. Prof. Moraes Rego, 1235, CEP 50.670-423, Recife, PE, Brazil.
[2] Institute for Maternal and Child Health – IRCCS "Burlo Garofolo", Trieste, Italy.
[3] Departamento de Bioquímica e Biofísica, Centro de Biociências, Universidade Federal de Pernambuco, Recife, PE, Brazil.
* Corresponding author: ana.iseppon@gmail.com

phytochemical extraction of molecules, several solvents of different polarities have been used, in addition to chromatographic techniques, applicable for the purification steps (Nelson and Cox 2014). Subsequently, the molecular weight of isolated products can be evaluated by electrospray mass spectrometry (ESMS), for instance, where the ratio between the mass and the charge of intact molecules can be determined (Kumar et al. 2020; Mann et al. 2001).

In addition to the traditional biochemical approach, sequence data deposited in omics data banks (e.g., genomics, transcriptomics, proteomics, etc.) can be used to prospect and identify biomolecules using bioinformatics tools in order to identify groups of molecules with antimicrobial capacity through *in silico* techniques. The identified genes/proteins can be used as targets to laboratory studies, including genomic or transcriptomic sequencing and subsequent validation of gene expression using Real-Time quantitative Polymerase Chain Reaction (RT-qPCR or qPCR). Subsequent structural analysis and simulation of molecular dynamics are generally used to assess the structural dynamics and stability of a given biological compound (reviewed by Santos-Silva et al. 2020).

The employment of traditional biochemical inference to find AMPs is indicated here as the 'bottom-up' approach, whereas the use of omics to predict AMP candidates is pointed as the 'top-down' approach. Both approaches have pros and cons and are crucial to characterize and validate antimicrobial compounds from plants. Thus this chapter assesses the strategies to identify, characterize, compare and validate new antimicrobial molecules, as well as their trajectory to become a potential therapeutic agent.

Bottom-up approach for isolation and characterization of antimicrobial compounds

Plants are rich in primary and secondary metabolic compounds that act on virus, bacteria and fungi, reducing their growth and survival (Coelho et al. 2018; Andleeb et al. 2020). Ethnobiological data regarding written or oral reports have been used as a starting point for choosing a given species besides plant availability in the field or in herbal markets (Benko-Iseppon and Crovella 2010).

There is an extensive collection of protocols for the purification of antimicrobial compounds in literature (Cen et al. 2021; Suda et al. 2016). The chosen method needs to be compatible with the target compound to preserve its activity during the extraction and purification processes. For this, one must keep in mind the physical-chemical characteristics linked to the molecular classes sought, such as hydrophobicity, residual charge, sensitivity to changes in pH, temperature and chemical solvents. Figure 10.1 outlines the main steps in prospecting for antimicrobial compounds covered in this chapter, including proteins, peptides and secondary metabolites.

Proteins and peptides

Extraction

The first step in the extraction process from solid tissue is cell disintegration aiming to increase the contact surface between the plant tissue and the extraction solution. The fragmentation method can be carried out using different tools as, for example, a mortar and pestle, blender, processor, mill, and electric crusher. Plant tissues that have a high fat or oil content (e.g., seeds) can be pretreated with nonpolar solvents to remove lipids that may interfere with the extraction, purification and Activity of Antimicrobial Peptides (AMPs) that have low molecular weight (Bard et al. 2016; Nazeer et al. 2019).

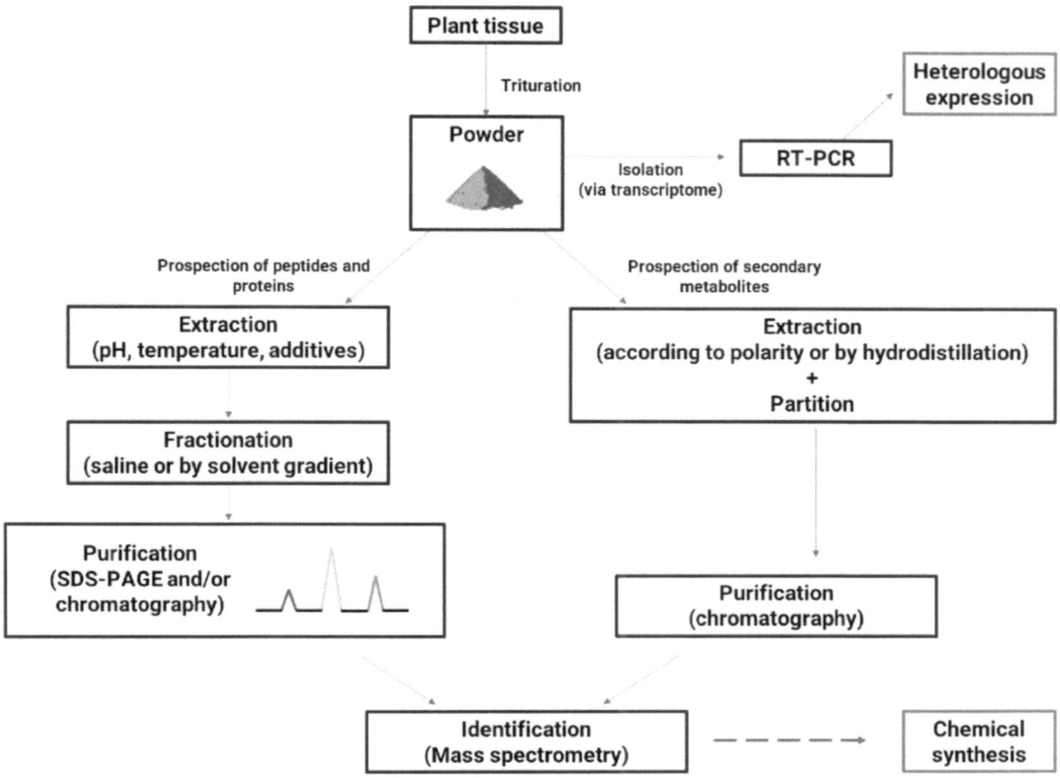

Figure 10.1. Schematic representation of the 'bottom-up' approach. The plant tissue of choice (e.g., roots, leaves, flowers, bark, fruits or the whole plant) is triturated to a powder and processed for the extraction of proteins or secondary metabolites, prior to purification and characterization of antimicrobial compounds.

Considering AMPs, the selection of the extraction solution will depend on the solubility of the targeted peptides. These compounds have degrees of hydrophobicity that can hinder their solubility in aqueous media. Acidic solutions or buffers are an alternative since AMPs are protonated when submitted to a pH range below their isoelectric point (pI). In this way, their solubility in an aqueous medium is increased (Finkina et al. 2016; D'Agostino et al. 2019).

Acetic acid and acidic buffers are effective in solubilizing AMPs. For instance, the latter was used to extract AMPs from *Ginkgo biloba* leaves and from *Coffea canephora* seeds (Liang et al. 2011; Bard et al. 2016; Wong et al. 2016). However, the acidic medium can be a problem for proteins and peptides sensitive to extremely low pH, as it may result in alteration of the native structure and loss of biological activity. In this sense, safe pH ranges for the extraction of AMPs are registered, as they were performed on *Pisum sativum* seeds, *Capsicum annuum* leaves and *Morinda citrifolia* seeds, which were subjected to extraction in CH_3COONH_4/NaCl buffer - pH 5.5, Tris-HCl - pH 7.0 and Tris-HCl/NaCl - pH 8.5, respectively (Bogdanov et al. 2016; Campos et al. 2016; Games et al. 2016; Souza et al. 2018).

Extracting proteins is extremely challenging, as the amino acid composition determines the protein's interaction with the medium. Several factors such as ionic strength, solvent polarity, pH and temperature can interfere with the protein's conformational structure and lead to a loss of its biological function (Gao et al. 2020; Ioerger et al. 2020; Ippoushi et al. 2020). Thus, it is necessary to define the experimental conditions in which the targeted protein is in its native conformation (or closer to it) to preserve its antimicrobial activity (Tanger et al. 2020).

An excellent criterion for selecting the solution for the extraction of globular proteins is the relationship between protein quantity and its biological activity. Thus, the protein purifier must avoid producing an inactive crude extract even if it contains a high protein concentration or an active extract, but with a minimum amount of the targeted protein. Protein detection can be done through absorbance at 280 nm or by colorimetric methods as Bradford, Lowry, Biuret, and bicinchoninic acid (Zheng et al. 2017; Silva et al. 2019; Licini et al. 2020).

The presence of phenolic compounds can result in extracts with strong coloring and high viscosity, which hinder the peptide purification process (Ramirez et al. 2021; Poletto et al. 2021). The choice of solvent is crucial for selective solubilization. Polar and nonpolar solvents are generally used in extraction processes of peptide molecules and secondary metabolites, respectively. The presence of non-protein contaminants can interfere with the antimicrobial activity of the peptides. Additionally, the extracts' clarification can be carried out by adsorption to activated carbon and silica, for example (Meric et al. 2019; Supong et al. 2020).

The stability of the biological activity must be observed during the extraction step. The isolation of a bioactive protein is not feasible if the extraction medium's conditions lead to the elimination of biological activity. Thus, the selection of the extraction conditions must be adjusted, aiming to produce an extract with stable biological activity, in detriment of one that presents greater activity but contains protein with an unstable structure.

One of the main causes of inactivity of crude extracts is the loss of the integrity of the protein's native structure due to the medium in which they are solubilized and/or by the action of proteases. The inclusion of ions in the extraction solution (e.g., 0.15 M NaCl or 0.05 M Tris-HCl) may be necessary to solubilize proteins whose structures are maintained by ions (Patriota et al. 2017; Costa et al. 2018). Besides, proteins may have their structure altered due to the formation of disulfide bridges between cysteine residues, making them more rigid and turning the extraction process more difficult (Santos-Silva et al. 2020). The addition of a reducing agent in the extraction solution can prevent the interaction between these amino acids. However, some authors indicate that disulfide bridges are of great importance for proteins' biological activity because they contribute to a compact and stable active structure against chemical and proteolytic degradation (Pokoj et al. 2010; Zhang et al. 2011; Tam et al. 2015; Santos-Silva et al. 2021). Thus, the addition of an alkaline buffer favors the formation of disulfide bridges (Zhang et al. 2011). The solubilization of constituents of an acidic or basic character can result in protein denaturation. In turn, the use of a buffer solution is essential for maintaining the pH of the medium. The absence of activity resulting from the action of proteases—that were extracted together—can be resolved by including non-protein protease inhibitors in the extraction solution, besides the selection of pH value and temperature that cause the blocking of the enzymatic action (Fibriana and Upaichit 2015; Bogdanov et al. 2016; Games et al. 2016; Wong et al. 2016; Wong et al. 2017).

Fractionation

The protein concentration and the volume of the extract determine the next stage of the purification process. For example, the production of a large volume of the extract with a low concentration of peptides requires methods that result in concentrated fractions of proteins. In contrast, the production of an extract rich in peptides may already lead the way for their separation. Peptide enriched preparations can be obtained by precipitation and ultrafiltration methods on membranes used for selective separation of peptides present in the precipitated fractions (Games et al. 2016; Wang et al. 2019). Solid-phase fractionation is also used and consists of separating peptides through a gradient of increasing concentration (v/v) of solvent in a polar or nonpolar matrix. For hydrophobic or amphipathic AMPs, reverse phase flash columns or Sep-Pak C18 matrix cartridges are used. Nonpolar amino acid residues adsorb to the matrix and are eluted by an increasing gradient of organic solvents such as ethanol at 20, 40, 60 and 80% or acetonitrile at 5, 40, and 80% (Chaparro and Silva 2016; Wong et al. 2016; Wong et al. 2017; Chaparro-Aguirre et al. 2019).

Proteins can be precipitated with acetone, trichloroacetic acid and polyethylene glycol (Bekale et al. 2015; Bose et al. 2019). However, the salting out method with ammonium sulfate is the most used method because it generally does not cause the protein's denaturation, besides the facility to eliminate the salt of fractions obtained through dialysis (Oliveira et al. 2016; Procópio et al. 2017a). Another advantage of saline precipitation is the possibility of promoting purification, that is, the possibility of obtaining a preparation rich in the protein of interest, free from other contaminants (Santana et al. 2012). The use of different salt concentrations can result in partial or total purification of the protein since highly soluble proteins need a higher concentration of salt to be removed from the solution than those less soluble. On the other hand, if the extract contains a mixture of proteins of similar solubility, it is better to use a high saturation of the salt and obtain a fraction containing all the proteins of the mixture in a reduced volume.

Purification

The purification of peptides and proteins from fractions or even crude extract can follow one or more chromatographic processes. Before performing a chromatographic procedure, the insoluble matrix and the chromatographic conditions of pH, ionic strength, volume of the chromatographic bed, amount of protein applied per mL of matrix, flow rate in the column, elution strategy and sample volume must be chosen (Paiva et al. 2010; Procópio et al. 2017b). Column chromatography occurs in the steps of equilibrating the column with the solution in which the process will take place, applying the sample and obtaining fractions.

Ion exchange chromatography, affinity and molecular exclusion are efficient in promoting the purification of active proteins because they explore structural characteristics such as charge, biospecificity and size without promoting denaturation (Feng et al. 2021; Benedini et al. 2019). Reverse-phase chromatography is advisable only for small molecules such as AMPs since the hydrophobic matrix and organic eluents used can result in the denaturation of larger proteins by the interaction with nonpolar amino acid residues (Feng et al. 2021). The reverse-phase procedure that separates based on hydrophobicity and uses denaturation conditions is used to determine the impurity profile and prepare samples for mass spectrometry (Soares et al. 2011).

To increase the chromatographic process's efficiency, the protein mixture must be in the same equilibrium solution as the column and it is important that the collected fractions are evaluated by absorbance at 280 nm and biological activity, the first time that the procedure is performed (Nelson and Cox 2014; Patriota et al. 2017). Proteins that are poor in phenylalanine, tyrosine and tryptophan residues can be followed rather by their activity than by absorption at 280 nm (Nelson and Cox 2014).

In ion exchange chromatography, proteins of opposite charge the matrix adsorb to it while proteins of the same charge do not interact. The interaction strength between the protein and the matrix depends on the number of charges of the protein. That is, proteins with a greater number of charges contrary to the matrix bind more strongly than those less charged (Nelson and Cox 2014; Procópio et al. 2017b). The first step is the choice of the matrix and the two types of ion exchangers that must be tested when the net charge of the protein is not known a priori. An excellent alternative to save time in selecting the matrix and the chromatographic conditions that favor the adsorption of the protein is to perform the steps of the chromatographic process in an Eppendorf. In this assay, the sample containing the protein mixture is added to the Eppendorf containing the opposite charge matrices balanced in buffers of different ionic strengths and pH values. After incubation and centrifugation, the collected supernatant is evaluated for absorption at 280 nm and biological activity. Subsequently, the matrices are washed with the buffer solution until the supernatant has no protein. The next step is the addition of 1.0 M NaCl and protein detection in the collected supernatant that indicates the matrix for protein isolation (Tan et al. 2009; Procópio et al. 2017b).

In ion exchange chromatography, the separation of proteins into distinct fractions occurs by selective desorption by altering the ionic strength or pH of the chromatographic medium (Nelson and Cox 2014; Procópio et al. 2017b; Patriota et al. 2017). For example, proteins strongly bound to the matrix will be eluted with a solution with greater ionic strength, while proteins that are less adsorbed will require less ionic strength to shut down. The use of saline gradients in the elution stage is very efficient in separating proteins, which—although having the same type of charge—differ in their amount. The gradient can be of the linear type, where the column is irrigated with a similar ionic strength buffer or of the ladder type, where the column is eluted with solutions with concentrations that differ in the order of 0.1 to 0.2 M (Procópio et al. 2017b; Patriota et al. 2017).

Affinity chromatography is extremely efficient in providing a high degree of purity because it explores a specific region of the protein, that is, the specificity of a binding site (Paiva et al. 2010; Nelson and Cox 2014; Procópio et al. 2017b). It is the chromatographic process of choice if the protein of interest forms a reversible complex with a ligand. For example, a protein that contains a glucose binding site can be isolated using a column containing a glucose polymer. An enzyme can be separated from other proteins using a matrix containing substrate or inhibitor and the enzyme inhibitor will bind to a matrix that has the enzyme immobilized (Coelho et al. 2012; Pontual et al. 2014; Ferreira et al. 2019). When the affinity matrix is not commercially available, it can be produced in the laboratory by immobilizing the binder to an insoluble material such as, for example, Sepharose 4B (Silva et al. 2020). It is important to emphasize that the ligand immobilization must be via functional groups that are not involved in the interaction with the protein of interest. Also the smaller the number of regions for nonspecific binding in the matrix, the greater is its selectivity, that is, its efficiency in selecting the protein among others present in the mixture.

The protein shutting down from an affinity matrix can occur by breaking interactions between the matrix and the binding site using a ligand solution (Gomes et al. 2012; Nelson and Cox 2014; Procópio et al. 2017b). Another way to recover the protein in the collected fractions is by breaking down interactions between the matrix and the binding site or other parts of the protein. In this type, the column is irrigated with a solution of ionic strength greater than the equilibrium solution or by decreasing the pH of the chromatographic medium. One strategy that can be adopted for proteins sensitive to low pH values is the addition of a solution collection tube that neutralizes the pH.

Molecular exclusion chromatography uses a porous matrix that acts as a sieve and separates proteins according to mass. Larger proteins are excluded from the spine more quickly than smaller proteins, thus separating them. There is no interaction of the protein with the matrix in this process because if this occurs, the protein's migration speed will not be according to its size (Lubomirsky et al. 2021).

Fast-acting liquid chromatography (FPLC) uses gel filtration or ion exchange matrices. In contrast, High-Performance Liquid Chromatography (HPLC) employs matrices that separate by charge, size or hydrophobicity and solvents that generally promote protein denaturation (Rocha-Filho et al. 2020). The FPLC and HPLC pump system provides high flow and efficiently controls the entry of solutions into the column, which results in the development of elution gradients with high precision, thus increasing the resolution of the process by limiting the diffusion of the protein.

Peptides and proteins may be isolated by the one-dimensional polyacrylamide gel electrophoresis in the presence of Sodium Dodecyl Sulfate (SDS) (Nawrot et al. 2017; Schmitt et al. 2018). The denaturation caused by the detergent compromises the biological activity; however, the method can be used to determine the sequence and predict the structure. Bands with different molecular weights are excised from the gel, and the retained compounds are cleaved with proteases to identify their sequences by Liquid Chromatography coupled to Mass Spectrometry (LC-MS) (Nawrot et al. 2017; Schmitt et al. 2018).

Identification and isolation

Peptides isolated by chromatographic methods have their sequence analyzed by Tandem Mass spectrometry (MS/MS), where larger sequences are initially cleaved by triptych digestion (Schmitt et al. 2018; Segura-Ramírez and Silva Júnior 2018; Offret et al. 2019). The generated files are analyzed by tools, such as MASCOT (Matrix Science), to assemble the compounds' primary structure. Peptides isolated by SDS-PAGE may have the sequences identified after band recovery of the gel, cleavage with proteases and analysis by mass spectrometry (Lima et al. 2018; Silva et al. 2020). *In silico* evaluation has been used to prove whether the sequences belong to the class of antimicrobial peptides through data on conserved domains, secondary structure and tertiary structure prediction (Nawrot et al. 2017; Schmitt et al. 2018).

After all the extraction, purification and identification steps, the amino acid sequences can be identified in the encoded cDNA sequence that can be used for cloning in heterologous systems. The first step in obtaining the cDNA is the total extraction of RNA from the studied tissue. The coding sequence of the target peptide is amplified and converted to cDNA by reverse transcription in a thermocycler. For this amplification, primers associated with the partial sequence of the identified peptide or its conserved sequence must be developed (Santos-Silva et al. 2021; Hong et al. 2018; Oh et al. 2018).

The determination of the primary protein sequence makes it possible to produce the recombinant protein, with the cloning and sequencing of DNA or RNA, resulting in a large number of proteins produced by this approach. Some proteins require disulfide and glycosylation bridges to mimic the folding of the protein in its native form and some expression systems (such as prokaryotes) do not perform these post-translational modifications naturally, so they may not produce soluble and functional proteins, even after the addition of buffers that favor these changes (Gazzaneo et al. 2017).

Some techniques allow to infer the secondary structure, such as circular dichroism spectroscopy that identifies changes in the secondary structure when determining the percentage of alpha-helix, beta sheet and turns. X-ray crystallography and nuclear magnetic resonance reveal the tertiary structure and the interaction of the protein with ligands. Fluorimetric methods indicate changes in the hydrophobic environment, while thermodynamic measurements reveal regions of the protein that differ in stability (Nelson and Cox 2014).

Extraction and identification of secondary metabolites

Secondary antimicrobial metabolites are identified, among many parameters, by solubility in water or organic solvents. The extraction of these compounds is carried out using a solvent that can be polar (such as organic acids), intermediate polarity (such as methanol and ethanol) or nonpolar (such as chloroform and hexane) (Paiva et al. 2010; Arachchige et al. 2021). For a larger amount of extracted compounds, successive extractions (at least three times) by replacing the same solvent type are recommended. After, for example, the first extraction for 5 to 24 hours, the solvent will be saturated, preventing the extraction of metabolites still present in the crushed plant tissue. Thus, the mixture must be filtered so that the retained solid residue is subjected to a second extraction (Long et al. 2018; Abdelgaleil et al. 2020; Zhao et al. 2020). All the crude extract obtained is concentrated by evaporation of the solvent, for example at low pressure, followed by resuspension in a polar or nonpolar solvent.

Although the choice of the solvent is based on the polarity of the antimicrobial compound of interest, other molecules of opposite polarity may be extracted. Partitioning (or liquid-liquid extraction) is an optional technique that separates the compounds from the crude extract according to their polarity. For this, a solvent that is immiscible or partially miscible is added to the suspended extract (Long et al. 2018). For example, when adding hexane, chloroform or ethyl acetate to a crude extract in an acidic medium (e.g., 3.6% HCl or 2.0% H_2SO_4), some simple polyphenols and alkaloids will be found in the aqueous phase, while terpenoids, hydrophobic and higher carbon chain compounds will partition into the organic phase (Abdelgaleil et al. 2020; Zhao et al. 2020).

The essential oils are also represented within the secondary metabolites, which are composed of a mixture of organic and volatile compounds (Tisserand and Young 2014). These compounds are extracted by hydrodistillation in two forms: the classic one, in which the plant material is boiled in water or the modern one, where steam passes through plant material (Tisserand and Young 2014). The volatile compounds are carried away by the water vapor and then condensed when passing through a cold chamber. In this way, two immiscible phases are formed: the aqueous and the oily. They are separated using a separating funnel. Through this technique, Brito et al. (2018) obtained essential oils from *Croton argyrophyllus* leaves with broad antibacterial activity. Oil samples from *Syagrus coronata* seeds (Santos et al. 2019) and from the aerial parts of *Osmunda regalis* (Bouazzi et al. 2018) were also extracted by hydrodistillation with antibacterial action against *Staphylococcus aureus* and antiviral against *Coxsackie B*, respectively.

Thin Layer Chromatography (TLC) is a qualitative technique that identifies groups of secondary metabolites by comparing their Retention factor (Rf) values with reference substances. The compounds are initially separated according to the affinity between the mobile phase (nonpolar solvent) and the stationary phase (e.g., silica or cellulose) (Akash and Rehman 2020; Kashid and Kadam 2021). At the end of the chromatography, chromogenic markers and excitation by UV light are used to demarcate classes of compounds. For example, Dragendorff's reagent allows the detection of alkaloids, while Folin reveals polyphenols such as isoflavones. Coumarins are detected by marking in 10% potassium hydroxide and excitation in UV 365 nm (Braz et al. 2012; Vieira et al. 2008). TLC can be used as the first step of qualitative identification of complex compounds, being an inexpensive technique, which does not require sophisticated equipment or even laborious sample preparation (Klimek-Turek et al. 2017).

Gas Chromatography (GC) is used to detect volatile secondary metabolites found, for example, in essential oils (Dagnino et al. 1991; Tisserand and Young 2014). One of the main characteristics of GC is its great power of detection and separation of different compounds. The sample is gradually heated and most volatile components are detected first. The association of this technique with mass spectrometry (GC/MS) allows identifying the compounds present in minimal quantities, even in mixtures presenting a high degree of complexity. These compounds are registered according to the characteristic retention time (Tisserand and Young 2014).

Applications of antimicrobial compounds

Humans have used plants for medicinal purposes since time immemorial and the growing interest in studies evaluating the medicinal properties of plant compounds is notable. Many medicines on the market originated from natural products, especially plants (Jamshidi-Kia et al. 2018). Infectious diseases are responsible for a high rate of morbidity and mortality worldwide, and the excessive and/or indiscriminate use of antibiotics has stimulated the appearance of multidrug-resistant strains. There are many uses and applications of antimicrobial compounds, including the food, cosmetic and pharmaceutical industry (Table 10.1).

Several pathogens are regarded as severe threats to public health. Besides, several side effects of the drugs currently used are described. This scenario encourages the search for new antimicrobial compounds of plant origin that have proved to be good alternatives (Yu et al. 2020). The rapid advance of biotechnology and molecular biology since the 1970s has accelerated the search for new drugs through plant origin (Anand et al. 2020). Understanding the mechanisms responsible for diseases at the molecular level, in addition to the ability to design proteins with the aid of new computational tools that were also developing rapidly, opened up the perspective of synthesizing new drugs with greater efficiency, starting only from the accumulated scientific knowledge (Sarris 2018).

Evaluation of *in vitro* antimicrobial activity

Various methods can be used to assess the *in vitro* sensitivity of microorganisms to antimicrobial agents, with emphasis on disk diffusion, macro- and microdilution assays.

The diffusion test is a physical method in which a microorganism is challenged, in a solid culture medium, against a biologically active substance. It is considered one of the simplest, most reliable susceptibility methods used by microbiological laboratories. Its basic principle is the diffusion of antimicrobials on the agar surface, from a disk impregnated with it, determining in millimeters the size of the growth inhibition zone around the disk (Schapoval 2005; Moradi et al. 2021). After measuring the halos, their interpretation is made, using the criteria established by the Clinical & Laboratory Standards Institute (CLSI) for the tested microorganism. Bacterial samples are thus categorized as sensitive, resistant or intermediate. This test is accepted by the Food and Drug Administration (FDA) and established as a standard by the National Committee for Clinical Laboratory Standards (NCCLS) (Barry and Thornsberry 1991). The agar disk diffusion test was described by Bauer et al. (1966) and provides qualitative results. Some advantages regard high reproducibility, easy execution and interpretation of results, besides flexibility concerning the choice of antimicrobial agents. Furthermore, in studies with resistance genes and

Table 10.1. Exemplary list of substances with antimicrobial activity, with source species and reported applications.

Antimicrobial agent	**Source**	**Biological activity**	**Applications/use**	**Reference**
Extract	Nut of *Carya illinoinensis*	Antimicrobial	Food preservative	Yemmireddy et al. 2020
	Inflorescences of *Tilia* spp.	Antimicrobial	Cosmetic industry, popular use in colds, pharyngitis and cough, diuretic and sedative action	Pavlović et al. 2020
	Roots of *Hypericum perforatum*	Antimicrobial, antidepressant, neuroprotective and antidiabetic	Used in depressive disorders, anxiety and/or nervous agitation	Tusevski et al. (2018)
Trypsin inhibitor	Cactus *Pilosocereus gounellei*	Antimicrobial	Potential use in the pharmaceutical industry	Rocha Filho et al. (2019)
	Solanum tuberosum	Antimicrobial	Food industry/emulsifiers	Cisneros et al. (2020)
Lectin	PgTeL– *Punica granatum* L. juice	Antimicrobial	Potential use in the pharmaceutical and food industry	Silva et al. (2019)
	PeRoL - *Portulaca elatior* root	Antimicrobial	Potential use in the pharmaceutical industry	Silva et al. (2019)
	WSMoL - *Moringa oleifera* seeds	Antimicrobial	Potential use in the pharmaceutical industry	Oliveira et al. (2017)
	CasuL - *Calliandra surinamensis* leaves	Antimicrobial	Potential use in the pharmaceutical industry	Procópio et al. (2017a)
Antimicrobial peptides	*Cryptostegia grandiflora* cysteine peptidase	Antimicrobial	Potential use in the pharmaceutical industry	Ramos et al. (2014)
	Hispidalina - *Benincasa hispida* seeds	Antimicrobial	Potential use in the pharmaceutical industry	Sharma et al. (2014)
	Lr-AMP1 -*L. rotundifolia* leaves	Antimicrobial	Potential use in the pharmaceutical industry	Tavares et al. (2020)
Essential oil	*Mentha longifolia*	Antibacterial and antioxidant	Food industry as a preservative agent	Sharma et al. (2020)
	Zingiber officinale	Antimicrobial and antioxidant	Food industry as a preservative agent	Preedy (2015)
	Litsea glaucescens	Antidepressant and sedative	Alternative medication	Guzmán-Gutiérrez et al. (2012)
	Citrus lemon	Antimicrobial and antioxidant	Cosmetic and pharmaceutical industry/fragrances and pharmaceutical preparations	Raspo et al. (2020)
	Plantago afra	Antimicrobial	Therapeutic and cosmetic applications. Has hypoglycemic and hypolipidemic action	Hammami et al. (2020)
	Eucalyptus globulus	Antimicrobial	Food preservative	Lee et al. (2020)

strains producing Extended-Spectrum Beta-Lactamase (ESBL) enzymes, controls need to be performed in a solid medium because PCR based techniques can identify resistance genes that are not always related to phenotypic resistance and microdilution systems may not detect ESBL production (Périllaud et al. 2019). However, there are limitations in this methodology, such as the lack of mechanization or automation, lack of standardization for some microorganisms and antimicrobial agents, short storage times for agar plates and the possibility of misinterpretations (Liu et al. 2016).

Many variables can affect the results of the disk diffusion test as (i) the preparation of the culture medium, (ii) control of the pH of the medium, (iii) adequate stock of the disks, (iv) rigor in the execution of the technique, (v) temperature and atmosphere of incubation, (vi) storage time of the plates, and (vii) thickness of the agar in the plate and the concentration inoculum. In addition, the culture medium must be homogeneously distributed to avoid the formation of superficial streaks and bubbles (Wang et al. 2016). Generally, the qualitative test is sufficient to guide the therapeutic choice. However in some situations, it is necessary to quantify the susceptibility (sensitivity or resistance) through macro- or microdilution tests and the consequent determination of Minimum Inhibitory Concentration (MIC) (Miranda et al. 2020).

Tanhaeian et al. (2020) analyzed the antibacterial, antibiofilm and *in vitro* antioxidant activities of a recombinant chimeric peptide in some bacterial strains of *Lactococcus lactis*, which cause food spoilage. Disk diffusion tests revealed that the analyzed peptides had considerable antimicrobial activity against these bacteria, with an inhibition halo varying between 9.1 and 28.6 mm. Using the disk diffusion method, Kameri et al. (2019) determined the antifungal efficacy of *Tanacetum vulgare* extract against *Candida albicans*. A synergistic effect was identified between the extract and the chlorhexidine antibiotic with an inhibition zone of 30 to 32.7 mm. Therefore, the extract may be useful as a natural bioactive agent with antifungal activity against this fungus. In another study using the disk diffusion test, Perim et al. (2019) reported the antimicrobial activity of the bark extract of *Davilla nitida* against multi-resistant bacteria, inhibiting the growth of *Pseudomonas* spp., *Enterobacter* spp., *Staphylococcus aureus*, *Streptococcus pneumoniae* and *Staphylococcus saprophyticus* with a zone of inhibition ranging between 9 and 14 mm.

As described above, it is necessary to determine the IMC values in many cases, which can be done through macrodilution and microdilution tests. In both cases, the inoculum can be prepared in a similar way. For bacteria, the inoculum must be prepared from colonies grown on agar plates. These colonies are transferred to the appropriate liquid culture medium and are incubated until the turbidity of a 0.5 solution on the McFarland scale is reached or exceeded, equivalent to approximately 1 to 2 × 10^8 Colony-Forming Units (CFU)/mL. The turbidity of the active growth culture can be adjusted with sterile saline solution (NaCl 0.9%) or culture medium. For fastidious organisms, it is recommended to prepare the inoculum by making a direct suspension, in saline or culture medium, of colonies grown on an agar plate for 18–24 hours (CLSI 2016; 2018). In the case of yeast fungus, the microorganisms must be subcultured in sterile tubes containing, for example, Sabouraud Dextrose Agar or Potato Dextrose Agar for 24 to 48 hours at 30°C. As with bacteria, colonies must be suspended in sterile saline and the cell density adjusted to 0.5 on the McFarland scale. The working suspension is produced by a 1:100 dilution, providing a standard yeast suspension containing from 1 to 5 × 10^6 CFU/mL (CLSI 2017a).

To prepare suspensions of sporangiospores or non-germinated conidia of filamentous fungi, the inoculum concentrations should be in the range of 0.4 to 5 × 10^4 CFU/mL. Fungi should be grown on Potato Dextrose Agar for seven days at 35°C. The inoculants are prepared in sterile saline. When necessary, a drop of Tween 20 can be added to facilitate solubilization. The supernatant is transferred to another sterile tube and the density of the conidial or sporangiospores suspensions should be adjusted to an Optical Density (OD) ranging from 0.09 to 0.17 (80 to 82% transmittance) for *Aspergillus* species and *Sporothrix schenckii* and from 0.15 to 0.17 (transmittance from 68 to 70%) for *Fusarium*, *Pseudallescheria boydii* and *Rhizopus arrhizus*, for example (CLSI 2017b).

For sensitivity tests on fast-growing aerobic bacteria, the use of the Broth Mueller Hinton medium is recommended. The pH of the medium is adjusted to 7.2 to 7.4, which corresponds to the physiological pH. If the concentrations of Ca^{+2} and Mg^{+2} cations are not ideal (20–25 and 10–12.5 mg, respectively), the medium must be supplemented with these ions. It is recommended for tests with fungi to use the medium RPMI-1640 (with glutamine, without bicarbonate and containing a red phenol indicator) (Fromtling et al. 1993; EUCAST 2020).

For the macrodilution test, dilutions must be prepared in the medium of the antimicrobial agent, with a minimum final volume of 1 mL for each dilution. The inoculum should be adjusted in saline solution, so that after inoculation, each test tube contains approximately 5×10^5 (bacteria), 0.5×10^2 to 2.5×10^3 (yeast fungi), or 0.4×10^4 at 5×10^4 (filamentous fungi) CFU/mL. The test tubes are incubated at 35°C between 16 to 20 h (bacteria), at 30°C for 46 to 74 hours (yeasts), or at 30°C for 26 to 74 hours (filamentous fungi). Fastidious bacteria should be incubated for 20 to 24 hours. It is recommended to check the purity of the inoculum suspension by subculture an aliquot on a non-selective agar plate for simultaneous incubation. To determine the IMC, which is the lowest concentration of antimicrobial agent that inhibits the growth of the organism in the tubes, a comparison of microbial growth in the treated tubes and in the control tube is performed (Feßler et al. 2018; Esaú et al. 2019; Eryılmaz et al. 2020).

The microdilution test is performed on flat-bottomed 96-well microtiter plates and each row of a microplate corresponds to a line. Successive dilutions (1:1) of the antimicrobial agent are performed in the culture medium. Then the same volume of inoculum is added. A well corresponding to 100% growth is prepared without the presence of the antimicrobial agent and a sterility control that has the culture medium and the antimicrobial agent. Then the plates should be incubated in the same conditions as described for the macrodilution test. Dissolved Oxygen (DO) should be measured at the beginning (zero time) and the end of the test using a microplate reader. The IMC will be determined as the lowest concentration of the sample capable of promoting a certain minimum value of OD reduction compared to the growth control. This value can be 50 or 90%, for example.

In both tests (macro- and microdilution) to determine the minimum bactericidal or fungicidal concentration (CMB/CMF), inocula from tubes or wells containing the sample in concentrations greater than or equal to CMI must be transferred to Petri dishes containing culture medium with agar. Then, the plates must be incubated under the same conditions as described. The CMB/CMF will correspond to the lowest concentration of the sample capable of reducing the number of CFU by 99.9% compared to the initial inoculum (Gezgin et al. 2019).

In addition to determining susceptibility, the growth kinetics of the microorganism in the presence of the antimicrobial agent may also be evaluated. Microbial growth involves the multiplication and population growth of a cell culture, in addition to the increase in biomass produced by such microorganisms. Several indirect methods to measure microbial biomass are being used, based on the measurement of some metabolic activity or some specific component of the biomass (Lekha and Lonsane 1994). The evaluation of the growth rate and the generation time allows to evaluate the concentration of the cells during the incubation time and to verify the cell growth in function of possible alterations of the environmental conditions, as the presence of an antimicrobial agent. Growth must be quantified by the spectrophotometric method of turbidimetry or nephelometry. In these methods, cell growth is monitored according to the turbidity of the solution. The result obtained is expressed in terms of absorbance as a function of time. The most used wavelength for turbidimetry/nephelometry readings is found in the red region (~ 660 nm), as wavelengths in this range generate less light scattering compared to shorter wavelengths (Abdallah et al. 2014; Kogawa et al. 2021).

Another applied methodology is digital image processing, which regards the extraction and treatment of quantitative data from digital images. This analysis allows faster and more accurate measurements, making it possible to carry out measurements that are difficult to perform manually. Digital image processing may represent a good alternative to estimate microbial growth in solid culture media, allowing a better evaluation of the microorganism's growth morphology and kinetics (Ganjdoost et al. 2021). Some of the digital image processing applications regard quantification of cell growth and viability, biomass determination, cellular sensitivity of antibiotics, assessment of adherence, morphology and microbial mobility (Wang et al. 2021; Yin et al. 2021).

Evaluation of antimicrobial activity *in vivo*

There are also different models for assessing *in vivo* antimicrobial activity. All procedures must be conducted following animal testing regulations and approved by institutional ethics committees. Some examples of these *in vivo* infection models are presented here.

In a murine pulmonary infection model, inoculum with the infectious agent (for example, *Pseudomonas aeruginosa*) is applied directly to the nasal passages and later (30 minutes) the antimicrobial agent is administered by the same route. It is important to note that before this injection, it is necessary to induce neutropenia, which can be done by injecting cyclophosphamide. To investigate the action of the compound in the *in vivo* model, animals are usually sacrificed between 20 and 24 hours (depending on the type of compound to be analyzed, the interaction kinetics of the compound, etc.). Further, the organs and tissues affected by infection are collected and analyzed (in this case, the lungs). Total lung weights must be recorded and the organs are then used to prepare homogenates in sterile saline and subsequent inoculation and quantification of the number of CFUs. Cytokine concentrations in pulmonary homogenates can also be measured (Martínez et al. 2020; Vairo et al. 2020).

Another murine model developed concerns gastric infection. After intragastric administration of saturated sodium bicarbonate for 30 minutes, *Helicobacter pylori* inoculum is administered three times, on alternate days, followed by a waiting period of 30 weeks. Then, the animals are treated intragastrically with the antimicrobial agent for the desired time. Forty-eight hours after the end of the treatment, the animals are sacrificed, the stomach is removed, which is then cut and the gastric contents removed by PBS washing. Subsequently, a section of the gastric tissue must be suspended in PBS and homogenized to evaluate bacterial colonization (Jiang et al. 2020).

Another murine model concerns corneal infection. Animals are immunosuppressed by injections of cyclophosphamide, and the left corneas are scratched by sterile needles and infected with the microbial inoculum. After one day of infection, the antimicrobial agent should be topically applied four times a day, in a continuous treatment of three days. The rats are then sacrificed, the corneas are collected, and the number of viable bacteria is determined (Lin et al. 2020). The action of antimicrobials on skin infections can also be assessed in rodents. The animals are anesthetized and the hair on the dorsal region is removed completely. The test area (2 cm^2) is disinfected and a bacterial suspension is applied to the dorsal area. The antimicrobial agent is administered topically to the contaminated site. After 2 minutes of treatment, the skin from the application site is removed under anesthesia, and placed on a plate containing the culture medium. Then, the number of UFCs is counted (Liu et al. 2017a). Porto et al. (2018) reported that the Guavanin 2 peptide was able to reduce the amount of bacteria in a skin infection model.

The last murine models presented here concerns acute lethal infection and bacteremia. In the first, microbial cultures are adequately diluted with bacteriological mucin and neutropenia is also induced by administration of cyclophosphamide. The infection is induced by intraperitoneal injection, which results in 100% mortality in untreated controls within 48 to 72 hours after infection. The antimicrobial agent should be administered between 10 and 15 minutes after infection, intravenously or subcutaneously. Mortality must be recorded daily. Animals that survive after seven days of treatment must be euthanized. The effective dose in reducing mortality by 50% (DE50) is calculated (Jabés et al. 2011). In the murine model of bacteremia, the bacterial suspension is also injected intraperitoneally into the animal. The antimicrobial agent is injected immediately after infection. After two hours, blood samples are collected, which are sown in plates containing culture medium, which will be incubated for later counting the number of CFUs. Animal mortality is also observed for up to seven days (Liu et al. 2017b).

Cows can also be used as an *in vivo* infection model to check the action of antimicrobial compounds against mastitis. Intramammary infusions of the antimicrobial agent are performed daily at the end of lactation. Samples of milk and breast secretions must be collected manually and aseptically. To check the antimicrobial action, samples should be sown immediately after collection in an appropriate medium and incubated at 37°C for 24 hours. Then, the number of UFCs is quantified (Lanctôt et al. 2017).

As an alternative to models that use mammals, assays have been developed that use insect larvae model, such as those from *Galleria mellonella*. This infection model is well established and offers many advantages, including reduced cost and fast speed of results. A previous *in vivo* toxicity assessment of the antimicrobial agent for the larvae is necessary to perform this assay. The antimicrobial capacity is confirmed by infecting the larva with lethal doses of the microorganism. There are two possibilities: (1) pre-treatment—in which the antimicrobial agent (in non-lethal doses) is injected first and after one hour when the infection with the microorganism is performed—or (2) post-treatment, where the larvae are infected before the treatment. The solutions and microorganisms are injected directly into the hemocele through the last pro-leg in the larvae. The larvae are incubated at 30°C and monitored for 24, 48, and 72 hours, with larval death determined by the lack of movement in response to the stimulus, along with cuticle discoloration (Jakobsen et al. 2018).

Evaluation of the toxicological effect of antimicrobial compounds

After confirmation of the antimicrobial potential, preclinical tests should be conducted to detect the new compounds' possible adverse effects at different concentrations on human health (Shegokar 2020). These tests provide information about *in vivo* and *in vitro* toxicology, including *in vivo* analysis of organ damage, individually and systemically, until cellular and DNA levels. If approved in preclinical trials, the new non-toxic compound can proceed to clinical trials in humans. The main preclinical tests based on current international legislation and their impacts on human health are explored here (Shegokar 2020).

The main preclinical toxicological tests that are requested internationally are based on the international guides of the International Council for Harmonization (ICH) number M3

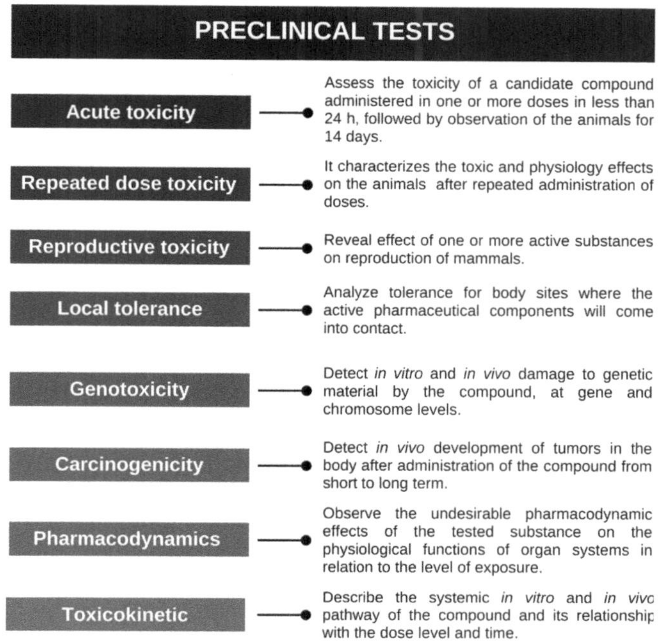

Figure 10.2. Preclinical toxicological tests of the antimicrobial candidate according to the guide ICH M3(R2): "Nonclinical Safety Studies for the Conduct of Human Clinical Trials and Marketing Authorization for Pharmaceuticals" (EMA 2013; FDA 2013).

(R2) from the Food and Drug Administration (FDA 2013) and of the European Medicine Agency (EMA 2013). Tests are conducted *in vitro* using cell culture or *in vivo* involving rodents and other mammals (Fig. 10.2). In general, preclinical trials assess the toxicity in different model organisms at diverse levels. Acute toxicity in a single dose or repeated doses and reproductive toxicity generally assess how the compound acts in various organ systems in mice, including reproductive organs (*in vivo*). Genotoxicity and carcinogenicity tests assess the compound's ability to induce DNA damage and develop a tumor, respectively. Local toxicology observes whether the drug induces an undesirable reaction in the contact region. Pharmacodynamics is an observation of the effects of the compound on organ systems, emphasizing the central nervous, cardiovascular and respiratory systems. Finally, toxicokinetic observes the pathway of the compound throughout the body, including absorption, metabolism and excretion. So far, genotoxicity and toxicokinetic tests can be conducted *in vitro* (Sjöberg and Jones 2013; Denny and Sterwart 2017; Madia et al. 2021).

For the use of plants in folk medicine, herbal preparations are established by international regulatory agencies as supplemental food, and toxicological tests become differentiated. Their analysis can be reduced to reproductive toxicology and genotoxicity assays (Poivre et al. 2017) and, if there is evidence of genotoxicity, carcinogenicity tests should be performed (EMA 2013; FDA 2015). Also, *in vitro* cytotoxicity tests must be conducted to provide toxicological information at the cellular level, in addition to indicate concentrations to be used in genotoxicity tests (Araújo et al. 2016).

Cytotoxicity

The cytotoxicity tests evaluate the potential of the candidate compound for damaging the function of the vital cell organelles, such as mitochondrial and

Table 10.2. List of cell lines recommended for cytotoxicity and genotoxicity tests.

Guideline	Cell lineage
ISO 10993-5-2009[1]	BALB/c 3T3, MRC-5, WI-38, Vero, V79, BHK-21, L929
OECD TG 473[2] and 487[3]	Primary human culture or mammalian peripheral blood lymphocytes, CHO, V79, CHL/IU e L5178Y e TK6
OECD TG 473 and 487*	HT29, Caco-2, HepaRG, HepG2, A549, embryonic Syrian hamster cells

[1] ISO - International Organization for Standardization 10993-5-2009 (2009) Tests for *in vitro* cytotoxicity; [2] OECD -Organization for Economic Co-operation and Development TG 473: *In vitro* Mammalian Chromosomal Aberration Test (2016a); [3] OECD TG 487 *In vitro* Mammalian Cell Micronucleus Test (2016b). * Cells under validation.

Table 10.3. Main tests used to assess cytotoxicity evaluation and respective cellular mechanism approached.

Assay	Mechanism
MTT*	Mitochondrial functionality
Trypan blue exclusion	Cell membrane integrity
Neutral red*	Selective permeability of the cell membrane
MTS	Functionality of the NAD(P)H-dependent dehydrogenase enzyme
Sulforamine B	Total cellular protein level
Lactate dehydrogenase	Cell membrane integrity
Colony formation*	Cell cycle
XTT*	Mitochondrial functionality

* Recommended by ISO 10993-5-2009 Tests for *in vitro* cytotoxicity (Miller et al. 2017).
MTT: 3-(4,5-dimethyl-2-thiazolyl)-2,5-diphenyl-2H-tetrazolium;
MTS: 3-(4,5-dimethylthiazol-2-yl)-5-(3-carboxymethoxyphenyl)-2-(4-sulfophenyl)-2H-tetrazolium;
XTT: 2,3-bis-(2-methoxy-4-nitro-5-sulfophenyl)-2H-tetrazolium-5-carboxanilide.

plasmatic membrane and/or induce mitosis interference. These tests are necessary to define concentration limits for *in vitro* tests (FDA 2013; EMA 2013; Aschrafi et al. 2021). Different cell lines are recommended for the scientific validation of cytotoxic assays by different guides (Table 10.2). Other lines, still undergoing validation and recognition are also recommended (OECD 2016a; 2016b).

The cytotoxicity assays can be divided into two types. The first type analyzes changes in cell integrity, apoptosis, necrosis and the cell cycle by different methodologies (Table 10.3), generally using dyes or chemical reagents acting as enzymatic markers (Araújo et al. 2016). Tests such as MTT [3-(4,5-dimethyl-2-thiazolyl)-2,5-diphenyl-2H-tetrazolium], neutral red, colony formation and XTT [2,3-bis-(2-methoxy-4-nitro-5-sulfophenyl)-2H-tetrazolium-5-carboxanilide] are recommended only by ISO 10993-5-2009 (Miller et al. 2017). The MTT colorimetric assay, for instance, is based on the mitochondrial function in reducing the soluble and yellow tetrazolium salt of MTT to an insoluble formazan violet product in viable cells (Vajrabhaya and Korsuwannawong 2018). On the other hand, the neutral red test assesses the selective permeability of the membrane. When the membrane is damaged, the dye passes through it and concentrates in the lysosomes (Ates et al. 2017).

In the second type of cytotoxic tests, there is an evaluation of the cell cycle by analyzing the following assessments: (1) Relative Increase in Cell Count (RICC), which measure the proportion of cells has divided in treated cultures; (2) Relative Population Doubling (RPD), which is similar to RICC but takes into account a log calculation for population doubling;

(3) Cytokinesis-Block Proliferation Index (CBPI), and (4) Replication Index (RI). The last two assays are related to Cytokinesis-Block Micronucleus assay (CBMN), a genotoxicity test described below. The CBPI analyzes the number of nuclei per cell as indicative of cell proliferation during the treatment, while RI calculates the cytostasis percentage (inhibition of cell multiplication) by comparing the number of cell cycles in treated and control cultures. The four assessments have a common characteristic: the analyzed cells undergo one or more cell cycles of exposure to the candidate compound. Measures to standardize the interpretation of test results and increase data reliability and acceptability were implemented by Organization for Economic Cooperation and Development (OECD) (OECD 2016a). The mentioned assessments are guided by the OECD and S2 (R1) TGs 487 and 473 guides recommended by FDA and EMA, determining standards for measuring cytotoxicity and the limit concentration for further tests (Fujita et al. 2017; Galloway 2017).

For the detection of the *in vitro* limit cytotoxic concentration, one out of four cytogenetic tests (RICC, RPD, CBPI or RI) with sufficient compound concentration to induce cytotoxicity is recommended by ICH and OECD (OECD 2016a). However, there are divergences in the selection of the maximum concentration used for *in vitro* tests. FDA and EMA S2 (R1) guides recommend maximum concentrations of 1 mM or 500 µg/mL (Kirkland and Fowler 2010), while OECD TGs 487 and 473 recommend 10 mM or 5000 µg/mL. Additionally, in some cases, as plant extracts, the composition of the tested substance is not well defined (OECD 2016a). On the other hand, all guides recommend that genotoxic analysis use concentrations that do not exceed 50% cytotoxicity (FDA 2012; OECD 2016a; OECD 2016b).

Several plant compounds with antimicrobial activities have been tested for cytotoxic potential. For instance, PaDef defensin from *Persea americana* Mill., with activity against *Escherichia coli* and *Staphylococcus aureus* (Guzmán-Rodríguez et al. 2013), exhibited no cytotoxicity action in human peripheral blood mononucleated cells (Flores-Alvarez et al. 2018). On the other hand, gallic acid with recognized activity against three gram-positive and five gram-negative bacteria (Özçelik et al. 2011) exhibited a cytotoxic potential at 25 µM over 3T3 cells in the MTT assay (Hsu et al. 2007).

An additional method to analyze the cytotoxicity is to determine the capacity of antimicrobial compounds to induce lysis of red blood cells (Pancotti et al. 2021). When the compounds possess an alternative route of administration, like intravenous, and are in contact with blood, they must be tested for potential hemolytic properties (Almizraq et al. 2017) by *in vitro* hemolytic assays, as preconized by the FDA regulation (2015).

Genotoxicity

The analysis of a compound's genotoxic potential is critical to verify its safety of use as a medicine. Genotoxicity tests assess the compound's ability to interact and induce changes in DNA as gene mutations or as numerical or structural chromosomal rearrangements. Such modifications can induce changes in germ and somatic cells (Habas et al. 2017; Taylor et al. 2019; Choudhuri et al. 2021), which may result in teratogenic and carcinogenic development (Soeteman-Hernández et al. 2016).

As mentioned earlier, genotoxicity can be assessed *in vitro* and *in vivo* models [Guideline M3 (R2)]. For *in vitro* assessments, two tests are required: (1) *Salmonella* (Ames) test; and (2) a test using mammalian cells. The latter can be one of the three possibilities: (1) assay of gene mutation in mouse lymphoma; (2) chromosomal aberrations; and (3) micronucleus test (Sekizawa et al. 2017). In the development of antimicrobials, the Ames test is often unfeasible because it involves bacterial strains of *Salmonella typhimurium*. In this case, more than one analysis can be carried out strategically on mammalian cells. On the other hand, for *in vivo* tests, the following assays are recommended: (1) the bone marrow micronucleus test; (2) chromosomal aberrations of the bone marrow; and (3) the micronucleus test in peripheral blood cells of mice or rats (EMA 2013).

Although there are limitations for predicting *in vivo* results based on *in vitro* assays, the tests are complementary since negative results *in vitro* also show negative results *in vivo* in most cases (Wilk-Zasadna et al. 2015). Besides reducing the use of animals, *in vitro* tests have been improved to increase the tests' sensitivity and the predictions of results (Corvi and Madia 2017).

The *in vitro* and *in vivo* micronucleus test is considered the gold standard for the analysis of genotoxicity. It analyzes the genotoxic potential by the frequency of nuclear changes resulting from chromosomal changes in the previous mitosis, including chromosome breaks or loss of whole chromosomes. Nuclear changes are mainly observed as micronucleus formation but also by the presence of nuclear buds and nucleoplasmic bridges (Fenech 2007), as well as the detection of apoptosis, necrosis and cytotoxicity. Regulatory agencies recommend that *in vitro* tests should include the step of blocking cytokinesis for the formation of binucleated cells, denominated cytokinesis-block micronucleus assay (CBMN) (EMA 2013; OECD 2016a), which allows the identification of the cells exposed to the candidate compound for one or more division cycles (Fenech 2007). The cytotoxicity (CBPI) by analyzing the number of cell cycles using CBMN assays is one of the tests recommended by international agencies (OECD 2016a; 2016b).

A second *in vitro* and *in vivo* test widely used in the analysis of candidate antimicrobial compounds is the chromosomal aberration test, by analyzing both structural and numerical changes scored in metaphase cells (FDA 2012; EMA 2013; Sekizawa et al. 2017). Structural chromosomal aberrations include chromosome gaps, breaks, deletion and other rearrangements like duplications, dicentrics, rings, isochromosomes, inversions and translocations. While the numerical changes result in disturbances or interferences in the chromosomal number at cell division, that can lead to cells bearing aneuploidy or/and polyploidy (Mosesso et al. 2019).

Many antimicrobial compounds from plants are considered less toxic, such as carvacrol, commonly found in aromatic plants (Vincenzi et al. 2004). This component possesses activity against *Staphylococcus aureus* (40 μg/mL) and *Escherichia coli* (125 μg/mL) (Marchese et al. 2018), with no genotoxic potential in the CBMN assay in human lymphocytes cells up to the concentration of 200 μg/mL (Aydin et al. 2013). On the other hand, certain terpenes like citral, which has good antimicrobial activity against *Bacillus cereus*, *Salmonella enterica* serovar Typhimurium, *Staphylococcus aureus*, and *Escherichia coli* (Guimarães et al. 2019), presented genotoxic potential above 0.625 μg/mL concentration on HepG2 and human leukocytes cells (Souza et al. 2020).

Thus, compounds with antimicrobial potential from plants must undergo a battery of toxicological tests at different stages to assess their risks of adverse effects. Several countries and agencies have harmonized toxicological tests to unify them worldwide and provide toxicological information covering the entire world population.

In silico approaches as new insights for the discovery of new antimicrobial agents—a top-down approach

The urgent demand to obtain alternatives against antibiotic-resistant pathogens has driven antimicrobial compounds into the spotlight (Eckert 2011). Since the design and development of a new drug candidate is an expensive process, the use of *in silico* approaches is frequently applied to improve the discovery of novel compounds and to optimize these molecules in clinical use (Anand et al. 2019; Charoenkwan et al. 2021).

The goal of modern plant-derived antimicrobial discovery is to find chemical compounds that have a specific biological effect. One of the main techniques used in drug discovery is structure-based high-throughput docking screening, in which automated assays are used to test large databases of thousands or millions of chemical compounds available in public databases for a desired biological activity (Alam and Khan 2018). This is a robust theoretical routine and requires substantial computational resources. The idea behind virtual docking screening is to test compounds *in silico* in a preliminary step to reduce the number of candidates to be screened experimentally, thus reducing the physical experiments time (Slater and Kontoyianni 2019). When the investigated compound is an approved commercial drug, this approach is called drug repositioning or repurposing, having advantages over the development of a new drug for a given indication.

Another valuable technique is the computational analysis of plant protein evolution and its conserved functions (Shafee et al. 2020). These findings constitute the basic assumption behind most of the available bioinformatics algorithms for alignment and protein structure prediction (Campos et al. 2018; Shafee et al. 2020). Among the more used available prediction methods (e.g., template-based, threading and *ab initio* modeling), those based on comparative modeling provide more accurate structures that can be used in a large variety of applications, including ligand binding sites prediction and virtual screening (Ehrt et al. 2016).

Biomolecular compounds are dynamic bodies with molecular rearrangements in their structure due to the environment (physiological conditions) and their targets (Reif and Zacharias 2019). Molecular Dynamics (MD) simulations use classical mechanics to describe these motions in biomolecules. This *in silico* technique is commonly used to assess structural dynamics and stability of biological compounds (Dror et al. 2012; Reif and Zacharias 2019).

The MD simulations of ligand-target systems provide valuable information (e.g., stability, structural dynamics, solvent accessibility and other features) that is not accessible in routines that use static structures, as cited earlier (Fradera and Babaoglu 2017). MD simulations allowed the visualization in full-atom of the interactions between antimicrobial molecules and various bacterial membranes and also other cellular structures, mimicking the pathogen and its physical-chemical features. Hence, MD simulations can reveal structural motions of accessible binding sites and provide details about dynamic receptor-ligand interactions (Hollingsworth and Dror 2018). Moreover, MD and free-energy calculations enable estimating binding affinity for ranking ligands according to their binding propensity to a specific partner (Alford et al. 2017). The use of MD simulations and free-energy calculations in plant-derived drug candidates are manifold. The combination of the previously cited *in silico* methods and machine-learning approaches has recently become increasingly popular and plays a prominent role in drug discovery (Vamathevan et al. 2019). In summary, these methods enable efficient strategies to discover novel antimicrobials aiming to control microbial diseases.

Computational tools for prospecting small molecules

Chemoinformatics tools for small molecule analysis play an essential role in many fields, impacting biochemistry, genomics, agronomy and drug discovery. For this, tools are applied to analyze structural similarities, physical-chemical properties, and bioactivity profiles of natural and synthetic compounds to predict their mechanisms of action in biological systems (Backman et al. 2011; Medina-Franco et al. 2021). Drug design based on molecular structure is becoming an essential tool for discovering compounds faster and more economically than the traditional method (Batool et al. 2019), here named bottom-up. The introduction of bioinformatics analysis in these surveys has led to an expansive collection of computational tools that assist in searching and designing proteins and enzymes (Copeland et al. 2012), i.e., a top-down approach.

The main functionalities of these tools fall into five main areas of application (Fig. 10.3): (i) database search and integration, (ii) comparison of structures prioritizing similar compounds, (iii) search for a compound in different ways (chemical formula or two-dimensional structure), (iv) the grouping of compounds correlating characteristics of biological activity through the conversion of a similarity matrix by a distance matrix and prediction of chemical properties, considering that physicochemical properties data are essential to predict bioactivity and (v) other properties of small molecules recognized using

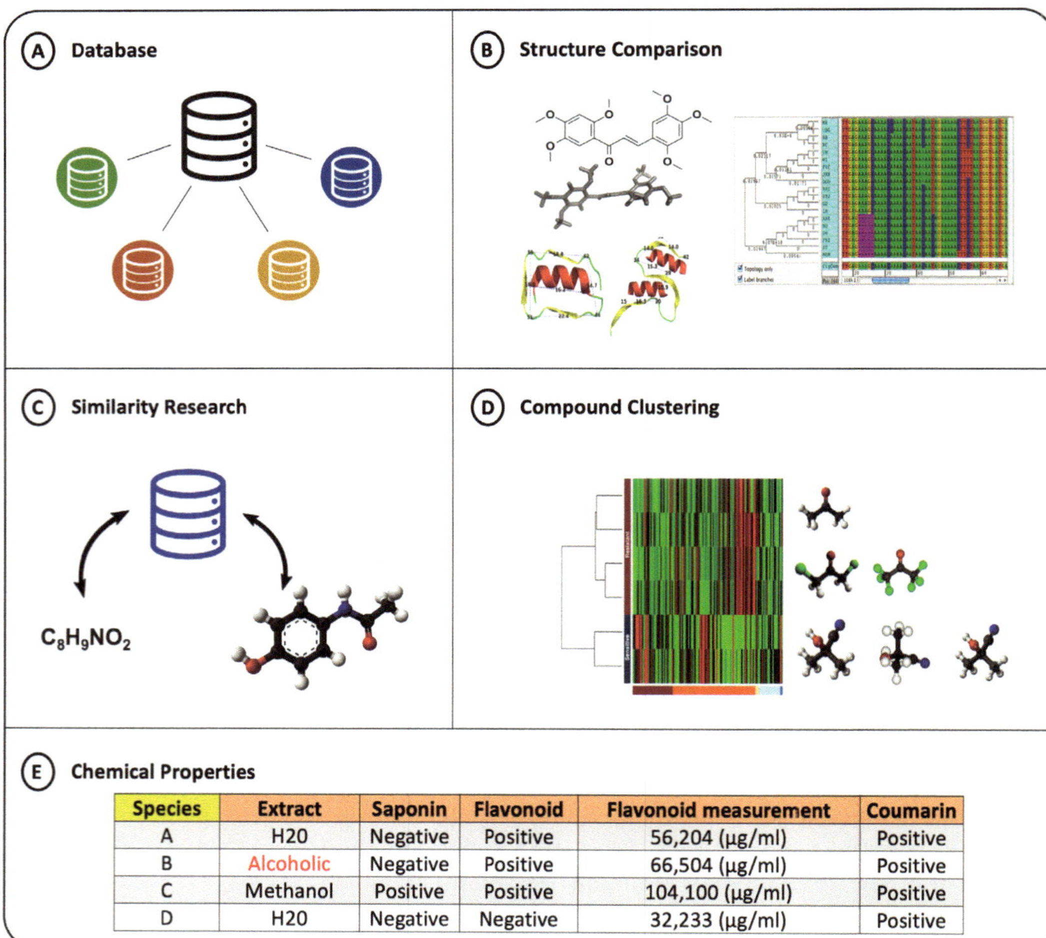

Species	Extract	Saponin	Flavonoid	Flavonoid measurement	Coumarin
A	H20	Negative	Positive	56,204 (μg/ml)	Positive
B	Alcoholic	Negative	Positive	66,504 (μg/ml)	Positive
C	Methanol	Positive	Positive	104,100 (μg/ml)	Positive
D	H20	Negative	Negative	32,233 (μg/ml)	Positive

Figure 10.3. Top-down approaches to search for new compounds, including (A) database search and integration, (B) comparison of features prioritizing structurally similar compounds, (C) search based on similarity (chemical formula or two-dimensional structure), (D) clustering of compounds by similarity/distance matrix and prediction of chemical properties (E) other properties of small molecules using machine learning.

Table 10.4. Main databanks with tools and resources frequently used in the identification, characterization and evaluation of antimicrobial compounds from plant origin. All repositories (see URLs) are freely available (no payment required).

Database	URL
PubChem	https://pubchem.ncbi.nlm.nih.gov
ZINC	http://zinc.docking.org
ChEMBL	https://www.ebi.ac.uk/chembl/
NCI	https://cactus.nci.nih.gov/ncidb2.2/
ChemDB	http://cdb.ics.uci.edu
ChemSpider	http://www.chemspider.com
BindingDB	http://www.bindingdb.org/bind/index.jsp
PDB-Bind	http://sw16.im.med.umich.edu/databases/pdbbind/index.jsp
PDBeChem	https://www.ebi.ac.uk/pdbe-srv/pdbechem/
KEGG	https://www.genome.jp/kegg/
HMDB	https://hmdb.ca
SMPDB	https://smpdb.ca
BIAdb	http://crdd.osdd.net/raghava/biadb/
DrugBank	https://www.drugbank.ca
HIT	https://bio.tools/hit
SuperNatural	http://bioinf-applied.charite.de/supernatural_new/index.php
NPACT	http://crdd.osdd.net/raghava/npact/
TTD	http://bidd.nus.edu.sg/group/cjttd/
PharmaGKB	https://www.pharmgkb.org
SuperDrug	http://cheminfo.charite.de/superdrug2
CRDD	http://crdd.osdd.net/chemdatabase.php

machine learning approaches (Backman et al. 2011). These procedures allowing a more in-depth understanding of chemical compounds' nature is highly relevant, promoting knowledge for both, to understand fundamental biological processes and to develop new therapeutic strategies (Dobson 2004).

Currently, there are more than 100 million small molecules available in open-access databases (Table 10.4), in addition to preliminary bioactivity data from hundreds of screening experiments, as well as evaluation of subcelular location (Pence and Williams 2010; Kim et al. 2019).

Computational techniques such as virtual structure-based screening, molecular docking and molecular dynamics simulation are the most commonly used methods to evaluate compounds to interact with a target molecule. These methods have numerous applications in the analysis of binding energy, ligand-protein interactions and evaluation of structural changes that occur during the binding process between these molecules (Kalyaanamoorthy and Chen 2011). The availability of supercomputers (High-Performance Computing, HPC), computer clusters and cloud computing has accelerated the identification and evaluation of these molecules (Puertas-Martín et al. 2020; Li et al. 2021). These computational resources serve as an efficient technology to accelerate the entire process, including various screening procedures, combinatorial chemistry and calculations of properties such as absorption, distribution, metabolism, excretion and toxicity (Batool and Choi 2017).

Despite that, it is worth mentioning that biological systems are increasingly being considered sets of interactions of several highly interconnected molecules (as shown, for example, by the emergence of systems biology) in contrast to the reductionist view of traditional biochemistry that aims only at singular interactions (Dobson 2004; Lai et al. 2017).

Metabolomics and the discovery of new therapeutic agents—uncovering 'the big picture'

One of the first steps towards the design of safer drugs is the characterization of their interactions with proteins and other bioactive compounds. For this, the omics sciences use biochemical analyzes, structural biology, enzymology, toxicology, etc. (Kaczor et al. 2018). Thus, the progress and the interaction of these areas allowed the consequent growth of the databases. Therefore, these analyses to prospect all molecules and their interactions are not manually feasible (Petrey and Honig 2014).

Interactomics is the branch of omics usually referred to in the field of protein-protein interaction analysis. However, this term may be applied to other types of molecule-molecule interactions (De Las Rivas and Prieto 2012; Schäfer and Ciaudo 2020). This field has been expanding and interacting increasingly with an omics area that performs analyzes in a broader context, metabolomics. This term is often related to a set of approaches that aim to describe all metabolites in a biological system (Fukushima et al. 2009). In other words, one can define metabolomics as the interactomics put in context.

Some databases have been dedicated to compiling information from these interactions between biomolecules that may come from different types of experiments (Pedamallu and Ozdamar 2014), the best known example of a repository that contains metabolic pathways regards the Kyoto Encyclopedia of Genes and Genomes (Kanehisa et al. 2017). Other databases like STRING are able to map interactions between proteins based on annotations of coexpression, co-occurrence, text mining, mapping of interactions already known by different experiments (Szklarczyk et al. 2017). Like STRING, other databases also compile such information and may add data from genes, and chemical associations, in addition to post-translational modifications; BioGRID is a well-cured repository that presents these characteristics (Oughtred et al. 2019). Another very useful tool to understand the interactions between proteins is IntAct (Kerrien et al. 2011), a database that groups information cured from literature and reports in the form of binary interactions. A thorough analysis of such banks can help understand and analyze the complexity of the different interactions between proteins and metabolic pathways.

In general, it is expected that a drug, be it a simple molecule or a peptide, will interact with several other biological structures. These same structures can form a complex cascade of reactions. Therefore, given this condition, the composition of a given network in a system, even if approximate, can be crucial in elucidating aspects from the toxicological and drug efficacy points of view (Petrey and Honig 2014; Luo et al. 2019).

Structural biology tools play a crucial role in the simulation of biological interactomes (Petrey and Honig 2014). The availability of protein structures has enabled researchers to use multiple drug design approaches, ranging from design optimization to mimicking interaction(s) of the target drug (Ivanov et al. 2013; Petrey and Honig 2014).

Pharmaceutical market, challenges and investments in antimicrobial compounds

The development of new antimicrobial substances is a challenge even for the scientific community. It requires great effort and time, ranging from the isolation of bioactive compounds, the phase of development of new drugs, to their acceptance by the pharmaceutical market.

The modest use, the limited time of treatment and the sporadic rate of patients in the general population, compared to treatments for chronic diseases (e.g., hypertension, diabetes) with continuous treatments, do not favor the investment and production of these drugs in the industry. Besides, complicated regulatory processes, large investments with low financial returns, and reduced market share, mean that the antimicrobials market is not prioritized (Spellberg 2014; 2015; Schulz et al. 2019).

For the use of antibiotics, doctors and experts recommend that broad-spectrum drugs with sensitivity to available multi-drug resistant microorganisms, be kept as a last resort for bacterial strains resistant to conventional treatments, which further complicates the financial return for companies developing new medicines. A well-known example is polymyxins, which are used to treat gram-negative strains. They are so toxic to kidney cells that they are used only as a last resource to treat infections caused by multi-resistant pathogens (Fernandes and Martens 2017; Moffatt et al. 2019).

In this sense, the pharmaceutical industry has shown greater interest in the development of drugs against methicillin-resistant *Staphylococcus aureus* (MRSA), which can be explained by the recurrent number of infections caused by this pathogen worldwide, thus creating a global market for these drugs concerning antibiotics for the treatment of gram-negative microorganisms (Abushaheen et al. 2020). At the same time, some recently approved antibiotics (such as fidaxomicin) are highly efficient compared to traditional antibiotics. However, they cannot compete in the market with existing drugs at lower prices, such as vancomycin, a widely used antibiotic to treat Gram-positive bacterial infections (Jeffres 2017).

Other difficulties encountered refer to some pipelines for obtaining new antimicrobials that are often scaffolds of pre-existing molecules, which are redirected to unique enzyme targets, that are quickly doomed to resistance. In contrast, new companies are adopting new approaches to drug development, using modern pipelines with new screening methods, using scaffolds of molecules of natural origin, expression of genes encoding small proteins (peptides) in heterologous hosts to produce new therapeutic agents, especially those chemically complex to synthesize and costly (Deng et al. 2017; Fernandes and Martens 2017; Gazzaneo et al. 2017). These companies are overcoming the challenge in developing new antimicrobials based on focus and persistence, relying on allies such as chemistry, computational tools and the bioavailability of raw material from nature.

In this sense, chemical compounds present in abundance in plants have been extensively investigated by the pharmaceutical industry as potential therapeutic molecules in recent years. Among these compounds, peptides have been extensively studied due to their increasing use and multifunctionality. Numerous functions can be found in peptides, such as antimicrobial activity, antibodies and vaccines. Also the use of peptides in the preparation of cosmetic products and as adjuvants' in treating infectious and inflammatory diseases or even rare or orphan diseases. Besides, their use in the food industry has been encouraged. Due to the ease of penetration into cells and the most varied mechanisms of action, peptides have also been used in cell targeting to find intracellular targets (Fosgerau and Hoffmann 2015; Negahdaripour et al. 2019; Buckton et al. 2021).

According to the repository of antimicrobial peptides (DRAMP) (Kang et al. 2019), until July 2020, there were 76 AMPs under development to be used as drugs in the different preclinical and clinical phases. Although interest in these molecules is growing, only a small part is available to the population as a drug; until 2018, only 1% of the studied peptides reached the earlier registration stage (Lau and Dunn 2018). Further growth is expected, as approximately 140 peptide drugs have been used in clinical trials and more than 600 peptides are in the pre-clinical and clinical testing phase. Most likely, other compounds will be approved by the FDA between 2019 and 2021, including cefiderocol, imipenem-relebactam, and lefamulin (Fosgerau and Hoffmann 2015; Schulz et al. 2019).

New technologies using bioactive compounds have helped the pharmaceutical industry innovate, allowing rapid diagnostic tools capable of selecting and directing therapies. A minimal number has been used clinically to the detriment of the higher costs that prevent them from being used on a large scale. Even so, many of them are being studied, as is the case of subunit vaccines that include fragments of pathogen proteins or even peptide epitopes of protective antigens. The latter shows advantages concerning stability, purity and lesser side effects, such as autoimmune responses and enabling responses to more than one antigen aimed at producing an immune response against more than two pathogens at the same time (Skwarczynski and Toth 2014; Negahdaripour et al. 2019). On the other hand, this technology has some flaws, as the induction of prolonged immune responses, probably due to its lower bioavailability due to rapid proteolytic degradation *in vivo* (Croft and Purcell 2011).

Development of antibacterial compounds

Bacteria evolve when subjected to selective pressure, thus managing to circumvent the mechanisms of action of several available drugs. Therefore the availability of antibacterial drugs in relation to the speed with which bacteria develop new resistance mechanisms has become obsolete.

Alarming data, such as the number of fatalities (about 700,000 people a year) as a result of infections by antibiotic-resistant bacteria, have led experts to estimate a future number of 10 million deaths in 2050 if new antibacterial drugs are not developed (Willyard 2017). In response to these figures, the Infectious Diseases Society of America (IDSA) at the European Center for Disease Prevention and Control (CDC) launched a bold proposal to develop 10 new antibacterial drugs by 2020 from new classes of antibiotics, as well as from existing ones. Concomitant to this, IDSA has stimulated the development of new diagnostic tests essential in the optimization and rational use of new drugs to identify pathogens that cause multidrug-resistant infections (IDSA 2010; Tamma et al. 2021).

Institutions that aim to promote public health care have proposed some incentives for investments in the production of new antibacterials, such as economic incentives, through public-private partnerships and government-supported collaboration programs. An example of an incentive was a European investment in which the European Union and the European Federation of Pharmaceutical Industries and Associations (EFPIA) donated a total of € 600 million planned to be spent by 2020 on developing new antibiotics (Kirby 2012).

Other incentives come from institutions that are also involved in the development of new antibiotics such as US federal government agencies, including the Food and Drug Administration (FDA), National Institute of Allergy and Infectious Diseases (NIAID), Department of Defense (DoD) and the Biomedical Authority for Advanced Research and Development (BARDA), as well as global scientific agencies, such as the World Health Organization (WHO) and the European Medicines Agency that are committed to recognizing the severe crisis of antibiotic resistance and fostering investments in new drugs (Talbot et al. 2019). More than half of the drugs in clinical use and approved by the FDA are derived and/or inspired by natural products (Newman and Cragg 2016; Barbosa et al. 2020).

Another proposal to encourage antimicrobials' production focuses on their regulatory approaches, essential for the safe use of medicines and their evaluation in the decision-making processes, ensuring that clinical studies can keep up with the growing threat of resistance (Gould and Bal 2013). Besides, a surveillance system for the sale and consumption of antimicrobials would be paramount in combating resistance, with the efficient administration of antibiotics necessary at all levels, from hospitals, pharmacies to communities.

Recognition steps and pharmacological records for antimicrobial products

In general, a drug discovery process consists of four main steps: the first is the discovery phase, where a potential therapeutic target and active ligands are identified. The second is the development phase, where the main hits are synthesized, optimized, tested *in vitro* in biochemical tests, and have their three-dimensional structure determined. The third is the clinical testing phase, and then the fourth and final phase is the registration for later distribution on the market for clinical use (Batool et al. 2019).

However, it is noteworthy that it is a process with retroactive takes, where resumes from previous phases are more common than one may imagine. When a potential candidate arrives in the clinical tests phase and presents undesirable characteristics, it is advisable to return to the previous phases to remodel the molecules and find plausible alternatives. Choosing the right biological target or a combination of targets is one of the main tasks for any successful drug discovery project. All subsequent efforts, whether small molecule identification, compound optimization, pharmacokinetic studies or a clinical trial, will be as effective as the initial decision to choose one target or another (Blaney 2012; Faust et al. 2021).

Here the focus is on developing an antimicrobial peptide (AMP), reporting on the recognition steps to become a new antimicrobial compound until acceptance by the pharmaceutical industry. From development to commercial approval, anew AMP can take years to be available on the market. These must go through a rigorous process that can be divided into: pre-clinical (*in vitro* and *in vivo* tests) and clinical phases (I, II, and III). In this last stage, aspects related to the safety and activity of the drug in humans are evaluated. One can also include phase IV or pre-registration that consists of monitoring the drug after launching and production on a large scale (Jiang et al. 2021).

Patent registration

An alternative to protect the stages of the antimicrobial drug discovery and commercialization process concerns patent registration. This register serves to protect the intellectual properties of inventors and companies (Farjadian et al. 2018).

The patent registration process can take years due to the time required to recognize and evaluate the new antimicrobial agent. This process begins with the evaluation carried out by technical teams from companies specialized in the field or by institutions responsible for discovering and producing future therapeutic agents. From there, the registration number of the patent to which the new product was submitted is obtained. Concomitant to this, the registration process is still pending while pre-clinical and clinical trials are carried out. However, taking into account the economic investments and ethical issues involved in trials for the development of pharmaceutical products, the issue of pre-clinical tests and clinical trials should be extended to the period of market exclusivity; that is, the company investing in the new product must bear the costs of testing if it intends to launch the product on the market (Bosetti and Vereeck 2012; Farjadian et al. 2018; Vandenplas et al. 2021).

Introducing a new drug to the market may require 10 to 20 years, depending on the country. This period decreases significantly when the product is under commercial exclusivity (Farjadian et al. 2018). In this sense, registration of a patent is essential for commercializing the new antimicrobial product and can signal safety and effectiveness for the market. The patents also provide credibility for shareholder companies, attracting even more investment (Bosetti and Vereeck 2011; Imran et al. 2021).

Current scenario in search for antiviral agents and COVID-19

Severe acute respiratory syndrome (SARS) caused by the new coronavirus (SARS-Cov-2) was initially diagnosed in China's Wuhan province in late 2019, spreading to countless countries worldwide. After the increasing number of cases, on March 11, 2020, this disease was recognized by the WHO as a pandemic. The disease's name comes from Coronavirus disease (COVID) since it belongs to the family Coronaviridaeae and as it was identified in 2019, it was named COVID-19 (Gupta 2020). Due to the pandemic character, infectious potential, mortality rate and lack of scientifically proven treatments to date, the search for antiviral therapies has been intensified to find bioactive compounds that help fight COVID-19 or assist in the treatment process. Currently, treatments for respiratory diseases caused by viruses present difficulties in their effectiveness, and the forms of prevention (such as vaccines) are still the best options.

The main targets used for the production of new therapeutic agents target the Spike (S) protein, contained in the viral surface, being among the responsible for the beginning of the connection between the virus and the Receptor-Binding Domain (RBD) of the Angiotensin-Converting Enzyme 2 (ACE2) that functions as a cellular receptor in the host. Virus S protein is considered a potential target for the design and development of a new drug or vaccine, as well as being essential for *in silico* tests in the search for new candidates with antiviral agents (Shang et al. 2020; Gupta 2020; Siriwattananon et al. 2021).

To obtain a new drug or vaccine, several parameters must be analyzed. The screening of antiviral molecules has been one of the first steps in searching for molecules against SARS-CoV2. For this purpose, bioinformatics provides essential tools to speed up the identification of new molecules. Also, other *in silico* analysis work as parameters to determine the choice between the candidates. Some techniques are essential in this search, such as virtual screening, used in the search for candidates available in databases (mining), characterization of the primary structure through the prediction of physicochemical properties, antimicrobial activity, toxicological, allergenic and immunogenic potential. The evaluation of the tertiary structure performed through molecular modeling, refinement of the model obtained and the evaluation of its quality are essential steps to obtain the most representative model for the next stages of molecular simulation (Odhar et al. 2020; Santos-Silva et al. 2020).

Finally, molecular docking analysis will represent the interaction between the antimicrobial model and the host cell's binding protein or receptor to evaluate the molecules' maximum interaction with the lowest energy. To verify the interaction force between the molecules, it is necessary to evaluate the simulation in molecular dynamics to test the interaction forces between these molecules against a force field and the interactions

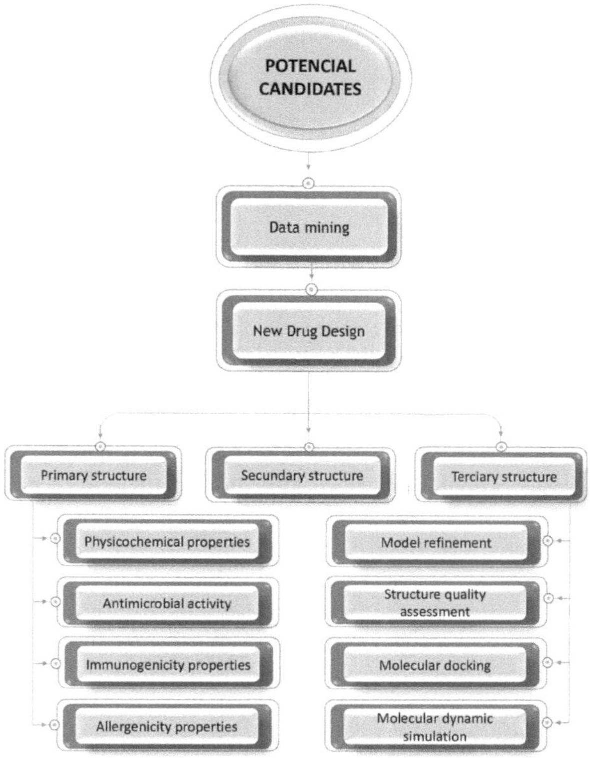

Figure 10.4. *In silico* steps for the evaluation of a potential antiviral candidate.

with a solvent (Hossain et al. 2021). A scheme of the *in silico* steps for evaluating a potential antiviral candidate is shown in Fig. 10.4.

Many studies have been developed to reveal the mechanism of action of viruses that cause respiratory diseases. A study carried out with phytocomposites from traditional Indian medicinal plants was carried out to verify multi-omics interactions and their actions concerning the immunological profile in the COVID-19 manifestation at the molecular level. The authors used immune transcriptomics data sets from healthy patients with COVID-19, besides searching for bioactive compounds and found that among the 259 phytocomposites available in the database (PubChem), 154 compounds targeted 13 reported immunity genes COVID-19 involved in several signaling pathways. Subsequently, active compounds were compared with drugs used against COVID-19 prescribed by the WHO, and 25 new phytocomposites were more efficient with higher bioactive scores than the drugs tested. Thus, the study indicated some signatures of SARS-CoV-2 virus genes that can serve as therapeutic targets of phytocomposites. However, it is necessary to validate these candidates to verify their potential to develop commercial drugs (Muthuramalingam et al. 2020).

Conclusions and perspectives

Not long ago, the process of discovering a new drug started from a natural product with an ethnobiological indication, based on its use by traditional communities. From this natural product (e.g., a plant part), a series of steps were initiated to identify, isolate, characterize and test new compounds, including those with antimicrobial potential.

In this sense, a series of biochemical and molecular protocols have been developed and improved, allowing for more specific and precise analysis.

The advent of omics, from the turn of the century, generated a new trend. Since then, massive amounts of omics data (DNA, RNA, proteins and metabolites) have been produced, culminating in an increased systemic view of biological systems.

Today it is possible to evaluate a large number of potential drugs against pathogen-specific targets simultaneously with their analysis in the human host context, predicting possible deleterious effects.

Although bioinformatics techniques allow the screening of a large amount of targets, posterior steps for the generation of a new commercial drug are identical to those of a compound isolated in the laboratory, involving *in vitro* and *in vivo* steps, prior to clinical tests.

These laboratory trials take time but are critical to patient safety. Many drugs are excluded after a long journey due to a fraction of the population with side effects. Such results highlight the importance of the so-called 'personalized medicine' where the same drug can be healing to an individual and poisonous to the other, demanding detailed mapping of the patient's clinical condition. Therefore, in the future drugs that have been disregarded may be reconsidered for more personalized applications.

It is also important to emphasize the urgency in identifying new antimicrobial compounds considering emerging pathogens (bacteria, fungi and viruses) that do not respond to existing drugs. In this context, antimicrobial peptides (AMPs) deserve special attention since they are most likely to escape the pathogen resistance.

Taken together, laboratory approaches of analytical chemistry, biochemistry, molecular biology and biotechnology can and shall successfully be complemented by omics data and bioinformatics, favoring the development of more precise, personalized and optimized medicine.

Acknowledgements

The authors thank Conselho Nacional de Desenvolvimento Científico Tecnológico (CNPq), Coordenação de Aperfeiçoamento de Pessoal de Nível Superior (CAPES) and Fundação de Amparo à Pesquisa do Estado de Pernambuco for fellowships and grants.

References

Abdallah, A.M., Alves, B.C., Gehrke, F.S., Sant'Ana, A.V., Azzalis, L.A., Junqueira, V.B., Adami, F., Perreira, E.C., Fonseca, F.L. (2014). Inability of turbidimetry method in detecting glycated hemoglobin to select diabetes mellitus patients according to their concentrations of blood glucose levels. J Clin Lab Anal 29: 312–316.

Abdelgaleil, S.A.M., Saad, M.M.G., Ariefta, N.R., Shiono, Y. (2020). Antimicrobial and phytotoxic activities of secondary metabolites from *Haplophyllum tuberculatum* and *Chrysanthemum coronarium*. S Afr J Bot 128: 35–41.

Abushaheen, M.A., Muzaheed, Fatani, A.J., Alosaimi, M., Mansy, W., George, M., Acharya, S., Rathod, S., Divakar, D.D., Jhugroo, C., Vellappally, S., Khan, A.A., Shaik, J., Jhugroo, P. (2020). Antimicrobial resistance, mechanisms and its clinical significance. Dis Mon Disease-a-Month 66: 100971.

Akash, M.S.H., Rehman, K. (2020). Thin layer chromatography. pp. 157–165. *In*: Essentials of Pharmaceutical Analysis. Springer, Singapore.

Alam, S., Khan, F. (2018). Virtual screening, Docking, ADMET and System Pharmacology studies on *Garcinia* caged Xanthone derivatives for anticancer activity. Sci Rep 8(1): 1–16.

Alford, R.F., Leaver-Fay, A., Jeliazkov, J.R., O'Meara, M.J., DiMaio, F.P., Park, H., Shapovalov, M.V., Renfrew, P.D., Mulligan, V.K., Kappel, K., Labonte, J.W., Pacella, M.S., Bonneau, R., Bradley, P., Dunbrack Jr., R.L., Das, R., Baker, D., Kuhlman, B., Kortemme, T., Gray, J.J. (2017). The Rosetta all-atom energy function for macromolecular modeling and design. J Chem Theory Comput 13(6): 3031–3048.

Almizraq, R.J., Yi, Q.L., Acker, J.P. (2017). Biomedical Excellence for Safer Transfusion (BEST) collaborative. Impact of technical and assay variation on reporting of hemolysis in stored red blood cell products. Clin Chim Acta 468: 90–97.

Anand, U., Jacobo-Herrera, N., Altemimi, A., Lakhssassi, N. (2019). A comprehensive review on medicinal plants as antimicrobial therapeutics: potential avenues of biocompatible drug discovery. Metabolites 9(11): 258.

Anand, U., Nandy, S., Mundhra, A., Das, N., Pandey, D.K., Dey, A. (2020). A review on antimicrobial botanicals, phytochemicals and natural resistance modifying agents from Apocynaceae family: Possible therapeutic approaches against multidrug resistance in pathogenic microorganisms. Drug Resist Updat 51: 100695–100695.

Andleeb, S., Alsalme, A., Al-Zaqri, N., Warad, I., Alkahtani, J., Bukhari, S.M. (2020). *In-vitro* antibacterial and antifungal properties of the organic solvent extract of *Argemone mexicana* L. J King Saud Univ Sci 32(3): 2053–2058.

Arachchige, N.I.B., Khan, F., Kim, Y.M. (2021). Antimicrobial properties of actively purified secondary metabolites isolated from different marine organisms. Curr Pharm Biotechnol 22(7): 920–944.

Araújo, S.S., Fernandes, T.C., Marin-Morales, M.A., Benko-Iseppon, A.M., Brasileiro-Vidal, A.C. (2016). Mutagenicity, genotoxicity and cytotoxicity assays of medicinal plants: first step for drug development. *In*: Rai, M., Duarte, M.C.T. (eds.). Therapeutic Medicinal Plants—From Lab to the Market. CRC Press 7: 130–153.

Aschrafi, A., Zupin, L., Vilela, L.M.B., Santos-Silva, C.A., Roldan-Filho, R.S., Lima, L.M., Lima, C.S.A., Petix, V., Tossi, A., Barbosa, L.L., Benko-Iseppon, A.M., Crovella, S. (2021). Antimicrobial and cytotoxic properties of extracts from plants traditionally used in North-East Brazil. Int J Pharmacol Phytochem Ethnomed 16: 21–32.

Ates, G., Vanhaecke, T., Rogiers, V., Rodrigues, R.M. (2017). Assaying cellular viability using the neutral red uptake assay. *In*: Gilbert, D.F., Friedrich, O. (eds.). Cell Viability Assays. Humana Press, New York, NY 1601: 19–26.

Aydin, E., Türkez, H., Geyikoğlu, F. (2013). Antioxidative, anticancer and genotoxic properties of α-pinene on N2a neuroblastoma cells. Biologia 68(5): 1004–1009.

Backman, T.W., Cao, Y., Girke, T. (2011). ChemMine tools: an online service for analyzing and clustering small molecules. Nucleic Acids Res 39(suppl_2): W486–W491.

Barbosa, F., Pinto, E., Kijjoa, A., Pinto, M., Sousa, E. (2020). Targeting antimicrobial drug resistance with marine natural products. Int J Antimicrob Agents 56(1): 106005.

Bard, G.C.V., Zottich, U., Souza, T.A.M., Ribeiro, S.F.F., Dias, G.B., Pireda, S., Cunha, M., Rodrigues, R., Pereira, L.S., Machado, O.L.T., Carvalho, A.O., Gomes, V.M. (2016). Purification, biochemical characterization, and antimicrobial activity of a new lipid transfer protein from *Coffea canephora* seeds'. Genet Mol Res 15(4): 1–16.

Barry, A.L., Thornsberry, C. (1991). Susceptibility test: diffusion test procedures. pp. 1526–1542. *In*: Balows, A., Hausler, W.J., Herramann, K.L., Isenberg, H.D., Shadomy, H.J. (eds.). Manual of Clinical Microbiology. ASM, Washington, DC A. SocietyMicrob.

Batool, M., Choi, S. (2017). Identification of druggable genome in *Staphylococcus aureus* multidrug resistant strain. *In*: 2017 IEEE Life Sciences Conference (LSC).

Batool, M., Ahmad, B., Choi, S. (2019). A structure-based drug discovery paradigm. Int J Mol Sci 20(11): 2783.

Bauer, A.W., Kirby, W.M.M., Sherris, J.C., Turck, M. (1966). Antibiotic susceptibility testing by a standardized single disc method. Am J Clin Pathol 45: 493–496.

Bekale, L., Agudelo, D., Tajmir-Riahi, H.A. (2015). The role of polymer size and hydrophobic end-group in PEG–protein interaction. Colloids Surf B 130: 141–148.

Benedini, L.J., Figueiredo, D., Cabrera-Crespo, J., Gonçalves, V.M., Silva, G.G., Campani, G., Zangirolami, T.C., Furlan, F.F. (2019). Modeling and simulation of anion exchange chromatography for purification of proteins in complex mixtures. J Chromatogr A 1613: 460685.

Benko-Iseppon, A.M., Crovella, S. (2010). Ethnobotanical bioprospection of candidates for potential antimicrobial drugs from Brazilian plants: state of art and perspectives. Curr Protein Pept Sci 11(3): 189–194.

Benko-Iseppon, A.M., Galdino, S.L., Calsa, T.Jr., Akio-Kido, E., Tossi, A., Belarmino, L.C., Crovella, S. (2010). Overview on plant antimicrobial peptides. Curr Protein Pept Sci 11(3): 181–188.

Benko-Iseppon, A.M., Crovella, S. (2017). Plant immunity and beyond: Signals from proteins & peptides. Curr Protein Pept Sci 18(4): 292–293.

Blaney, J. (2012). A very short history of structure-based design: how did we get here and where do we need to go? J Comput Aided Mol Des 26(1): 13–14.

Bogdanov, I.V., Shenkarev, Z.O., Finkina, E.I., Melnikova, D.N., Rumynskiy, E.I., Arseniev, A.S., Ovchinnikova, T.V. (2016). A novel lipid transfer protein from the pea *Pisum sativum*: isolation, recombinant expression, solution structure, antifungal activity, lipid binding, and allergenic properties'. BMC Plant Biol 16(1): 107.

Bose, U., Broadbent, J.A., Byrne, K., Hasan, S., Howitt, C.A., Colgrave, M.L. (2019). Optimisation of protein extraction for in-depth profiling of the cereal grain proteome. J Proteomics 197: 23–33.

Bosetti, R., Vereeck, L. (2011). Future of nanomedicine: obstacles and remedies. Nanomedicine 6(4): 747–755.

Bosetti, R., Vereeck, L. (2012). The impact of effective patents on future innovations in nanomedicine. Pharm Pat Anal 1(1): 37–43.

Bouazzi, S., Jmii, H., El Mokni, R., Faidi, K., Falconieri, D., Piras, A., Jaïdane, H., Porcedda, S., Hammami, S. (2018). Cytotoxic and antiviral activities of the essential oils from Tunisian Fern, *Osmunda regalis*. S Afr J Bot 118: 52–57.

Braz, R., Wolf, L.G., Lopes, G.C., Mello, J.C.P. (2012). Quality control and TLC profile data on selected plant species commonly found in the Brazilian market. Rev Bbras fFarmacogn 22(5): 1111–1118.

Brito, S.S.S., Silva, F., Malheiro, R., Baptista, P., Pereira, J.A. (2018). *Croton argyrophyllus* Kunth and *Croton heliotropiifolius* Kunth: Phytochemical characterization and bioactive properties. Ind Crops Prod 113: 308–315.

Buckton, L.K., Rahimi, M.N., McAlpine, S.R. (2021). Cyclic peptides as drugs for intracellular targets: The next frontier in peptide therapeutic development. Chem Eur J 27(5): 1487–1513.

Campos, D.C.O., Costa, A.S., Lima, A.D.R., Silva, F.D.A., Lobo, M.D.P., Monteiro-Moreira, A.C.O., Moreira, R.A., Leal, L.K.A.M., Miron, D., Vasconcelos, I.M., Oliveira, H.D. (2016). First isolation and antinociceptive activity of a lipid transfer protein from noni (*Morinda citrifolia*) seeds. Int J Biol Macromol 86: 71–79.

Campos, M.L., Lião, L.M., Alves, E.S.F., Migliolo, L., Dias, S.C., Franco, O.L. (2018). A structural perspective of plant antimicrobial peptides. Biochem J 475(21): 3359–3375.

Cen, Q.W., Wang, Z.Y., Tang, Z.X., Zhang, Y., Chen, T., Xue, D.W., Xu, M.F., Bai, X.L., Zhou, T., Shi, L.E. (2021). Initial purification of antimicrobial fermentation metabolites from *Paecilomyces cicadae* and its antimicrobial mechanism. LWT - Food Sci Technol LWT 148: 111785.

Chaparro, E., Silva, P.I. (2016). Lacrain: the first antimicrobial peptide from the body extract of the Brazilian centipede *Scolopendra viridicornis*. Int J Antimicrob Agents 48(3): 277–285.

Chaparro-Aguirre, E., Segura-Ramírez, P.J., Alves, F.L., Riske, K.A., Miranda, A., Silva Júnior, P.I. (2019). Antimicrobial activity and mechanism of action of a novel peptide present in the ecdysis process of centipede *Scolopendra subspinipes subspinipes*'. Sci Rep 9(1): 13631.

Charoenkwan, P., Anuwongcharoen, N., Nantasenamat, C., Hasan, M., Shoombuatong, W. (2021). *In silico* approaches for the prediction and analysis of antiviral peptides: A review. Curr Pharm Des 27(18): 2180–2188.

Choudhuri, S., Kaur, T., Jain, S., Sharma, C., Asthana, S. (2021). A review on genotoxicity in connection to infertility and cancer. Chem Biol Interact 345: 109531.

Cisneros, J.S., Cotabarren, J., Parisi, M.G., Vasconcelos, M.W., Obregón, W.D. (2020). Purification and characterization of a novel trypsin inhibitor from *Solanum tuberosum* subsp. *andigenum* var. overa: Study of the expression levels and preliminary evaluation of its antimicrobial activity. Int J Biol Macromol 158: 1279–1287.

CLSI - Clinical and Laboratory Standards Institute. (2016). Methods for Antimicrobial Dilution and Disk Susceptibility Testing of Infrequently Isolated or Fastidious Bacteria, 3rd edition, CLSI Guideline M45.

CLSI - Clinical and Laboratory Standards Institute. (2017a). Reference Method for Broth Dilution Antifungal Susceptibility Testing of Yeasts, 4th Edition, CLSI Guideline M27.

CLSI - Clinical and Laboratory Standards Institute. (2017b). Reference Method for Broth Dilution Antifungal Susceptibility Testing of Filamentous Fungi, 3rd Edition, CLSI Guideline M38.

CSLI - Clinical and Laboratory Standards Institute. (2018). Methods for Dilution Antimicrobial Susceptibility Tests for Bacteria That Grow Aerobically, 11th Edition, CLSI Guideline M07.

Coelho, L.C., Santos, A.F., Napoleão, T.H., Correia, M.T., Paiva, P.M. (2012). Protein purification by affinity chromatography. *In*: Rizwan Ahmad. (Orged.). Protein Purific. Rijeka: InTechOpen 3: 53–72.

Coelho, L.C.B.B., Silva, P.M.S., Oliveira, W.F.O., Moura, M.C., Pontual, E.V., Gomes, F.S., Paiva, P.M.G., Napoleão, T.H., Correia, M.T.S. (2018). Lectins as antimicrobial agents. J Appl Microbiol 125(5): 1238–1252.

Copeland, W.B., Bartley, B.A., Chandran, D., Galdzicki, M., Kim, K.H., Sleight, S.C., Maranas, C. D. Sauro, H.M. (2012). Computational tools for metabolic engineering. Metab Eng 14(3): 270–280.

Corvi, R., Madia, F. (2017). *In vitro* genotoxicity testing—Can the performance be enhanced? Food Chem Toxicol 106: 600–608.

Costa, R.B., Campana, P.T., Chambergo, F.S., Napoleão, T.H., Paiva, P.M.G., Pereira, H.J.V., Oliva, M.L.V., Gomes, F.S. (2018). Purification and characterization of a lectin with refolding ability from *Genipa americana* bark. Int J Biol Macromol 119: 517–523.

Croft, N.P., Purcell, A.W. (2011). Peptidomimetics: modifying peptides in the pursuit of better vaccines. Expert Rev Vaccines 10(2): 211–226.

D'Agostino, N., Buonanno, M., Ayoub, J., Barone, A., Monti, S.M., Rigano, M.M. (2019). Identification of non-specific lipid transfer protein gene family members in *Solanum lycopersicum* and insights into the features of Sola I 3 protein. Sci Rep 9(1): 1607.

Dagnino, D., Schripsema, J., Peltenburg, A., Verpoorte, R., Teunis, K. (1991). Capillary gas chromatographic analysis of indole alkaloids: investigation of the indole alkaloids present in *Tabernaemontana divaricata* cell suspension culture. J Nat Prod 54(6): 1558–1563.

De Las Rivas, J., Prieto, C. (2012). Protein interactions: mapping interactome networks to support drug target discovery and selection. *In*: Larson, R. (eds.). Bioinformatics and Drug Discovery. Methods Mol Biol, Humana Press, Protein Interactions: Mapping Interactome Networks to Support Drug Target Discovery and Selection. Bioinf Drug Discov 910: 1–13.

Deng, T., Ge, H., He, H., Liu, Y., Zhai, C., Feng, L., Yi, L. (2017). The heterologous expression strategies of antimicrobial peptides in microbial systems. Protein Expr Purif 140: 52–59.

Denny, K.H., Stewart, C.W. (2017). Acute, subacute, subchronic, and chronic general toxicity testing for preclinical drug development. pp. 109–127. *In*: Faqi, A.S. (ed.). A Comprehensive Guide to Toxicology in Nonclinical Drug Development. 2nd edn. Academic Press, London.

Dobson, C.M. (2004). Chemical space and biology. Nature 432(7019): 824–828.

Dror, R.O., Dirks, R.M., Grossman, J.P., Xu, H., Shaw, D.E. (2012). Biomolecular simulation: a computational microscope for molecular biology. Annu Rev Biophys 41: 429–452.

Eckert, R. (2011). Road to clinical efficacy: challenges and novel strategies for antimicrobial peptide development. Fut Microbiol 6(6): 635–651.

Ehrt, C., Brinkjost, T., Koch, O. (2016). Impact of binding site comparisons on medicinal chemistry and rational molecular design. J Med Chem 59(9): 4121–4151.

EMA – European Medicines Agency. (2013). ICH guideline M3(R2) Non-clinical safety studies for the conduct of human clinical trials for pharmaceuticals and marketing authorization for pharmaceuticals. https://www.ema.europa.eu/en/documents/scientific-guideline/ich-guideline-m3r2-non-clinical-safety-studies conduct-human-clinical-trials-marketing-authorisation_en.pdf. Accessed on July 22, 2020.

Eryılmaz, S., Çelikoğlu, E.T., İdil, Ö., İnkaya, E., Kozak, Z., Mısır, E., Gül, M. (2020). Derivatives of pyridine and thiazole hybrid: Synthesis, DFT, biological evaluation via antimicrobial and DNA cleavage activity. Bioorg Med Chem 95: 103476.

Esaú, L.J.L., Rodolfo, R.G.C., Melissa, H.D., Adriana, C.C.C., Rodolfo, G.C., Rafael, F.C. (2019). An alternative disk diffusion test in broth and macrodilution method for colistin susceptibility in Enterobacteriales. J Microbiol Methods 167: 105765.

EUCAST - European Committee on Antimicrobial Susceptibility Testing. (2020). Method for the determination of broth dilution minimum inhibitory concentrations of antifungal agents for yeasts. EUCAST definitive document E.DEF 7.3.2. *In*: Arendrup, M.C., Meletiadis, J., Mouton, J.W., Lagrou, K., Hamal, P., Guinea, J. (eds.). ESCMID. Available at: http://www.eucast.org. Accessed on December 19 2020.

Farjadian, F., Ghasemi, A., Gohari, O., Roointan, A., Karimi, M., Hamblin, M.R. (2018). Nanopharmaceuticals and nanomedicines currently on the market: challenges and opportunities. Nanomedicine 14(1): 93–126.

Faust, T.B., Donovan, K.A., Yue, H., Chamberlain, P.P., Fischer, E.S. (2021). Small-molecule approaches to targeted protein degradation. Annu Rev Cancer Biol 5: 181–201.

FDA – Food and Drug Administration. (2012). Guidance for Industry S2(R1) Genotoxicity testing and data interpretation for pharmaceuticals intended for human use. Department of Health and Human Services Food and Drug. Administration Center for Drug Evaluation and Research (CDER) EUA, Rockville, U.S.

FDA – Food and Drug Administration. (2013). M3(R2) Nonclinical Safety Studies for the Conduct of Human Clinical Trials and Marketing Authorization for Pharmaceuticals. Department of Health and Human Services Food and Drug. Administration Center for Drug Evaluation and Research (CDER) EUA, Rockville, U.S.

FDA – Food and Drug Administration. (2015). Nonclinical Safety Evaluation of Reformulated Drug Products and Products Intended for Administration by an Alternate Route Guidance for Industry and Review Staff. Department of Health and Human Services Food and Drug. Administration Center for Drug Evaluation and Research (CDER), Rockville, U.S.

Fenech, M. (2007). Cytokinesis-block micronucleus cytome assay. Nat Protoc 2(5): 1084.

Feng, K., Wang, S., Han, L., Qian, Y., Li, H., Li, X., Jia, L., Hu, Y., Wang, H., Liu, M., Hu, W., Guo, D. Yang, W. (2021). Configuration of the ion exchange chromatography, hydrophilic interaction chromatography, and reversed-phase chromatography as off-line three-dimensional chromatography coupled with high-resolution quadrupole-Orbitrap mass spectrometry for the multicomponent characterization of *Uncaria sessilifructus*. J Chromatogr A 1649: 462237.

Fernandes, P., Martens, E. (2017). Antibiotics in late clinical development. Biochem Pharmacol 133: 152–163.

Ferreira, R.S., Napoleão, T.H., Silva-Lucca, R.A., Silva, M.C.C., Paiva, P.M.G., Oliva, M.L.V. (2019). The effects of *Enterolobium contortisiliquum* Serine protease inhibitor on the survival of the termite *Nasutitermes corniger*, and its use as affinity adsorbent to purify termite proteases. Pest Manag Sci 75(3): 632–638.

Feßler, A.T., Schug, A.R., Geber, F., Scholtzek, A.D., Merle, R., Brombach, J., Hensel, V., Meurer, M., Michael, G.B., Reinhardt, M., Speck, S., Truyen, U., Schwarz, S. (2018). Development and evaluation of a broth macrodilution method to determine the biocide susceptibility of bacteria. Vet Microbiol 223: 59–64.

Fibriana, F., Upaichit, A. (2015). Proteases from latex of *Euphorbia* spp. and its application on milk clot formation. Biosaintifika: J Biol Biol Educat 7(2): 1–8.

Finkina, E.I., Melnikova, D.N., Bogdanov, I.V., Ovchinnikova, T.V. (2016). Lipid transfer proteins asas components of the plant innate immune system: structure, functions, and applications. Acta Naturae 8(2): 47–61.

Flores-Alvarez, L.J., Guzmán-Rodríguez, J.J., López-Gómez, R., Salgado-Garciglia, R., Ochoa-Zarzosa, A., López Meza, J.E. (2018). PaDef defensin from avocado (*Persea americana* var. *drymifolia*) is cytotoxic to K562 chronic myeloid leukemia cells through extrinsic apoptosis. Int J Biochem Cell B 99: 10–18.

Fosgerau, K., Hoffmann, T. (2015). Peptide therapeutics: Current status and future directions. Drug Discov Today 20(1): 122–128.

Fradera, X., Babaoglu, K. (2017). Overview of methods and strategies for conducting virtual small molecule screening. Curr Prot Chem Biol 9(3): 196–212.

Fromtling, R.A., Galgiani, J.N., Pfaller, M.A., Espinel-Ingroff, A., Bartizal, K.F., Bartlett, M.S., Body, B.A., Frey, C., Hall, G., Roberts, G.D., Nolte, F.B., Odds, F.C., Rinaldi, M.G., Sugar, A.M., Roberts, G.D. (1993). Multicenter evaluation of a broth macrodilution antifungal susceptibility test for yeasts. Antimicrob Agents Chemother 37: 39–45.

Fujita, Y., Kasamatsu, T., Ikeda, N., Nishiyama, N., Honda, H. (2016). A retrospective evaluation method for *in vitro* mammalian genotoxicity tests using cytotoxicity index transformation formulae. Mut Res/Gen Toxicol Environ Mut 796: 1–7.

Fukushima, A., Kusano, M., Redestig, H., Arita, M., Saito, K. (2009). Integrated omics approaches in plant systems biology. Curr Opin Chem Biol 13(5-6): 532–538.

Galloway, S.M. (2017). International regulatory requirements for genotoxicity testing for pharmaceuticals used in human medicine, and their impurities and metabolites. Environ Mol Mut 58(5): 296–324.

Games, P.D., Silva, E.Q.G., Barbosa, M.O., Almeida-Souza, H.O., Fontes, P.P., Magalhães-Jr, M.J., Pereira, P.R.G., Prates, M.V., Franco, G.R., Faria-Campos, A., Campos, S.V.A., Baracat-Pereira, M.C. (2016). Computer aided identification of a Hevein-like antimicrobial peptide of bell pepper leaves for biotechnological use. BMC Genomics 17(S12): 999.

Ganjdoost, M., Aboonajmi, M., Mirsaeedghazi, H., Asefpour Vakilian, K. (2021). Effects of power ultrasound treatment on the shelf life of button mushrooms: Digital image processing and microbial counting can reveal the effects. Food Sci Nutr 9: 3538–3548.

Gao, J., Weng, W., Yan, Y., Wang, Y., Wang, Q. (2020). Comparison of protein extraction methods from excess activated sludge. Chemosphere 249: 126107.

Gazzaneo, R.S.L., Pandolfi, V., Jesus, A.L.S., Crovella, S., Benko-Iseppon, A.M., Freitas, A.C. (2017). Heterologous expression systems for plant defensin expression: examples of success and pitfalls. Curr Protein Pept Sci 18(4): 391–399.

Gezgin, Y., Kazan, A., Ulucan, F., Yesil-Celiktas, O. (2019). Antimicrobial activity of propolis and gentamycin against methicillin-resistant *Staphylococcus aureus* in a 3D thermo-sensitive hydrogel. Ind Crop Prod 139: 111588.

Gomes, F.S., Procópio, T.F., Napoleão, T.H., Coelho, L.C.B., Paiva, P.M.G. (2012). Antimicrobial lectin from *Schinus terebinthifolius* leaf. J Appl Microbiol 114(3): 672–679.

Gould, I.M., Bal, A.M. (2013). New antibiotic agents in the pipeline and how they can help overcome microbial resistance. Virulence 4(2): 185–191.

Guimarães, A.C., Meireles, L.M., Lemos, M.F., Guimarães, M.C.C., Endringer, D.C., Fronza, M., Scherer, R. (2019). Antibacterial activity of terpenes and terpenoids present in essential oils. Molecules 24(13): 2471.

Gupta, A., Madhavan, M.V., Sehgal, K., Nair, N., Mahajan, S., Sehrawat, T.S., Bikdeli, B., Ahluwalia, N., Ausiello, J.C., Wan, E.Y., Freedberg, D.E., Kirtane, A.J., Parikh, S.A., Maurer, M.S., Nordvig, A.S., Accili, D., Bathon, J.M., Mohan, S., Bauer, K.A., Leon, M.B., Krumholz, H.M., Uriel, N., Mehra, M.R., Elkind, M.S.V., Stone, G.W., Schwartz, A., Ho, D.D., Bilezikian, J.P., Landry, D.W. (2020). Extrapulmonary manifestations of COVID-19. Nat Med 26(7): 1017–1032.

Guzmán-Gutiérrez, S.L., Gómez-Cansino, R., García-Zebadúa, J.C., Jiménez-Pérez, N.C., Reyes-Chilpa, R. (2012). Antidepressant activity of Litsea glaucescens essential oil: identification of β-pinene and linalool as active principles. J Ethnopharmacol 143(2): 673–679.

Guzmán-Rodríguez, J.J., López-Gómez, R., Suárez-Rodríguez, L.M., Salgado-Garciglia, R., Rodríguez-Zapata, L.C., Ochoa-Zarzosa, A., López-Meza, J.E. (2013). Antibacterial activity of defensin PaDef from avocado fruit (*Persea americana* var. *drymifolia*) expressed in endothelial cells against *Escherichia coli* and *Staphylococcus aureus*. BioMed Res Int 2013: 986273.

Habas, K., Brinkworth, M.H., Anderson, D. (2017). *In vitro* responses to known *in vivo* genotoxic agents in mouse germ cells. Environ Mol Mutagen 58(2): 99–107.

Hammami, S., Debbabi, H., Jlassi, I., Joshi, R.K., El Mokni, R. (2020). Chemical composition and antimicrobial activity of essential oil from the aerial parts of *Plantago afra* L. (*Plantaginaceae*) growing wild in Tunisia. S Afr J Bot 132: 410–414.

Hollingsworth, S.A., Dror, R.O. (2018). Molecular dynamics simulation for all. Neuron 99(6): 1129–1143.

Hong, S.-Y., Kim, D.-G., Kim, Y.-O., Park, J.Y., Seo, J.-K., Nam, B.-H., Hong, Y.-K. (2018). Purification and cDNA cloning of the antimicrobial peptide apMolluscidin from the pen shell, Atrina pectinata. Fish Shellfish Immun 81: 408–415.

Hossain, M.U., Bhattacharjee, A., Emon, M.T.H., Chowdhury, Z.M., Mosaib, M.G., Mourin, M., Das, K.C., Keya, C.A., Salimullah, M. (2021). Recognition of plausible therapeutic agents to combat COVID-19: an Omics data based combined approach. Gene 771: 145368.

Hsu, C.L., Lo, W.H., Yen, G.C. (2007). Gallic acid induces apoptosis in 3T3-L1 pre-adipocytes via a Fas-and mitochondrial-mediated pathway. J Agr Food Chem 55(18): 7359–7365.

IDSA - Infectious Diseases Society of America. (2010). The 10 × '20 Initiative: Pursuing a Global Commitment to Develop 10 New Antibacterial Drugs by 2020. Clin Infect Dis 50(8): 1081–1083.

Imran, M., A.S., A., Tauseef, M., Khan, S.A., Hudu, S.A., Abida. (2021). *Mucormycosis medications:* a patent review. Expert Opin Ther Pat 1–16.

Ioerger, B.P., Bean, S.R., Tilley, M., Lin, H. (2020). An improved method for extraction of sorghum polymeric protein complexes. J Cereal Sci 91: 102876.

Ippoushi, K., Tanaka, Y., Wakagi, M., Hashimoto, N. (2020). Evaluation of protein extraction methods for β-conglycinin quantification in soybeans and soybean products. LWT-Food Sci Technol 132: 109871.

ISO 10993-5:2009. (2009). Biological evaluation of medical devices—Part 5: Tests for *in vitro* cytotoxicity. https://www.iso.org/standard/36406.html. Accessed on July 29, 2020.

Ivanov, A.A., Khuri, F.R., Fu, H. (2013). Targeting protein–protein interactions as an anticancer strategy. Trends Pharmacol Sci 34(7): 393–400.

Jabés, D., Brunati, C., Candiani, G., Riva, S., Romanó, G., Donadio, S. (2011). Efficacy of the new lantibiotic NAI-107 in experimental infections induced by multidrug-resistant Gram-positive pathogens. Antimicrob Agents Chemother 55: 1671–1676.

Jakobsen, V., Viganor, L., Blanco-Fernández, A., Howe, O., Devereux, M., McKenzie, C.J., McKee, V. (2018). Tetrameric and polymeric silver complexes of the omeprazole scaffold; synthesis, structure, *in vitro* and *in vivo* antimicrobial activities and DNA interaction. J Inorg Biochem 186: 317–328.

Jamshidi-Kia, F., Lorigooini, Z., Amini-Khoei, H. (2018). Medicinal plants: Past history and future perspective. J Herbmed Pharmacol 7(1): 1–7.

Jeffres, M.N. (2017). The whole price of vancomycin: toxicities, troughs, and time. Drugs 77(11): 1143–1154.

Jiang, M., Ma, L., Huang, Y., Wu, H., Dou, J., Zhou, C. (2020). Antimicrobial activities of peptide Cbf-K16 against drug-resistant *Helicobacter pylori* infection *in vitro* and *in vivo*. Microb Pathog 138: 103847.

Jiang, Y., Chen, Y., Song, Z., Tan, Z., Cheng, J. (2021). Recent advances in design of antimicrobial peptides and polypeptides toward clinical translation. Adv Drug Deliv Rev 170: 261–280.

Kaczor, A.A., Bartuzi, D., Stępniewski, T.M., Matosiuk, D., Selent, J. (2018). Protein–protein docking in drug design and discovery. Methods Mol Biol 1762: 285–305.

Kalyaanamoorthy, S., Chen, Y-P.P. (2011). Structure-based drug design to augment hit discovery. Drug Discov Today 16: 831–839.

Kameri, A., Koçani, F., Hashani, Z., Kurteshi, K., Kamberi, B., Kurti, A., Haziri, A. (2019). Antifungal and synergistic effects of the ethyl acetate extract of *Tanacetum vulgare* (L.) against *Candida albicans*. Med Sci Monit Basic Res 25: 179.

Kanehisa, M., Miho, F., Mao, T., Yoko, S., Kanae, M. (2017). KEGG: New perspectives on genomes, pathways, diseases and drugs. Nucleic Acids Res 45(D1): D353–D361.

Kang, X., Dong, F., Shi, C., Liu, S., Sun, J., Chen, J., Li, H., Xu, H., Lao, X., Zheng, H. (2019). DRAMP 2.0, an updated data repository of antimicrobial peptides. Sci Data 6(1): 1–10.

Kashid, A.M., Kadam, R.R. (2021). Stability-indicating high-performance thin-layer chromatography method for the simultaneous estimation of emtricitabine and tenofovir alafenamide fumarate. J Planar Chromatogr–Mod TLC 1: 1–9.

Kerrien, S., Aranda, B., Breuza, L., Bridge, A., Broackes-Carter, F., Chen, C., Duesbury, M., Dumousseau, M., Feuermann, M., Hinz, U., Jandrasits, C., Jimenez, R., Khadake, J., Mahadevan, U., Masson, P., Pedruzzi, I., Pfeiffenberger, E., Porras, P., Raghunath, A., Roechert, B., Orchard, S., Hermjakob, H. (2011). The IntAct molecular interaction database in 2012. Nucleic Acids Res 40(D1): D841–D846.

Kim, S., Chen, J., Cheng, T., Gindulyte, A., He, J., He, S., Li, Q., Shoemaker, B.A., Thiessen, P.A., Yu, B., Zaslavsky, L., Zhang, J., Bolton, E.E. (2019). PubChem 2019 update: improved access to chemical data. Nucleic Acids Res 47(D1): D1102–D1109.

Kirby, T. (2012). Europe to boost development of new antimicrobial drugs. The Lancet 379(9833): 2229–2230.

Kirkland, D., Fowler, P. (2010). Further analysis of Ames-negative rodent carcinogens that are only genotoxic in mammalian cells *in vitro* at concentrations exceeding 1 mM, including retesting of compounds of concern. Mutagenesis 25(6): 539–553.

Klimek-Turek, A., Sikora, E., Dzido, T.H. (2017). Solvent front position extraction procedure with thin-layer chromatography as a mode of multicomponent sample preparation for quantitative analysis by instrumental technique. J Chromatogr A 1530: 204–210.

Kogawa, A.C., Mendonça, J.N., Lopes, N.P., Salgado, H.R.N. (2021). Eco-friendly pharmaceutical analysis of rifaximin in tablets by HPLC-MS and microbiological turbidimetry. J Chromatogr Sci 59(7): 597–605.

Kumar, P.P., Siva, B., Katragunta, K., Beedkar, S.D., Ramakrishna, S., Barua, C.C., Suresh Babu, K. (2020). Investigation of cytotoxic constituents from seed-pulp of Entada Phaseoloides and metabolite profiling using UPLC-QTOF-MSE. J AOAC Intern 104(3): 827–835.

Lai, J.K., Ambia, J., Wang, Y., Barth, P. (2017). Enhancing structure prediction and design of soluble and membrane proteins with explicit solvent-protein interactions. Structure 25(11): 1758–1770.

Lanctôt, S., Fustier, P., Taherian, A.R., Bisakowski, B., Zhao, X., Lacasse, P. (2017). Effect of intramammary infusion of chitosan hydrogels at drying-off on bovine mammary gland involution. J Dairy Sci 100: 2269–2281.

Lau, J.L., Dunn, M.K. (2018). Therapeutic peptides: Historical perspectives, current development trends, and future directions. Bioorg Med Chem 26(10): 2700–2707.

Lee, S., Kim, H., Beuchat, L.R., Kim, Y., Ryu, J.H. (2020). Synergistic antimicrobial activity of oregano and thyme thymol essential oils against *Leuconostoc citreum* in a laboratory medium and tomato juice. Food Microbiol 90: 103489.

Lekha, P.K., Lonsane, B.K. (1994). Comparative titres, location and properties of tannin acyl hydrolase produced by *Aspergillus niger* PKL 104 in solid-state, liquid surface and submerged fermentations. Process Biochem 29: 497–503.

Li, Z., Li, H., Yu, K., Luo, H.B. (2021). Perspective of drug design with high-performance computing. Natl Sci Rev 0: 105–107.

Liang, Y., Guan, R., Huang, W., Xu, T. (2011). Isolation and identification of a novel inducible antibacterial peptide from the skin mucus of Japanese eel, *Anguilla japonica*. Protein J 30(6): 413–421.

Licini, C., Montalbano, G., Ciapetti, G., Cerqueni, G., Vitale-Brovarone, C., Mattioli-Belmonte, M. (2020). Analysis of multiple protein detection methods in human osteoporotic bone extracellular matrix: From literature to practice. Bone 137: 115363.

Lima, T.A., Dornelles, L.P., Oliveira, A.P.S., Guedes, C.C., Souza, S.O., Sá, R.A., Zingalic, R.B., Napoleão, T.H., Paiva, P.M. (2018). Binding targets of termiticidal lectins from the bark and leaf of *Myracrodruon urundeuva* in the gut of *Nasutitermes corniger* workers. Pest Manag Sci 74(7): 1593–1599.

Lin, S., Zhu, C., Li, H., Chen, Y., Liu, S. (2020). Potent *in vitro* and *in vivo* antimicrobial activity of semisynthetic amphiphilic γ-mangostin derivative LS02 against Gram-positive bacteria with destructive effect on bacterial membrane. Biochim Biophys Acta Biomembr 1862: 183353.

Liu, B., Huang, H., Yang, Z., Liu, B., Gou, S., Zhong, C., Han, X., Zhang, Y., Ni, J., Wang, R. (2017b). Design of novel antimicrobial peptide dimer analogues with enhanced antimicrobial activity *in vitro* and *in vivo* by intermolecular triazole bridge strategy. Peptides 88: 115–125.

Liu, H., Taylor, T.H., Pettus, K., Johnson, S., Papp, J.R., Trees, D. (2016). Comparing the disk-diffusion and agar dilution tests for *Neisseria gonorrhoeae* antimicrobial susceptibility testing. Antimicrob Resist Infect Control 5(1): 1–6.

Liu, S., Ono, R.J., Wu, H., Teo, J.Y., Liang, Z.C., Xu, K., Zhang, M., Zhong, G., Tan, J.P.K., Ng, M., Yang, C., Chan, J., Ji, Z., Bao, C., Kumar, K., Gao, S., Lee, A., Fevre, M., Yang, Y.Y. (2017a). Highly potent antimicrobial polyionenes with rapid killing kinetics, skin biocompatibility and *in vivo* bactericidal activity. Biomaterials 127: 36–48.

Long, S.-Y., Li, C.-L., Hu, J., Zhao, Q.-J., Chen, D. (2018). Indole alkaloids from the aerial parts of *Kopsia fruticosa* and their cytotoxic, antimicrobial and antifungal activities. Fitoterapia 129: 145–149.

Lubomirsky, E., Khodabandeh, A., Preis, J., Susewind, M., Hofe, T., Hilder, E.F., Arrua, R.D. (2021). Polymeric stationary phases for size exclusion chromatography: A review. Anal Chim Acta 1151: 338244.

Luo, M., Jiao, J., Wang, R. (2019). Screening drug target combinations in disease-related molecular networks. BMC Bioinformatics 20(7): 129–138.

Madia, F., Pillo, G., Worth, A., Corvi, R., Prieto, P. (2021). Integration of data across toxicity endpoints for improved safety assessment of chemicals: the example of carcinogenicity assessment. Arch Toxicol 95(6): 1971–1993.

Mann, M., Hendrickson, R.C., Pandey, A. (2001). Analysis of proteins and proteomes by mass spectrometry. Annu Rev Biochem 70(1): 437–473.

Marchese, A., Arciola, C.R., Coppo, E., Barbieri, R., Barreca, D., Chebaibi, S., Sobarzo-Sánchez, E., Nabavi, S.F., Daglia, M. (2018). The natural plant compound carvacrol as an antimicrobial and anti-biofilm agent: mechanisms, synergies and bio-inspired anti-infective materials. Biofouling 34(6): 630–656.

Martínez, M., Polizzotto, A., Flores, N., Semorile, L., Maffía, P.C. (2020). Antibacterial, anti-biofilm and *in vivo* activities of the antimicrobial peptides P5 and P6. 2. Microb Pathog 139: 103886.

Medina-Franco, J.L., Sánchez-Cruz, N., López-López, E., Díaz-Eufracio, B.I. (2021). Progress on open chemoinformatic tools for expanding and exploring the chemical space. J Comput-Aided Mol Des 1: 1–14.

Meric, G.G., Gegic, F., Leblebici, S., Gungor, S., Degirmenci, L. (2019). Clarification of *Stevia rebaudiana* Bertoni extract by Ca based silica microspheres. Heliyon 5(7): e02062.

Miller, F., Hinze, U., Chichkov, B., Leibold, W., Lenarz, T., Paasche, G. (2017). Validation of eGFP fluorescence intensity for testing *in vitro* cytotoxicity according to ISO 10993-5. J Biomed Mater Res B 105(4): 715–722.

Miranda, C.D., Rojas, R., Contreras-Lynch, S., Vega, A. (2020). Evaluation of the correlation between minimum inhibitory concentrations (MIC) and disk diffusion data of *Flavobacterium psychrophilum* isolated from outbreaks occurred in Chilean salmonid farms. Aquaculture 530: 735811.

Moffatt, J.H., Harper, M., Boyce, J.D. (2019). Mechanisms of polymyxin resistance. pp. 55–71. *In*: Li, J., Nation, R.L., Kaye, K.S. (eds.). Polymyxin A: From Laboratory Bench to Bedside.

Moradi, M., Kousheh, S.A., Razavi, R., Rasouli, Y., Ghorbani, M., Divsalar, E., Tajik, H., Guimaraes, J.T., Ibrahim, S.A. (2021). Review of microbiological methods for testing protein and carbohydrate-based antimicrobial food packaging. Trends Food Sci Technol 111: 595–609.

Mosesso, P., Cinelli, S., Natarajan, A.T., Palitti, F. (2019). *In vitro* cytogenetic assays: chromosomal aberrations and micronucleus tests. *In*: Genotoxicity Assessment. Humana Press, New York, 123-146: 79–104.

Muthuramalingam, P., Jeyasri, R., Valliammai, A., Selvaraj, A., Karthika, C., Gowrishankar, S., Pandiana, S.K., Ramesh, M., Chen, J.-T. (2020). Global multi-omics and systems pharmacological strategy unravel the multi-targeted therapeutic potential of natural bioactive molecules against COVID-19: An *in silico* approach. Genomics 112(6): 4486–4504.

Nawrot, R., Józefiak, D., Sip, A., Kuźma, D., Musidlak, O., Goździcka-Józefiak, A. (2017). Isolation and characterization of a non-specific lipid transfer protein from *Chelidonium majus* L. latex. Int J Biol Macromol 104: 554–563.

Nazeer, M., Waheed, H., Saeed, M., Ali, S.Y., Choudhary, M.I., Ul-Haq, Z., Ahmed, A. (2019). Purification and characterization of a nonspecific lipid transfer protein 1 (nsLTP1) from Ajwain (*Trachyspermum ammi*) seeds. Sci Rep 9(1): 4148.

Negahdaripour, M., Owji, H., Eslami, M., Zamani, M., Vakili, B., Sabetian, S., Nezafat, N., Ghasemi, Y. (2019). Selected application of peptide molecules as pharmaceutical agents and in cosmeceuticals. Expert Opin Biol Ther 19(12): 1275–1287.

Nelson, D.L., Cox, M.M. (2014). Amino acids, peptides, and proteins. *In*: Lehninger Principles of Biochemistry. 4th edition. Macmillan 262: 1680–1685.

Newman, D.J., Cragg, G.M. (2016). Natural products as sources of new drugs from 1981 to 2014. J Nat Prod 79(3): 629–661.

Odhar, H.A., Ahjel, S.W., Humadi, S.S. (2020). Towards the design of multiepitope-based peptide vaccine candidate against SARS-CoV-2. BioRxiv 45: 546–557.

OECD - Organization for Economic Co-operation and Development. (2016a). TG 473: *In vitro* Mammalian Chromosomal Aberration Test. https://www.oecd-ilibrary.org/docserver/9789264264649-en.pdf?expires=1593694339&id=id&accname=guest&checksum=1005A77848BF954DF24E669817571BB0. Accessed on August 30, 2020.

OECD - Organization for Economic Co-operation and Development. (2016b). TG 487: *In vitro* Mammalian Cell Micronucleus Test. https://www.oecd-ilibrary.org/docserver/9789264264861-en.pdf?expires=1593694259&id=id&accname=guest&checksum=C401D5563DF80463BBC3E538FEB204C6. Accessed on August 30, 2020.

Offret, C., Fliss, I., Bazinet, L., Marette, A., Beaulieu, L. (2019). Identification of a novel antibacterial peptide from Atlantic mackerel belonging to the GAPDH-related antimicrobial family and its *in vitro* digestibility. Mar Drugs 17(7): 413.

Oh, R., Lee, M.J., Kim, Y.O., Nam, B.H., Kong, H.J., Kim, J.W., Park, J.Y., Seo, J.K., Kim, D.G. (2018). Purification and characterization of an antimicrobial peptide mytichitin-chitin binding domain from the hard-shelled mussel, *Mytilus coruscus*. Fish & Shellfish Immun 18: 1050–1093.

Oliveira, A.P.S., Santana Silva, L.L., Albuquerque Lima, T., Pontual, E.V., Lima Santos, N.D., Coelho, L.C.B.B., Navarro, D.M.A.F., Zingali, R.B., Napoleão, T.H.P., Paiva, P.M.G. (2016). Biotechnological value of *Moringa oleifera* seed cake as source of insecticidal lectin against *Aedes aegypti*. Process Biochem 51(10): 1683–1690.

Oliveira, C.F.R., Moura, M.C., Napoleão, T.H., Paiva, P.M.G., Coelho, L.C.B.B., Macedo, M.L.R. (2017). A chitin-binding lectin from *Moringa oleifera* seeds (WSMoL) impairs the digestive physiology of the Mediterranean flour larvae, Anagasta kuehniella. Pestic Biochem Physiol 142: 67–76.

Oughtred, R., Stark, C., Breitkreutz, B.-J., Rust, J., Boucher, L., Chang, C., Kolas, N., O'Donnell, L., Leung, G., McAdam, R., Zhang, F., Dolma, S., Willems, A., Coulombe-Huntington, J., Chatr-aryamontri, A., Dolinski, K. Tyers, M. (2019). The BioGRID interaction database: 2019 update. Nucleic Acids Res 39: D698–D704.

Özçelik, B., Kartal, M., Orhan, I. (2011). Cytotoxicity, antiviral and antimicrobial activities of alkaloids, flavonoids, and phenolic acids. Pharma Biol 49(4): 396–402.

Paiva, P.M.G., Gomes, F.S., Napoleão, T.H., Sá, R.A., Correia, M.T.S., Coelho, L.C.B.B. (2010). Antimicrobial activity of secondary metabolites and lectins from plants. Curr Research Tech Educ Top App Microbiol Microb Biotech 1(2): 396–406.

Pancotti, A., Santos, D.P., Morais, D.O., de Barros Souza, M.V., Lima, D.R., Vulcani, V.A.S., Martins, A., Landers, R., Braoios, A. (2021). Synthesis, characterization and *in vitro* cytotoxicity study of Co and Ni ferrite nanoparticles prepared by sol-gel method. SN Applied Sciences Sci 3(7): 1–13.

Patriota, L.L., Procópio, T.F., Santana Brito, J., Sebag, V., Oliveira, A.P.S., Araújo Soares, A.K., Moreira, L.R., Lima, T.A., Soares, T., Silva, T.D., Paiva, P.M.G., Lorena, V.M.B., Melo, C.M.L., Albuquerque, L.P., Napoleão, T.H. (2017). *Microgramma vacciniifolia* (Polypodiaceae) fronds contain a multifunctional lectin with immunomodulatory properties on human cells. Int J Biol Macromol 103: 36–46.

Pavlović, T., Dimkić, I., Andrić, S., Milojković-Opsenica, D., Stanković, S., Janaćković, P., Gavrilović, M., Ristivojević, P. (2020). Linden tea from Serbia–an insight into the phenolic profile, radical scavenging and antimicrobial activities. Ind Crops Prod 154: 112639.

Pedamallu, C.S., Ozdamar, L. (2014). A review on protein-protein interaction network databases. pp. 511–519. *In*: Modeling, Dynamics, Optimization and Bioeconomics I. Springer, Cham.

Pence, H.E., Williams, A. (2010). ChemSpider: An online chemical information resource. J Chem Educ 87: 1123–1124.

Périllaud, C., Pilmis, B., Diep, J., Ponfilly, G.P., Vidal, B., Couzigou, C., Mizrahi, A., Lourtet-Hascoët, J., Le Monnier, A., Van, J.C.N. (2019). Prospective evaluation of rapid antimicrobial susceptibility testing by disk diffusion on Mueller-Hinton rapid-SIR directly on blood cultures. Diagn Microbiol Infect Dis 93(1): 14–21.

Perim, M.C., Borges, J.D.C., Silva, E.M.L., Araújo, T.A.D.S., Silva, A.C.O., Silva, V.C., Correia, S.C., Cunha, A.F., Pranchevicius, M.C.D.S. (2019). *In vitro* antibacterial and time-kill assay of ethanolic extract of *Davilla nitida* bark on multidrug resistant bacteria isolated from diabetic foot lesions. Nat Prod Res 33: 2383–2388.

Petrey, D., Honig, B. (2014). Structural bioinformatics of the interactome. Annu Rev Biophys 43(1): 193–210.

Poivre, M., Nachtergael, A., Bunel, V., Philippe, O.N., Duez, P. (2017). Genotoxicity and carcinogenicity of herbal products. pp. 179–215. *In*: Pelkonen, O., Duez, P., Vuorela P.M., Vuorela, H. (eds.). Toxicology of Herbal Products. Springer, Switzerland.

Pokoj, S., Lauer, I., Fötisch, K., Himly, M., Mari, A., Enrique, E., Miguel-Moncin, M.M.S., Lidholm, J., Vieths, S., Scheurer, S. (2010). *Pichia pastoris* is superior to *E. coli* for the production of recombinant allergenic non-specific lipid-transfer proteins. Protein Expres Purif 69(1): 68–75.

Poletto, P., Álvarez-Rivera, G., López, G.D., Borges, O.M., Mendiola, J.A., Ibáñez, E., Cifuentes, A. (2021). Recovery of ascorbic acid, phenolic compounds and carotenoids from acerola by-products: An opportunity for their valorization. Food Sci Technol LWT 146: 111654.

Pontual, E.V., Santos, N.D., Moura, M.C., Coelho, L.C.B.B., Navarro, D.M.A.F., Napoleão, T.H., Paiva, P.M.G. (2014). Trypsin inhibitor from *Moringa oleifera* flowers interferes with survival and development of *Aedes aegypti* larvae and kills bacteria inhabitant of larvae midgut. Parasitol Res 113(2): 727–733.

Porto, W.F., Irazazabal, L., Alves, E.S., Ribeiro, S.M., Matos, C.O., Pires, Á.S., Fensterseifer, I.C.M., Miranda, V.J., Haney, E.F., Humblot, V., Torres, M.D.T., Hancock, R.E.W., Liao, L.M., Ladram, A., Lu, T.K., Fuente-Nunez, C.L., Franco, O.L. (2018). *In silico* optimization of a guava antimicrobial peptide enables combinatorial exploration for peptide design. Nat Commun 9(1): 1–12.

Preedy, V.R. (ed.). (2015). Essential Oils in Food Preservation, Flavor and Safety. Academic Press.

Procópio, T.F., Patriota, L.L.S., Moura, M.C., Silva, P.M., Oliveira, A.P.S., Carvalho, L.V.N., Lima, T.A., Soares, T., Silva, T.D., Coelho, L.C.B.B., Pitta, M.G., Rêgo, M.J.B., Figueiredo, R.C.B.Q., Paiva, P.M.G., Napoleão, T.H.

(2017a). CasuL: A new lectin isolated from *Calliandra surinamensis* leaf pinnulae with cytotoxicity to cancer cells, antimicrobial activity and antibiofilm effect. Int J Biol Macromol 98: 419–429.

Procópio, T.F., Moura, M.C., Albuquerque, L.P., Gomes, F.S., Santos, N.D., Coelho, L.C.B.B., Pontual, E.V., Paiva, P.M.G., Napoleão, T.H. (2017b). Antibacterial lectins: action mechanisms, defensive roles and biotechnological potential. pp. 69–90. *In*: Colins, E. (ed.). Antibacterials: Synthesis, Properties and Biological Activities. New York: Nova Sci. Pub.

Puertas-Martín, S., Banegas-Luna, A.J., Paredes-Ramos, M., Redondo, J.L., Ortigosa, P.M., Brovarets', O.H.O., Pérez-Sánchez, H. (2020). Is high performance computing a requirement for novel drug discovery and how will this impact academic efforts? Expert Opin Drug Discov 15(9): 981–985.

Ramirez, D.A., Altamirano, J.C., Camargo, A.B. (2021). Multi-phytochemical determination of polar and non-polar garlic bioactive compounds in different food and nutraceutical preparations. Food Chem 337: 127648.

Ramos, M.V., Souza, D.P., Gomes, M.T.R., Freitas, C.D.T., Carvalho, C.P.S., Júnior, P.A.V.R., Salas, C.E. (2014). A phytopathogenic cysteine peptidase from latex of wild rubber vine *Cryptostegia grandiflora*. Protein J 33(2): 199-209.

Raspo, M.A., Vignola, M.B., Andreatta, A.E., Juliani, H.R. (2020). Antioxidant and antimicrobial activities of citrus essential oils from Argentina and the United States. Food Biosci 36: 100651.

Reif, M., Zacharias, M. (2019). Computer modelling and molecular dynamics simulation of biomolecules. Biomol Bioanal Tech: TMA 54: 501–535.

Rocha Filho, C.A.A., Amorim, P.K., de Albuquerque Lima, T., da Silva, P.M., de Moura, M.C., Coelho, L.C.B.B., Zingali, R.B., Pontual, E.V., Napoleão, T.H., Paiva, P.M.G. (2019). PgTI, the first bioactive protein isolated from the Cactus Pilosocereus gounellei, is a Trypsin inhibitor with antimicrobial activity. Adv Res 20(3): 1–11.

Rocha-Filho, C.A.A., Araújo, L.C.C., Silva, P.M., Machado, J.C.B., Assunção Ferreira, M.R.A., Soares, L.A.L., Silva, M.V., Silva, T.G., Napoleão, T.H., Paiva, P.M.G. (2020). Phytochemical analysis and evaluation of acute toxicity and antioxidant, antibacterial, and anti-inflammatory activities of ethanolic extracts from branches and leaves of *Jacaranda rugosa*. S Afr J Bot 135: 233–239.

Santana, G.M.S., Albuquerque, L.P., Napoleão, T.H., Souza, S.R., Coelho, L.C.B.B., Paiva, P.M.G. (2012). Electrochemical potential of *Microgramma vaccinifolia* rhizome lectin. Bioelectrochem 85: 56–60.

Santos, B.S., Bezerra-Filho, C.M., Nascimento-Junior, J.A.A., Brust, F.R., Bezerra-Silva, P.C., Lino da Rocha, S.K., Krogfeltd, K.A., Navarroc, D.M.A.F., Correia, M.T.S., Napoleão, T.H., Silva, L.C.N., Macedo, A.J., Silva, M.V., Paiva, P.M.G. (2019). Anti-staphylococcal activity of *Syagrus coronata* essential oil: Biofilm eradication and *in vivo* action on *Galleria mellonela* infection model. Microb Pathog 131: 150–157.

Santos-Silva, C.A., Zupin, L., Oliveira-Lima, M., Vilela, L.M.B., Bezerra-Neto, J.P., Ferreira-Neto, J.R., Cavalcanti-Ferreira, J.D., Oliveira-Silva, R.L., Jesús-Pires, C., Aburjaile, F.F., Oliveira, M.F., Akio-Kido, E., Crovella, S., Benko-Iseppon, A.M. (2020). Plant antimicrobial peptides: state of the art, *in silico* prediction and perspectives in the omics era. Bioinform Biol Insights 14: 1–22.

Santos-Silva, C.A., Vilela, L.M.B., Oliveira-Silva, R.L., Silva, J.B., Machado, A.R., Bezerra-Neto, J.P., Crovella, S., Benko-Iseppon, A.M. (2021). Cassava (*Manihot esculenta*) defensins: Prospection, structural analysis and tissue-specific expression under biotic/abiotic stresses. Biochimie 186: 1–12.

Sarris, J. (2018). Herbal medicines in the treatment of psychiatric disorders: 10-year updated review. Phytother Res 32(7): 1147–1162.

Schäfer, M., Ciaudo, C. (2020). Prediction of the MiRNA interactome—established methods and upcoming perspectives. Comput Struct Biotechnol 18: 548–57.

Schapoval, E.E.S. (2005). Controle biológico de qualidade de produtos farmacêuticos, correlatos e cosméticos. Rev Bras Cienc Farm 41: 279–280.

Schmitt, A.J., Sathoff, A.E., Holl, C., Bauer, B., Samac, D., Carter, C.J. (2018). The major nectar protein of *Brassica rapa* is a non-specific lipid transfer protein, BrLTP2.1, with strong antifungal activity. J Exp Bot 140: 170531.

Schulz, L.T., Kim, S.Y., Hartsell, A., Rose, W.E. (2019). Antimicrobial stewardship during a time of rapid antimicrobial development: Potential impact on industry for future investment. Diagn Microbiol Infect Dis 95(3): 114857.

Segura-Ramírez, Silva-Júnior, P. (2018). *Loxosceles gaucho* spider venom: an untapped source of antimicrobial agents. Toxins 10(12): 522.

Sekizawa, S.I., Hoshino, Y., Takasu, A. (2017). *In vitro* genotoxicity test package of antibiotics for human use submitted to the Japanese regulatory agency during 2004–2015. Fundam Toxicol Sci 4(5): 241–245.

Shafee, T., Bacic, A., Johnson, K. (2020). Evolution of sequence-diverse disordered regions in a protein family: order within the chaos. Mol Biol Evol 10: 1093–1121.

Shang, J., Ye, G., Shi, K., Wan, Y., Luo, C., Aihara, H., Geng, Q., Auerbach, A., Li, F. (2020). Structural basis of receptor recognition by SARS-CoV-2. Nature 581(7807): 221–224.

Sharma, K., Guleria, S., Razdan, V.K., Babu, V. (2020). Synergistic antioxidant and antimicrobial activities of essential oils of some selected medicinal plants in combination and with synthetic compounds. Ind Crops Prod 154: 112569.

Sharma, S., Verma, H.N., Sharma, N.K. (2014). Cationic bioactive peptide from the seeds of Benincasa hispida. Int J Pept 2014: 12.

Shegokar, R. (2020). Preclinical testing—Understanding the basics first. pp. 19–32. *In*: Shegokar, R. (eds.). Drug Deliv. Aspects. Elsevier, Germany.

Silva, J.D.F., Silva, S.P., Silva, P.M., Vieira, A.M., de Araújo, L.C.C., de Albuquerque Lima, T., Silva, A.P., Carvalho, L.V.N., Da Rocha Pita, M.G., Rêgo, M.J.B., Pinheiro, I.O., Zingale, R.B., Cavalcanti, M.S.M., Napoleão, T.H., Paiva, P.M.G. (2019). *Portulaca elatior* root contains a trehalose-binding lectin with antibacterial and antifungal activities. Int J Biol Macromol 126: 291–297.

Silva, L.L., Santos, N.D., Silva, S.C.C., Lima, T., Coelho, L.C.B.B., Zingali, R.B., Navarro, D.M.A.F., Napoleão, T.H., Paiva, P.M.G. (2020). Investigation of the ability of the oviposition-stimulant lectin from *Moringa oleifera* seeds (WSMoL) to bind with membrane proteins present in the legs of *Aedes aegypti*. Int J Biol Macromol 162: 657–662.

Siriwattananon, K., Manopwisedjaroen, S., Kanjanasirirat, P., Budi Purwono, P., Rattanapisit, K., Shanmugaraj, B., Smith, D.R., Borwornpinyo, S., Thitithanyanont, A., Phoolcharoen, W. (2021). Development of plant-produced recombinant ACE2-Fc fusion protein as a potential therapeutic agent against SARS-CoV-2. Front Plant Sci 11: 2117.

Sjöberg, P., Jones, D.R. (2013). Non-clinical safety studies for the conduct of human clinical trials for pharmaceuticals: ICH M3 and M3 (R2). pp. 299–309. *In*: Global Approach in Safety Testing. Springer, New York.

Skwarczynski, M., Toth, I. (2014). Recent advances in peptide-based subunit nanovaccines. Nanomedicine 9(17): 2657–2669.

Slater, O., Kontoyianni, M. (2019). The compromise of virtual screening and its impact on drug discovery. Expert Opin Drug Discovery 14(7): 619–637.

Soares, T., Ferreira, F.R.B., Gomes, F.S., Coelho, L.C.B.B., Torquato, R.J.S., Napoleão, T.H., Cavalcanti, M.S.M., Tanaka, A.S., Paiva, P.M.G. (2011). The first serine protease inhibitor from *Lasiodora* sp. (Araneae: Theraphosidae) hemocytes. Process Biochem 46: 2317–2321.

Soeteman-Hernández, L.G., Johnson, G.E., Slob, W. (2016). Estimating the carcinogenic potency of chemicals from the *in vivo* micronucleus test. Mutagenesis 31(3): 347–358.

Souza, A.A., Costa, A.S., Campos, D.C.O., Batista, A.H.M., Sales, G.W.P., Nogueira, N.A.P., Alves, K.M.M., Coelho-de-Souza, A.N., Oliveira, H.D. (2018). Lipid transfer protein isolated from noni seeds displays antibacterial activity *in vitro* and improves survival in lethal sepsis induced by CLP in mice. Biochim 149: 9–17.

Souza, A.C., Silva, L.K., Queiroz, T.B., Marques, E.S., Hiruma-Lima, C.A., Gaivão, I.O., Maistro, E.L. (2020). Citral presents cytotoxic and genotoxic effects in human cultured cells. Drug Chem Toxicol 43(4): 435–440.

Spellberg, B. (2014). The future of antibiotics. Crit Care 18(3): 228.

Spellberg, B., Bartlett, J., Wunderink, R., Gilbert, D.N. (2015). Novel approaches are needed to develop tomorrow's antibacterial therapies. Am J Respir Crit Care Med 191(2): 135–140.

Suda, S., Field, D., Barron, N. (2016). Antimicrobial peptide production and purification. Prot Chromatog 401–410.

Supong, A., Bhomick, C.P., Karmaker, R., Ezung, S.L., Jamir, L., Bora Sinha, U., Sinha, D. (2020). Experimental and theoretical insight into the adsorption of phenol and 2,4-dinitrophenol onto *Tithonia diversifolia* activated carbon. Appl Surf Sci 147046.

Szklarczyk, D., Morris, J.H., Cook, H., Kuhn, M., Wyder, S., Simonovic, M., Santos, A., Doncheva, N.T., Roth, A., Bork, P., Jensen, L.J., von Mering C. (2017). The STRING database in 2017: quality-controlled protein–protein association networks, made broadly accessible. Nucleic Acids Res 45(D1): D362–D368.

Talbot, G.H., Jezek, A., Murray, B.E., Jones, R.N., Ebright, R.H., Nau, G.J., Rodvold, K., Newland, J.G., Boucher, H.W., Infectious Diseases Society of America. (2019). The Infectious Diseases Society of America's 10 × '20 initiative (10 new systemic antibacterial agents US Food and Drug Administration approved by 2020): is 20 × '20 a possibility? Clin Infect Dis 69(1): 1–11.

Tam, J.P., Wang, S., Wong, K.H., Tan, W.L. (2015). Antimicrobial peptides from plants. Pharmaceut 8(4): 711–757.

Tamma, P.D., Aitken, S.L., Bonomo, R.A., Mathers, A.J., van Duin, D., Clancy, C.J. (2021). Infectious Diseases Society of America guidance on the treatment of extended-spectrum β-lactamase producing Enterobacterales (ESBL-E), carbapenem-resistant Enterobacterales (CRE), and *Pseudomonas aeruginosa* with difficult-to-treat resistance (DTR-*P. aeruginosa*). Clin Infect Dis 72(7): e169–e183.

Tan, S.C., Yiap, B.C. (2009). DNA, RNA, and protein extraction: the past and the present. J Biomed Biotechnol 1–10.

Tanger, C., Engel, J., Kulozik, U. (2020). Influence of extraction conditions on the conformational alteration of pea protein extracted from pea flour. Food Hydrocoll 107: 105949.

Tanhaeian, A., Mirzaii, M., Pirkhezranian, Z., Sekhavati, M.H. (2020). Generation of an engineered food-grade *Lactococcus lactis* strain for production of an antimicrobial peptide: *in vitro* and *in silico* evaluation. BMC Biotechnol 20: 1–13.

Tavares, L.S., de Souza, V.C., Nunes, V.S., Silva, O.N., de Souza, G.T., Marques, L.F., Goliatt, P.V.Z.C., Viccinia, L.F., Franco, O.L., Oliveira Santos, M. (2020). Antimicrobial peptide selection from *Lippia* spp. leaf transcriptomes. Peptid 129: 170317.

Taylor, J.D., Baumgartner, A., Schmid, T.E., Brinkworth, M.H. (2019). Responses to genotoxicity in mouse testicular germ cells and epididymal spermatozoa are affected by increased age. Toxicol Lett 310: 1–6.

Tisserand, R., Young, R. (2014). Essential oil composition. pp. 5–22. *In*: Essential Oil Safety. Elsevier.

Tusevski, O., Krstikj, M., Stanoeva, J.P., Stefova, M., Simic, S.G. (2018). Phenolic profile and biological activity of *Hypericum perforatum* L.: Can roots be considered as a new source of natural compounds? S Afr J Bot 117: 301–310.

Vairo, C., Basas, J., Pastor, M., Palau, M., Gomis, X., Almirante, B., Gainza, E., Hernández, R.M., Igartua, M., Gavaldà, J., Gainza, G. (2020). *In vitro* and *in vivo* antimicrobial activity of sodium colistimethate and amikacin-loaded nanostructured lipid carriers (NLC). Nanomedicine: NBM, 102259.

Vajrabhaya, L.O., Korsuwannawong, S. (2018). Cytotoxicity evaluation of a Thai herb using tetrazolium (MTT) and sulforhodamine B (SRB) assays. J Anal Sci Technol 9(1): 15.

Vamathevan, J., Clark, D., Czodrowski, P., Dunham, I., Ferran, E., Lee, G., Li, B., Madabhushi, A., Shah, P., Spitzer, M., Zhao, S. (2019). Applications of machine learning in drug discovery and development. Nat Rev Drug Discov 18(6): 463–477.

Vandenplas, Y., Simoens, S., Van Wilder, P., Vulto, A.G., Huys, I. (2021). Off-patent biological and biosimilar medicines in Belgium: a market landscape analysis. Front Pharmacol 12: 684147.

Vieira, I.J.C., Medeiros, W.L.B., Monnerat, C.S., Souza, J.J., Mathias, L., Braz-Filho, R., Pinto, A.C., Sousa, P.M., Rezende, C.M., Epifanio, R.D.A. (2008). Two fast screening methods (GC-MS and TLC-ChEI assay) for rapid evaluation of potential anticholinesterasic indole alkaloids in complex mixtures. An Acad Bras Ciênc 80: 419–426.

Vilela, L.M.B., Santos-Silva, C.A., Roldan-Filho, R.S., Araújo, S.S., Ferreira-Neto, J.R.C., Oliveira, W.D., Amorim, L.L.B., Pandolf, V., Benko-Iseppon, A.M. (2021). *Cenostigma pyramidale*: ethnomedicinal properties and perspectives on a legume tree highly adapted to semiarid 'Caatinga' region. pp. 218–233. *In*: Mahendra, R., Bhattarai, S., Feitosa, C.M. (eds.). Ethnopharmacol Wild Plants. CRC Press.

Vincenzi, M., Stammati, A., Vincenzi, A., Silano, M. (2004). Constituents of aromatic plants: carvacrol. Fitoterapia 75(7-8): 801–804.

Wang, L.F., Luo, F., Xue, C.R., Deng, M., Chen, C., Wu, H. (2016). Antibacterial effect and shear bond strength of an orthodontic adhesive cement containing *Galla chinensis* extract. Biomed Rep 4: 507–511.

Wang, X., Yu, H., Xing, R., Liu, S., Chen, X., Li, P. (2019). Preparation and identification of antioxidative peptides from Pacific Herring (*Clupea pallasii*). Prot Mol 24(10): 1946.

Wang, Z., Guo, Y., Zhang, P. (2021). A rapid quantitation of cell attachment and spreading based on digital image analysis: application for cell affinity and compatibility assessment of synthetic polymers. Mater Sci Eng C 128: 112267.

Wilk-Zasadna, I., Bernasconi, C., Pelkonen, O., Coecke, S. (2015). Biotransformation *in vitro*: An essential consideration in the quantitative *in vitro*-to-*in vivo* extrapolation (QIVIVE) of toxicity data. Toxicology 332: 8–19.

Willyard, C. (2017). The drug-resistant bacteria that pose the greatest health threats. Nature 543(7643): 15.

Wong, K.H., Tan, W.L., Serra, A., Xiao, T., Sze, S.K., Yang, D., Tam, J.P. (2016). Ginkgotides: Proline-rich hevein-like peptides from gymnosperm *Ginkgo biloba*. Front Plant Sci 7: 1639.

Wong, K.H., Tan, W.L., Kini, S.G., Xiao, T., Serra, A., Sze, S.K., Tam, J.P. (2017). Vaccatides: Antifungal glutamine-rich hevein-like peptides from *Vaccaria hispanica*. Front Plant Sci 8: 1100.

Yemmireddy, V.K., Cason, C., Moreira, J., Adhikari, A. (2020). Effect of pecan variety and the method of extraction on the antimicrobial activity of pecan shell extracts against different foodborne pathogens and their efficacy on food matrices. Food Control 112: 107098.

Yin, F., Cheng, S., Liu, S., Ma, C., Wang, L., Zhao, R., Lin, J.-M., Hu, Q. (2021). A portable digital optical kanamycin sensor developed by surface-anchored liquid crystal droplets. J Hazard Mater 128: 112266017.

Yu, Z., Tang, J., Khare, T., Kumar, V. (2020). The alarming antimicrobial resistance in ESKAPEE pathogens: Can essential oils come to the rescue? Fitoterapia 140: 104433.

Zhang, L., Chou, C.P., Moo-Young, M. (2011). Disulfide bond formation and its impact on the biological activity and stability of recombinant therapeutic proteins produced by *Escherichia coli* expression system. Biotechnol Adv 29(6): 923–929.

Zhao, B., Ablajan, N., Zhao, J.-Y., Kodirova, D.R., Sagdullaev, Sh.Sh., Aisa, H.A. (2020). Two new C19-diterpenoid alkaloids from *Aconitum smirnovii*. Phytochem Lett 38: 96–100.

Zheng, K., Wu, L., He, Z., Yang, B., Yang, Y. (2017). Measurement of the total protein in serum by biuret method with uncertainty evaluation. Measurement 112: 16–21.

11

Plant-Based Essential Oils in The Treatment of Microbial Infections

Hercília Maria Lins Rolim,[1,*] *Alessandra Braga Ribeiro,*[2] *Monalisa de Alencar Lucena,*[3] *Andressa Barros Ibiapina*[4] and *Thais Cruz Ramalho*[1]

Introduction

Just a few decades ago, researches from around the world presented the large possibilities of native flora for providing a sustainable source of bioactive compounds to develop new drugs and herbal medicines. In this context, Brazil possesses the status of the greatest and more diverse flora around the world, representing 20% of the total flora, with more than 40,000 of different species (Oliveira et al. 2012). However, despite the great potential of Brazilian native flora, most plants remain commercially underexploited considering their exquisite phytochemical and biological profile (Fraga et al. 2020).

Globally, the actual trend in the industrial field is signaling for natural compounds capable of reducing or even replacing synthetic chemical compounds (Almeida et al. 2018). In this way, ingredients and active compounds based on plants have been intensively searched for application in several fields, such as cosmetic, food and pharmaceutical industries. Additionally, several plant-based drugs are already included in the national health care in many countries, mainly due to their interesting aspects, namely lower cost, higher tolerance and fewer side effects compared to traditional synthetic drugs (Jardim et al. 2015; Uttra et al. 2018).

Most plants used for developing new products present a complex chemical composition composed mainly by secondary metabolites, which exerts a fundamental role in plant survival, protecting them against infectious diseases, also acting as an insecticidal or even promoting the healing after damage caused by external factors (Sharifi-Rad et al. 2017;

[1] Graduate Program in Pharmaceutical Sciences, Federal University of Piauí, Teresina, PI, 64049-550, Brazil.
[2] Universidade Católica Portuguesa, CBQF – Centro de Biotecnologia e Química Fina – Laboratório Associado, Escola Superior de Biotecnologia, Rua Diogo Botelho 1327, 4169-005 Porto, Portugal.
[3] Graduate Program of Science and Engineering of Materials, Federal University of Piauí, Teresina, PI, 64049-550, Brazil.
[4] Graduate Program in Science and Health, Federal University of Piauí, Teresina, PI, 64049-550, Brazil.
* Corresponding author: hercilia.rolim@gmail.com

Silva et al. 2020; Tariq et al. 2019). In the range of secondary metabolites produced by plants, essential oils configure in one of the most important and effectively used groups of compounds in many different fields.

Essential Oils (EO) are liquids that present a complex mixture of volatile compounds, with a characteristic odor, color that varies between colorless and slightly yellow, lower density than water and solubility in organic solvents. In general, the chemical composition of the essential oil can be influenced by many factors such as part of the plant, method performed to extract the essential oil, external factors as environmental conditions, soil type, altitude, period of the day that the plant was collected, seasonality and genetic varieties (Veras et al. 2017; Sciarrone et al. 2017). The EO is commonly extracted from plants by hydrodistillation and in this process different parts of plants can be used, such as leaves, flowers, seeds, trunks and roots (Mancarz et al. 2019; Meroni et al. 2020).

These natural compounds have in their composition a complex mixture of several molecules, which have low molecular weight and concentrations that vary due to the influence of numerous factors, as mentioned earlier. The main compounds present in EO are monoterpenes, sesquiterpenes and their oxygenated derivatives, and can have diterpenes, phenylpropanoids and isothiocyanates. Generally a few of these molecules represent the highest percentage of compounds present in EO (major compounds), with the others being present in low concentrations. The major compounds, alone or in combination with the other components, are responsible for the biological properties attributed to the EO (De Souza et al. 2017; Sharifi-Rad et al. 2017).

Essential oils are used in several applications, whether in cosmetics (in perfumes and soaps) and food (as flavorings and additives) industry, or as pesticide and insecticide agents in the agro-industrial sector (Yu et al. 2020). In addition to these applications, of late EOs have received great attention for their pharmacological properties, which has increased their application in the medical field and in the development of alternative treatments for numerous diseases. The bioactive compounds present in its composition can act as antioxidant and anti-inflammatory agents, as well as antimicrobial agents, due to their lipophilic nature, which facilitates the interaction with microbial cells resulting in their destruction. Due to the intensive use of antimicrobial drugs, an increase in the number of resistant strains has been observed in recent years, which calls attention to the development of new natural and effective antimicrobial compounds, such as essential oils (Mancarz et al. 2019; Ragno et al. 2020).

Antimicrobial activity of Brazilian native plants essential oils

Brazil has a large distribution of potential therapeutic compounds in their native flora, that plays a fundamental role in traditional medicine. The use of medicinal plants by the Brazilian population for the treatment of different diseases is traditionally old and has increased in the last years, arousing the interest of researchers and pharmaceutical companies to gain knowledge about the biological activities from natural compounds of native medicinal plants (Duarte et al. 2007; Jardim et al. 2015; Dutra et al. 2016).

Among these natural compounds obtained from plants, essential oils stand out for having several biological activities due to their bioactive compounds, as described earlier. As Brazil has the highest biodiversity in the world, its native plants are important sources for obtaining essential oils (Dutra et al. 2016). Considering the above, in this chapter the main studies related to the antimicrobial activity of essential oils obtained from native Brazilian plants belonging to the genus *Lippia*, *Cordia*, *Croton*, *Copaífera*, *Eugenia*, *Baccharis*, *Hedyosmum*, *Schinus*, and *Aniba* will be discussed.

Lippia

The genus *Lippia* belongs to the family Verbenaceae and includes a large number of species distributed throughout tropical Africa, Central, and South America. The vast majority of known species present in the Brazilian territory and are widely used in food and drinks, in addition to their use in traditional medicine for having a wide range of biological properties (Funari et al. 2012; De Melo et al. 2020). Among the native Brazilian species, *Lippia sidoides*, *Lippia origanoides*, *Lippia grandis*, *Lippia thymoides*, and *Lippia lasiocalycina* are rich in aromatic essential oils that present important biological activities, with emphasis on antimicrobial activity.

Lippia sidoides Cham., whose popular name is *"alecrim-pimenta"*, is a perennial bushy plant native to *Caatinga* and widely cultivated in Northeast Brazil (Cavalcanti et al. 2010; Medeiros et al. 2011). The aromatic stem and leaves of this plant are used in therapeutic treatments and the extracted essential oil is used to treat skin lesions, as an oral antiseptic, and in the treatment of oral and vaginal infections (Medeiros et al. 2011; De Farias et al. 2012). Regarding the chemical composition, *L. sidoides* EO contains major constituent thymol (2-Isopropyl-5-methylphenol), a monoterpene phenolic which is responsible for potent antibacterial and antifungal activities, in addition to other biological activities such as larvicidal action, antioxidant, anti-inflammatory and analgesic (Veras et al. 2013; Saraiva et al. 2020). Lobo et al. (2014) reported the potential activity of *L. sidoides* EO against *Streptococcus mutans*, a Gram positive bacteria from the saliva of children with cavities, corroborating with results obtained in the study by Veras et al. (2017), who evaluated the antibacterial activity of *L. sidoides* EO against *S. mutans* and other Gram-positive and Gram-negative bacteria (Table 11.1). In both studies, thymol and carvacrol (a structural molecular isomer of thymol) were considered the major compound, corresponding to approximately 90% of EO chemical composition, which explains the antimicrobial activity shown by the EO under study.

Lippia origanoides Kunt., popularly known as *"alecrim-pimenta"*, *"alecrim-do-nordeste"* or *"estrepa-cavalo"* (Queiroz et al. 2014), is a synonym of *L. sidoides* and has several uses, either in food (as seasoning), or as a medicinal plant where leaf infusion is used to treat gastrointestinal (stomach pain, flatulence and indigestion) and respiratory diseases (Vicuña et al. 2010; Teles et al. 2014). Its essential oil has thymol, carvacrol and p-cymene as major compounds, which are also responsible for their biological activities, as observed in the study by Oliveira et al. (2007) that showed the potential antimicrobial activity of EO from *L. origanoides* against bacteria (Gram-negative and Gram-positive) and yeast (Table 11.1).

Lippia grandis Schauer, commonly known as *"erva-do-marajó"*, is an aromatic shrub whose leaves and branches are widely used in food (giving flavor) and in the production of medicinal teas for the treatment of stomach and liver diseases (Damasceno et al. 2011). Sarrazin et al. (2012) evaluated EO activity obtained from *L. grandis* against Gram-positive, Gram-negative bacteria and against *C. albicans* recovered from clinical samples. It was observed that EO showed potential activity against most of the tested microorganisms and this antimicrobial activity seems to be related to phenolic compounds as carvacrol and thymol, acting synergistically with p-cymene and other minority compounds present in EO composition. These compounds have shown activity against *Candida* species related to envelope damage causing depletion of biofilm formation (Braga et al. 2008).

Lippia thymoides Mart. & Schauer, also known as *"alecrim-do-mato"* or *"alecrim-do-campo"*, is a shrub popularly used in religious rituals and in medicine for respiratory problems (bronchitis), in the treatment of rheumatism, as well as headaches and weakness (Silva et al. 2016). The analysis of the chemical composition of the essential oil

Table 11.1. Antimicrobial activity of essential oils from plants of the genus *Lippia, Cordia, Croton, Copaífera, Eugenia, Baccharis, Hedyosmum, Schinus*, and *Aniba*.

Genus	Specie/part of the plant	Yield (% EO)/Major compounds	Methodology for AM activity determination	Antimicrobial activity		Reference
				Antibacterial	Antifungal	
Lippia	*Lippia sidoides* (leaves)	Yield: NM Thymol/carvacrol (93.36%) Caryophyllene (3.59%)	Essential oil was incorporated in toothpaste, gel and mouthwash; *In vivo* test (microbiological analysis of saliva)	Toothpaste promoted lower levels of salivarius *Streptococcus mutans* than chlorhexidine	NT	Lobo et al. (2014)
		Yield: 1.06% Thymol (84.9%) Ethyl-methyl-carvacrol (5.33%) p-cymene (3.01%)	Microdilution assay (MIC)	*Staphylococcus aureus* (128 µg/mL) *S. mutans* (256 µg/mL) *Enterococcus faecalis* (512 µg/mL) *Escherichia coli* (512 µg/mL) *Enterobacter cloacae* (256 µg/mL) *Klebsiella pneumoniae* (256 µg/mL) *Pseudomonas aeruginosa* (512 µg/mL) *Providencia rettgeri* (256 µg/mL)	NT	Veras et al. (2017)
	Lippia origanoides (leaves)	Yield: 1.0% (v/w) Oxygenated monoterpenes (66%) Carvacrol (38.6%) Thymol (18.5%)	Drop agar diffusion (inhibition halo diameter)	*S. aureus* (25 mm) *Staphylococcus aureus* MRSA (25 mm) *S. mutans* (26 mm)	*Candida albicans* Serotype B ATCC 36802 (25 mm) *C. albicans* (27 mm) *Candida guilliermondii* (40 mm) *Candida parapsilosis* (35 mm) *Cryptococcus neoformans* (24 mm) *Trichophytum rubrum* (30 mm)	Oliveira et al. (2007)
	Lippia grandis (leaves)	Yield: 2.7% Carvacrol (37.12%) p-cymene (11.64%) Thymol (7.8%)	-Disk diffusion assay (inhibition halo diameter) -Microdilution assay (MIC)	*S. aureus* ATCC 25923 (33.9 mm; 1.15 mg/mL) *E. faecalis* ATCC 51299 (24.5mm; 0.57 mg/mL) *E. faecalis* ATCC 29212 (13.0 mm; 0.57 mg/mL) *E. coli* ATCC 35218 (29.3 mm; 1.15 mg/mL) *E. coli* ATCC 25922 (22.7 mm; 1.15 mg/mL) *K. pneumoniae* ATCC 700603 (9.8 mm; 1.15 mg/mL)	NA	Sarrazin et al. (2012)
	Lippia thymoides (leaves)	Yield: 1.63% Thymol (88.56%)	Disk diffusion assay (inhibition halo diameter)	*S. aureus* (11 mm)	*C. albicans* (10 mm) *C. tropicalis* (8 mm)	Silva et al. (2020)
	Lippia lasiocalycina (leaves)	Yield: ranged from 0.40 to 0.52% Piperitenone oxide (57.55%), limonene (20.69%)	Microdilution assay (MIC)	NA	*C. albicans* ATCC 10231 (512 µg/mL)	Almeida et al. (2018)

Cordia	*Cordia verbenacea* (aerial parts and leaves)	Yield: 0.19% (aerial parts) and 0.23% (leaves) α-pinene (29.69%), *Trans*-caryophyllene (25.27%), alloaromadendrene (9.99%)	Disk diffusion assay (inhibition halo diameter)	*Bacillus cereus* (23 mm) *Bacillus subtilis* (24 mm) *S. aureus 5051* (48 mm) *S. aureus 5164* (41 mm) *S. aureus* ATCC13150 (27 mm) *S. aureus* ATCC 6534 (21 mm) *Staphylococcus epidermidis* (31 mm) *S. epidermidis* ATCC 27626 (39 mm) *Proteus mirabilis* 1821 (15 mm) *Proteus vulgaris* ATCC 13315 (22 mm)	*C. albicans* 4241 (48 mm) *C. albicans* 14 (22 mm) *C. albicans* 14.1 (15 mm) *C. albicans* ICB11 (48 mm) *C. albicans* ICB58 (29 mm) *Candida dublinenesis* 558 (48 mm) *Candida glabrata* (48 mm) *C. guilliermondii* 158 (37 mm) *Candida lusitaniae* (48 mm) *C. parapsilosis* 006 (48 mm) *C. stellatoidea* 10C (36 mm) *C. tropicalis* ATCC 13803 (17 mm) *Candida tropicalis* 75 (20 mm) *Cryptococcus* sp. 202 (48 mm)	de CarvalhoJr et al. (2004)
		Yield: 0.37% NT	Rat periodontitis model-topical application of essential oil (detection of microorganisms by PCR)	Significant decline in *Porphyromonas gingivalis*	Reduction of *Actinobacillus actinomycetemcomitans*	Pimentel et al. (2012)
		Yield: ranged from 0.166% to 0.366% Sabinene (15.69–69.68%), α-pinene (5.57–16.16%), β-phellandrene (2.95–11.02%), γ-elemene (6.49–12.57%), δ-elemene (3.3–12.65%), β-caryophyllene (5.58–11.49%), γ-caryophyllene (5.03–15.65%), Germacrene B (2.89–13.85%)	Microdilution assay (MIC)	-EO combinated with antibiotics: a potential modulator (synergistic effect), reducing the MIC of the antibiotic tested against all bacterial strains	NT	Matias et al. (2016)

Table 11.1 contd. ...

...Table 11.1 contd.

Genus	Specie/part of the plant	Yield (% EO)/Major compounds	Methodology for AM activity determination	Antimicrobial activity		Reference
				Antibacterial	Antifungal	
Croton	*Croton cajucara* Benth (leaves)	Yield: NM Linalool*	-Microdilution assay (MIC) -Antimicrobial assays in artificial biofilms	-MIC: *Lactobacillus casei* ATCC 4646 (22.3 μg/mL) *Streptococcus sobrinus* ATCC 27609 (13.8 μg/mL) *S. mutans* ATCC 25175 (40.1 μg/mL) *P. gingivalis* ATCC 43146 (31.2 μg/mL) *S. aureus* ATCC 49456 (33.4 μg/mL) -EO inhibited all the microorganisms tested (artificial biofilms)	*C. albicans* ATCC 51501 (13.4 μg/mL)	Alviano et al. (2005)
		Yield: 0.65% (average) α-Pinene (EO1: 7.5%; EO2: 24.7%; EO3: 0.1%; EO4: 0.5%; EO5: less than 0.1%), linalool (EO1: 6.3%; EO2: 11.6%; EO3: 11.0%; EO4: 9.9%; EO5: 13.2%), 7-Hydroxycalamenene (EO1: 37.5%; EO2: ND; EO3: 28.4%; EO4: 30.9%; EO5: 32.9%), β-Caryophyllene (EO1: 4.1%; EO2: 5.7%; EO3: 2.8%; EO4: 4.0%; EO5: 2.6%)	-Drop agar diffusion method (inhibition zones); -Microdilution assay (MIC)	-Inhibition zones (mm): *S. aureus* (MRSA-BMB9393) (EO1: 7; EO2: 5; EO3: 7; EO4: 13; EO5: 38) *E. faecalis* (EO1: 10; EO2: 6; EO3: 8; EO4: 6; EO5: 9) *S. epidermidis* (EO1: 6; EO2: 9; EO3: 7; EO4: 8; EO5: 20) *L. casei* (EO1: 6; EO2: 9; EO3: 7; EO4: 8; EO5: 20) *Mycobacterium smegmatis* (00061) (EO1: 10; EO2: 24; EO3: 12; EO4: 18; EO5: 18) -MIC (μg/mL): *M. smegmatis* (EO1: 39.06; EO2: 5000; EO3: 78.12; EO4: 156.25; EO5: 156.25); *Mycobacterium tuberculosis* H37Rv ATCC 27294 (all EO samples: 4.88); *S. aureus* MRSA (EO1: 0.019; EO2: NA; EO3: 0.019; EO4: 0.004; EO5: 0.001).	-Inhibition zones (mm): *C. albicans* ATCC 24433 (EO1: 6; EO2: 10; EO3: 7; EO4: 8; EO5: 8) *Aspergillus fumigatus* ATCC 16913 ATCC 24433 (EO1: 4.8; EO2: 3.2; EO3: 4.5; EO4: 4.6; EO5: 5) *Aspergillus niger* (EO1: 5.7; EO2: 3.4; EO3: 5.6; EO4: 5.9; EO5: 6) *Aspergillus ochraceus* ATCC 22947 (EO1: 4.5; EO2: 3.1; EO3: 4.5; EO4: 4.7; EO5: 5) *Mucor circinelloides* LIKA0066 (EO1: 8; EO2: 15; EO3: 6; EO4: 10; EO5: 10) *Rhizopus oryzae* UCP1506 (EO1: 6; EO2: 12; EO3: 5; EO4: 8; EO5: 10); *T. rubrum* (T544) (EO1: 5.8; EO2: 0; EO3: 5.5; EO4: 5.7; EO5: 6); *Fusarium solani* (EO1: 3.1; EO2: 2.2; EO3: 3.4; EO4: 3.5; EO5: 3.5); *Microsporum gypseum* (EO1: 14; EO2: 3; EO3: 13; EO4: 14; EO5: 14). -MIC (μg/mL): *C. albicans* (EO1: 1.22; EO2: 1250; EO3: 156.25; EO4: 0.001; EO5: 0.038); *M. circinelloides* (EO1: NT; EO2: NT; EO3: NT; EO4: NT; EO5: 3.63 x 10^{-8}); *R. oryzae* (EO1: NT; EO2: NT; EO3: NT; EO4: NT; EO5: 0.152).	Azevedo et al. (2013)

	Croton campestris (leaves; branches)	Yield: 0.04% (leaves) and 0.02% (branches) -Leaves: β-caryophyllene (17%), bicyclogermacrene (16.2%) -Branches: Bicyclogermacrene (11.3%), spathulenol (14.7%), β-caryophyllene (8.2%)	Microdilution assay (MIC)	*S. aureus* ATCC 12624 (leaves: ≥ 512 μg/mL; branches: ≥ 128 μg/mL) *B. cereus* ATCC 33018 (leaves: NA; branches: ≥ 256 μg/mL) *E. coli* ATCC 25922 (leaves and branches: ≥ 512 μg/mL) *P. aeruginosa* ATCC 15442 (leaves: NA; branches: ≥ 512 μg/mL)	NT	de Almeida et al. (2013)
	Croton heliotropiifolius (leaves; steams)	Yield: 0.60% (winter), 0.24% (spring), 0.36% (summer), 0.16% (autumn) β-caryophyllene (28.61–46.99%), bicyclogermacrene (8.47–22.60%), germacrene-D (3.06–7.80%), limonene (1.76–9.56%) 1,8-cineole (1.12–8.66%)	Microdilution assay (MIC)	-MIC (μg/mL): *B. cereus* ATCC 11778 (winter: NA; spring: NA; summer: 500; autumn: NA) *E. faecalis* ATCC 19433 (winter: 500; spring: 500; summer: 62.5; autumn: 125) *E. coli* ATCC 25922 (winter: 500; spring: 500; summer: 500; autumn: NA) *Salmonella enterica* ATCC 10708 (winter: NA; spring: 500; summer: 500; autumn: NA) *Serratia marcescens* ATCC 13880 (winter: 500; spring: 500; summer: 500; autumn: 500) *Shigella flexneri* ATCC 12022 (winter: NA; spring: NA; summer: 500; autumn: 500)	NT	de Alencar Filho et al. (2017)
		Yield: 0.17% (E)-caryophyllene (23.85%), γ-muurolene (10.52%), viridiflorene (8.08%)	Microdilution assay (MIC)	*S. aureus* ATCC 6538 (500.0 μg/mL) *B. subtilis* ATCC 6833 (62.5 μg/mL)	NT	Araújo et al. (2017)

Table 11.1 contd. ...

...Table 11.1 contd.

Genus	Specie/part of the plant	Yield (% EO)/Major compounds	Methodology for AM activity determination	Antimicrobial activity		Reference
				Antibacterial	Antifungal	
Copaifera	*Copaifera multijuga* Hayne (oilsresin from trunk)	Yield: 75% β-cariophylene (57.29%), cariophylene oxide (10.34%), α-humuleno (9.11%), trans-α-bergamotene (5.31%), α-copaene (3.29%), γ-muurolene (1.63%), β-bisabolene (1.08%)	-Adapted disk diffusion assay (inhibition halo diameter) -Microdilution assay (MIC)	NT	*Aspergillus flavus* (IOC-3974) (19.5 mm; 0.08 mg/mL) *A. niger* (IOC-200) (9.5 mm; 0.1 mg/mL) *Aspergillus tamarii* (IOC-186) (9.0 mm; 0.5 mg/mL) *A. tamarii* (IOC-187) (12.5 mm; 0.3 mg/mL) *Aspergillus terreus* (IOC-217) (11.5 mm; 0.3 mg/mL[1]) *Candida guilliermondii* (IOC-2889) (9.5 mm; 0.1 mg/mL) *C. tropicalis* (IOC-3610) (10.0; 0.5 mg/mL) *C. parapsilosis* (IOC-2882) (16.0 mm; 0.1 mg/mL)	Deus et al. (2011)
	Copaifera langsdorffii (oilsresin from trunk)	Yield: 36.4% β-Bisabolene (32.0%), trans-α-Bergamotene (15.1%), β-Caryophyllene (9.7%), β-Famesene (3.9%), α-Bisabolene (3.3%), caryophyllene oxide (3.0%), β-Sesquiphellandrene (2.8%)	Microdilution assay (MIC)	*S. aureus* ATCC 6538 (125 μg/mL) *P. aeruginosa* ATCC 15442 (500 μg/mL) *S. choleraesuis* ATCC 10708 (250 μg/mL)	*C. albicans* ATCC 10231 (250 μg/mL) *C. tropicalis* ATCC 750 (15.6 μg/mL) *Candida krusei* ATCC 6258 (62.5 μg/mL)	Ribeiro et al. (2019)
		Yield: NM Cis-thujopsene (46.96%), seychellene (8.04%), α-copaene (7.75), β-sesquiphellandrene (7.34%), caryophyllene (6.71%)	Clinical study: Evaluation of the evolution of surface affected with acne from ten volunteers (photographs)	After 21 days of treatment, there was a highly significant decrease in the extent of area affected by acne (*Propionibacterium acnes*) in the region treated with the active gel (containing *Copaifera langsdorffii* EO)	NT	Gomes da Silva et al. (2012)

Eugenia	*Eugenia uniflora* (leaves)	Yield: NM Selina-1,3,7(11)-trien-8-one (36.37%), selina-1,3,7(11)-trien-8-one epoxide (27.32%)	Microdilution assay with spectrophotometric readings (IC_{50})	NT	*C. albicans* INCQS 40006 (1892.47 μg/mL) *C. tropicalis* INCQS 40042 (4511.82 μg/mL) *C. krusei* INCQS 40095 (12491.80 μg/mL)	dos Santos et al. (2018)
		Yield: 1.08% Isoflurane-germacrene (65.80%), germacra 3,7,9-trien-6-one (16.19%), β-elemenone (4.47%), γ-elemene (3.97%), germacrene B (2.19%)	Microdilution assay (MIC)	*S. aureus* 99 ATCC 25923 and *S. aureus* SA 358 (surgical wound) (> 256 μg/ml)	NT	Pereira et al. (2017)
		Yield: NM Germacrene B (21.2%), seline-1,3,7-(11)-trien-8-one oxide (19.3%), β-caryophyllene (12.6%), germacrene A (11.6%), germacrene C (11.4%), seline-1,3,7-(11)-trien-8-one (9.7%), curzerene (3.9%)	-Agar diffusion method (inhibition halo diameter) -Microdilution assay (MIC)	*L. monocytogenes* 138 ATCC 19117 (18 mm) *S. aureus* 139 ATCC 27664 (26 mm)	*C. albicans* (208.3 μg/mL) *C. parapsilosis* (208.3 μg/mL) *C. guilliermondii* (109.4 μg/mL) *Candida globose* (187.5 μg/mL) *Candida lipolytica* (93.7 μg/mL) *Cryptococcus laurentii* (208.3 μg/mL) *Trichosporon asahii* (312.5 μg/mL)	Victoria et al. (2012)
Baccharis	*Baccharis dracunculifolia* D.C. (leaves)	Yield: 0.31% (fresh) and 0.43% (dried) Germacrene D (18.4%) (E)-nerolidol (14.0%), spathulenol (11%) β-Pinene (9.5%)	Microdilution assay (MIC)	*S. aureus* ATCC 25923 (102 μg/mL) -Resistant to antibiotics: *S. aureus* SA10 (512 μg/mL) *P. aeruginosa* PA24 (813 μg/mL)	NT	Salazar et al. (2018)

Table 11.1 contd. ...

...Table 11.1 contd.

Genus	Specie/part of the plant	Yield (% EO)/Major compounds	Methodology for AM activity determination	Antimicrobial activity		Reference
				Antibacterial	Antifungal	
Hedyosmum	*Hedyosmum brasiliense* (leaves; flowers)	Yield: 0.5% (v/w) α-terpineol (10.2%), curzerene (8.9%), pinocarvone (8.4%), β-thujene (7.1%)	Agar dilution method (MIC)	MIC: % (vol/vol) *B. subtilis* ATCC 23858 (0.312) *S. aureus* ATCC 25923 (0.312) *Staphylococcus saprophyticus* ATCC 35552 (0.312)	MIC: % (vol/vol) *C. albicans* ATCC 10231 (0.312) *C. parapsilosis* ATCC 22019 (0.312) *Microsporum canis* C112 (0.125) *M. gypseum* C115 (0.312) *Trichophyton mentagrophytes* ATCC 9972 (0.312) *T. rubrum* C137 (0.125)	Kirchner et al. (2010)
		Yield: 0.24% (male flowers) and 0.38% (female flowers) -Leaves: Sabinene (16%), β-pinene (5%), 1,8-cineole (3–7%), methyl eugenol (2–5%), curzerene (17–18%), carotol (6%) -Flowers: Sabinene (8–10%), α-phellandrene (1–8%), 1,8-cineole (2–7%), germacrene D (4–6%), curzerene (11%), carotol (9%)	Adapted Bioautography (evaluation of clear inhibition zones on pre-coated Thin Layer Chromatography - TLC - plates sprayed with a suspension containing fungus spores)	NT	Female flowers EO: presented the strongest activity against *Cladosporium cladosporioides* CCIBt 140 and *Cladosporium sphaerospermum* Penz CCIBt 491) in all tested concentrations. Other oils: weaker inhibition against both fungi tested.	Murakami et al. (2017)
Schinus	*Schinus mole* (leaves; fruits)	Yield: 1.2% (w/w) β-pinene (25.23%), Epi α cadinol (21.29%), α-pinene (18.72%), Myrcene (11.54%), Sabinene (5.02%)	Microdilution assay (MIC)	*P. aeruginosa* ATCC 27853 (> 5000 µg/mL) *E. coli* ATCC 43895 (> 5000 µg/mL) *Acinetobacter baumannii* ATCC 19600 (5000 µg/mL) *S. aureus* (MRSA) ATCC 33591 (5000 µg/mL) *E. faecalis* ATCC 10100 (2500 µg/mL)	*C. neoformans* ATCC 90012 (625 µg/mL) *Paracoccidioides brasiliensis* (strain T17LM2) (39.06 µg/mL) *Trichophyton quinckeanum* ATCC 11480 (625 µg/mL)	do Prado et al. (2019)
		Yield: NM Sabinene (34.77%), germacrene-D (7.06%), terpinen-4-ol (5.50%), α-pinene (5.32%), β-pinene (4.50%)	-Diffusion in cylindrical agar cavities (Inhibition halo); -Macrodilution broth assay (MIC)	*S. aureus* ATCC 25923 (38.2 mm; 2.81 mg/mL) *E. coli* ATCC 25922 (22.0 mm; 11.25 mg/mL) *P. aeruginosa* ATCC 27853 (16.0 mm; 11.25 mg/mL)	NT	Rocha et al. (2012)

		Yield: 1.09 ± 0.22% (leaf) and 0.91 ± 0.09% (fruit) -Leaf: α-phellandrene (25.9%), limonene (11.7%), β-myrcene (11.1%), β-phellandrene (10.5%) -Fruit: β-myrcene (51.3%), limonene (14.1%), α-phellandrene (14.0%), β-phellandrene (11.0%)	- Agar diffusion method (inhibition halo diameter); - Macrodilution broth assay (MIC)	-Inhibition growth zone (mm): *S. epidermidis* ATCC 12228 (leaf: 11.3; fruit: 6.7) *S. aureus* ATCC 29213 (leaf: 9.7; fruit: 10.0) *E. coli* ATCC 25922 (leaf: 7.7; fruit: 6.3) *P. aeruginosa* ATCC 27853 (leaf: 6.3; fruit: 6.3) *E. faecalis* (leaf: 9.7; fruit: 8.3) *P. mirabilis* (leaf: 7.3; fruit: 6.3) *S. enteritidis* (leaf: 7.7; fruit: 6.3) *Salmonella entiritidis serovar Typhimurium* (leaf: 6.3; fruit: 6.3) -MIC (μg/mL): *S. epidermidis* ATCC 12228 (leaf: 66; fruit: 125) *S. aureus* ATCC 29213 (leaf: 125; fruit: 500) *E. coli* ATCC 25922 (leaf: 1000; fruit: 1000) *P. aeruginosa* ATCC 27853 (leaf: > 1000; fruit: > 1000) *E. faecalis* (leaf: 500; fruit: 1000) *P. mirabilis* (leaf: 1000; fruit: > 1000) *S. enteritidis* (leaf: 750; fruit: 1000) *S. entiritidis serovar Typhimurium* (leaf: > 1000; fruit: > 1000)	- Inhibition growth zone (mm): *A. niger* (leaf: 6.3; fruit: 6.3) *Aspergillus japonicus* (leaf: 7.3; fruit: 9.3) *Aspergillus oryzae* (leaf: 7.3; fruit: 6.7) *Fusarium oxysporum* (leaf: 6.3; fruit: 6.3) *R. oryzae* (leaf: 7.7; fruit: NA) *Rhizopus stolonifer* (leaf: 8.7; fruit: NA) - MIC (μg/mL): *A. niger* (leaf: 1000; fruit: 750) *A. japonicus* (leaf: 750; fruit: 250) *A. oryzae* (leaf: > 1000; fruit: > 1000) *F. oxysporum* (leaf: > 1000; fruit: > 1000) *R. oryzae* (leaf: 750) *R. stolonifera* (leaf: 500)	Martins et al. (2014)
Aniba	*Aniba rosaeodora* Ducke (aerial parts - leaves and thin stems)	Yield: 1.0% Linalool (88.6%)	-Disk diffusion assay (inhibition halo diameter in mm) -Microdilution assay (MIC in μL/mL)	*E. coli* ATCC 35218 (13.2; > 10) *K. pneumoniae* ATCC 13883 (11.6; > 10) *S. aureus* ATCC 25923 (26.7; 1.3) *E. faecalis* ATCC 29212 (8.80; 5) *S. pyogenes* ATCC 19615 (38.4; 5) *S. epidermidis* ATCC 12228 (> 40; 1.3)	NT	Sarrazin et al. (2016)

NM: not mentioned; NT: not tested; NA: no activity; ND: not detected; MIC: minimum inhibitory concentration; MBC: minimum bactericidal concentration; * percentage not mentioned.

extracted from *L. thymoides* carried out by Silva et al. (2019) demonstrated that thymol is the main compound present in EO extracted from flowers, leaves and branches (19.34–66.33%), which corroborates with the results obtained by Silva et al. (2020). The authors obtained thymol (88.56%) as the major EO compound of *L. thymoides*. In addition, they evaluated antimicrobial activity of EO that showed activity against Gram-positive bacteria (*S. aureus*) and yeasts (*C. albicans* and *C. tropicalis*) (Table 11.1), however, it showed no activity against Gram-negative bacteria (*E. coli*). In this study, thymol (88.56%) was also the major compound present in the EO.

In the research carried out by Almeida et al. (2018), the EO of *Lippia lasiocalycina* Cham. was evaluated for its antimicrobial activity against bacteria (Gram-negative, Gram-positive) and fungus (*C. albicans*) (Table 11.1). The analysis of the EO chemical composition of this species was carried out by Santos et al. (2020), who found the main components β-myrcene (31.17%), (E)-ocimenone (24.10%), p-cymene (7.17%) and (Z)-ocimenone (6.51%), this differed from the results obtained by Almeida et al. (2018), who identified piperitenone oxide (57.55%) and limonene (20.69%) as major compounds in the EO of *L. lasiocalycina*. Despite the great interest in research on the biological properties of natural compounds obtained from the genus *Lippia*, studies related to the species *L. lasiocalycina* are still scarce, which emphasizes the search for more in-depth studies related to the antimicrobial activity of this EO.

In view of the studies carried out to evaluate the essential oil of *Lippia* spp., it is observed that the major phenolic compounds (thymol and carvacrol) show antimicrobial activity against a vast number of microorganisms important for human health. The mechanism of action of these monoterpenes is related to the direct interaction of these hydrophobic compounds and the cell membrane, causing loss of integrity or even disintegration, which releases lipopolysaccharides (LPS) altering the selective permeability of cell membrane by escaping internal vital components (Helander et al. 1998; Chorianopoulos et al. 2004; Sarrazin et al. 2012).

Cordia

The genus *Cordia* belongs to the family Boraginaceae and is distributed in tropical and subtropical regions (Matias et al. 2010). The species *Cordia verbenacea* D.C. (synonym *Varronia curassavica* Jacq.) is popularly known as *"erva-baleeira"*, *"baleeira"*, *"maria-milagrosa"* or *"erva-preta"* and has leaves with a strong and characteristic odor. This plant has traditionally been used in popular medicine to treat rheumatism, inflammation and ulcers, since they have numerous biological activities, including anti-inflammatory, anti-allergic, antioxidant and antiulcerogenic (Nizio et al. 2015; Sciarrone et al. 2017). The EO from this plant is commonly extracted from the aerial part and leaves, presenting a low yield that varies from 0.17 to 0.37% according to data obtained in studies described in the literature (Table 11.1).

Regarding the chemical compounds present in the EO of *C. verbenacea*, it was observed that α-pirene (~ 30%), sabinene (15.69–69.68%), and β-caryophyllene (5.58–11.49%) are the major compounds (Table 11.1). Pimentel et al. (2012) showed a promising application of essential oil of *C. verbenacea* in periodontitis, using a topical application in a rat model, the authors showed a significant reduction of *P. gingivalis* and *A. actinomycetemcomitans*, which are important putative period onto pathogens. In another study by de Carvalho et al. (2004), *C. verbenacea* EO showed strong antimicrobial activity against Gram-positive bacteria and yeast, however, most Gram-negative bacteria were resistant to the action of EO. The analysis of the antimicrobial activity of the EO of *C. verbenacea* performed in a study by Rodrigues et al. (2012) showed a strong and similar activity against Gram-positive bacteria (*S. aureus* ATCC 12692 and *B. cereus*) and Gram-negative bacteria (multi-drug resistant *E. coli* 27), presenting MICs of 64 µg/mL. In addition, this study obtained as major compounds present in the EO the sesquiterpenes (83.66%), as ß-caryophyllene (25.4%), bicyclogermacrene (11.3%) d δ-cadinene (9.4%) and the monoterpene α-pinene (9.5%). In addition, the *C. verbenace* EO also acts synergistically with antibiotics, decreasing the MIC values obtained by the antibiotics used alone, as shown in the study by Matias et al. (2016).

The excellent antimicrobial activities of EO of *Cordia* genus are directly related to major compounds, especially α-pinene, ß-phellandrene, ß-caryophyllene and germacrene B. Numerous studies have shown the effects of these compounds against microorganisms, as in the study by Nissen et al. (2010), who found the excellent activity of α-pirene against Gram-positive and Gram-negative bacteria, as well as yeast species. It was observed that this compound was found in significant amounts in most EO obtained from *C. verbancea*, thus being one of the main compounds responsible for the antimicrobial activity of the EO, synergistically with the other major compounds. In the study by Stojković et al. (2008) α-pirene showed high activity against three tested strains of *Actinomadura madurae* (MICs of 5 µg/mL), demonstrating the excellent antimicrobial activity of this compound, whose mechanism of action is related to its oxidation by microorganism, which results in the formation of acyclic metabolites and consequently death of microbial cells.

Croton

Considered one of the main genus of the Euphorbiaceae family, the genus *Croton* is widely distributed in the Northeast region of Brazil. Many species of this genus have tremendous economic importance due to the great natural compounds diversity, with several pharmacological properties, emphasizing on anti-inflammatory, gastroportiva, antiulcerogenic, antioxidant and antimicrobial activities, which justifies the wide use in traditional medicine by the local population. Most species of this genus, such as *Croton cajucara, Croton campestris* and *Croton heliotropiifolius*, are rich in essential oils with complex chemical composition formed by monoterpenes, sesquiterpenes and phenylpropanoids (Da Costa et al. 2013; Martins et al. 2017).

The species *Croton cajucara* Benth. is commonly known as "*sacaca*" or "*cajuçara*" and grows mainly in the Amazon and the Northeast region of Brazil. Leaf tea is used in traditional medicine to treat diarrhea, diabetes, gastrointestinal disorders, as well as fever, liver problems and malaria (Da Silva et al. 2012; Freires et al. 2015). In addition, the leaves of *C. cajucara* are rich in essential oil, which presents as major compounds linalool (~ 11% average) and 7-hydroxycalamenene (~ 30%) (Table 11.1). Alviano et al. (2005) reported the antimicrobial activity of *C. cajucara* EO against *C. albicans, Lactobacillus casei, S. aureus, S. sobrinus, Porphyromonas gingivalis* and *S. mutans* cell suspensions, that are associated with oral cavity infections. The results show excellent antimicrobial and antifungal activity of EO, with MICs ranging from 13.4–40.1 µg/mL (Table 11.1). These results were confirmed by the analysis of the antimicrobial activities of the chemical compounds present in the EO (earlier separated by chromatography), which show linalool as responsible for the activity against *C. albicans* and other polar compounds present in the composition of EO was responsible for antibacterial activity. In the study by Azevedo et al. (2013), five samples of EO obtained from leaves of the red morphotype *C. cajucara* were analyzed, and an excellent antimicrobial activity against pathogenic microorganisms was observed, both by the inhibition of growth (inhibition halos) and by the evaluation of the MICs. It was observed that the EO showed significant activities against Methicillin Resistant *S. aureus* (MRSA), *Enterococcus faecalis*, *Mycobacterium tuberculosis*, *M. smegmatis*, *Mucor circinelloides* and *Rhizopus oryzae* (Table 11.1). In this case, in addition to linalool, EO samples showed large amounts of 7-hydroxyicalamenene (28.4–37.5%), which was analyzed in isolation and demonstrated activity on a wide spectrum of microorganisms, thus being responsible for antimicrobial activity synergistically with linalool.

In another study by Liu et al. (2020), linalool alone showed strong antibacterial activity against *P. aeruginosa* (MIC = 431 μg/mL), and it was observed through Scanning Electron Microscopy (SEM) that the mechanism of action of this compound is related to the permeability and integrity of cell membrane, resulting in lysis of the bacterial cell wall.

Croton campestris A. St.-Hil, which is known commonly as *"velame do campo"*, is a native plant from Brazil, mainly from the Southeast and Northeast regions and is traditionally used in folk medicine to treat many disorders, such as general pain, constipation, bronchitis, intestinal problems, among others. In relation the biological activities, antimicrobial, antiulcerogenic and antiinflammatory activities are the ones that stand out. Regarding the chemical composition of the EO obtained from this plant, it was observed that β-caryophyllene is present as a major compound, with the biological activities attributed to this sesquiterpene (De Morais et al. 2018; Monteiro et al. 2019). In the antibacterial activity evaluation of the of *C. campestris* EO, de Almeida et al. (2013) reported that the EO extracted from the leaves showed activity against Gram-positive and Gram-negative strains (*S. aureus* and *E. coli* with MICs = 512 μg/mL) and the EO extracted from the branches was active against all bacterial strains (the best MIC was 128 μg/mL against *S. aureus*) (Table11.1). Babili et al. (2009) showed that the EO obtained from the leaves also exhibited antifungal activity, with a MIC of 6.25 μg/mL for *C. albicans*. In this case, the caryophyllene oxide (29.9%) was found as the major compound and the β-caryophyllene presented 1.6% of the composition of the analyzed EO.

Another species from this genus is *Croton heliotropiifolius* Kunth., that is popularly known as *"velame-preto"* and is widely distributed in the Northeast region of Brazil, mainly in the *Caatinga* vegetation. The infusion of the roots of this species is popularly used in the treatment of pain in general, inflammation, as well as in the treatment of influenza infection and dermatitis. *C. heliotropiifolius* is considered a medicinal plant of great economic interest, since it is rich in bioactive compounds that confer numerous activities, such as antimicrobial, antioxidant, among others (Angélico et al. 2014; Saraiva et al. 2015; De Sousa et al. 2020).

In literature it was found that the major compounds present in EO extracted from leaves and stems were β-caryophyllene (maximum found was 46.99%), in addition to γ-muurolene and germacrene-D (Table 11.1). As for antimicrobial activity, de Alencar Filho et al. (2017) evaluated EO samples obtained in different seasons and observed that the samples showed weak to moderate activity (most had MICs = 500 μg/ml) for Gram-positive and Gram-negative bacteria. EO obtained in the summer showed the best activity compared with the ones obtained from other seasons, with activity against all bacteria tested and lower MIC against *E. faecalis* (62.5 μg/ml). In the study reported by Araujo et al. (2017) activity against Gram-positive bacteria was also observed, for the species *B. subtilis* (MIC = 62.5 $\mu g.mL^{-1}$) and *S. aureus* (MIC = 500.0 $\mu g.mL^{-1}$) (Table 11.1). This result was similar to the observed by Angélico et al. (2014), where the *C. heliotropiifolius* EO showed inhibitory activity for the multiresistant lineage of *S. aureus* (MR 358) with MIC of 512 μg/mL. In this case, the major compounds found in the EO were eucalyptol (16.9%), β-caryophyllene (15.9%) and germacrene-D (14.5%).

Copaifera

Belonging to the Leguminosae Juss. family (Fabaceae) and subfamily Caesalpinoideae Kunth, the genus *Copaifera* L. is widely distributed in the tropical regions of Latin America, and can also be found in Asia and Africa. However, it is in Brazil that the largest number of species of this genus occurs, mainly in the central regions and in the Amazon area (Junior et al. 2007; Furtado et al. 2018; Da Trindade et al. 2018). Plants of this genus are commonly called *"copaíba"*, *"copaibeira"* or *"pau-de-óleo"* (Arruda et al. 2019) and are traditionally used in popular medicine, mainly by the Amazonian people, in the treatment of cystitis, urinary incontinence, some respiratory diseases (bronchitis, pneumonia, sinusitis), in the treatment of skin and mucosal wounds, among others (Da Trindade et al. 2018). These uses are related to oil extracted from tree trunks, which are called oil-resin due the composition formed by resin acids (resin fraction) in diterpenes and sesquiterpenes (volatile fraction, which corresponds to most of the oil-resin mass). These oils have numerous biological activities, including antimicrobial, anti-inflammatory, analgesic activities, among others responsible for their application in various treatments, as mentioned earlier (Kobayashi et al. 2011; Santos et al. 2012; Diefenbach et al. 2018).

Among the copaifera species native to Brazil, *Copaifera multijuga* Hayne has received great prominence for the antimicrobial activity derived from its oil. This species is found mainly in the western Amazon, in the south of Roraima and in the north of Mato Grosso (Mendonça et al. 2009). In the study by Deus et al. (2011), the antifungal activity of the fraction of essential oil present in the oil-resin extracted from *C. multijuga* trunk was analyzed and it was observed that EO inhibited the growth of all species of *Aspergillus* and *Candida* tested, in addition to showing MIC values ranging from 0.08 to 0.5 $mg.mL^{-1}$ (Table 11.1), indicating its potential antifungal activity. In this case, the volatile fraction of the oil-resin present as major compounds the β-caryophyllene (57.29%), caryophyllene oxide (10.34%) and α-humulene (9.11%), which are related to the observed antimicrobial activity.

Copaifera langsdorffii Desf. is another species of the genus *Copaifera* that has great commercial potential and is commonly used in traditional medicine, due to the biological properties present in the extracts of the leaves and in the oil-resin extracted from the trunks, including antimicrobial, leishmanicidal and antimalarial activity (Lemos et al. 2015). Ribeiro et al. (2019) showed that the essential oil extracted from the oil-resin of *C. langsdorffii* presented β-bisabolene (32.0%) as a major compound and a small amount of β-caryophyllene (9.7%). The activity was observed against *S. aureus*, *P. aeruginosa* and *S. choleraesuis* (MICs between 125 and 500 μg/mL), in addition to antifungal activity against *Candida* species, with the greatest activity against *C. tropicalis* (MIC = 15.6 μg/mL) (Table 11.1). Gomes da Silva et al. (2012) conducted a clinical study with formulations containing *C. langsdorffii* EO for the treatment of acne and reported a significant decrease in the surface affected with acne (*Propionibacterium acnes*) after application of the formulation, with cis-thujopsene (46.96%) as the major compound, and also the presence of caryophyllene (6.71%) (Table 11.1).

As can be seen, β-caryophyllene is a chemical compound present in essential oils of various plant species and numerous studies have reported its potent antimicrobial activity against different species of microorganisms, as this compound is capable of interfering in the metabolism of microbial cells (Sharma et al. 2016). In the study by Yoo and Jwa (2018), it was observed that β-caryophyllene penetrates very well into exopolysaccharide (EPS) of *S. mutans* biofilm, resulting in decreased microbial growth. These results demonstrate that this compound has a great potential to become a strong ally against infectious microorganisms important for human health.

Eugenia

Considered one of the broadest genera of the Myrtaceae family, the genus *Eugenia* has hundreds of species distributed mainly in South America. *Eugenia uniflora* L., popularly known as "*pitanga* cherry" or "*pitangueira*", consists of a fruit species found throughout the Brazilian territory and is one of the main representative species of this genus, attracting important economic interest because of its health benefits. Infusions of *E. uniflora* leaves have been reported to be widely used in folk medicine to treat inflammatory problems, as hypoglycemiant and in intestinal pain. This species contains numerous components with biological activity, such as essential oils rich in monoterpenes, which gives anti-inflammatory, antimicrobial and antioxidant properties, among others (Ogunwande et al. 2005; Amorim et al. 2009; De Araújo et al. 2019).

Victoria et al. (2012) reported, as the major compounds of EO of *E. uniflora*, seline-1,3,7-(11)-trien-8-one oxide (19.3%), β-caryophyllene (12.6%), germacrene (A, B and C) (11.4–21.2%) and seline-1,3,7-(11)-trien-8-one (9.7%) (Table 11.1). In accordance with the earlier results, Dos Santos et al. (2018) showed the predominance of seline-1,3,7(11)-trien-8-one (36.37%) and seline-1,3,7(11)-trien-8-one epoxide (27.32%). In relation the biological activity, both studies observed the EO present activity against *Candida* species (Table 11.1), however, some variations in the results were observed since they used different methodologies to evaluate the antimicrobial activity. In addition, the EO showed excellent activity against *L. monocytogenes* and *S. aureus*, with growth inhibition halos of 18 ± 3.2 mm and 26 ± 7.0 mm respectively (Victoria et al. 2012). These results corroborate with the study by Pereira et al. (2017), who evaluated the antibacterial activity *E. uniflora* EO against two strains of *S. aureus*, obtaining MIC > 256 µg/mL. In the latter case, the EO presented isoflurane-germacrene (65.80%) as a major compound (Table 11.1). Considering the results, it is possible to observe once again that the influence of the EO chemical composition on antimicrobial activity of EO. According to Sousa et al. (2015), oxygenated compounds, such as seline-1,3,7-(11)-trien-8-one oxide, give EO more pronounced antimicrobial activities in relation to non-oxygenated compounds, which may explain the results obtained.

Baccharis

The genus *Baccharis* L. belongs to the Asteraceae family and has numerous species distributed throughout South America that are commonly used in popular medicine to treat headaches, skin wounds, liver problems, among others. Native to Brazil, the *Baccharis dracunculifolia* D.C. species, commonly known as "*vassourinha*" or "*alecrim do campo*", has received great importance due to its biological properties attributed to the presence of active compounds, such as the extracted essential oil from the leaves (diterpene and triterpene) and flavonoids, responsible for the anti-inflammatory, anti-carcinogenic, antimicrobial and immunostimulatory activities (Sobrinho et al. 2016; Bonin et al. 2020).

Salazar et al. (2018) evaluated the antibacterial activity of EO extracted from the leaves of *B. dracunculifolia* and a potential activity was observed against Gram-negative bacteria *P. aeruginosa* (MIC = 813 µg/mL), however, it showed more effective activity against Gram-positive bacteria *S. aureus*, presenting MIC for standard and multi-resistant strains of 102 µg/mL and 512 µg/mL, respectively (Table 11.1). In this case, germacrene D (18.4%), (E)-nerolidol (14.0%), spathulenol (11%) and β-pinene (9.5%) were obtained as major compounds, which have a mechanism of action against microorganisms associated with destructing of microbial membranes and their functions, since they penetrate cells due to their lipophilic character and results in their destruction.

Hedyosmum

Hedyosmum brasiliense is included in one of 48 species of the genus *Hedyosmum* (Chloranthaceae) spread in American countries, such as Brazil (Radice et al. 2019). In folk medicine, this plant is used in treatment of migraine, stomach pain, rheumatism, ovarian dysfunction and other conditions due to its analgesic indications. Popularly known as "chá de bugre", "cidrão" or "cidreira", *H. brasiliense* presents essential oil from male and female leaves and flowers with low yield, both not exceeding 0.5% (Kirchner et al. 2010; Murakami et al. 2017). Monoterpenes, sesquiterpenes and phenylpropanoids are the major compounds of this EO. Regarding antimicrobial activity, it was found that the activity was limited against Gram-positive bacteria (*S. aureus*, *S. saprophyticus* and *B. subtilis*), fungal dermatophytes, opportunistic fungi responsible for onychomycosis, probably caused by α-terpineol (Kirchner et al. 2010).

Schinus

Schinus molle belongs to the genus *Schinus* L., from the Anacardiaceae family, which is native of South America (Barbosa et al. 2007). It is a species popularly known as *"aroeira-periquita"* which occurs in the Brazilian region from Pernambuco to Rio Grande do Sul (Dos Santos et al. 2010). *S. molle* has been used in treatment of different conditions including respiratory and urinary infections. This plant provides essential oil from leaves and fruits, both with low yield (1.09 ± 0.22% and 0.91 ± 0.09%, respectively). *S. molle* EO is composed of monoterpenes and sesquiterpenes. Leaf EO and fruit EOs have a large spectrum of antibacterial activity, in particular important activity against Gram-positive bacteria (Do Rosário Martins et al. 2014). It was suggested that monoterpenes, such as pinenes and sabinene, act synergistically in Gram-positive strain (Rocha et al. 2012). *S. molle* EO has a promising application due to activity against *Paracoccidioides brasiliensis*, the etiologic agent of paracoccidioidomycosis (Do Prado et al. 2019). Nonetheless, *S. molle* EO showed cytotoxicity in lymphocytes and macrophages, probably in relation of the major compound α-pinene because of oxidative stress, but the mechanism of cytotoxicity of α-pinene remains unclear (Türkez and Aydin 2016; Duarte et al. 2018).

Aniba

The species *Aniba rosaeodora* Ducke, popularly known as *"pau-rosa"* (rosewood), is considered one of the main representatives of the genus *Aniba* (Lauraceae) and is distributed mainly in the Amazon region. This plant has an essential oil rich in linalool which is widely used in the cosmetic industry (perfumery) and in popular medicine, as it has numerous pharmacological properties with emphasis on its analgesic, antidepressant and anticonvulsant effect, and is also used in tissue regeneration (Simić et al. 2004; May and Barata 2004; Sarrazin et al. 2016). Another important biological property attributed to the EO of *A. rosaeodora* is its antimicrobial activity, which was evaluated in the study by Sarrazin et al. (2016). The EO showed activity against Gram-positive and Gram-negative bacteria, with the highest activity observed against *S. pyogenes* (inhibition zone > 40 mm and MIC of 1.3 μL/mL) (Table 11.1). In the study by Donaldson et al. (2005), in addition to antibacterial activity, EO showed antifungal activity of *A. rosaeodora* against *C. albicans* (inhibition zone of 20 mm). The antimicrobial activity is directly related with the presence of linalool in the oil composition, whose mechanism of action has been discussed earlier.

Promising plants and research gaps

Despite the increase in research related to the evaluation of the biological activities of essential oils extracted from native Brazilian plants, there are still an unknown number of EO plants that have not been studied in terms of the antimicrobial activity. Thus, there are many research possibilities to explore the huge diversity of the Brazilian flora using a sustainable approach, which will allow in the future, the discovery of new potent drugs to treat infectious diseases.

The EO from some species, that are already commonly used in popular medicine, have shown an interesting profile regarding their antimicrobial activity, with emphasis on the species *Annona foetida* (*"araticum caatinga"*), *Annona muricata* Lin., *Hypericum* spp. (the species *H. brasiliense* being the most common in Brazil), *Rhaphiodon echinus* (Nees & Mart.) Schauer (*"betônica"*) and *Vanillosmopsis arborea* Barker (*"candeeiro"*) (Costa et al. 2009; França et al. 2009; Kuete et al. 2016; Costa et al. 2017; Rodrigues et al. 2018). The few studies on these species showed the unexploited potential antimicrobial activity of their bioactive compounds. In this point of view, further investigation of their EO must be carried out in order to obtain new sources of natural compounds for application in various industrial sectors, including pharmaceutical fields.

Conclusion

The chemical composition of essential oils from species of the genus *Lippia*, *Cordia*, *Croton*, *Copaífera*, *Eugenia*, *Baccharis*, *Hedyosmum*, *Schinus* and *Aniba* is represented mainly by an aromatic class of compounds, namely terpenes. Many of the essential oils from plants showed a broad spectrum of action against important pathogens for human health, including multi-drug resistant strains of *S. aureus* and fungus such as *Paracoccidioides brasiliensis*, although their activity is, in general, stronger against Gram positive bacteria. However, despite the great antimicrobial potential of essential oils from native Brazilian flora, it is evident that there is a great necessity to conjugate the efforts from research centers, universities and industries aiming to develop efficient, low cost and sustainable new natural antimicrobial drugs.

References

Almeida, W.S., de Lima, S.G., Barreto, H.M., de Sousa Andrade, L.M., Fonseca, L., Sobrinho, C.A., Muratori, M.C.S. (2018). Chemical composition and antimicrobial activity of the essential oil of *Lippia lasiocalycina* Cham. (Verbenaceae). Ind Crops Prod 125: 236–240. DOI: 10.1016/j.indcrop.2018.09.007.

Alviano, W.S., Mendonça-Filho, R.R., Alviano, D.S., Bizzo, H.R., Souto-Padrón, T., Rodrigues, M.L., Souza, M.M.G. (2005). Antimicrobial activity of *Croton cajucara* Benth linalool-rich essential oil on artificial biofilms and planktonic microorganisms. Oral Microbiol Immunol 20(2): 101–105. DOI: 10.1111/j.1399-302X.2004.00201.x.

Amorim, A.C.L., Lima, C.K.F., Hovell, A.M.C., Miranda, A.L.P., Rezende, C.M. (2009). Antinociceptive and hypothermic evaluation of the leaf essential oil and isolated terpenoids from *Eugenia uniflora* L. (Brazilian Pitanga). Phytomedicine 16(10): 923–928. DOI: 10.1016/j.phymed.2009.03.009.

Angélico, E.C., Rodrigues, O.G., da Costa, J.E.G., Maria de F&atima, A.L., Neto, V.Q., de Medeiros, R.A.S. (2014). Chemical characterization and antimicrobial activity of essential oils and *Crotons* varieties modulator in the Brazilians Northeast semiarid. Afr J Plant Sci 8(7): 392–397. DOI: 10.5897/AJPS2014.1198.

Araújo, F.M., Dantas, M.C., e Silva, L.S., Aona, L.Y., Tavares, I.F., de Souza-Neta, L.C. (2017). Antibacterial activity and chemical composition of the essential oil of *Croton heliotropiifolius* Kunth from Amargosa, Bahia, Brazil. Ind Crops Prod 105: 203–206. DOI: 10.1016/j.indcrop.2017.05.016.

Arruda, C., Mejía, J.A.A., Ribeiro, V.P., Borges, C.H.G., Martins, C.H.G., Veneziani, R.C.S., Bastos, J.K. (2019). Occurrence, chemical composition, biological activities and analytical methods on *Copaifera* genus—A review. Biomed Pharmacother 109: 1–20. DOI: 10.1016/j.biopha.2018.10.030.

Azevedo, M., Chaves, F., Almeida, C.A., Bizzo, H.R., Duarte, R.S., Campos-Takaki, G.M., Alviano, D.S. (2013). Antioxidant and antimicrobial activities of 7-hydroxy-calamenene-rich essential oils from *Croton cajucara* Benth. Molecules 18(1): 1128–1137. DOI: 10.3390/molecules18011128.

Babili, F.E., Fouraste, I., Moulis, C., Bessiere, J.M., Roques, C., Haddioui, L. (2009). Essential oil of leaves of *Croton campestris* St. Hilaire, its secretory elements, and its biological activity. J Essent Oil Res 21(3): 272–275. DOI: 10.1080/10412905.2009.9700168.

Barbosa, L.C.A., Demuner, A.J., Clemente, A.D., Paula, V.F.D., Ismail, F. (2007). Seasonal variation in the composition of volatile oils from *Schinus terebinthifolius* Raddi. Quim Nova 30(8): 1959–1965. DOI: 10.1590/S0100-40422007000800030.

Bonin, E., Carvalho, V.M., Avila, V.D., dos Santos, N.C.A., Benassi-Zanqueta, É., Lancheros, C.A.C., do Prado, I.N. (2020). *Baccharis dracunculifolia*: Chemical constituents, cytotoxicity and antimicrobial activity. LWT 120: 108920. DOI: 10.1016/j.lwt.2019.108920.

Braga, P.C., Culici, M., Alfieri, M., Dal Sasso, M. (2008). Thymol inhibits *Candida albicans* biofilm formation and mature biofilm. Int J Antimicrob Agents 31(5): 472–477. DOI: 10.1016/j.ijantimicag.2007.12.013.

Cavalcanti, S.C.H., Niculau, E.D.S., Blank, A.F., Câmara, C.A.G., Araújo, I.N., Alves, P.B. (2010). Composition and acaricidal activity of *Lippia sidoides* essential oil against two-spotted spider mite (*Tetranychus urticae* Koch). Bioresour Technol 101(2): 829–832. DOI: 10.1016/j.biortech.2009.08.053.

Chorianopoulos, N., Kalpoutzakis, E., Aligiannis, N., Mitaku, S., Nychas, G.J., Haroutounian, S.A. (2004). Essential oils of *Satureja, Origanum*, and *Thymus* species: chemical composition and antibacterial activities against foodborne pathogens. J Agric Food Chem 52(26): 8261–8267. DOI: 10.1021/jf049113i.

Costa, A.R., de Lima Silva, J., Lima, K.R.R., Rocha, M.I., Barros, L.M., da Costa, J.G.M., e Menezes, I.R.A. (2017). *Rhaphiodon echinus* (Nees & Mart.) Schauer: Chemical, toxicological activity and increased antibiotic activity of antifungal drug activity and antibacterial. Microb Pathog 107: 280–286. DOI: 10.1016/j.micpath.2017.04.001.

Costa, E.V., Pinheiro, M.L.B., Silva, J.R.D.A., Maia, B.H.L.D.N.S., Duarte, M.C.T., Amaral, A.C.F., Leon, L.L. (2009). Antimicrobial and antileishmanial activity of essential oil from the leaves of *Annona foetida* (Annonaceae). Quim Nova 32(1): 78–81. DOI: 10.1590/S0100-40422009000100015.

Da Costa, A.C.V., do Amarante Melo, G.F., Madruga, M.S., da Costa, J.G.M., Junior, F.G., Neto, V.Q. (2013). Chemical composition and antibacterial activity of essential oil of a *Croton rhamnifolioides* leaves Pax & Hoffm. Semina: Ciênc Agrár 34(6): 2853–2863. DOI: 10.5433/1679-0359.2013v34n6p2853.

Da Silva, F.R., Junior, A.W., Cechinel Filho, V., Nunes, D.S. (2012). Chemical composition of essential oil from the bark of *Croton cajucara* Bentham. Acta Sci Technol 34(3): 325–329. DOI: 10.4025/actascitechnol.v34i3.11712.

Da Trindade, R., Da Silva, J.K., Setzer, W.N. (2018). *Copaifera* of the Neotropics: A review of the phytochemistry and pharmacology. Int J Mol Sci 19(5): 1511. DOI: 10.3390/ijms19051511.

Damasceno, E.I.T., Silva, J.K.R., Andrade, E.H.A., Sousa, P.J.C., Maia, J.G.S. (2011). Antioxidant capacity and larvicidal activity of essential oil and extracts from *Lippia grandis*. Rev Bras Farmacogn 21(1): 0–0. DOI: 10.1590/S0102-695X2011005000013.

De Alencar Filho, J.M., Araújo, L.D.C., Oliveira, A.P., Guimarães, A.L., Pacheco, A.G., Silva, F.S., Araújo, E.C.D.C. (2017). Chemical composition and antibacterial activity of essential oil from leaves of *Croton heliotropiifolius* in different seasons of the year. Rev Bras Farmacogn 27(4): 440–444. DOI: 10.1016/j.bjp.2017.02.004.

De Almeida, T.S., Rocha, J.B.T., Rodrigues, F.F.G., Campos, A.R., da Costa, J.G.M. (2013). Chemical composition, antibacterial and antibiotic modulatory effect of *Croton campestris* essential oils. Ind Crops Prod 44: 630–633. DOI: 10.1016/j.indcrop.2012.09.010.

De Araújo, F.F., Neri-Numa, I.A., de Paulo Farias, D., da Cunha, G.R.M.C., Pastore, G.M. (2019). Wild Brazilian species of *Eugenia* genera (Myrtaceae) as an innovation hotspot for food and pharmacological purposes. Food Res Int 121: 57–72. DOI: 10.1016/j.foodres.2019.03.018.

De Carvalho Jr, P.M., Rodrigues, R.F.O., Sawaya, A.C.H.F., Marques, M.O.M., Shimizu, M.T. (2004). Chemical composition and antimicrobial activity of the essential oil of *Cordia verbenacea* DC. J Ethnopharmacol 95(2-3): 297–301. DOI: 10.1016/j.jep.2004.07.028.

De Farias, E.M.F.G., Ximenes, R.M., Magalhães, L.P.M., de Andrade Chiappeta, A., de Albuquerque, J.F.C. (2012). Antifungal activity of *Lippia sidoides* Cham. (Verbenaceae) against clinical isolates of *Candida* species. J Herb Med 2(3): 63–67. DOI: 10.1016/j.hermed.2012.06.002.

De Melo, A.R.B., Higino, T.M.M., da Rocha Oliveira, A.D.P., Fontes, A., da Silva, D.C.N., de Castro, M.C.A.B., de Figueiredo, R.C.B.Q. (2020). *Lippia sidoides* and *Lippia origanoides* essential oils affect the viability, motility and ultrastructure of *Trypanosoma cruzi*. Micron 129: 102781. DOI: 10.1016/j.micron.2019.102781.

De Morais Oliveira-Tintino, C.D., Pessoa, R.T., Fernandes, M.N.M., Alcântara, I.S., da Silva, B.A.F., de Oliveira, M.R.C., da Costa, J.G.M. (2018). Anti-inflammatory and anti-edematogenic action of the *Croton campestris* A. St.-Hil (Euphorbiaceae) essential oil and the compound β-caryophyllene in *in vivo* models. Phytomedicine 41: 82–95. DOI: 10.1016/j.phymed.2018.02.004.

De Sousa, C.B.D.C., dos Anjos, G.L., Nóbrega, R.S., Magaton, A.D.S., de Miranda, F.M., Dias, F.D.S. (2020). Greener ultrasound-assisted extraction of bioactive phenolic compounds in *Croton heliotropiifolius* Kunth leaves. Microchem J 105525. DOI: 10.1016/j.microc.2020.105525.

De Souza, L.I.O., Bezzera-Silva, P.C., Navarro, D.M.D.A.F., da Silva, A.G., dos Santos Correia, M.T., da Silva, M.V., de Figueiredo, R.C.B.Q. (2017). The chemical composition and trypanocidal activity of volatile oils from Brazilian Caatinga plants. Biomed Pharmacother 96: 1055–1064. DOI: 10.1016/j.biopha.2017.11.121.

Deus, R.J.A., Alves, C.N., Arruda, M.S.P. (2011). Avaliação do efeito antifúngico do óleo resina e do óleo essencial de copaíba (*Copaifera multijuga* Hayne). Rev Bras de Plantas Medicinais 13(1): 01–07. DOI: 10.1590/S1516-05722011000100001.

Diefenbach, A.L., Muniz, F.W.M.G., Oballe, H.J.R., Rösing, C.K. (2018). Antimicrobial activity of copaiba oil (*Copaifera* ssp.) on oral pathogens: systematic review. Phytother Res 32(4): 586–596. DOI: 10.1002/ptr.5992.

Do Prado, A.C., Garces, H.G., Bagagli, E., Rall, V.L.M., Furlanetto, A., Fernandes Junior, A., Furtado, F.B. (2019). *Schinus molle* essential oil as a potential source of bioactive compounds: antifungal and antibacterial properties. J Appl Microbiol 126(2): 516–522. DOI: 10.1111/jam.14157.

Do Rosário Martins, M., Arantes, S., Candeias, F., Tinoco, M.T., Cruz-Morais, J. (2014). Antioxidant, antimicrobial and toxicological properties of *Schinus molle* L. essential oils. J Ethnopharmacol 151(1): 485–492. DOI: 10.1016/j.jep.2013.10.063.

Donaldson, J.R., Warner, S.L., Cates, R.G., Gary Young, D. (2005). Assessment of antimicrobial activity of fourteen essential oils when using dilution and diffusion methods. Pharm Biol 43(8): 687–695. DOI: 10.1080/13880200500384932.

Dos Santos, A.C.A., Rossato, M., Serafini, L.A., Bueno, M., Crippa, L.B., Sartori, V.C., Moyna, P. (2010). Efeito fungicida dos óleos essenciais de *Schinus molle* L. e *Schinus terebinthifolius* Raddi, Anacardiaceae, do Rio Grande do Sul. Rev Bras Farmacogn 20(2): 154–159.

Dos Santos, J.F.S., Rocha, J.E., Bezerra, C.F., do Nascimento Silva, M.K., de Matos, Y.M.L.S., de Freitas, T.S., de Brito, E.S. (2018). Chemical composition, antifungal activity and potential anti-virulence evaluation of the *Eugenia uniflora* essential oil against *Candida* spp. Food Chem 261: 233–239. DOI: 10.1016/j.foodchem.2018.04.015.

Duarte, J.A., Zambrano, L.A.D.B., Quintana, L.D., Rocha, M.B., Schmitt, E.G., Boligon, A.A., Machado, M.M. (2018). Immunotoxicological evaluation of *Schinus molle* L. (Anacardiaceae) essential oil in lymphocytes and macrophages. Evid Based Complement Alternat Med. DOI: 10.1155/2018/6541583.

Duarte, M.C.T., Leme, E.E., Delarmelina, C., Soares, A.A., Figueira, G.M., Sartoratto, A. (2007). Activity of essential oils from Brazilian medicinal plants on *Escherichia coli*. J Ethnopharmacol 111(2): 197–201. DOI: 10.1016/j.jep.2006.11.034.

Dutra, R.C., Campos, M.M., Santos, A.R., Calixto, J.B. (2016). Medicinal plants in Brazil: Pharmacological studies, drug discovery, challenges and perspectives. Pharmacol Res 112: 4–29. DOI: 10.1016/j.phrs.2016.01.021.

Fraga, S., Gonçalves, D., Nasário, F., Pereira, E., Pontes, P., Ribas, L., Sampaio, K.A. (2020). Sequential high-pressure extraction of caffeine and bioactive compounds from caferana seeds (*Bunchosia glandulifera*). J Supercrit Fluids 165: 104958. DOI: 10.1016/j.supflu.2020.104958.

França, H.S., Kuster, R.M., Rito, P.D.N., Oliveira, A.P.D., Teixeira, L.A., Rocha, L. (2009). Atividade antibacteriana de floroglucinóis e do extrato hexânico de *Hypericum brasiliense* Choysi. Quim Nova 32(5): 1103–1106. DOI: 10.1590/S0100-40422009000500004.

Freires, I.A., Denny, C., Benso, B., De Alencar, S.M., Rosalen, P.L. (2015). Antibacterial activity of essential oils and their isolated constituents against cariogenic bacteria: a systematic review. Molecules 20(4): 7329–7358. DOI: 10.3390/molecules20047329.

Funari, C.S., Gullo, F.P., Napolitano, A., Carneiro, R.L., Mendes-Giannini, M.J.S., Fusco-Almeida, A.M., Silva, D.H.S. (2012). Chemical and antifungal investigations of six *Lippia* species (Verbenaceae) from Brazil. Food Chem 135(3): 2086–2094. DOI: 10.1016/j.foodchem.2012.06.077.

Furtado, R.A., de Oliveira, P.F., Senedese, J.M., Ozelin, S.D., de Souza, L.D.R., Leandro, L.F., Ambrósio, S.R. (2018). Assessment of genotoxic activity of oleoresins and leaves extracts of six *Copaifera* species for prediction of potential human risks. J Ethnopharmacol 221: 119–125. DOI: 10.1016/j.jep.2018.04.002.

Gomes da Silva, A., de Freitas Puziol, P., Nunes Leitão, R., Gomes, T.R., Scherer, R., Lacerda Lopes Martins, M., Cavalcanti, L.C. (2012). Application of the essential oil from copaiba (*Copaifera langsdorffii* Desf.) for acne vulgaris: a double-blind, placebo controlled clinical trial. Altern Med Rev 17(1): 69–75.

Helander, I.M., Alakomi, H.L., Latva-Kala, K., Mattila-Sandholm, T., Pol, I., Smid, E.J., Von Wright, A. (1998). Characterization of the action of selected essential oil components on Gram-negative bacteria. J Agric Food Chem 46(9): 3590–3595. DOI: 10.1021/jf980154m.

Jardim, A.C.G., Igloi, Z., Shimizu, J.F., Santos, V.A.D.F.F.M.D., Felippe, L.G., Mazzeu, B.F., Rahal, P. (2015). Natural compounds isolated from Brazilian plants are potent inhibitors of hepatitis C virus replication *in vitro*. Antivir Res 115: 39–47. DOI: 10.1016/j.antiviral.2014.12.018.

Junior, V.V., Rosas, E.C., Carvalho, M.V.D., Henriques, M.D.G.M.D.O., Pinto, A.C. (2007). Chemical composition and anti-inflammatory activity of copaiba oils from *Copaifera cearensis* Huber ex Ducke, *Copaifera reticulata* Ducke and *Copaifera multijuga* Hayne—A comparative study. J Ethnopharmacol 112(2): 248–254. DOI: 10.1016/j.jep.2007.03.005.

Kirchner, K., Wisniewski Jr, A., Cruz, A.B., Biavatti, M.W., Netz, D.J. (2010). Chemical composition and antimicrobial activity of *Hedyosmum brasiliense* Miq., Chloranthaceae, essential oil. Rev Bras Farmacogn 20(5): 692–699. DOI: 10.1590/S0102-695X2010005000005.

Kobayashi, C., Fontanive, T.O., Enzweiler, B.G., de Bona, L.R., Massoni, T., Apel, M.A., Suyenaga, E.S. (2011). Pharmacological evaluation of *Copaifera multijuga* oil in rats. Pharm Biol 49(3): 306–313. DOI: 10.3109/13880209.2010.515595.

Kuete, V., Dzotam, J.K., Voukeng, I.K., Fankam, A.G., Efferth, T. (2016). Cytotoxicity of methanol extracts of *Annona muricata, Passiflora edulis* and nine other Cameroonian medicinal plants towards multi-factorial drug-resistant cancer cell lines. Springerplus 5(1): 1666. DOI: 10.1186/s40064-016-3361-4.

Lemos, M., Santin, J.R., Mizuno, C.S., Boeing, T., Sousa, J.P.B.D., Nanayakkara, D., Andrade, S.F.D. (2015). *Copaifera langsdorffii*: evaluation of potential gastroprotective of extract and isolated compounds obtained from leaves. Rev Bras Farmacogn 25(3): 238–245. DOI: 10.1016/j.bjp.2015.05.005.

Liu, X., Cai, J., Chen, H., Zhong, Q., Hou, Y., Chen, W., Chen, W. (2020). Antibacterial activity and mechanism of linalool against *Pseudomonas aeruginosa*. Microb Pathog 141: 103980. DOI: 10.1016/j.micpath.2020.103980.

Lobo, P.L.D., Fonteles, C.S.R., Marques, L.A.R.V., Jamacaru, F.V.F., da Cruz Fonseca, S.G., de Carvalho, C.B.M., de Moraes, M.E.A. (2014). The efficacy of three formulations of *Lippia sidoides* Cham. essential oil in the reduction of salivary *Streptococcus mutans* in children with caries: A randomized, double-blind, controlled study. Phytomedicine 21(8-9): 1043–1047. DOI:10.1016/j.phymed.2014.04.021.

Mancarz, G.F.F., Laba, L.C., Silva, T.A.M., de Santi Pazzim, M., de Souza, D., Prado, M.R.M., Mello, R.G. (2019). Chemical composition and biological activity of *Liquidambar styraciflua* L. leaf essential oil. Ind Crops Prod 138: 111446. DOI: 10.1016/j.indcrop.2019.06.009.

Martins, A.O.B.P.B., Rodrigues, L.B., Cesário, F.R.A.S., de Oliveira, M.R.C., Tintino, C.D.M., Castro, F.F., de Sousa Araújo, A.A. (2017). Anti-edematogenic and anti-inflammatory activity of the essential oil from *Croton rhamnifolioides* leaves and its major constituent 1, 8-cineole (eucalyptol). Biomed Pharmacother 96: 384–395. DOI: 10.1016/j.biopha.2017.10.005.

Martins, M.R., Arantes, S., Candeias, F., Tinoco, M.T., Cruz-Morais, J. (2014). Antioxidant, antimicrobial and toxicological properties of *Schinus molle* L. essential oils. J Ethnopharmacol 151(1): 485–492. DOI: 10.1016/j.jep.2013.10.063.

Matias, E.F., Santos, K.K., Almeida, T.S., Costa, J.G., Coutinho, H.D. (2010). Enhancement of antibiotic activity by *Cordia verbenacea* DC. Lat Am J Pharm 29(6): 1049–1052.

Matias, E.F., Alves, E.F., Silva, M.K., Carvalho, V.R., Figueredo, F.G., Ferreira, J.V., Costa, J.G. (2016). Seasonal variation, chemical composition and biological activity of the essential oil of *Cordia verbenacea* DC (Boraginaceae) and the sabinene. Ind Crops Prod 87: 45–53. DOI: 10.1016/j.indcrop.2016.04.028.

May, P.H. and Barata, L.E. (2004). Rosewood exploitation in the Brazilian Amazon: options for sustainable production. Econ Bot 58(2): 257–265. DOI: 10.1663/0013-0001(2004)058[0257:REITBA]2.0.CO;2.

Medeiros, M.D.G.F., Da Silva, A.C., Citó, A.M.D.G.L., Borges, A.R., De Lima, S.G., Lopes, J.A.D., Figueiredo, R.C.B.Q. (2011). *In vitro* antileishmanial activity and cytotoxicity of essential oil from *Lippia sidoides* Cham. Parasitol Int 60(3): 237–241. DOI: 10.1016/j.parint.2011.03.004.

Mendonça, D.E., Onofre, S.B. (2009). Antimicrobial activity of the oil-resin produced by copaiba *Copaifera multijuga* Hayne (Leguminosae). Rev Bras Farmacogn 19(2B): 577–581. DOI: 10.1590/S0102-695X2009000400012.

Meroni, G., Cardin, E., Rendina, C., Herrera Millar, V.R., Soares Filipe, J.F., Martino, P.A. (2020). *In vitro* efficacy of essential oils from *Melaleuca alternifolia* and *Rosmarinus officinalis,* Manuka honey-based gel, and propolis as antibacterial agents against canine *Staphylococcus Pseudintermedius* strains. Antibiotics 9(6): 344. DOI: 10.3390/antibiotics9060344.

Monteiro, P.A., Maccari Zelioli, Í.A., de Oliveira Sousa, I.M., Ruiz, A.L.T.G., Vendramini-Costa, D.B., Foglio, M.A., Carvalho, J.E.D. (2019). Chemical composition and antiproliferative activity of *Croton campestris* A. St.-Hil. essential oil. Nat Prod Res 33(4): 580–583. DOI: 10.1080/14786419.2017.1399377.

Murakami, C., Cordeiro, I., Scotti, M.T., Moreno, P.R.H., Young, M.C.M. (2017). Chemical composition, antifungal and antioxidant activities of *Hedyosmum brasiliense* Mart. Ex Miq. (Chloranthaceae) essential oils. Medicines 4(3): 55. DOI: 10.3390/medicines4030055.

Nissen, L., Zatta, A., Stefanini, I., Grandi, S., Sgorbati, B., Biavati, B., Monti, A. (2010). Characterization and antimicrobial activity of essential oils of industrial hemp varieties (*Cannabis sativa* L.). Fitoterapia 81(5): 413–419. DOI: 10.1016/j.fitote.2009.11.010.

Nizio, D.A.C., de Andrade Brito, F., Sampaio, T.S., de Oliveira Melo, J., da Silva, F.L.S., Gagliardi, P.R., Blank, A.F. (2015). Chemical diversity of native populations of *Varronia curassavica* Jacq. and antifungal activity against *Lasiodoplodia theobromae*. Ind Crops Prod 76: 437–448. DOI: 10.1016/j.indcrop.2015.07.026.

Ogunwande, I.A., Olawore, N.O., Ekundayo, O., Walker, T.M., Schmidt, J.M., Setzer, W.N. (2005). Studies on the essential oils composition, antibacterial and cytotoxicity of *Eugenia uniflora* L. Int J Aromather 15(3): 147–152. DOI: 10.1016/j.ijat.2005.07.004.

Oliveira, D.R., Leitao, G.G., Bizzo, H.R., Alviano, D.S., Alviano, C.S., Leitão, S.G. (2007). Chemical and antimicrobial analyses of essential oil of *Lippia origanoides* HBK. Food Chem 101(1): 236–240. DOI: 10.1016/j.foodchem.2006.01.022.

Oliveira, V.B., Yamada, L.T., Fagg, C.W., Brandão, M.G. (2012). Native foods from Brazilian biodiversity as a source of bioactive compounds. Food Res Int 48(1): 170–179. DOI: 10.1016/j.foodres.2012.03.011.

Pereira, N.L., Aquino, P.E., Júnior, J.G., Cristo, J.S., Vieira Filho, M.A., Moura, F.F., Cunha, F.A. (2017). *In vitro* evaluation of the antibacterial potential and modification of antibiotic activity of the *Eugenia uniflora* L. essential oil in association with led lights. Microb Pathog 110: 512–518. DOI: 10.1016/j.micpath.2017.07.048.

Pimentel, S.P., Barrella, G.E., Casarin, R.C.V., Cirano, F.R., Casati, M.Z., Foglio, M.A., Ribeiro, F.V. (2012). Protective effect of topical *Cordia verbenacea* in a rat periodontitis model: immune-inflammatory, antibacterial and morphometric assays. BMC Complement Altern Med 12(1): 1–8. DOI: 10.1186/1472-6882-12-224.

Queiroz, M.R.A., Almeida, A.C., Andrade, V.A., Lima, T.S., Martins, E.R., Figueiredo, L.S., Careli, R.T. (2014). Avaliação da atividade antibacteriana do óleo essencial de *Lippia origanoides* frente à *Staphylococcus* sp. isolados de alimentos de origem animal. Rev Bras de Plantas Medicinais 16(3): 737–743. DOI: 10.1590/1983-084x/13_083.

Radice, M., Tasambay, A., Pérez, A., Diéguez-Santana, K., Sacchetti, G., Buso, P., Baldisserotto, A. (2019). Ethnopharmacology, phytochemistry and pharmacology of the genus *Hedyosmum* (Chlorantaceae): A review. J Ethnopharmacol 244: 111932. DOI: 10.1016/j.jep.2019.111932.

Ragno, R., Papa, R., Patsilinakos, A., Vrenna, G., Garzoli, S., Tuccio, V., Artini, M. (2020). Essential oils against bacterial isolates from cystic fibrosis patients by means of antimicrobial and unsupervised machine learning approaches. Sci Rep 10(1): 1–11. DOI: 10.1038/s41598-020-59553-8.

Ribeiro, V.P., Arruda, C., da Silva, J.J.M., Aldana Mejia, J.A., Furtado, N.A.J.C., Bastos, J.K. (2019). Use of spinning band distillation equipment for fractionation of volatile compounds of *Copaifera* oleoresins for developing a validated gas chromatographic method and evaluating antimicrobial activity. Biomed Chromatogr 33(2): e4412. DOI: 10.1002/bmc.4412.

Rocha, P.M., Rodilla, J.M., Díez, D., Elder, H., Guala, M.S., Silva, L.A., Pombo, E.B. (2012). Synergistic antibacterial activity of the essential oil of aguaribay (*Schinus molle* L.). Molecules 17(10): 12023–12036. DOI: 10.3390/molecules171012023.

Rodrigues, F.F., Oliveira, L.G., Rodrigues, F.F., Saraiva, M.E., Almeida, S.C., Cabral, M.E., Costa, J.G.M. (2012). Chemical composition, antibacterial and antifungal activities of essential oil from *Cordia verbenacea* DC leaves. Pharmacogn Res 4(3): 161. DOI: 10.4103/0974-8490.99080.

Rodrigues, F.F.G., Colares, A.V., Nonato, C.D.F.A., Galvão-Rodrigues, F.F., Mota, M.L., Braga, M.F.B.M., da Costa, J.G.M. (2018). *In vitro* antimicrobial activity of the essential oil from *Vanillosmopsis arborea* Barker (Asteraceae) and its major constituent, α-bisabolol. Microb Pathog 125: 144–149. DOI: 10.1016/j.micpath.2018.09.024.

Salazar, G.J.T., de Sousa, J.P., Lima, C.N.F., Lemos, I.C.S., da Silva, A.R.P., de Freitas, T.S., Deschamps, C. (2018). Phytochemical characterization of the *Baccharis dracunculifolia* DC (Asteraceae) essential oil and antibacterial activity evaluation. Ind Crops Prod 122: 591–595. DOI: 10.1016/j.indcrop.2018.06.052.

Santos, D.R., Oliveira, L.M., Lucchese, A.M., de Freitas Espeleta, A., da Cruz, J.D., Lordelo, M.S. (2020). Insecticidal activity of essential oils of species from the genus *Lippia* against *Nasutitermes corniger* (Motschulsky) (Isoptera: Termitidae). Sociobiology 67(2): 292–300. DOI: 10.13102/sociobiology.v67i2.4992.

Santos, R.C.V., dos Santos Alves, C.F., Schneider, T., Lopes, L.Q.S., Aurich, C., Giongo, J.L., de Almeida Vaucher, R. (2012). Antimicrobial activity of Amazonian oils against *Paenibacillus species*. J Invertebr Pathol 109(3): 265–268. DOI: 10.1016/j.jip.2011.12.002.

Saraiva, A.G.Q., Saraiva, G.D., Albuquerque, R.L., Nogueira, C.E.S., Teixeira, A.M.R., Lima, L.B., de Sousa, F.F. (2020). Chemical analysis and vibrational spectroscopy study of essential oils from *Lippia sidoides* and of its major constituent. Vib Spectrosc 110: 103111. DOI: 10.1016/j.vibspec.2020.103111.

Saraiva, M.E., de Alencar Ulisses, A.V.R., Ribeiro, D.A., de Oliveira, L.G.S., de Macêdo, D.G., de Sousa, F.D.F.S., de Almeida Souza, M.M. (2015). Plant species as a therapeutic resource in areas of the savanna in the state of Pernambuco, Northeast Brazil. J Ethnopharmacol 171: 141–153. DOI: 10.1016/j.jep.2015.05.034.

Sarrazin, S.L.F., Oliveira, R.B., Barata, L.E.S., Mourão, R.H.V. (2012). Chemical composition and antimicrobial activity of the essential oil of *Lippia grandis* Schauer (Verbenaceae) from the western Amazon. Food Chem 134(3): 1474–1478. DOI: 10.1016/j.foodchem.2012.03.058.

Sarrazin, S.L.F., Oliveira, R.B., Maia, J.G.S., Mourão, R.H.V. (2016). Antibacterial activity of the rosewood (*Aniba rosaeodora* and *A. parviflora*) linalool-rich oils from the Amazon. European J Med Plants 12: 1–9. DOI: 10.9734/EJMP/2016/22901.

Sciarrone, D., Giuffrida, D., Rotondo, A., Micalizzi, G., Zoccali, M., Pantò, S., Mondello, L. (2017). Quali-quantitative characterization of the volatile constituents in *Cordia verbenacea* DC essential oil exploiting advanced chromatographic approaches and nuclear magnetic resonance analysis. J Chromatogr A 1524: 246–253. DOI: 10.1016/j.chroma.2017.10.007.

Sharifi-Rad, J., Sureda, A., Tenore, G.C., Daglia, M., Sharifi-Rad, M., Valussi, M., Sharifi-Rad, R. (2017). Biological activities of essential oils: From plant chemoecology to traditional healing systems. Molecules 22(1): 70. DOI: 10.3390/molecules22010070.

Sharma, C., M Al Kaabi, J., M Nurulain, S., N Goyal, S., Amjad Kamal, M., Ojha, S. (2016). Polypharmacological properties and therapeutic potential of β-caryophyllene: a dietary phytocannabinoid of pharmaceutical promise. Curr Pharm Des 22(21): 3237–3264.

Silva, F.S., Menezes, P.M.N., Sá, P.G.S.D., Oliveira, A.L.D.S., Souza, E.A.A., Almeida, J.R.G.D.S., Lucchese, A.M. (2016). Chemical composition and pharmacological properties of the essential oils obtained seasonally from *Lippia thymoides*. Pharm Biol 54(1): 25–34. DOI: 10.3109/13880209.2015.1005751.

Silva, S.G., da Costa, R.A., de Oliveira, M.S., da Cruz, J.N., Figueiredo, P.L.B., Brasil, D.D.S.B., Andrade, E.H.D.A. (2019). Chemical profile of *Lippia thymoides*, evaluation of the acetylcholinesterase inhibitory activity of its essential oil, and molecular docking and molecular dynamics simulations. PloS One 14(3): e0213393. DOI: 10.1371/journal.pone.0213393.

Silva, S.G., de Oliveira, M.S., Cruz, J.N., da Costa, W.A., da Silva, S.H.M., Maia, A.A.B., de Aguiar Andrade, E.H. (2020). Supercritical CO_2 extraction to obtain *Lippia thymoides* Mart. & Schauer (Verbenaceae) essential oil rich in thymol and evaluation of its antimicrobial activity. J Supercrit Fluids 105064. DOI: 10.1016/j.supflu.2020.105064.

Simić, A., Soković, M.D., Ristić, M., Grujić-Jovanović, S., Vukojević, J., Marin, P.D. (2004). The chemical composition of some Lauraceae essential oils and their antifungal activities. Phytother Res 18(9): 713–717. DOI: 10.1002/ptr.1516.

Sobrinho, A.C.N., de Souza, E.B., Rocha, M.F.G., Albuquerque, M.R.J.R., Bandeira, P.N., dos Santos, H.S., dos Santos Fontenelle, R.O. (2016). Chemical composition, antioxidant, antifungal and hemolytic activities of essential oil from *Baccharis trinervis* (Lam.) Pers. (Asteraceae). Ind Crops Prod 84: 108–115. DOI: 10.1016/j.indcrop.2016.01.051.

Sousa, R.M.F., de Morais, S.A., Vieira, R.B., Napolitano, D.R., Guzman, V.B., Moraes, T.S., do Nascimento, E.A. (2015). Chemical composition, cytotoxic, and antibacterial activity of the essential oil from *Eugenia calycina* Cambess. leaves against oral bacteria. Ind Crops Prod 65: 71–78. DOI: 10.1016/j.indcrop.2014.11.050.

Stojković, D., Soković, M., Glamočlija, J., Džamić, A., Ristić, M., Fahal, A., Petrović, S. (2008). Susceptibility of three clinical isolates of *Actinomodura madurae* to α-pinene, the bioactive agent of *Pinus pinaster* turpentine oil. Arch Biol Sci 60(4): 697–701. DOI: 10.2298/ABS0804697S.

Tariq, S., Wani, S., Rasool, W., Shafi, K., Bhat, M.A., Prabhakar, A., Rather, M.A. (2019). A comprehensive review of the antibacterial, antifungal and antiviral potential of essential oils and their chemical constituents against drug-resistant microbial pathogens. Microb Pathog 134: 103580. DOI: 10.1016/j.micpath.2019.103580.

Teles, S., Pereira, J.A., de Oliveira, L.M., Malheiro, R., Machado, S.S., Lucchese, A.M., Silva, F. (2014). Organic and mineral fertilization influence on biomass and essential oil production, composition and antioxidant activity of *Lippia origanoides* HBK. Ind Crops Prod 59: 169–176. DOI: 10.1016/j.indcrop.2014.05.010.

Türkez, H. and Aydın, E. (2016). *In vitro* assessment of cytogenetic and oxidative effects of α-pinene. Toxicol Ind Health 32(1): 168–176. DOI: 10.1177/0748233713498456.

Uttra, A.M., Ahsan, H., Hasan, U.H., Chaudhary, M.A. (2018). Traditional medicines of plant origin used for the treatment of inflammatory disorders in Pakistan: A review. J Tradit Chin Med 38(4): 636–656. DOI: 10.1016/S0254-6272(18)30897-5.

Veras, H.N., Araruna, M.K., Costa, J.G., Coutinho, H.D., Kerntopf, M.R., Botelho, M.A., Menezes, I.R. (2013). Topical antiinflammatory activity of essential oil of *Lippia sidoides* Cham: possible mechanism of action. Phytother Res 27(2): 179–185. DOI: 10.1002/ptr.4695.

Veras, H.N., Rodrigues, F.F., Botelho, M.A., Menezes, I.R., Coutinho, H.D., Costa, J.G. (2017). Enhancement of aminoglycosides and β-lactams antibiotic activity by essential oil of *Lippia sidoides* Cham. and the Thymol. Arab J Chem 10: S2790–S2795. DOI: 10.1016/j.arabjc.2013.10.030.

Victoria, F.N., Lenardão, E.J., Savegnago, L., Perin, G., Jacob, R.G., Alves, D., da Silva Nascente, P. (2012). Essential oil of the leaves of *Eugenia uniflora* L.: antioxidant and antimicrobial properties. Food Chem Toxicol 50(8): 2668–2674. DOI: 10.1016/j.fct.2012.05.002.

Vicuña, G.C., Stashenko, E.E., Fuentes, J.L. (2010). Chemical composition of the *Lippia origanoides* essential oils and their antigenotoxicity against bleomycin-induced DNA damage. Fitoterapia 81(5): 343–349. DOI: 10.1016/j.fitote.2009.10.008.

Yoo, H.J., Jwa, S.K. (2018). Inhibitory effects of β-caryophyllene on *Streptococcus mutans* biofilm. Arch Oral Biol 88: 42–46. DOI: 10.1016/j.archoralbio.2018.01.009.

Yu, Z., Tang, J., Khare, T., Kumar, V. (2020). The alarming antimicrobial resistance in ESKAPEE pathogens: Can essential oils come to the rescue? Fitoterapia 140: 104433. DOI: 10.1016/j.fitote.2019.104433.

12

Terpenes: A Source of Novel Antimicrobials, Applications and Recent Advances

Nawal M. Al Musayeib,[1] *Musarat Amina*[1,*] and *Farah Maqsood*[2]

Introduction

Antimicrobial agents are composed of naturally occurring components, including phytochemicals and essential oils (Moo et al. 2019). They can exist in nature either as synthetic or semi-synthetic compounds (Rudramurthy et al. 2016). Natural phytochemicals with a molecular weight ≤ 500 g/mol may have ability to serve as antimicrobial adjuvants and exert synergistic effects (Langeveld et al. 2014; Barbieri et al. 2017). The exploration of new antimicrobial agents through microbial modification is an important alternative (Yu et al. 2010). The combination of an antimicrobial agent with a low molecular weight natural compound such as terpene and terpene derivatives has shown promising effects due to their ability to remove bacterial and fungal biofilm production (Zacchino et al. 2017). Terpenes and terpene derivatives are secondary metabolites commonly present in essential oils and have exhibited excellent antimicrobial effects against susceptible and resistant microorganisms (Thapa et al. 2015). Combination therapy between antimicrobial drugs and natural products may recover the lost function of existing drugs and potentiating drug action (Lewis and Kontoyiannis 2001). The potentiation of antimicrobial drugs was accomplished by various mechanisms in combinatorial therapy, multi-targeted pharmacokinetic potential, allowing simultaneous destruction of existing resistance mechanisms in a specific microorganism.

Essential Oils (EOs) are secondary metabolites naturally produced by aromatic plants. They usually exist in liquid form, are volatile and easily soluble in less dense solvents such as lipids and organic solvents. They are extracted from different parts of plants including seed, flowers, buds, leaves, stem, bark, twigs, wood and roots by several extraction techniques such as solvent extraction (organic solvent, subcritical water, supercritical

[1] Department of Pharmacognosy, Pharmacy College, King Saud University, Riyadh 11451, Saudi Arabia.
[2] Department of Optometry and Vision Science, College of Applied Medical Sciences, King Saud University, Riyadh 11451, Saudi Arabia.
* Corresponding author: mamina@ksu.edu.sa, musarat.org@gmail.com

CO_2), distillation (hydrodistillation, steam, hydrodiffusion), solvent combination (solvent and steam) and solvent free (microwave) technique (Tongnuanchan and Benjakul 2014). Plants generally store their EOs in the specific cellular chamber such as cavities, secretory cells and glandular trichomes. The primary function of these EOs is to protect plants from harmful pests and infections (Helander et al. 1998). In particular, EOs are reported as eminent antimicrobial, insecticidal and antioxidant agents of plants that are significantly involved in inhibition of microbial biofilm production and growth of bacteria, molds and yeast (Liu et al. 2015). A list of reports summarized in Table 12.1, indicates various antimicrobial properties possessed by EOs. This chapter will mainly focus on one particular group of compounds commonly found in EOs, namely the terpenes and terpenoids, their antimicrobial potential and possible mechanism of action.

Antimicrobial terpenes and their derivatives

Terpenes or isoprenoids are the largest and most diverse class of aromatic natural products. They are widely outspread in nature, found in nearly all forms of life and perform various functions ranging from participation in the primary cell structure (steroids and cholesterol in cell membranes) to contribution in functions of the cell (carotenoids in photosynthesis, retinal in vision, quinones in electron transport) (Paduch et al. 2007; Oldfield and Lin 2012). They are also abundantly present in fruits, flowers and vegetables. Foliage and reproductive structures of plants are mainly rich in terpenes throughout and immediately after flowering. Terpenes are responsible for the common aroma, taste, pigments of numerous plants and major components of herbal resins (Paduch et al. 2007). Terpenes are composed of isoperene units (C_5H_8) and classified on the basis of organization and number of isoprene residues present in a molecule. Monoterpenes and sesquiterpenes are commonly found terpenes which differ in number of isoperene units, while as long chains including diterpenes and triterpenes also exist in nature. Terpenes play a vital role in plants such as growth and development, defense against herbivores and environmental stress (Touani et al. 2014). There are about 50,000 terpenes isolated from different species of plants. Terpenes are reported to exhibit a wide range of biological properties such as antimicrobial, anticancer and anti-inflammatory effects (Zaidi et al. 2005; Davies and Davies 2010). Terpenoids are classified into eight different classes, including, hemiterpenoids, monoterpenoids, sesquiterpenoids, diterpenoids, triterpenoids, tetraterpinoids and polyterpenoids based on the number of isoprene units. A recent study reported that 67% of an available terpenoids are monoterpenes and sesquiterpenes (Zacchino et al. 2017). Furthermore, among the discovered potentiators of antibacterial drugs, 75% were terpenes such as mono-, di- and tri-terepenes (Zacchino et al. 2017). Although the mechanism of antibacterial action of terpenes largely remains unknown, Griffin et al. (1999) described in their study that most of the terpenoids have the ability to inhibit two crucial processes which are important for the survival of microorganisms, this includes oxidative phosphorylation and the uptake of oxygen. Oxygen is required by aerobic microbes to produce energy for their growth and development. Earlier studies have shown that low concentration of oxygen caused a restriction in respiration rate of bacteria (Griffin et al. 1999). Oxidative phosphorylation is an essential biochemical process which is responsible for cellular respiration taking place in the cytoplasmic cell membrane (Zengin and Baysal 2014). Additionally, carbonylation of terpenoids believed to enhance the bacteriostatic effect, but not certainly the bactericidal effect. A bacteriostatic agent is an agent that inhibits the growth of microbe, whereas a bactericidal agent is responsible for the microbial demise. Terpenoids have also shown a significant antiseptic effect depending on their solubility in water. The presence of

Table 12.1. Essential oils extracted from different plants against tested microorganism.

Plants species	Essential oil	Tested microorganism	MIC/Sensitivity/ Inhibition Zone	References
Origanum vulgare	Oregano	*Burkholderia cepacia* complex	ES	Maida et al. (2014)
Eugenia caryophyllata	Clove	*B. cepacia* complex	ES	Maida et al. (2014)
Thymus vulgaris	Thyme	*B. cepacia* complex *Streptococcus pyogenes*	ES 1 mg mL^{-1}	Maida et al. (2014) Rasooli et al. (2009)
Cymbopogon citratus	Lemongrass	*Acinetobacter baumannii* *Fusarium oxysporum* f. sp. *lycopersici* *Candida albicans*	0.65% (v/v) 250 ppm 360 μg mL^{-1}	Adukwu et al. (2016) Sharma et al. (2017) De Andrade et al. (2013)
Mentha spicata	Spearmint	*Escherichia coli* *Salmonella thyphi* *Staphylococcus aureus*	21 mm at 150 μL 13 mm at 150 μL 12 mm at 150 μL	Shrigod et al. (2017)
Eucalyptus camadulensis	Eucalyptus	*Fusarium oxysporum* f. sp. *lycopersici* *Streptococcus pyogenes*	15.93% to 72.5% 2 mg mL^{-1}	Sharma et al. (2017) Rasooli et al. (2009)
Laurus nobilis	Bay laurel	*E. coli* *Candida* spp.	1% (v/v) 250 to 500 μg mL^{-1}	Peixoto et al. (2017)
Syzygium aromaticum	Clove	*Fusarium oxysporum* f. sp. *lycopersici* *B. cepacia* complex	125 ppm 0.4% (v/v)	Sharma et al. (2017) Vasireddy et al. (2019)
Melaleuca alternifolia	Tea tree	*Campylobacter* spp.	0.00%	Kurekci et al. (2013)
Backhousia citriodora	Lemon myrtle	*Campylobacter* spp. *S. aureus*	0.01% 2 mg mL^{-1}	Kurekci et al. (2013)
Lavandula angustifolia	Lavender	*Pseudomonas aeruginosa* *Candida albicans*	2 mg mL^{-1} 3 mg mL^{-1}	De Rapper et al. (2016)
Leptospermum petersonii	Manuka	*Campylobacter* spp.	0.01%	Kurekci et al. (2013)
Pelargonium graveolens	Geranium	*B. cepacia* complex *Prototheca zopfii* *Salmonella typhimurium*	0.4% (v/v) 3.5 to 4.0 μg mL^{-1} 3% (v/v)	Vasireddy et al. (2019) Grzesiak et al. (2016) Rafiq et al. (2016)
Origanum majorana	Marjoram	*Micrococcus luteus* *Vibrio alginolyticus*	0.097 mg mL^{-1} 0.39 mg mL^{-1}	Hajlaoui et al. (2016)
Mentha piperita	Peppermint	*Clostridium perfringens* *Fusarium oxysporum* f. sp. *lycopersici*	10 mg mL^{-1} 500 ppm	Swamy et al. (2016) Sharma et al. (2017)
Foeniculum vulgare	Fennel	*Candida* spp.	1.56 to 12.48 mg mL^{-1}	Garzoli et al. (2018)
Chamaemelum nobile	Roman chamomile	*Porphyromonas gingivali*	20.5 ± 0.5 mm	Al-Snafi (2016)
Cedrus atlantica	Cedarwoood	*Bacillus subtilis* *Bacillus cereus*	0.2 μg mL^{-1} 0.4 μg mL^{-1}	Zrira and Ghanmi (2016)
Aniba rosaeodora	Rosewood	*Trichophyton mentagrophytes*	0.002 M	El Omari et al. (2016)
Pinus sylvestris	Pine	*Pseudomonas* spp. *E. coli*	4.33 ± 0.58 mm 0.4 μg mL^{-1}	Kacániová et al. (2017)

hydroxyl groups, lipophilic and hydrobhobic properties of terepenes are the responsible factors for their antibacterial potential (Peixoto et al. 2017). Terpenes have shown affect on lipid membrane activity by interacting with intermembrane lipophilic tails and polar head groups which affected the pathways of lipodial intermembrane and polar transmembrane in skin barrier-associated treatments (Chen et al. 2016). Several studies have shown that various terpenes and their derivatives act as an imperative defense against herbivores and pathogens (Lee and Ho 2008; Togashi et al. 2010; Catteau et al. 2018).

Hemiterpenes and Hemiterpenoids

Hemiterpenes are among the simplest terpenoids. Isoprene is the most prominent hemiterpene isolated from the leaves of different herbs and plant species, including conifers, poplars, oaks and willows (Shakeri et al. 2017). However, oxygen containing derivatives such as isovaleric, tiglic, angelic, senecioic acids and isoamyl alcohol found in plants is known as hemiterpenoids (Touani et al. 2014). These are mainly components of oils and have shown antibacterial properties.

Monoterpenes

Monoterpenes are comprised of two isoprene units and found in various plants. Griffin et al. (1999) reported that monoterpenes possesses antibacterial effects. For instance, thymol, carvacrol, geraniol and menthol have shown antibacterial potential against Gram-positive and Gram-negative bacterial strains. Geraniol has been claimed to increase efficacy against *Enterobacter aerogenes* a Multi-Drug Resistant (MDR) by becoming a potent efflux pump inhibitor (Laxminarayan et al. 2013). Three monoterpenes menthol, linalyl actetate and thymol have been reported to possess antibacterial potential against *S. aureus* and *E. coli*. Other constituents such as trans-cinnamaldehyde, carvacrol and (+)-carvone have shown potent inhibitory effect against *S. typhimurium* and *E. coli*. However halogenated monoterpenes are reported to exert antimalarial, antialgal and cytotoxic potentials, while as a monocyclic monoterpenes have shown potent insecticidal and antifungal effects (Lorenzi et al. 2009). One of the most important monoterpene that has been used as anti-infective therapy is carvone (Paduch et al. 2007). It has been reported that (4R)-(−)-carvone was active against *E. coli*, *E. faecium* and *Campylobacter jejuni*, while as (4S)-(+)-carvone was found effective against *L. monocytogenes* (De Inés et al. 2004). Both the optical isomers have shown antifungal activity towards several pathogenic fungi. Carvone generally inhibits the transformation of cellular shape of *C. albicans* yeast to the filamentous form which is responsible for the pathogenicity of the fungus (Paduch et al. 2007). Another antifungal monoterpene is thymol which has displayed potent activity towards clinical isolates of *Candida glabrata*, *Candida krusei* and *C. albicans*, alone and in the combination with antibiotic fluconazole. The MIC values of thymol were 70, 51.25 and 49.37 $\mu g\ mL^{-1}$ for *C. krusei*, *C. glabrata* and *C. albicans* pathogenic strains, respectively. Thymol has also shown synergistic effect against all the tested species of *Candida* in combination with drug fluconazole (De Carvalho and Da Fonseca 2006). Due to their potent antifungal properties Abbaszadeh et al. (2014) suggested that thymol and carvacrol along with eugenol and menthol can serve as good natural alternatives of synthetic fungicides in food industries. They also reported that these chemical constituents were active in different extents towards many food-decaying fungi such as *Aspergillus fumigatus*, *Aspergillus niger*, *Aspergillus flavus*, *Aspergillus ochraceus*, *Alternaria alternata*, *Botrytis cinerea*, *Cladosporium* spp., *Fusarium oxysporum*, *Penicillium chrysogenum*, *Penicillium citrinum* and *Rhizopus oryzae* (Abbaszadeh et al. 2014).

Althunibat et al. (2016) reported that thymol and carvacrol are the major components found in *Thymus capitatus*. The antibacterial effects of these compounds were tested against *E. coli*, *S. aureus*, *P. aeruginosa* and *Enterobacter aerogenes*. The MIC values for thymol and carvacol obtained were in the range 0.005–0.008 and 0.007–0.008 mg mL^{-1}, respectively (Althunibat et al. 2016). Both the compounds have decreased the microbial count on polypropylene surface during the formation of biofilm in Salmonella spp. (*Salmonella enteritidis*, *Salmonella saintpaul* and *Salmonella typhimurium*) at subinhibitory concentration (1–2 log), but for the established biofilms around 1–5 log concentration were observed for MIC or 2 × MIC (Amaral et al. 2015). Although Chauhan and Kang (2014) suggested that "disruption of membrane integrity" as the main responsible mechanism of thymol towards *S. typhimurium* (Chauhan and Kang 2014), however, Miladi et al. (2016) hypothesized that the two major phenolic monoterpenic, thymol and carvacrol, acts through the suppression of EP in a dose-dependent manner. They observed that these natural constituents enhanced the ethidium bromide (EtBr) accumulation in foodborne microbes via inhibition of EtBr cell efflux (Miladi et al. 2016). Kurita et al. (1979) documented 13 monoterpenes namely; α-terpinene, γ-terpinene, terpinolene, α-pinene, (+)-β-pinene, 1,8,cinaole α-cymene, α-myrecene, (+)-limonene, (±)-linalool, α-terpinoel, and α-phellandrene which showed strong antifungal effects towards 14 fungal pathogens (Kurita et al. 1979).

Phenolic monoterpenes like eugenol, thymol and carvacrol have shown significant activity against bacteria (Zaidi et al. 2005). Carvacrol has been also reported to inhibit the development of biofilm of *S. aureus* and *S. typhimurium* bacterial strains (Dorman and Deans 2000). Previous literature survey claimed that 21 phenolic monoterpenes such as carvacrol, borneol, carvacrol methyl ester, cis/trans citral, geraniol, eugenol, methone, geranyl acetate, R(+) limonene, α-pinene, β-pinene, cis-hex-3-en-1-ol, (+)sabinene, thymol, (-)-thujone, α-terpinene, α-terpineole, terpinen-4-ol, d-3-carene and (2)-linalool have shown antibacterial properties against 25 microbial strains (Knowles et al. 2005). Eugenol and cinnamaldehyde are two popular phenolic terpenoids that are naturally found in essential oils of various plants and demonstrated to be active towards a vast spectrum of microbes (Liu et al. 2008). After an extensive research study conducted on 30 strains of *H. pylori* as one of the major human pathogens responsible for gastric and duodenal ulcers as well asgastric cancer, Ali et al. showed that these two bioactive (eugenol and cinnamaldehyde) constituents could inhibit the growth of *H. pylori* without showing any resistance against these compounds (Ali et al. 2005). Eugenol also has displayed a remarkable bactericidal effect against biofilms of clinical isolates of MRSA and MSSA strains. According to the study by Yadav et al. (2015) eugenol inhibits the formation of biofilm, interrupts the cell to cell communication, destroys the pre-established biofilms and squashs the pathogens in biofilms, equally for both MRSA and MSSA strains. These antibacterial effects of eugenol were attributed to the destruction of microbial cell membrane and cell content leakage. Yadav et al. (2015) further revealed that eugenol decreased the gene expression that are related to the biofilm and the enterotoxin formation at a sub-inhibitory concentration. Rathinam et al. (2017) reported that eugenol showed comparable effects on the development of biofilm and synthesis of virulence factor in *P. aeruginosa* (Rathinam et al. 2017). The electron scanning microscope was used to study cinnamaldehyde mechanism of action against *E. coli* and *S. aureus* that the cinnamaldehyde anchored microbial cell membrane and damaged the cell membrane, decreased the membrane potential and affected the metabolic activity of cells that has finally resulted in inhibition of bacterial growth (Zhang et al. 2015).

A recent study reports that linalool isolated from lavender essential oil exhibited strong antibacterial activity against resistant *K. pneumoniae* and suggested the mechanism of action for linalool which disrupts the bacterial membrane (Yang et al. 2019). Alcoholic monoterpenes are considered, more as bactericidal agents rather than bacteriostatic agents. A study conducted by Bhatti et al. (2014) showed that monoterpenes alcohol such as α-terpineole, terpinen-4-ol, linalool and 1,8,cineole exerted good antifungal potential and suggested that alcohol moieties were responsible for antifungal properties. Additionally, myrcene an acyclic monoterpene alcohol displayed a negative response against fungal strains, indicating that the cyclic structure of monoterpenes may be determinants of this response (Smid et al. 1995).

Togashi et al. (2010) screened the inhibitory potential of several terpene alcohols with a different length of aliphatic carbon chains such as linalool (C6), geraniol (C8), nerolidol (C10), plaunotol (C11), farnesol (C12) and geranylgeraniol and phytol (C16) (the numbering is from the first carbon bonded to the hydroxyl group) on the growth of *S. aureus*. Among all the tested compounds, only farnesol and nerolidol showed strong antibacterial activity with MBC as 20 and 40 μg/mL, respectively. They also investigated the interaction of these terpene alcohols with the cell membrane of bacteria by evaluating the leakage of intracellular K^+ ion. They proposed that K^+ leakage from the cells reflects the antibacterial potency of the membrane disturbing compounds. The initial rate of leakage was considered as the damage to the cell membranes while the total amount of leaked K^+ was evaluated as the antibacterial activity. Again, farnesol and nerolidol were the most effective compounds. Finally, they concluded that the length of the hydrocarbon chain connected to the hydroxyl group, plays an important role in antibacterial and cell membrane disrupting activity and it should be between C10 and C12 for appropriate effect against *S. aureus* (Togashi et al. 2010).

Ketonic monoterpenes and monoterpenes containing aldehydes were found to possess potent antimicrobial activities due to their high electronegativity created by carbon double bond arrangements in their structure (Smid et al. 1995). Based on this observation, reported that nine monoterpenes including α-pinene, α-terpinene, α-terpineol, γ-terpinene, δ-cymene, 1, 8-cineole, thymol, terpinen-4-ol, and citral showed antimicrobial potential against Herpes Simplex Virus type 1 (HSV-1). Dunkic et al. (2010) listed a few more monoterpenes (1, 8-cineole, borneol, thujone, bornyl acetate, isoborneol and camphor) displaying a considerable effect against HSV-1. Carvacrol and thymol were also reported to exhibit a powerful antimicrobial effect against the *Tobacco mosaic* and *Cucumber mosaic* virus.

Seven essential oils were tested for biological activity and their eight major constituents against bacteria (Gram-negative and Gram-positive), filamentous fungi and protozoans. The antimicrobial properties were examined by determination of the Minimal Inhibitory Concentration (MIC) for *Escherichia coli*, *Staphylococcus aureus*, *Bacillus cereus*, *Listeria monocytogenes*, *Shigella sonnei*, *Salmonella Typhimurium*, *Fusarium oxysporum*, *Alternaria alternata*, *Aspergillus ochraceus* and *Aspergillus niger*, the half-maximal inhibitory concentration (IC_{50}) for *Leishmania mexicana* and *Trypanosoma cruzi*, and the median lethal dose (LD_{50}) for *Giardia lamblia*. The results revealed that oregano essential oil exhibited the strong antibacterial effect (66–100 μg mL^{-1}), while cinnamon essential oil showed promising fungicidal effect (66–116 μg mL^{-1}), and both had excellent antiprotozoal potential (22–108 μg mL^{-1}). Two main monoterpenoids, carvacrol and thymol were the best antimicrobials (23–200 μg mL^{-1}), and cinnamaldehyde was a good antifungal component (41–75 μg mL^{-1}) (Andrade-Ochoa et al. 2021).

Sesquiterpenes

Sesquiterpenes are terpenes with a $C_{15}H_{24}$ chemical formula and are much larger and more stable compounds in comparison to monoterpenes. Sesquiterpenes have been recognized for long to possess antimicrobial properties (Lin et al. 2000). Torres-Romero et al. (2011) isolated and identified a dihydro-β agarofuran sesquiterpenes named as 1α-acetoxy-6β, 9β-dibenzoyloxy-dihydro-β-agarofuran, which has been reported to inhibit the growth of *Bacillus* spp. (Torres-Romero et al. 2011). A natural acyclic isoprenoid sesquiterpene alcohol known as farnesol isolated from *Stellera chamaejasme*, showed moderate antibacterial effects against *Streptococcus mutans* and *Streptococcus sobrinus* biofilm production (Koo et al. 2002). It also displayed antibacterial potential against *S. epidermidis* and *S. aureus*, whereby it suppressed the biofilm formation (Gomes et al. 2009). Masako (2005) conducted two consecutive studies and reported that the combination of farnesol with xylitol showed positive effects towards the atopic dermatitis caused by *S. aureus* without altering the flora of bacteria and successfully suppressed the biofilm development of *S. aureus* (Masako et al. 2005). The sesquiterpenes were also used in the combinatorial therapy using existing antibiotics. Castelo-Branco et al. (2016) reported the potentiation effects of farnesol in combination with amoxicillin, ceftazidime, sulfamethoxazole-trimethoprim and doxycycline in combination therapy against *B. pseudomallei* (Castelo-Branco et al. 2016). A phenolic sesquiterpene, xanthorrhizol isolated from the rhizomes of *Curcuma xanthorrhiza* has been reported to reduce 60% of cell adherence ability of *Staphylococcus mutans* and inhibited the growth of *Mycobacterium smegmatis* (Rukayadi and Hwang 2006; Jin et al. 2010). It was discovered that sesquiterpenes showed potent antibiotic enhancement against MRSA and Gram-negative bacteria (Langeveld et al. 2014). Gonçalves et al. (2011) showed that the zone of inhibition was increased after the incorporation of sesquiterpenes into the antibiotic disk. The study was performed against MDR microbial strains of *S. aureus* with a combination of available sets of antibiotics such as penicillin, tetracycline, vancomycin and erythromycin (Gonçalves et al. 2011). A recent study has shown that farnesol endogenously produced by *C. albicans* and plant terpenoids, cuminaldehyde and carvacrol were tested individually or in combination on amphotericin-B-resistant *C. albicans* in either single or mixed infections. The results showed that combination of farnesol with carvacrol exhibited synergistic inhibitory effects not only on *C. albicans* yeast and hyphae, but also on biofilms formed from single and mixed-species and reduced doses (Touil et al. 2020).

Diterpenes

Diterpenes are naturally found chemical constituents containing the $C_{20}H_{32}$ molecular formula and an important class of terpene with a broad spectrum of interesting biological properties (Yang et al. 2012). About 60 previously reported terpenoids were examined for their Minimum Inhibitory Concentration (MIC) against *E. coli, S. aureus, C. albicans* and *P. aeruginosa*. The terpenoids were divided into five groups to estimate their activity patterns. Hydrogen bond was found to be the determinant factor for the positive antimicrobial potential, while low water solubility was found to be responsible for antimicrobial inactivity (Griffin et al. 1999). The probable mechanism of action of diterpenes for antimicrobial activity was that they restricted the microbial oxygen uptake and oxidative phosphorylation. Another study reported that ent-kaurane and ent-pimarane diterpene derivatives have ability to inhibit the growth of pathogens of dental caries. The antibacterial potential was confirmed by MIC values (2–10 mg mL^{-1}) (Ambrosio et al. 2006). Róz˙alski et al. (2007) extracted a diterpenoid salvipisone from the hairy roots of *Salvia sclarea* L. which prevented the cell adherence and development of biofilm of *S. epidermidis* and *S. aureus* (Rózalski et al. 2007).

Diterpenes also serve as a potential antibiotic enhancer against MRSA and are widely used in combination therapy with several antibiotics. For instance, a clerodane diterpenoid 16-α-hydroxycleroda-3, 13 (14)-Z-dien-15, 16-olide (CD) isolated from *Polyathia longifolia* leaves reported to enhanced the efficacy of tetracycline, linezolid, oxacillin and daptomycin against clinical MRSA. The MICs of all the antibiotic was dropped significantly between 4–16, 2–4, 10–80 and 2–8-folds respectively when mixed with CD (Gupta et al. 2016). Gupta et al. (2016) further suggested the *in vivo* mechanism of action of CD by reversing the resistance of clinical MRSA isolates. Similar clinical isolates of MRSA were investigated against CD combined with ciprofloxacin, ofloxacin and norfloxacin. qRT-PCR analysis revealed that cells treated with CD alone and in combination with antibiotics showed significant modulation in expression of genes for efflux pumps. The results of time-kill assay clearly showed that the MIC of CD in combination with norfloxacin was half as compared to MIC of individual CD and norfloxacin, undoubtedly reducing the bacterial cell viability (Gupta et al. 2016). Despite the promising potential of CD, limited sourcing of CD has become the bottleneck for further investigation (Zacchino et al. 2017). Two anti-biofilm diterpenoids, salvipisone and aethiopinone extracted from *Salvia sclarea* roots expressed antibacterial and anti-biofilm effects against *E. faecalis, S. epidermidis* and *S. aureus* (Walencka et al. 2007). Both the diterpenes were screened for synergistic effect when combined with antibiotics such as oxacillin, linezolid and vancomycin, against clinical isolates of MRSA and Methicillin Resistant *S. epidermidis* (MRSE). It was noticed that they displayed either bactericidal or bacteriostatic effects against planktonic cultures of investigated MRSA and MRSE. MIC was attained with 50% decrease in the dose of antimicrobial drug when combined with diterpenoids (Feio et al. 1999). Another diterpenoid compound with promising antibacterial effect against *S. aureus* is dehydroabietic acid, which is an abietane diterpenoid containing abieta-8,11,13-triene substituted with a carboxyl group at position 18. It plays a role as a metabolite as well as an allergen (Tapia et al. 1997). Various derivatives of dehydroabietic acid also have shown antibacterial activities (Savluchinske-Feio et al. 2006).

Cannabinoids

Cannabinoids are a C21 terpene phenolic group of compounds, including their analogs and transformation products, predominantly isolated from cannabis plants (Elsohly and Slade 2005). They are detected in different parts of these plants; however, they are largely found in the secretory cavity of the glandular trichomes of the female flowers (Andre et al. 2016). Cannabinoids are biosynthesized in their carboxylated form (acid forms) in fresh cannabis plant while during storage and smoking, they are non-enzymatically decarboxylated into their neutral forms. Cannabinoids can be categorized into 11 major structural types, including, Δ8-tetrahydrocannabinol, Δ9-tetrahydrocannabinol, cannabigerol, cannabidiol, cannabichromene, cannabinodiol, cannabielsoin, cannabinol, cannabicyclol, cannabitriol and miscellaneous types The first study addressing the quantification of antimicrobial effect of the pure cannabinoids (Δ9-tetrahydrocannabinol and cannabidiol) isolated from *C. sativa* was conducted by VanKlingeren and Ten Ham in 1976. They entrenched a bactericidal and bacteriostatic effect of Δ9-tetrahydrocannabinol and cannabidiol against *S. aureus* and *Streptococcus* spp. in the range of 1–5 μg mL^{-1}, but found no effects towards Gram-negative bacteria. They also showed a decrease in antibiotic activity in the presence of horse serum (Van Klingeren et al. 1976).

Later, Turner and Elsohly (1981) investigated the biological properties of cannabichromene and its related compounds. The results of their study showed that these compounds exhibited significant anti-inflammatory and antimicrobial effects. The cannabichromene and its homologs and isomers displayed strong antibacterial effects, whereas mild to moderate antifungal effects. Besides, this research group screened the antimicrobial potential of different cannabinoids towards fungi, Gram-positive and Gram-negative bacteria. Among all these compounds, cannabichromene and cannabigerol, their homologs and isomers were found most active components against fungi and bacteria assayed (Eisohly et al. 1982). A study reports the isolation of nine new cannabinoids and three of them namely 8-hydroxycannabinol, 4-acetoxy-2-geranyl-5-hydroxy-3-n-pentylphenol, and 5-acetyl-4-hydroxycannabigerol exerted significant antifungal (*C. albicans*), antibacterial (MRSA and *S. aureus*) and antileishmanial (*Leishmania donovani*) potential, respectively (Radwan et al. 2009). Another study has shown that pure cannabidiol, cannabichromene, cannabinol, cannabigerol and cannabigerolic acid were more effective in reducing the colony count of bacterial strains associated with dental plaque in comparison to commercial oral care products (Stahl and Vasudevan 2020).

Keeping in mind the resistant bacteria, five major cannabinoids including cannabidiol, cannabichromene, cannabigerol, Δ9-tetrahydrocannabinol and cannabinodiol and related compounds were tested against various multi-resistant *S. aureus* strains by Appendino and co-workers. All cannabinoids exhibited significant antibacterial activity, with MIC values in the range of 0.5–2 μg mL^{-1}. Moreover, a relation between chemical structure of cannabinoids and antibiotic potential, indicating a specific interaction to a bacterial target (Appendino et al. 2008). These results indicated that cannabinoids have significant antibacterial effect towards MRSA, inhibition of MRSA biofilm formation, elimination of preformed biofilms and stationary phase cells persistent to antibiotics (Farha et al. 2020). Furthermore, the endocannabinoid anandamide and the endocannabinoid-like, arachidonoyl serine were capable of inhibition of biofilm formation, the decreased metabolic effect of preformed biofilms, adjusting biofilm-associated virulence factors including cell aggregation, hydrophobicity and spreading MRSA abilities (Feldman et al. 2020). Recent studies illustrated that antibacterial property of cannabidiol, cannabidiolic acid and cannabichromenic acid towards Gram-positive bacteria such as MRSA. Cannabinoids has improved the antibacterial properties of conventional antibiotics towards resistant bacteria and vice versa. Some antibiotics have extended and enhanced the antimicrobial properties of cannabinoids (Kosgodage et al. 2019; Wassmann et al. 2020). Cannabidiol has significantly increased the antibacterial potential of rifampicin and erythromycin against *E. coli* VCS257 and enhanced the antibacterial potential of kanamycin towards *S. aureus* subsp. *aureus* Rosenbach (Kosgodage et al. 2019). Cannabidiol strengthens the effect of bacitracin towards MRSA, *Staphylococcus epidermidis* (MRSE), *L. monocytogenes*, and *Enterococcus faecalis* and the combination decreased the MIC value of bacitracin by 64-fold. Cannabidiol and bacitracin showed bactericidal and synergistic effects (Wassmann et al. 2020). The antimicrobial property of five major cannabinoids towards *E. coli* has increased in the presence of polymyxin B at sublethal concentration. Polymyxin B in combination with cannabigerol enhanced the antimicrobial potential towards multidrug-resistant clinical isolates of Gram-negative microorganisms including *E. coli, P. aeruginosa, Acinetobacter baumannii,* and *K. pneumoniae* (Farha et al. 2020). Interactions of Poly-L-lysine (natural antimicrobial agent), endocannabinoids and endocannabinoids-like components were also reported. Endocannabinoid anandamide and oleoyl ethanol amide with poly-L-lysine showed synergistic antimicrobial effect against *Streptococcus mutans* by effecting bacterial growth. Interaction of endocannabinoid anandamide and poly-L-lysine also exhibited antibiofilm effect (Feldman et al. 2020). Recently, cannabinodiol in combination with different types of antibiotics showed an indifferent effect against MRSA (Martinenghi et al. 2020).

Triterpenes

Triterpenes are made up of three or six isoprene units and contain the $C_{30}H_{48}$ chemical formula which includes sterols and steroids with saqualeneas the biological precursor for these tritepenes. They are produced naturally by plants, animals and microorganisms. Pandanaceae containing triterpenes such as 24, 24 dimethyl-5β-tirucall-9 and 25-dien-3-one has shown promising antimicrobial potential against tubercular strains (Saleem et al. 2010). Oleanic acid a common component of essential oil reported to possess potent antimicrobial effect against various pathogens such as *Mycobacterium tuberculosis*. Oleanic acid had also shown promising synergistic effect in combination with rifampicin, ethambutol and isoniazide against MDR with significant reduction in MIC by 128–16, 128–16 and 32–4 fold (Barbieri et al. 2017). Two new triterpenoids, bonianic acid A and B isolated from Radermacheraboniana exhibited antimicrobial activity against *M. tuberculosis*. Around six known triterpenoids including AO, ursolic acid and ergosterol peroxide found active against *M. tuberculosis*. In fact, a combination of ursolic acid and ergosterol peroxide expressed synergistic effect against *M. tuberculosis* (Jimenez-Arellanes et al. 2013). Another study conducted by Cunha et al. showed that oleanic acid and ursolic acid extracted from *Miconia ligustroides*, displayed potent antibacterial properties against five selected (*B. cereus, S. choleraesuis, S. pneumoniae, V. cholerae*, and *K. pneumoniae*) bacterial strains (Cunha et al. 2010). The MIC values of ursolic acid and oleanic acid were 20 μg mL^{-1} and 80 μg mL^{-1} when treated individually against *B. cereus*. However, MIC values were reduced when treated with ursolic acid and oleanic acid in combination. Zhou et al. (2013) reported that ursolic acid and oleanic acid were also active against planktonic cariogenic microbes and their biofilms (Zhou et al. 2013). The research was expanded by Liu et al. (2015) and depicted the combinatorial effects of ursolic acid and xylitol against the development of biofilm in *S. mutans* and *S. sobrinus* (Liu et al. 2015). Furthermore, ursolic acid and oleanic acid are reported to alleviate antimicrobial properties against *Listeria monocytogenes* without affecting the secretion of toxin; thereby influencing the virulence factors of *L. monocytogenes* and decreasing the biofilm production abilities from these microbes (Kurekci et al. 2013).

Oleanic acid also showed strong interactions with gentamicin and kanamycin aminoglycoside and exhibited strong antimicrobial activity against *A. baumanii*, but did not show good results with other antibiotics such as chloramphenicol, ampicillin, tetracycline, rifampicin and norfloxacin. The bactericidal activity of gentamicin was significantly increased in combination oleanic acid calculated by the time-kill assay (Zacchino et al. 2017). Three known triterpenoids amyrin, betulinaldehyde and betulinic acid isolated from the bark of *Callicarpa farinose* Roxb showed potent antibacterial potential against two clinical isolates methicillin-susceptible (MSSA) and methicillin-resistant (MRSA) with MIC values ranging from 2 to 512 μg mL^{-1} (Chung et al. 2014). However, there is no specific mode of action reported in literature for terpenoids until the mechanisms of action shown by other phytochemicals have been suggested for them. The phytochemicals act either by disruption of the cell membrane of bacteria, modulation of efflux pump of bacteria, suppression of biofilm development in bacteria and inhibition of some virulence factors which include enzyme and toxin (Barbieri et al. 2017). Carvacrol is among one of the responsible components for sub-lethal injury to microbial cells due to amendments in the composition of fatty acids. Other studies have reported in literature that carvacrol and thymol caused dissolution of outer cell membrane and disruption of cytoplasmic cell membrane of Gram-negative bacteria (Helander et al. 1998). Antimicrobial activity mechanism of two natural pentacyclic triterpenes, ursolic acid and α-amyrin was studied by Broniatowski et al. (2015) and demonstrated the broad-spectrum of antimicrobial potential. The interaction of these two compounds with the inner layer of the cell membrane of *E. coli* was studied by the Langmuir monolayer technique. Both the triterpenes showed disorganizing effects on the membrane of an applied model of *E. coli* (Broniatowski et al. 2015). The ethanol extract of leaves of *Syzygium szemaoense* resulted in the isolation of new six terpenoids (syzygiumursanolides A–F) and two phloroglucinol glucosides. Among the isolates, two compounds namely syzygiumursanolides A and B were unusual 28-norursane type triterpenoids with 19(18→17)-abeo spirocyclic skeleton. The antimicrobial bioassay showed that syzygiumursanolide D exhibited the most significant antifungal effect (MIC values, 6.25–25 μgmL^{-1}) towards a panel of fungi (Xu et al. 2020). Three terpenoid chloro-dihydroquinones were isolated from the actinomycete *Streptomyces* sp. All the extracted compounds showed significant antibiotic activity against methicillin-resistant *Staphylococcus aureus* (MRSA) and Vancomycin-Resistant Enterococcus Faecalis (VREF) with MIC value in the range of 1.90–15.6 $\mu g\ mL^{-1}$ (Socha et al. 2006).

Four bisanthraquinones extracted from *Streptomyces* sp. A43472 were first reported in a Japanese patent in 1996 as antitumor agents. These constituents were assigned as BE-43472A, BE-43472B, BE-43472C and BE-43472D in this patent (Socha et al. 2006). Several years later, bisanthraquinone metabolites isolated from streptomycete strain (N1-78-1) were reported to possess antibacterial activity (Cheng et al. 2013; Kumar et al. 2013). These constituents showed potent antibacterial effect against MRSA, MSSA and VRE. BE-43472B was the strongest compound towards MRSA and VRE (MIC range, 0.11–0.45 µM). BE-43472C and D exhibited lower potency in contrast to other two metabolites. Moreover, they showed no bactericidal effect. These results illustrated the deleterious effect of a C-1′ methoxy substituent. However, VRE isolates were clearly less sensitive to the C-1′ methoxy analogues (MICs range 30–233-fold higher) (Kumar et al. 2013).

Napyradiomycins A–F and B2–B4 terpenoids isolated from two marine-derived *Streptomyces* strains. Among all the isolates, napyradiomycin A, B and B2–B4 were active against MRSA, while as active analogues napyradiomycin 1 and B3 showed strong activity against MRSA with MIC values 16 $\mu g\ mL^{-1}$ and 2 $\mu g\ mL^{-1}$, respectively (Al-Footy et al. 2015).

Sarcophyton trocheliophorum, a soft coral collected from the Red Sea resulted in isolation of pyrane-based cembranoid type structure, sarcotrocheliol and sarcotrocheliol acetate, along with sarcophine, cembranoid cembrene C and the aromadendrene sesquiterpenoid palustrol. All the isolated derivative compounds except cembranoid cembrene C demonstrated antibacterial activity against various bacterial strains, including MRSA. Sarcotrocheliol and sarcotrocheliol acetate displayed the potent antibacterial activity with MIC between 4.34 µM and 3.06 µM (Buckle 2015).

Bioassay-guided isolation *Penicillus capitatus* (Halimedaceae) from the tropical green alga led to the identification of capisterones A and B, two cycloartanone triterpene sulfates. These two compounds have increased the fluconazole activity in two efflux pump-overexpressing strains. Two *S. cerevisiae* strains overexpressing the *C. albicans* CDR1 (DSY 415) and MDR1 (DSY 415) efflux pumps were used in a study of their fluconazole reversal activity, along with the null pump strain (DSY 390). Capisterone A was inactive towards the three strains, whilst capisterone B showed some intrinsic antifungal activity. Both the compounds significantly alleviated fluconazole effect in the CDR1 and MDR1 efflux pump-overexpressing strains (FICI < 0.5, synergistic). These secondary metabolites did not show any inherent antifungal effect when tested towards several opportunistic pathogens and were not cytotoxic. Conclusively, capisterones A and B appear to be promising for combination therapy of fungal infections caused by azole-resistant strains (Pettit et al. 1997).

In conclusion, BE-43472-B was the most active terpenoid, with MICs in the nanomolar range. Furthermore, capisterones A and B, received large attention due to their synergistic effect with conventional antifungals towards drug-resistant strains owing to their inhibitory effect on efflux pumps. These terpenoids were extracted from various marine sources such as algae, sponge, coral, and mostly from bacteria of *Streptomyces* sp.

Tetraterpenes

Tetraterpenes commonly known as carotenoids with $C_{40}H_{56}$ molecular formula and can be in the category of terpenes as they are made up of isoprene units. They are highly unsaturated carotenoids and thus are very difficult to isolate and purify. They are found in plants, bacteria and fungi and are the responsible for red, orange or yellow, fat soluble pigments of a plant and animal. One of the most common and crucial tetraterpene is β-carotene, which is responsible for the yellow color in carrots. It is an important component for mammals as it is a precursor for the production of vitamin A and other important terpenoids.

Terpenes of higher order showed increased thermotolerance. The thylakoid membrane permeability increases at higher temperatures due to increased cyclic photophosphorylation in photosystem II (Singsaas 2000). The continuous rise in temperature, the photophosphorylation system loses its ability to stop protons leaking, that resulted in a decrease of transmembrane gradient and ATP synthesis. All these events can potentially cause lowering in the Rubisco activation state due to an inhibition of RuBP regeneration (Friedman et al. 2006). They possess psychoactive effects and are used against various infectious diseases (Lewis and Haba 2019). The antimicrobial properties of some terpenoids are listed in Table 12.2. Chemical structures of some important terepenes and terpenoids (Fig. 12.1).

Possible mode of action terpene/terpenoids on antibiotic resistant microorganisms

There are various targets mentioned for terpenes/terpenoids, including, destabilization of microbial membranes, damage of membrane proteins, proton motive force depletion and release of cell contents (Ultee et al. 1999; Ultee et al. 2001; Trombetta et al. 2005; Turgis et al. 2009; Astani et al. 2010). Most times essential oil confers antimicrobial effect by damaging the cell wall and membranes, that lead to lysis of cell and leakage of cell components (Devi et al. 2010). Many reports suggested that the antimicrobial mode of action of essential oils and corresponding compounds depends on their chemical composition and on the quantity of single compounds. The bioactivity of the molecules is also affected by the presence and position of functional groups (Bajpai et al. 2013). Acyclic components showed higher antimicrobial potential as compared to cyclic compounds. With regard to the acyclic compounds, the difference between antimicrobial properties of citronellol and citronellal is the replacement of an alcohol by an aldehyde group. This replacement and consequently the absence of a double bond in citronellol were the major reasons for their higher antimicrobial effect against bacteria strains. However, presence of carbonyl group in carvone instead of a hydroxyl group at position 3 of carveol appears to be responsible for the lack of antimicrobial activity of carvone towards *S. aureus*. However, this difference did not affect the effect against *E. coli*.

Table 12.2. Antimicrobial potential of some extracted terpenoids from different plant against tested microbial strains.

Terpenoids class	Phytochemicals	Pathogen tested	Antimicrobial activity	References
Hemiterpenes	Isoprene Isovaleric Tiglic Angelic Senecioic acids Isoamyl alcohol	*E. coli*	Growth inhibition	Touani et al. (2014)
Monoterpenes	Carvacrol Trans-cinnamaldehyde (+)-Carvone	*E. coli* *S. typhimurium*	Growth inhibition	Helander and Alakomi (1998)
	Carvacrol Thymol Geraniol	Resistant *Enterobacter aerogenes*	Efflux pump inhibition	Lorenzi et al. (2009)
	Linalyl acetate (+)-Menthol Thymol	*S. aureus* *E. coli*	Growth inhibition	Trombetta et al. (2005)
	(+)-Terpinen-4-ol γ-Terpinene α-Terpinene Terpinolene α-Pinene 1,8-Cineole γ-Cymene (+)-Limonene β-Myrcene (+)-β-Pinene (±)-Linalool α-Phellandrene α-Terpinoel	*Trichophyton mentagrophytes* *Trichophyton violaceum* *Microsporium gypseum* *Histoplasma capsulatum* *Blastomyces dermatitidis*, etc.	Growth inhibition	Kurita et al. (1979)
	Eugenol Carvacrol Thymol	*Acinetobacter calcoaceticus* *Aeromonas hydrophila* *Bacillus subtilis*	Growth inhibition	Dorman and Deans (2000)
	Borneol Carvacrol Carvacrol methyl d-3-Carene ester cis/trans Citral Eugenol Geraniol cis-hex-3-en-1-ol R(+)Limonene Geranyl acetate (2)-Linalool Menthone Nerol α-Pinene β-Pinene (-)-Thujone Thymol (+)sabinene α-Terpinene Terpinen-4-ol α-terpineole	*E. coli* *S. aureus* *Salmonella typhia* *S. typhimurium* *Shigella* sp. *Salmonella enteritidis* *Aeromonas hydrophila* *Yersinia* sp. *Vibrio anguillarum* *Vibrio parahaemolyticus* *C. albicans* *Penicillium expansum* *Aspergillus niger*....., etc.	Growth inhibition	Liu et al. (2008)

Table 12.2 contd. ...

...Table 12.2 contd.

Terpenoids class	**Phytochemicals**	**Pathogen tested**	**Antimicrobial activity**	**References**
	Linalool	Resistant *Klebsiella pneumoniae carbapenemase* (KPC)	Cell membrane disruption	Yang et al. (2019)
	Carvacrol	*S. aureus* *S. typhimurium*	Biofilm inhibition	Knowles et al. (2005)
	Terpinen-4-ol α-Terpineol 1, 8-Cineole Linalool	*Aspergillus niger* *Botrytis cinerea*	Growth inhibition	Bhatti et al. (1995)
	α-Terpinene Terpinen-4-ol α-Terpineol γ-Terpinene Thymol α-Pinene γ-Cymene Citral 1, 8-Cineole Camphor Borneol Bornyl acetate Isoborneol Thujone	*Herpes simplex* virus type 1 (HSV-1)	Growth inhibition	Dunkic et al. (2010)
	Carvacrol Thymol	*Tobacco mosaic* virus (TMV) *Cucumber mosaic* virus (CMV)	Growth inhibition	Astani et al. (2010)
	1α-Acetoxy-6β, 9β-dibenzoyloxydihydro-β-agarofuran	*Bacillus* spp.	Growth inhibition	Torres-Romero et al. (2011)
Sesquiterpenes	Farnesol	*Streptococcus mutans* *Streptococcus sobrinus*	Biofilm formation inhibition	Koo et al. (2002)
		S. aureus *Staphylococcus epidermidis*		Gomes et al. (2009) Masako et al. (2005)
		Burkholderia pseudomallei	Potentiation effect-combination therapy	Castelo-Branco et al. (2016)
	Xanthorrhizol	*Staphylococcus mutans*	Reduction of cell adherence ability	Rukayadi and Hwang (2006)
		Mycobacterium smegmatis	Growth inhibition	Jin et al. (2010)
	Ent-kaurane Ent-pimarane	Dental carries pathogens	Growth inhibition	Souza et al. (2011)
	Thymol (-)-Carvone Carvacrol Dihydrocarveol (-)-Perilla alcohol (-)-Carveol, etc.	*Pseudomonas aeruginosa* *E. coli* *S. aureus* *C. albicans*	Growth inhibition	Griffin et al. (1999)

Table 12.2 contd. ...

...Table 12.2 contd.

Terpenoids class	Phytochemicals	Pathogen tested	Antimicrobial activity	References
	Salvipisone	*Staphylococcus epidermidis* *S. aureus*	Bacterial cell adherence prevention Biofilm development inhibition	Rózalski et al. (2007)
Diterpenes	16α-Hydroxycleroda-3, 13 (14)-Z-dien-15, 16-olide (CD)	MRSA	Antibiotic potentiation E_ux pump modulation	Gupta et al. (2016)
	Salvipisone Aethiopinone	*S. aureus* *S. epidermidis* *Enterococcus faecalis*	Biofilm production inhibition	Rózalski et al. (2007)
		MRSA MRSE	Synergistic activity alongside antibiotic	Walencka et al. (2007)
	24, 24-Dimethyl-5β-tirucall-9	Tubercular strains	Growth inhibition	Saleem et al. (2010)
	Bonianic acid A Bonianic acid B 25-Dien-3-one Oleanic acid	*Mycobacterium tuberculosis*	Synergistic activity alongside antibiotic	Barbieri et al. (2017)
	Oleanic acid Ergosterol peroxide Ursolic acid		Synergistic Effect-combination therapy	
Triterpenes	Oleanic acid Ursolic acid	*Bacillus cereus* *Vibrio cholerae* *Salmonella choleraesuis* *K. pneumoniae* *S. pneumoniae*	Growth inhibition	Cunha et al. (2010)
		Planktonic cariogenic microorganism *Streptococcus mutans* *Streptococcus sobrinus* *Listeria monocytogenes*	Biofilm inhibition	Zhou et al. (2013) Liu et al. (2015) Kurekci et al. (2013)
	Oleanic acid	*Acinetobacter baumanii*	Antibiotic potentiation	Zacchino et al. (2017)
	Amyrin Betulinaldehyde Betulinic acid	MRSA MMSA	Growth inhibition	Chung et al. (2014)

In general, essential oils have the ability to change physicochemical surface properties of microbial strains such as hydrophobicity and surface tension of the bacterial membrane. It is well known that EOs are hydrophobic nature, which permit them to penetrate bacterial cells and causes disruption of the structure of cell membranes and impairment of functions of cells. This leads to an increase in the cell permeability due to the inability of bacterial cell membranes to separate the EOs from its surface (Gallucc et al. 2009). Thus, it is possible to hypothesize that the hydrophobicity changes of microbial membranes, after treatment with

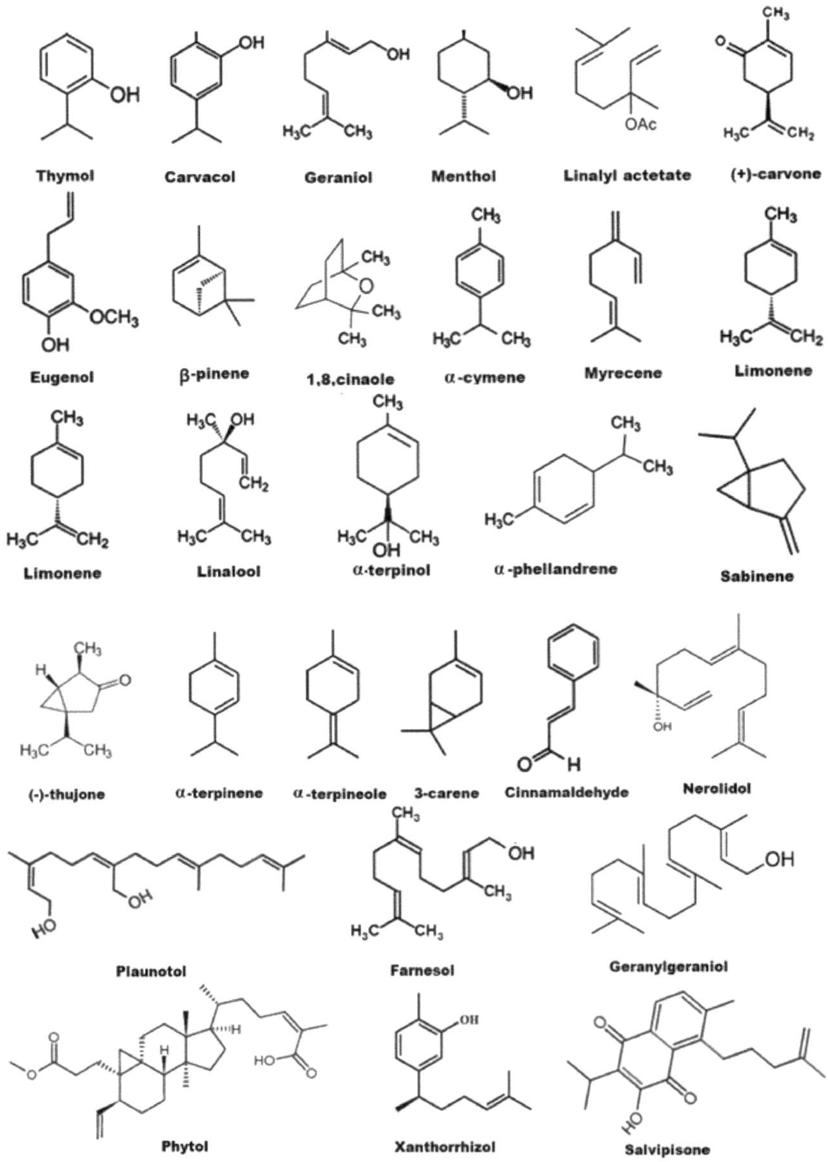

Figure 12.1. Chemical structure of important antimicrobial terpenes.

EOs components, may result in destabilization of the phospholipid bilayer of cytoplasmic membrane of Gram-positive bacteria. In addition to the interaction with the cytoplasmic membrane, these components may have also affected the hydrophobic character of LPS from the OM. As cell membranes provide a fence that is essential for various cellular processes undergoing within the cells, their damage warrant adverse effects that can result in cell inactivation and/or demise.

Another important characteristic that plays a crucial role in the pathogenic balance and resistance to antibiotics is the charge on the surface of cell (Burt 2004). Under normal physiological conditions microbial cells have a negative charge on the surface due to the presence of anionic groups (e.g., carboxyl and phosphate) on the cell membrane (Gilbert et al. 1991; Nazzaro et al. 2013; Borges et al. 2013). However, the extent of the charge differs with the species and are affected by various conditions such as ionic strength, pH, and age of the culture (Lerebour et al. 2004; Palmer et al. 2007). The majority of reported studies suggested that EOs and their constituents are more effective towards Gram-positive bacteria, because Gram negative bacteria cease the penetration of hydrophobic molecules (Gallucc et al. 2009). Moreover, the shape of the cell also contributes in the distinct susceptibility of microbial strains. The rod shaped bacterial cells are more sensitive to EOs than coccoid cells (Ahimou et al. 2002).

Furthermore, the cytoplasmic membrane plays a vital role in the maintenance of cellular homeostasis, as it directly controls the input and output of intracellular content. The internal matrix of cells is rich in K^+ and their presence in the extracellular medium is a clear indication of consequential and irreversible damage of cytoplasmic membrane (Kim et al. 1995; Hajlaoui et al. 2009). Some EOs are known to have membrane active properties against several microbial strains, leading to leakage of cell components such as K^+ (Carson et al. 2002; Newman and Cragg 2012). Cox and collaborators reported the antibacterial mode of action of tea tree oil towards *E. coli* and *S. aureus* strains and claimed that with the increase of PI uptake and K^+ leakage increases (Cox et al. 2000). In this study release of K^+ was verified for carveol, citronellol and citronellal treated *S. aureus*. However, no release of K^+ was observed for *E. coli* with all these components. These results of this study indicated that the selected components of EOs were able to cause alterations only in the outer envelope of the cell structure without promoting release content of the cell. Moreover, these changes on the cell surface were adequate to induce the death of the cell. Although the constituents of EOs did not cause any K^+ leakage, but they promoted the uptake of PI. Thus, it can be hypothesized that the absence of K^+ in the supernatant of *E. coli* was due to retention by the thin layer of the cell wall. In fact, with increased PI permeability, it seems unexpected that the cytoplasmic membrane stands K^+ impermeable.

Thus, the antimicrobial effect of an agent is credited mainly to two mechanisms, which include chemical interference in the synthesis or function of the important constituents of microbe, and/or avoiding the conventional mechanisms of antimicrobial resistance. Figure 12.2 depicts these mechanisms and as it can be noticed, there are multiple targets available for action of antimicrobial agents that are comprised of (I) biosynthesis of bacterial protein; (II) biosynthesis of bacterial cell-wall; (III) destruction of bacterial cell membrane; (IV) replication of bacterial DNA and repair, and (V) metabolic pathway inhibition.

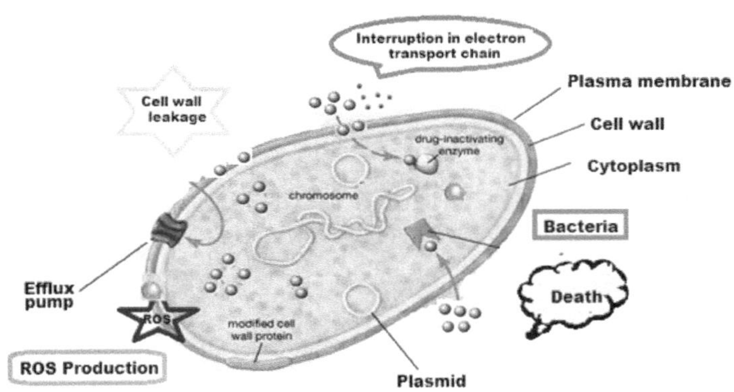

Figure 12.2. Cell-wall biosynthesis is inhibited by the various classes of antibiotics and multiple antibiotic resistance mechanisms in bacteria.

Challenges and future perspectives

The shift of *in vitro* experiments to *in vivo* studies and finally to human clinical trials is the major challenge for the development of new bioactive constituents. In the development of natural antimicrobial agent, the problem is even more serious since various factors can influence the potency of the natural product such as cell penetration, maximum concentration of plasma can be attained and the bioavailability. For instance, natural essential oil and their constituents are readily anchored to the surface of the cell membrane that dramatically affects their cell penetration and maximum plasma concentration. The spread of antibiotic-resistant microbe are a huge threat to successful therapy of microbial diseases. Thus, there is an urgent need to develop new approaches to fight antibiotic resistance. Bioactive terpenes are characterized by vast chemical structures with promising antimicrobial properties are captivating therapeutic alternatives for discovering natural bioactive products in the coming years. However, continued detailed researches are required to better understand the exact mechanism of action, pharmacodynamics and pharmacokinetic properties of the bioactive terpenes.

Conclusions

It is clear in this chapter that there is much evidence suggesting that some terpenes and their derivatives have proven to be significant antimicrobial agents towards various drug resistant microbes which are mainly bacteria and fungi. Specific mechanisms of action for each class of terpenes have also been highlighted and in general, terpenes follow a probable mitigation route for AMR and negotiate the dead end of the diminishing antibiotic pipeline. Therefore, a correct match between terpenoids and existing available antimicrobial agents may deliver ideal therapeutic alternatives for AMR-associated infections.

References

Abbaszadeh, S., Sharifzadeh, A., Shokri, H., Khosravi, A.R., Abbaszadeh, A. (2014). Antifungal efficacy of thymol, carvacrol, eugenol and menthol as alternative agents to control the growth of food-relevant fungi. J Mycol Med 24: e51.

Adukwu, E.C., Bowles, M., Edwards-Jones, V., Bone, H. (2016). Antimicrobial activity, cytotoxicity and chemical analysis of lemongrass essential oil (*Cymbopogon flexuosus*) and pure citral. Appl Microbiol Biotechnol 100: 9619.

Ahimou, F., Denis, F.A., Touhami, A., Dufrêne, Y.F. (2002). Probing microbial cell surface charges by atomic force microscopy. Langmuir 18: 9937.

Al-Footy, K.O., Alarif, W.M., Asiri, F., Aly, M.M., Ayyad, S.E.N. (2015). Rare pyrane-based cembranoids from the red sea soft coral *Sarcophyton trocheliophorum* as potential antimicrobial–antitumor agents. Med Chem Res 24: 505.

Al-Snafi, A.E. (2016). Medical importance of Anthemis nobilis—A review. Asian J Pharm Sci Technol 6: 89.

Ali, S.M., Khan, A.A., Ahmed, I., Musaddiq, M., Ahmed, K.S., Polasa, H., Rao, L.V., Habibullah, C.M., Sechi, L.A., Ahmad, N. (2005). Antimicrobial activities of Eugenol and Cinnamaldehyde against the human gastric pathogen *Helicobacter pylori*. Ann Clin Microbiol Antimicrob 4: 1.

Althunibat, O.Y., Qaralleh, H., Al-Dalin, S.Y.A., Abboud, M., Khleifat, K., Majali, I.S., Aldal'in, H.K., Rayyan, W.A., Jaafraa, A. (2016). Effect of thymol and carvacrol, the major components of *Thymus capitatus* on the growth of *Pseudomonas aeruginosa*. J Pure Appl Microbiol 10: 367.

Amaral, V.C.S., Santos, P.R., Silva, A.F.D., Santos, A.R.D., Machinski, M., Mikcha, J.M.G. (2015). Effect of carvacrol and thymol on salmonella spp. biofilms on polypropylene. Int J Food Sci Technol 50: 2639.

Ambrosio, S.R., Tirapelli, C.R., da Costa, F.B., de Oliveira, A.M. (2006). Kaurane and pimarane-type diterpenes from the Viguiera species inhibit vascular smooth muscle contractility. Life Sci 79: 925.

Andrade-Ochoa, S., Chacon-Vargas, K.F., Sanchez-Torres, L.E., Rivera-Chavira, B.E., Nogueda-Torres, B., Nevarez-Moorillon, G.V. (2021). Differential antimicrobial effect of essential oils and their main components: Insights based on the cell membrane and external structure. Membranes 11: 405.
Andre, C.M., Hausman, J.F., Guerriero, G. (2016). *Cannabis sativa*: the plant of the thousand and one molecule. Front Plant Sci 7: 19.
Appendino, G., Gibbons, S., Giana, A., Pagani, A., Grassi, G., Stavri, M., Smith, E., Rahman, M.M. (2008). Antibacterial cannabinoids from *Cannabis sativa*: A structure-activity study. J Nat Prod 71: 1427.
Astani, A., Reichling, J., Schnitzler, P. (2010). Comparative study on the antiviral activity of selected monoterpenes derived from essential oils. Phytother Res 24: 673.
Bajpai, V.K., Sharma, A., Baek, K.H. (2013). Antibacterial mode of action of *Cudrania tricuspidata* fruit essential oil, affecting membrane permeability and surface characteristics of food-borne pathogens. Food Control 32: 582.
Barbieri, R., Coppo, E., Marchese, A., Daglia, M., Sobarzo-Sánchez, E., Nabavi, S.F. (2017). Phytochemicals for human disease: An update on plant-derived compounds antibacterial activity. Microbiol Res 196: 44.
Bhatti, H.N., Khan, S.S., Khan, A., Rani, M., Ahmad, V.U., Choudhary, M.I. (2014). Biotransformation of monoterpenoids and their antimicrobial activities. Phytomed 21: 1597.
Borges, A., Ferreira, C., Saavedra, M.J., Simoes, M. (2013). Antibacterial activity and mode of action of ferulic and gallic acids against pathogenic bacteria. Microb Drug Resist 19: 256.
Broniatowski, M., Mastalerz, P., Flasiński, M. (2015). Studies of the interactions of ursane-type bioactive terpenes with the model of *Escherichia coli* inner membrane-Langmuir monolayer approach. Biochim Biophys Acta Biomembr 1848: 469.
Buckle, J. (2015). Clinical Aromatherapy. 3th ed. London, UK: Elsevier Health Sciences.
Burt, S. (2004). Essential oils: their antibacterial properties and potential applications in foods—a review. Int J Food Microbiol 94: 223.
Carson, C.F., Mee, B.J., Riley, T.V. (2002). Mechanism of action of Melaleuca alternifolia (tea tree) oil on *Staphylococcus aureus* determined by time-kill, lysis, leakage, and salt tolerance assays and electron microscopy. Antimicrob Agents Chemother 46: 1914.
Castelo-Branco, D.S.C.M., Riello, G.B., Vasconcelos, D.C., Guedes, G.M.M., Serpa, R., Bandeira, T.J.P.G., Monteiro, A.J., Cordeiro, R.A., Rocha, M.F., Sidrim, J.J.C., Brilhante, R.S.N. (2016). Farnesol increases the susceptibility of Burkholderia pseudomallei biofilm to antimicrobials used to treat melioidosis. J Appl Microbiol 120: 600.
Catteau, L., Zhu, L., Van-Bambeke, F., Quetin-Leclercq, J. (2018). Natural and hemisynthetic pentacyclic triterpenes as antimicrobials and resistance modifying agents against *Staphylococcus aureus*: a review. Phytochem Rev 17: 1129.
Chauhan, A.K., Kang, S.C. (2014). Thymol disrupts the membrane integrity of *salmonella ser. Typhimurium in vitro* and recovers infected macrophages from oxidative stress in an *ex vivo* model. Res Microbiol 165: 559.
Chen, J., Jiang, Q.D., Chai, Y.P., Zhang, H., Peng, P., Yang, X.X. (2016). Natural terpenes as penetration enhancers for transdermal drug delivery. Molecules 12: 1709.
Cheng, Y.B., Jensen, P.R., Fenical, W. (2013). Cytotoxic and antimicrobial napyradiomycins from two marine-derived MAR 4 Streptomyces strains. European J Org Chem 2013.
Chung, P.Y., Chung, L.Y., Navaratnam, P. (2014). Potential targets by pentacyclic triterpenoids from Callicarpa farinose against methicillin-resistant and sensitive *Staphylococcus aureus*. Fitoterapia 94: 48.
Cox, S.D., Mann, C.M., Markham, J.L., Bell, H.C., Gustafson, J.E., Warmington, J.R., Wyllie, S.G. (2000). The mode of antimicrobial action of the essential oil of *Melaleuca alternifolia* (tea tree oil). J Appl Microbiol 88: 170.
Cunha, W.R., De Matos, G.X., Souza, M.G.M., Tozatti, M.G., Andrade, E., Silva, M.L., Martins, C.H.G., Silva, R.D., Da Silva Filho, A.A. (2010). Evaluation of the antibacterial activity of the methylene chloride extract of *Miconia ligustroides*, isolated triterpene acids, and ursolic acid derivatives. Pharm Biol 48: 166.
Davies, J., Davies, D. (2010). Origins and evolution of antibiotic resistance. Microbiol Mol Biol Rev 3: 417.
De Andrade, F.B., Midena, R.Z., Koga-Ito, C.Y., Duarte, M.A. (2013). Conventional and natural products against oral infections. Microbial pathogens and strategies for combating them: Science, technology and education. pp. 1574–1583. *In*: Méndez-Vilas, A. (ed.). Pathogens and Strategies for Combating them: Science, Technology and Education. FORMATEX Research Center: Badajoz, Spain.
De Carvalho, C.C., Da Fonseca, M.M.R. (2006). Carvone: why and how should one bother to produce this terpene. Food Chem 95: 413.
De Rapper, S., Viljoen, A., Van Vuuren, S. (2016). The *in vitro* antimicrobial effects of *Lavandula angustifolia* essential oil in combination with conventional antimicrobial agents. Evidence-Based Complement Altern Med 2016.
De Inés, C., Argandoña, V.H., Rovirosa, J., San-Martín, A., Díaz-Marrero, A.R., Cueto, M., González-Coloma, A. (2004). Cytotoxic activity of halogenated monoterpenes from *Plocamium cartilagineum*. Zeitschrift fur Naturforsch-Sect C J Biosci 59: 339.

Devi, K.P., Nisha, S.A., Sakthivel, R., Pandian, S.K. (2010). Eugenol (an essential oil of clove) acts as an antibacterial agent against *Salmonella typhi* by disrupting the cellular membrane. J Ethnopharmacol 130: 107.

Dorman, H.J.D., Deans, S.G. (2000). Antimicrobial agents from plants: Antibacterial activity of plant volatile oils. J Appl Microbiol 88: 308.

Dunkic, V., Bezic, N., Vuko, E., Cukrov, D. (2010). Antiphytoviral activity of *Satureja montana* L. ssp. variegata (host) P. W. Ball essential oil and phenol compounds on CMV and TMV. Molecules 15: 6713.

Elsohly, M.A., Slade, D. (2005). Chemical constituents of marijuana: the complex mixture of natural cannabinoids. Life Sci 78: 539.

Eisohly, H.N., Turner, C.E., Clark, A.M., Eisohly, M.A. (1982). Synthesis and antimicrobial activities of certain cannabichromene and cannabigerol related compounds. J Pharm Sci 71: 1319.

El Omari, K., Hamze, M., Alwan, S., Jama, C., Chihib, N.E. (2016). Antifungal activity of the essential oil of *Micromeria barbata* an endemic *lebanese micromeria* species collected at North Lebanon. J Mater Environ Sci 7: 4158.

Farha, M.A., El-Halfawy, O.M., Gale, R.T., MacNair, C.R., Carfrae, L.A., Zhang, X., Jentsch, N.G., Magolan, J., Brown, E.D. (2020). Uncovering the hidden antibiotic potential of Cannabis. ACS Infect Dis 6: 338.

Feio, S.S., Gigante, B., Roseiro, J.C., Marcelo-Curto, M. (1999). Antimicrobial activity of diterpene resin acid derivatives. J Microbiol Methods 35: 201.

Feldman, M., Smoum, R., Mechoulam, R., Steinberg, D. (2020). Potential combinations of endocannabinoid/endocannabinoid-like compounds and antibiotics against methicillin-resistant *Staphylococcus aureus*. PLoS One 15: e0231583.

Feldman, M., Sionov, R., Smoum, R.M.R., Ginsburg, I., Steinberg, D. (2020). Comparative evaluation of combinatory interaction between endocannabinoid system compounds and poly-L-lysine against *Streptococcus mutans* growth and biofilm formation. Biomed Res Int 2020: 1.

Friedman, H., Pross, S., Klein, T.W. (2006). Addictive drugs and their relationship with infectious diseases. Immunol Med Microbiol 47: 330.

Gallucci, MN., Oliva, M., Casero, C., Dambolena, J., Luna, A., Zygadlo, J., Demo, M. (2009). Antimicrobial combined action of terpenes against the food-borne microorganisms *Escherichia coli, Staphylococcus aureus* and *Bacillus cereus*. Flavour Fragr J 24: 348.

Garzoli, S., Božovi´c, M., Baldisserotto, A., Sabatino, M., Cesa, S., Pepi, F., Vicentini, C.B., Manfredini, S., Ragno, R. (2018). Essential oil extraction, chemical analysis and anti-Candida activity of *Foeniculum vulgare* Miller—new approaches. Nat Prod Res 32: 1254.

Gilbert, P., Evans, D.J., Evans, E., Duguid, I.G., Brown, MR. (1991). Surface characteristics and adhesion of *Escherichia coli* and *Staphylococcus epidermidis*. J Appl Bacteriol 71: 72.

Gomes, F.I.A., Teixeira, P., Azeredo, J., Oliveira, R. (2009). Effect of farnesol on planktonic and biofilm cells of *Staphylococcus epidermidis*. Curr Microbiol 59: 118.

Gonçalves, O., Pereira, R., Gonçalves, F., Mendo, S., Coimbra, M.A., Rocha, S.M. (2011). Evaluation of the mutagenicity of sesquiterpenic compounds and their influence on the susceptibility towards antibiotics of two clinically relevant bacterial strains. Mutat Res 723: 18.

Griffin, S.G., Wyllie, S.G., Markham, J.L., Leach, D.N. (1999). The role of structure and molecular properties of terpenoids in determining their antimicrobial activity. Flavour Fragr J 14: 322.

Grzesiak, B., Głowacka, A., Krukowski, H., Lisowski, A., Lassa, H., Sienkiewicz, M. (2016). The *in vitro* efficacy of essential oils and antifungal drugs against *Prototheca zopfii*. Mycopathologia 181: 609.

Gupta, V.K., Tiwari, N., Gupta, P., Verma, S., Pal, A., Srivastava, S.K., Darokar, M.P. (2016). A clerodane diterpene from *Polyalthia longifolia* as a modifying agent of the resistance of methicillin resistant *Staphylococcus aureus*. Phytomedicine 23: 654.

Hajlaoui, H., Trabelsi, N., Noumi, E., Snoussi, M., Fallah, H., Ksouri, R., Bakhrouf, A. (2009). Biological activities of the essential oils and methanol extract of tow cultivated mint species (*Mentha longifolia* and *Mentha pulegium*) used in the Tunisian folkloric medicine. World J Microbiol Biotechnol 25: 2227.

Hajlaoui, H., Mighri, H., Aouni, M., Gharsallah, N., Kadri, A. (2016). Chemical composition and *in vitro* evaluation of antioxidant, antimicrobial, cytotoxicity and anti-acetylcholinesterase properties of Tunisian *Origanum majorana* L. essential oil. Microb Pathog 95: 86.

Helander, I.M., Alakomi, H.L., Latva-Kala, K., Mattila-Sandholm, T., Pol, I., Smid, E.J., Gorris, L.G., Wright, A. (1998). Characterization of the action of selected essential oil components on Gram-negative bacteria. J Agric Food Chem 46: 3590.

Jimenez-Arellanes, A., Luna-Herrera, J., Cornejo-Garrido, J., López-García, S., Castro-Mussot, M.E., Meckes-Fischer, M., Mata-Espinosa, D., Marquina, B., Torres, J., Hernández-Pando, R. (2013). Ursolic and oleanolic acids as antimicrobial and immunomodulatory compounds for tuberculosis treatment. BMC Complement Altern Med 13: 258.

Jin, J., Guo, N., Zhang, J., Ding, Y., Tang, X., Liang, J., Li, L., Deng, X., Yu, L. (2010). The synergy of honokiol and fluconazole against clinical isolates of azole-resistant *Candida albicans*. Lett Appl Microbiol 51: 351.
Kacaniova, M., Terentjeva, M., Vukovic, N., Puchalski, C., Roychoudhury, S., Kunová, S., Kluga, A., Tokar, M., Kluz, M., Ivanisova, E. (2017). The antioxidant and antimicrobial activity of essential oils against *Pseudomonas* spp. isolated from fish. Saudi Pharm J 25: 1108.
Kim, J., Marshall, M.R., Wei, C.I. (1995). Antibacterial activity of some essential oil components against five foodborne pathogens. J Agric Food Chem 43: 2839.
Knowles, J.R., Roller, S., Murray, D.B., Naidu, A.S. (2005). Antimicrobial action of carvacrol at different stages of dual-species biofilm development by *Staphylococcus aureus* and *Salmonella enterica serovar typhimurium*. Appl Environ Microbiol 71: 797.
Koo, H., Pearson, S.K., Scott-Anne, K., Abranches, J., Cury, J.A., Rosalen, P.L., Park, Y., Marquis, R.E., Bowen, W.H. (2002). Effects of apigenin and tt-farnesol on glucosyltransferase activity, biofilm viability and caries development in rats. Oral Microbiol Immunol 17: 337.
Kosgodage, U.S., Matewele, P., Awamaria, B., Kraev, I., Warde, P., Mastroianni, G., Nunn, A.V., Guy, G.W., Bell, J.D., Inal, J.M., Lange, S. (2019). Cannabidiol is a novel modulator of bacterial membrane vesicles. Front Cell Infect Microbiol 9: 324.
Kumar, R., Subramani, R., Aalbersberg, W. (2013). Three bioactive sesquiterpene quinones from the Fijian marine sponge of the genus Hippospongia. Nat Prod Res 27: 1488.
Kurekci, C., Padmanabha, J., Bishop-Hurley, S.L., Hassan, E., Al Jassim, R.A.M., McSweeney, C.S. (2013). Antimicrobial activity of essential oils and five terpenoid compounds against *Campylobacter jejuni* in pure and mixed culture experiments. Int J Food Microbiol 166: 450.
Kurita, N., Miyaji, M., Kuraney, R., Takahara, Y., Ichimura, K. (1979). Antifungal activity and molecular orbital energies of aldehyde compounds from oils of higher plants. Agric Biol Chem 43: 2365.
Langeveld, W.T., Veldhuizen, E.J.A., Burt, S.A. (2014). Synergy between essential oil components and antibiotics: A review. Crit Rev Microbiol 40: 76.
Laxminarayan, R., Duse, A., Wattal, C., Zaidi, A.K.M., Wertheim, H.F.L., Sumpradit, N., Vlieghe, E., Hara, G.L., Gould, I.M., Goossens, H., Greko, C. (2013). Antibiotic resistance-the need for global solutions. Lancet Infect Dis 13: 1057.
Lee, NH., Ho, J.W. (2008). Celastrol and terpenes as anti-infective agents. Anti-Infect Agents Med Chem Formerly Curr Med Chem Anti Infect Agents 7: 97.
Lerebour, G., Cupferman, S., Bellon-Fontaine, M.N. (2004). Adhesion of *Staphylococcus aureus* and *Staphylococcus epidermidis* to the Episkin® reconstructed epidermis model and to an inert 304 stainless steel substrate. J Appl Microbiol 97: 7.
Lewis, M.A., Haba, S.R. (2019). Biotech Institute llc, Cannabis plant named 'OG citron'. U.S. Patent Application 16/501,446.
Lewis, R.E., Kontoyiannis, D.P. (2001). Rationale for combination antifungal therapy. Pharmacotherapy 21: 149S.
Lin, CM., Preston, JF. 3rd, Wei, CI. (2000). Antibacterial mechanism of allyl isothiocyanate. J Food Prot 63: 727.
Liu, Q., Niu, H., Zhang, W., Mu, H., Sun, C., Duan, J. (2015). Synergy among thymol, eugenol, berberine, cinnamaldehyde and streptomycin against planktonic and biofilm-associated food-borne pathogens. Lett Appl Microbiol 60: 21.
Liu, X., Dong, M., Chen, X., Jiang, M., Lv, X., Zhou, J. (2008). Antimicrobial activity of an endophytic Xylaria sp. YX-28 and identification of its antimicrobial compound 7-amino-4-methylcoumarin. Appl Microbiol Biotechnol 78: 241.
Liu, X., Li, Y., Guo, Y., Zeng, Z., Li, B., Wood, T.K., Cai, X., Wang, X. (2015). Physiological function of rac prophage during biofilm formation and regulation of rac excision in *Escherichia coli* K-12. Sci Rep 5: 1.
Lorenzi, V., Muselli, A., Bernardini, A.F., Berti, L., Pagès, J.M., Amaral, L., Bolla, J.L. (2009). Geraniol restores antibiotic activities against multidrug-resistant isolates from Gram-negative species. Antimicrob Agents Chemother 53: 2209.
Maida, I., Lo Nostro, A., Pesavento, G., Barnabei, M., Calonico, C., Perrin, E., Chiellini, C., Fondi, M., Mengoni, A., Maggini, V., Vannacci, A. (2014). Exploring the anti-*Burkholderia cepacia* complex activity of essential oils: A preliminary analysis. Evidence-Based Complement Altern Med 2014.
Masako, K., Yusuke, K., Hideyuki, I., Atsuko, M., Yoshiki, M., Kayoko, M., Makoto, K. (2005). Corrigendum to A novel method to control the balance of skin microflora. Part 2. A study to assess the effect of a cream containing farnesol and xylitol on atopic dry skin. J Dermatol Sci 39: 197.
Martinenghi, L.D., Jonsson, R., Lund, T., Jenssen, H. (2020). Isolation, purification, and antimicrobial characterization of cannabidiolic acid and cannabidiol from *Cannabis sativa* L. Biomol 10: 900.
Miladi, H., Zmantar, T., Chaabouni, Y., Fedhila, K., Bakhrouf, A., Mahdouani, K., Chaieb, K. (2016). Antibacterial and efflux pump inhibitors of thymol and carvacrol against food-borne pathogens. Microb Pathog 99: 95.

Moo, C.L., Yang, S.K., Yusoff, K., Ajat, M., Thomas, W., Abushelaibi, A., Lim, S.H., Lai, K.S. (2019). Mechanisms of antimicrobial resistance (AMR) and alternative approaches to overcome AMR. Curr Drug Discov Technol 16.

Nazzaro, F., Fratianni, F., De Martino, L., Coppola, R., De Feo, V. (2013). Effect of essential oils on pathogenic bacteria. Pharmaceuticals 6: 1451.

Newman, D.J., Cragg, G.M. (2012). Natural products as sources of new drugs over the 30 years from 1981 to 2010. J Nat Prod 75: 311.

Oldfield, E., Lin, F.Y. (2012). Terpene biosynthesis: modularity rules. Angew Chem Int 51 ed. 1124.

Paduch, R., Kandefer-Szerszen, M., Trytek, M., Fiedurek, J. (2007). Terpenes: substances useful in human healthcare. Arch Immunol Ther Exp 55: 315.

Palmer, J., Flint, S., Brooks, J. (2007). Bacterial cell attachment, the beginning of a biofilm. J Industrial Microbiol Biotechnol 34: 577.

Peixoto, L.R., Rosalen, P.L., Ferreira, G.L.S., Freires, I.A., de Carvalho, F.G., Castellano, L.R., Castro, R.D. (2017). Antifungal activity, mode of action and anti-biofilm effects of *Laurus nobilis* Linnaeus essential oil against Candida spp. Arch Oral Biol 73: 179.

Pettit, R.K., McAllister, S.C., Pettit, G.R., Herald, C.L., Johnson, J.M., Cichacz, Z.A. (1997). A broad-spectrum antifungal from the marine sponge Hyrtios erecta. Int J Antimicrob Agents 9: 147.

Radwan, M.M., Elsohly, M.A., Slade, D., Ahmed, S.A., Khan, I.A., Ross, S.A. (2009). Biologically active cannabinoids from high-potency *Cannabis sativa*. J Nat Prod 72: 906.

Rafiq, R., Hayek, S., Anyanwu, U., Hardy, B., Giddings, V., Ibrahim, S., Tahergorabi, R., Kang, H. (2016). Antibacterial and antioxidant activities of essential oils from *Artemisia herba-alba* Asso., *Pelargonium capitatum* radens and *Laurus nobilis* L. Foods 5: 28.

Rasooli, I., Shayegh, S., Astaneh, S.D.A. (2009). The effect of *Mentha spicata* and *Eucalyptus camaldulensis* essential oils on dental biofilm. Int J Dent Hyg 7: 196.

Rathinam, P., Vijay Kumar, H.S., Viswanathan, P. (2017). Eugenol exhibits anti-virulence properties by competitively binding to quorum sensing receptors. Biofouling 33: 624.

Rozalski, M., Walencka, E., Roz alska, B., Wysokin, ska, H. (2007). Antimicrobial activity of diterpenoids from hairy roots of *Salvia sclarea* L.: Salvipisone as a potential anti-biofilm agent active against antibiotic resistant *Staphylococci*. Phytomedicine 14: 31.

Rudramurthy, G.R., Swamy, M.K., Sinniah, U.R., Ghasemzadeh, A. (2016). Nanoparticles: Alternatives against drug-resistant pathogenic microbes. Molecules 21: 836.

Rukayadi, Y., Hwang, J.K. (2006). Effect of coating the wells of a polystyrene microtiter plate with xanthorrhizol on the biofilm formation of *Streptococcus mutans*. J Basic Microbiol 46: 410.

Saleem, A.F., Ahmed, I., Mir, F., Ali, S.R., Zaidi, A.K.M. (2010). Pan-resistant *Acinetobacter* infection in neonates in Karachi, Pakistan. J Infect Dev Ctries 4: 30.

Savluchinske-Feio, S., Curto, M.J.M., Gigante, B., Roseiro, J.C. (2006). Antimicrobial activity of resin acid derivatives. Appl Microbiol Biotechnol 72: 430.

Shakeri, A., Akhtari, J., Soheili, V., Taghizadeh, S.F., Sahebkar, A., Shaddel, R. Asili, J. (2017). Identification and biological activity of the volatile compounds of *Glycyrrhiza triphylla* Fisch. &C.a.Mey. Microb Pathog 109: 39.

Sharma, A., Rajendran, S., Srivastava, A., Sharma, S., Kundu, B. (2017). Antifungal activities of selected essential oils against *Fusarium oxysporum* f. sp. *lycopersici* 1322, with emphasis on *Syzygium aromaticum* essential oil. J Biosci Bioeng 123: 308.

Shrigod, N.M., Swami Hulle, N.R., Prasad, R.V. (2017). Supercritical fluid extraction of essential oil from mint leaves (*Mentha spicata*): Process optimization and its quality evaluation. J Food Process Eng 40: 12488.

Singsaas, E.L. (2000). Terpenes and the thermotolerance of photosynthesis. *In*: New Phytologist. Cambridge University Press, Cambridge.

Smid, E.J., de Witte, Y., Gorris, L.G.M. (1995). Secondary plant metabolites as control agents of postharvest *Penicillium rot* on tulip bulbs. Postharvest Biol Technol 6: 303.

Stahl, V., Vasudevan, K. (2020). Comparison of efficacy of cannabinoids versus commercial oral care products in reducing bacterial content from dental plaque: a preliminary observation. Cureus 12: e6809.

Socha, A.M., Garcia, D., Sheffer, R., Rowley, D.C. (2006). Antibiotic bisanthraquinones produced by a streptomycete isolated from a cyanobacterium associated with *Ecteinascidia turbinata*. J Nat Prod 69: 1070.

Socha, A.M., LaPlante, K.L., Rowley, D.C. (2006). New bisanthraquinone antibiotics and semi-synthetic derivatives with potent activity against clinical *Staphylococcus aureus* and *Enterococcus faecium* isolates. Bioorg Med Chem 14: 8446.

Souza, A.B., Martins, C.H.G., Souza, M.G.M., Furtado, N.A.J.C., Heleno, V.C.G., De Sousa, J.P.B., Rocha, E.M., Bastos, J.K., Cunha, W.R., Veneziani, R.C., Ambrósio, S.R. (2011). Antimicrobial activity of terpenoids from *Copaifera langsdorffii* Desf. against cariogenic bacteria. Phythother Res 25: 215.

Swamy, M.K., Akhtar, M.S., Sinniah, U.R. (2016). Antimicrobial properties of plant essential oils against human pathogens and their mode of action: An updated review. Evid Based Complement Altern Med 2016.

Tapia, A.A., Vallejo, M.D., Gouiric, S.C., Feresin, G.E., Rossomando, P.C., Bustos, D.A. (1997). Hydroxylation of dehydroabietic acid by fusarium species. Phytochem 46: 131.

Thapa, D., Louis, P., Losa, R., Zweifel, B., Wallace, R.J. (2015). Essential oils have different effects on human pathogenic and commensal bacteria in mixed faecal fermentations compared with pure cultures. Microbiology 161: 441.

Togashi, N., Hamashima, H., Shiraishi, A., Inoue, Y., Takano, A. (2010). Antibacterial activities against *Staphylococcus aureus* of terpene alcohols with aliphatic carbon chains. J Essent Oil Res 22: 263.

Tongnuanchan, P., Benjakul, S. (2014). Essential oils: Extraction, bioactivities, and their uses for food preservation. J Food Sci 7.

Torres-Romero, D., Jiménez, I.A., Rojas, R., Gilman, R.H., López, M., Bazzocchi, I.L. (2011). Dihydroydro agarofuran sesquiterpenes isolated from *Celastrus vulcanicola* as potential anti-Mycobacterium tuberculosis multidrug-resistant agents. Bioorganic Med Chem 19: 2182.

Touani, F.K., Seukep, A.J., Djeussi, D.E., Fankam, A.G., Noumedem, J.A.K., Kuete, V. (2014). Antibiotic-potentiation activities of four Cameroonian dietary plants against multidrug-resistant Gram-negative bacteria expressing efflux pumps. BMC Complement Altern Med 1: 258.

Touil, H.F., Boucherit, K., Boucherit-Otmani, Z., Kohder, G., Madkour, M., Soliman, S.S. (2020). Optimum inhibition of amphotericin-B-resistant candida albicans strain in single- and mixed-species biofilms by candida and non-candida terpenoids. Biomolecules 10: 342.

Trombetta, D., Castelli, F., Sarpietro, M.G., Venuti, V., Cristani, M., Daniele, C., Saija, A., Mazzanti, G., Bisignano, G. (2005). Mechanisms of antibacterial action of three monoterpenes. Antimicrob Agents Chemother 49: 2474.

Turgis, M., Han, J., Caillet, S., Lacroix, M. (2009). Antimicrobial activity of mustard essential oil against *Escherichia coli* O157 : H7 and *Salmonella typhi*. Food Control 20: 1073.

Turner, C.E., Elsohly, M.A. (1981). Biological activity of cannabichromene, its homologs and isomers. J Clin Pharmacol 21: 283S.

Ultee, A., Kets, E.P., Smid, E.J. (1999). Mechanisms of action of carvacrol on the food-borne pathogen *Bacillus cereus*. Appl Environ Microbiol 65: 4606.

Ultee, A., Smid, E.J. (2001). Influence of carvacrol on growth and toxin production by *Bacillus cereus*. Int J Food Microbiol 64: 373.

VanKlingeren, B., Ten Ham, M. (1976). Antibacterial activity of delta9-tetrahydrocannabinol and cannabidiol. Antonie Van Leeuwenhoek 42: 9.

Vasireddy, L., Bingle, L.E.H., Davies, M.S. (2018). Antimicrobial activity of essential oils against multidrug-resistant clinical isolates of the *Burkholderia cepacia* complex. PLOS ONE.

Walencka, E., Rozalska, S., Wysokinska, H., Rozalski, M., Kuzma, L., Rozalska, B. (2007). Salvipisone and aethiopinone from *Salvia sclarea* hairy roots modulate staphylococcal antibiotic resistance and express anti-biofilm activity. Planta Med 73: 545.

Wassmann, C.S., Højrup, P., Klitgaard, J.K. (2020). Cannabidiol is an effective helper compound in combination with bacitracin to kill Gram-positive bacteria. Sci Rep 10: 4112.

Xu, W., Tan, J., Mu, Y., Zheng, D., Huang, X., Li, L. (2020). New antimicrobial terpenoids and phloroglucinol glucosides from *Syzygium szemaoense*. Bioorg Chem 103: 104242.

Yadav, M.K., Chae, S.W., Im, G.J., Chung, J.W., Song, J.J. (2015). Eugenol: a phyto-compound effective against methicillin-resistant and methicillin-sensitive *Staphylococcus aureus* clinical strain biofilms. PLoS ONE 10(3): e0119564.

Yang, S.K., Yusoff, K., Ajat, M., Thomas, W., Abushelaibi, A., Akseer, R., Lim, S.E., Lai, K.S. (2019). Disruption of KPC-producing *Klebsiella pneumoniae* membrane via induction of oxidative stress by cinnamon bark (Cinnamomum verum J. Presl) essential oil. PLoS ONE 1–20.

Yang, B., Zhou, X.F., Lin, X.P., Liu, J., Peng, Y., Yang, X.W., Liu, Y. (2012). Cembrane diterpenes chemistry and biological properties. Curr Org Chem 16(12): 1512–39.

Yu, H., Zhang, L., Li, L., Zheng, C., Guo, L., Li, W., Sun, P., Qin, L. (2010). Recent developments and future prospects of antimicrobial metabolites produced by endophytes. Microbiol Res 165(6): 437–49.

Zacchino, S.A., Butassi, E., Cordisco, E., Svetaz, L.A. (2017). Hybrid combinations containing natural products and antimicrobial drugs that interfere with bacterial and fungal biofilms. Phytomed 37: 14–26.

Zaidi, A.K.M., Huskins, W.C., Thaver, D., Bhutta, Z.A., Abbas, Z., Goldmann, D.A. (2005). Hospital-acquired neonatal infections in developing countries. Lancet 365: 1175–1188.

Zengin, H., Baysal, A. (2014). Antibacterial and antioxidant activity of essential oil terpenes against pathogenic and spoilage-forming bacteria and cell structure-activity relationships evaluated by SEM microscopy. Molecules 11: 17773–17798.

Zhang, Y.B., Liu, X.Y., Jiang, P.P., Li, W.D., Wang, Y.F. (2015). Mechanism and antibacterial activity of cinnamaldehyde against *Escherichia coli* and *Staphylococcus aureus*. Mod Food Sci Technol 31(5): 31–5.

Zhou, L., Ding, Y., Chen, W., Zhang, P., Chen, Y., Lv, X. (2013). The *in vitro* study of ursolic acid and oleanolic acid inhibiting cariogenic microorganisms as well as biofilm. Oral Dis 19: 494–500.

Zrira, S., Ghanmi, M. (2016). Chemical composition and antibacterial activity of the essential of *Cedrus atlantica* (Cedarwood oil). J Essent Oil-Bearing Plants 19: 1267–1272.

Section III

Biobased Antimicrobials from Mushrooms and Microbial Sources

13

Streptomyces:
A Potential Source of Natural Antimicrobial Drug Leads

Mahmoud A. Elfaky,[1] *Hanaa Nasr,*[2] *Ilham Touiss*[3] and *Mohamed L. Ashour*[4,5,*]

Introduction

In the early 1940s, Waksman and Henrici first described the genus *Streptomyces* (Williams et al. 1983). It was included in the family Streptomycetaceae (Arai 1997) based on physiological and morphological characteristics. Furthermore, some other features, such as the cell wall chemical composition, including the phospholipids/peptidoglycan type, fatty acid chains and the 16S rRNA sequence, are very useful in the taxonomy of this genus (Kämpfer et al. 2008). The genus *Streptomyces* represents the only member of the family Streptomycetaceae that belongs to Actinobacteria phylum and is classified in Actinomycetales order placed in the class Actinobacteria (Anderson and Wellington 2001). However, *Streptomyces* is among the richest taxonomic components of known actinomycetes in terms of the number of discovered species (Bhattacharyya et al. 1998), with more than 500 species producing approximately two thirds of the known natural antibiotics (Mohanraj and Sekar 2013). *Streptomyces* are aerobic, Gram-positive, but not acid-fast bacteria that are also characterized by high guanine/cytosine content (> 70%) (Reza Dehnad et al. 2010) and can grow in different environments (Maleki et al. 2013).

The *Streptomyces* produces a layer of hyphae that makes aerial mycelium, which is rarely rammed, although it differentiates into chains of three or more spores when mature. There are large varieties between the species among the genus (Holt et al. 1994). Most of these species are found in soil worldwide and have much ecological importance for the environment (Madigan et al. 2000). Differentiation of *Streptomyces* morphologically involves a mechanism distinctive of Gram-positive bacteria. It is divided through the aerial hyphae into non-mobile spores. A particular and coordinated metabolism is needed

[1] Department of Natural Products, Faculty of Pharmacy, King Abdulaziz University, Jeddah 21589, Saudi Arabia; Centre for Artificial Intelligence in Precision Medicines, King Abdulaziz University, Jeddah 21589, Saudi Arabia.
[2] Central Administration for Pharmaceutical Products, Egyptian Drug Authority, El-Manial, Cairo 11553, Egypt.
[3] Laboratory of Bioresources, Biotechnologies, Ethnopharmacology and Health, Faculty of Sciences, University Mohamed I, 60 000, Oujda, Morocco.
[4] Department of Pharmaceutical Sciences, Pharmacy Program, Batterjee Medical College, Jeddah 21442, Saudi Arabia.
[5] Department of Pharmacognosy, Faculty of Pharmacy, Ain-Shams University, Abbasia, Cairo 11566, Egypt.
* Corresponding author: mohamed.ashour@bmc.edu.sa

for this process (Robinson 2014). *Streptomyces* has a complex growth cycle in the soil that begins with producing a multi-hyphae substratum mycelium. The hyphae emerge through the ground to look for nutrients through a tip expansion. These systems secrete several enzymes that hydrolyze many organic polymers, such as chitin and cellulose, into substituted sugars that are taken by different ABC transporters.

The dynamic multicellular growth of the genus *Streptomyces*, in which spores germinate and form hyphae, with a multi-nuclear mycelium that regularly forms septa at certain intervals, forming an aggregate of uninuclear spores, is a highly significant and notable feature. The germ tube forms and the hyphae grow under optimal temperature, nutritional and moisture conditions. The existence of spores in *Streptomyces* is most likely due to the favorable conditions, which may necessitate the requirement to live outside of plants and invertebrates or in severe conditions (Robinson 2014).

The spore germination and outgrowth of mycelia are the first stages of the *Streptomyces* developmental cycle. The species produce spores that produce reproductive aerial hyphae. The dietary requirements and environmental factors influence the shape of aerial hyphae. While antibiotic development is often associated with species differentiation, not all antibiotics produced from *Streptomyces* are biosynthesized in the stationary phase (Seipke et al. 2012).

A large number of members of *Streptomyces* produce a plethora of bioactive secondary metabolites. As a result of the therapeutic benefits of these metabolites as antimicrobials and immune-suppressants agents, the genus *Streptomyces* plays an essential role in pharmaceutical industries (Kumagai et al. 2010). It is worth mentioning that although many microbes develop resistance to the antibiotics, it usually remains susceptible to antibiotics produced by other *Streptomycetes* species.

In addition, most of the *Streptomyces* species produce volatile substances in wet soil. These volatile components have a significant role in soil ecology, especially in the hydrolysis of many organic substances present in the soil. *Streptomyces* spp. have recently been found to be free-living in the soil and to have evolved to exist in symbiosis with plants, fungi and animals. Most *Streptomyces* are saprophytes; however, some infectious strains cause many severe human diseases, including wound infection and abscess development (Robinson 2014).

Secondary metabolites production by *Streptomyces*

Actinobacteria are Gram-positive bacteria known for their high potential to produce secondary metabolites (Jose et al. 2021).

Secondary metabolites produced by *Streptomyces* have diverse biological activities that cover the plant, the microbes and humans. They can be used as antibiotics, antiprotozoal, antibacterial, antifungal and antiviral agents, anthelmintic, herbicides/pesticides (Kariminik and Baniasadi 2010), plant growth-promoting rhizobia (Bibb 2005), anticancer drugs, immune modulators (Mann 2001).

Since the discovery of Streptothricin in 1942 (Sanglier et al. 1993), followed by the discovery of streptomycin in 1943 (Schatz et al. 1944), the need for finding other antibiotics from the genus *Streptomyces* has intensified, where the majority of the antibiotic findings occurred between 1945 and 1960 (Watve et al. 2001). Rachel Brown isolated Nystatin from *Streptomyces noursei* in 1949 as the first antibiotic to treat fungal disease (Straits et al. 2009). *Streptomyces* spp. produces more than two-thirds of clinically effective natural antibiotics, which represent 80% of all known antibiotics (Kieser et al. 2000).

The development of antibiotics, is related to sporulation (a process that is self-triggered by nutrient depletion) and maybe a method to prevent the growth of other organisms competing with *Streptomyces* cells for limited resources. This will cause *Streptomyces* to finish the sporulation process and form a dormant system with better survival chances (Madigan et al. 2000)

It is essential to report that cefoxitin, daptomycin, rifamycin and Thiostrepton A1 are antibacterial compounds produced by *Streptomyces lactamdurans, S. roseosporus, S. mediterranei* and *S. laurentii*, respectively. *Streptomyces* spp. provide antifungal agents such as nystatin (*S. noursei*), amphotericin B (*S. nodosus*), natamycin (*S. natalensis*) and rapamycin (*S. hygroscopicus*) (Liras and Demain 2009) (Morin and Sello 2010; Murphy et al. 2010; Sadhasivam et al. 2010; Li et al. 2011; Song et al. 2012).

Due to the fact that the majority of antibiotics are species-specific, these secondary metabolites are required for *Streptomyces* spp. to compete with other bacteria, even those belonging to the same genus (de Lima Procópio et al. 2012). Despite advances in antibiotic discovery, numerous studies have revealed that infectious diseases continue to be the second largest cause of mortality worldwide, with bacterial infections killing over 17 million people each year, the majority of whom are children and the elderly.

Due to their broad range of action and potency, aminoglycosides are useful antibiotics. They are Actinomycetes-derived natural or hemi-synthetic heterosides. Aminoglycosides work by preventing bacteria from making proteins. They were started in 1944 with the isolation of streptomycin from *Streptomyces griseus* (Krause et al. 2016). It continues releasing a sequence of milestone compounds that have become a cornerstone in antibacterial chemotherapy. Additionally, aminoglycoside antibiotics share several characteristics due to their chemical similarity, including a broad antibacterial spectrum, partial or complete cross-resistance, potentiation of their bactericidal action in a slightly alkaline medium, poor gastrointestinal absorption, and weak glomerular filtration and ototoxicity. The diffusion potential of aminoglycosides into bone tissue is moderate.

Aminoglycosides bind to the 30S subunit of bacterial ribosomes in the cytosol and interact with the translation of messenger RNAs into proteins, leaving the ribosome inaccessible for translation and resulting in cell death. Many aminoglycosides bind to 16S ribosomal RNA at the decoding site (site A). While this binding does not impede the forming of the peptide synthesis initiation complex, it does disrupt the elongation of the nascent chain by interfering with the proofreading mechanism that controls translational precision. The erroneous proteins may be injected into the cell membrane, altering permeability and stimulating aminoglycoside transport even further (Krause et al. 2016). Aminoglycosides may function on bacterial cells by more than one process, inducing misreading of the RNA code and/or inhibiting amino acid polymerization.

Bactericidal action of aminoglycosides against Gram-positive and Gram-negative cells. They are particularly effective against many members of the Enterobacteriaceae family such as *E. coli, Enterobacter aerogenes, Entero. cloacae, Klebsiella pneumoniae, K. oxytoca, Providencia* spp., *Morganella* spp., *Proteus* spp., and *Serratia* spp. Furthermore, they are effective against the causative agents of plague and tularemia, *Yersinia pestis* and *Francisella tularensis*, respectively. The class is also effective against *Staphylococcus aureus*, including methicillin-resistant and vancomycin-intermediate and -resistant isolates and *Pseudomonas aeruginosa* and, to a lesser degree, *Acinetobacter baumannii. Mycobacterium tuberculosis, Mycobacterium fortuitum, Mycobacterium chelonae,* and *Mycobacterium avium* are all vulnerable to aminoglycosides. However, since aminoglycosides' absorption into cells involves effective electron transfer, anaerobic bacteria are completely resistant to aminoglycosides. Resistance to aminoglycosides is common in most *Burkholderia* spp. and *Stenotrophomonas* spp., as well as *Streptococcus* spp. and *Enterococcus* spp. (Krause et al. 2016).

Amikacin, isepamicin, gentamicin (pINN), kanamycin, neomycin, netilmicin, paromomycin, sisomicin, streptomycin, dihydrostreptomycin and tobramycin are representatives of aminoglycoside antibiotics that have been and still are clinically important (Dominguez-Godinez et al. 2018).

Antibiotic resistance has increased, necessitating the development of new antibiotics. A part of secondary metabolite pathways is involved in actinomycete laboratory cultures, and there are several strains with untapped ability to manufacture bioactive compounds. At the beginning of the 21st century and after the whole genome sequencing became possible, it revealed the presence of several silent or mysterious biosynthetic gene clusters in actinomycetes that could encode antibiotic-like substances but are not, or only weakly, expressed in laboratory growth conditions. Only a subset of the 20–50 biosynthetic pathways for natural products found in each *Streptomyces* strain has been identified. The biggest problem is figuring out how to reopen these channels to resume normal secondary metabolite screening (Manteca and Yagüe 2019).

History of antibiotics

The term antibiotic was first used by Selman Waksman (1888–1973) to describe any small molecule formed by a microbe that can kill or prevent the development of other microbes. Generally, antibiotics have their specific potent action on bacteria at relatively small therapeutic concentrations while exhibiting reduced toxicity (Ribeiro da Cunha et al. 2019). It is often assumed that human resistance to antibiotics is related to the recent 'antibiotic period.' However, tetracycline traces were found in human skeletal remains from Nubia and late Roman times in Egypt, which may be one of the reasons for the low infectious disease prevalence in the Nubia population (Aminov 2010).

It was always thought salvarsan was the first antimicrobial agent used against African Trypanosomiasis in 1910s (Hutchings et al. 2019). However, two great scientists Paul Ehrlich and Alexander Fleming, have always been linked with the new antibiotic phase. Ehrlich's concept of a "silver bullet" was based on discovering that aniline and other synthetic dyes could stain specific microbes but not others hosting cells. This can explain the phenomenon that antibiotics can selectively kill disease-causing bacteria while leaving the host cells unaffected. While Fleming's (1929) accidental discovery of penicillin is familiar to all of us (Aminov 2010). It took almost a decade for purification and the first therapeutic application of penicillin. By 1943, the US-UK collaboration to develop penicillin had proven to be a great success that had been used in Europe during WWII (Gaynes 2017). It is worth mentioning that, during Fleming's study, he warned that bacteria could develop antibiotic resistance to penicillin whether they were used in small doses or for the short improper regime (Aminov 2010).

In 1935, a chemical derivative from sulfamidochrysoidine was discovered by the German bacteriologist Gerhard Domagk who named it prontosil. This compound has been used successfully in mice and humans against streptococcal infections. Simultaneously, in France, it was found that prontosil had no *in vitro* effect on the *Streptococci*. It was discovered that prontosil should be metabolized in the human body to produce the active half, sulfanilamide, that possesses antimicrobial activity (Gaynes 2017).

Selman Waksman initiated a systematic study in the late 1930s on many microorganisms to work as antimicrobial producers. That led to the discovery of many antibiotics, such as neomycin and streptomycin, from the soil-inhabited *Actinomycetes*. His work was pioneering in identifying the genus *Streptomyces* as prolific producers of (natural products) that can be used as antimicrobial agents (Hutchings et al. 2019).

The golden age of antibiotic research lasted from the 1940s to the 1960s. During this time, the most effective antibacterial antibiotic was discovered. Antibiotics greatly decreased the mortality associated with infectious diseases during the 20th century, and antimicrobial chemotherapy made impressive strides during this time, leading to the falsely positive view that contagious diseases will be defeated in the immediate future (Vickers 2017).

According to many reports, more than 2 million North Americans are infected with antibiotic-resistant microbes per year, resulting in at least 23,000 deaths. Currently, at least 700,000 cases of antibiotic-resistant infections result in over 33,000 deaths per year in Europe, at the cost of more than €1.5 billion. Infectious diseases now account for about 20% of all deaths globally, even though antibiotic intake grew by 36% between 2000 and 2010. This condition worsened when hospital infections became a leading cause of morbidity and death, resulting in longer hospital stays and higher healthcare costs. Furthermore, multidrug-resistant bacteria cause more than 15% of hospital infections that are resistant to many antimicrobials. Antibiotic-resistant diseases, for example, are expected to kill 10 million people annually by 2050, according to a government study in the United Kingdom (Ribeiro da Cunha et al. 2019).

Since the discovery of daptomycin and linezolid in the 1980s, no new antibiotic classes have been found. The newly discovered antibiotics are nothing more than a refinement of a mixture of previously discovered compounds (Vickers 2017). Due to the contributions of emerging technology such as gene mining in the development of new natural products with complex bioactivities, the future of antibiotic discovery seems to be promising (Fig. 13.1 and Table 13.1) (Hutchings et al. 2019).

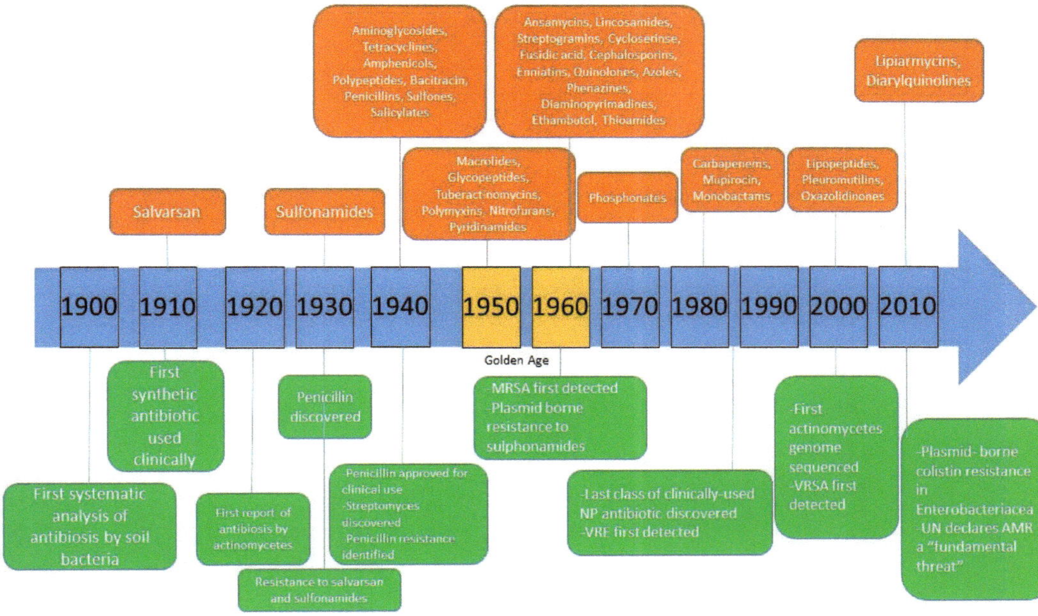

Figure 13.1. A schematic sketch for the most relevant key findings and dates of antibiotics discoveries.

Table 13.1. A summary of the antibiotics produced by genus *Streptomyces*, their mechanisms of action and uses

***Streptomyces* spp.**	**Antibiotic**	**Structure**	**Mechanism of action and uses**
S. venezuelae	Chloramphenicol	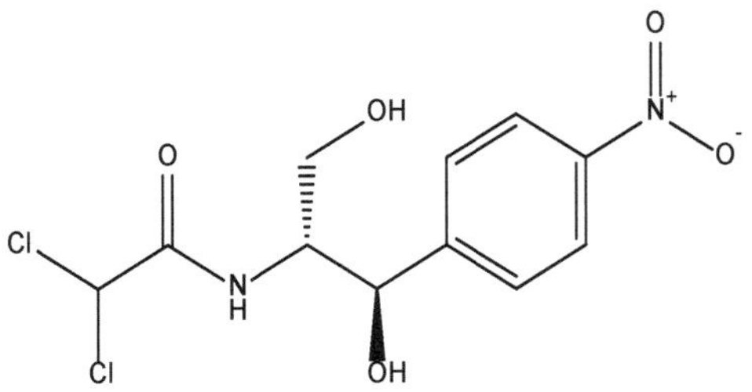	Chloramphenicol is a broad-spectrum antibiotic that works against both Gram-positive and negative bacteria. Spirochetes, Rickettsiae, anaerobic bacteria, *Streptococcus pneumoniae* and *Neisseria meningitides* are included. It is also used to treat staphylococcal brain abscesses and meningitis because of its excellent penetration of the blood brain barrier. In addition, infections caused by tetracycline-resistant cholera and VRE are usually treated. Its mechanism of action involves preventing protein chain elongation by inhibiting peptidyl transferase, resulting in a high-potency inhibition of bacterial protein synthesis (Brook 2016; Dinos et al. 2016; Kumar 2017).
S. aureofaciens	Chlortetracycline	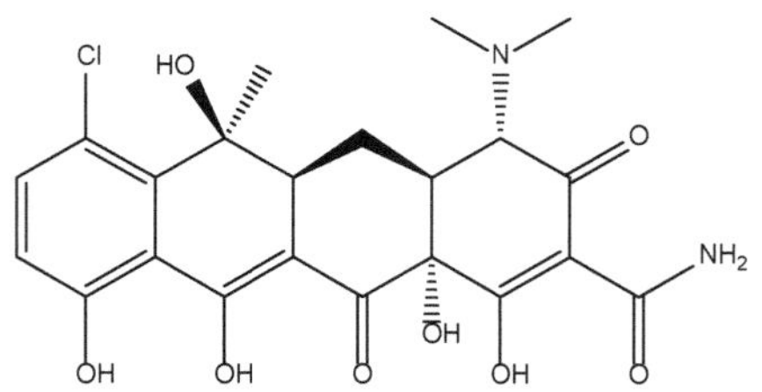	It inhibits protein synthesis by blocking the binding of aminoacyl-tRNA to the ribosomal acceptor (A) site. It has action against gram-positive and gram-negative bacteria, chlamydiae, mycoplasmas, rickettsiae and protozoan parasites, among others (Chopra and Roberts 2001).
S. aureofaciens	Tetracycline	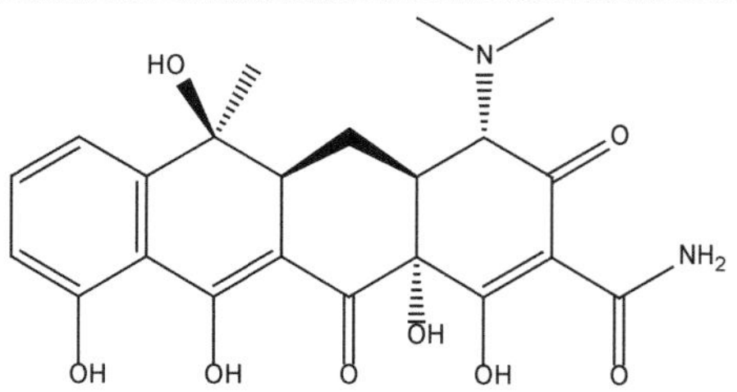	Same as chlortetracycline (Chopra and Roberts 2001).

Table 13.1 contd. ...

...*Table 13.1 contd.*

***Streptomyces* spp.**	**Antibiotic**	**Structure**	**Mechanism of action and uses**
S. platensis	Platensimycin	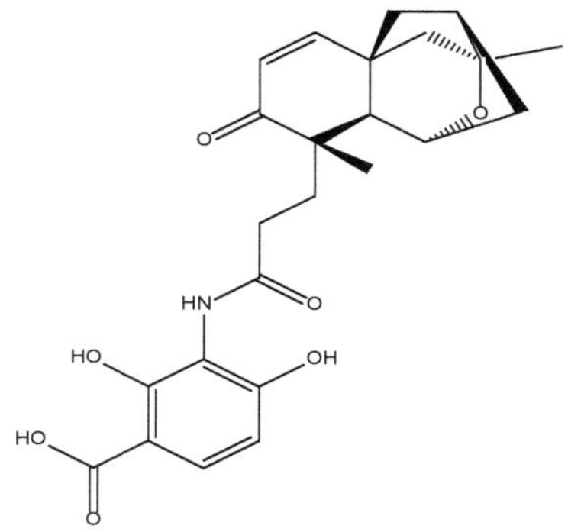	Its activity includes pathogenic Gram-positive bacteria, including antibiotic resistant strains and *M. tuberculosis* through inhibition of the elongation-condensing enzyme FabF (Martens and Demain 2011).
S. platensis	Platencin	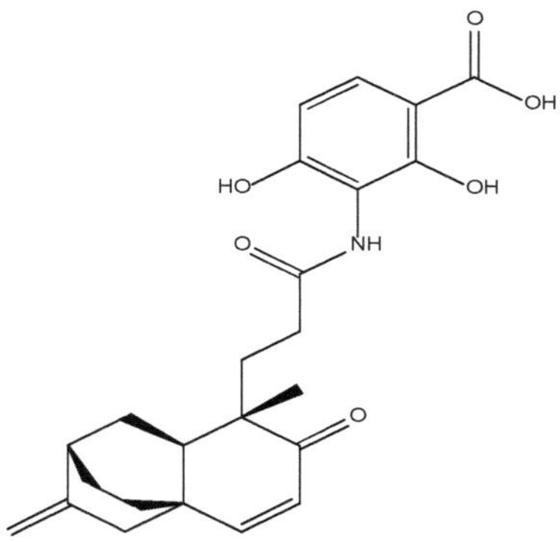	Similar activity like platensimycin but it can inhibit both FabF and the initiation condensing enzyme FabH (Martens and Demain 2011).
S. roseosporus	Daptomycin	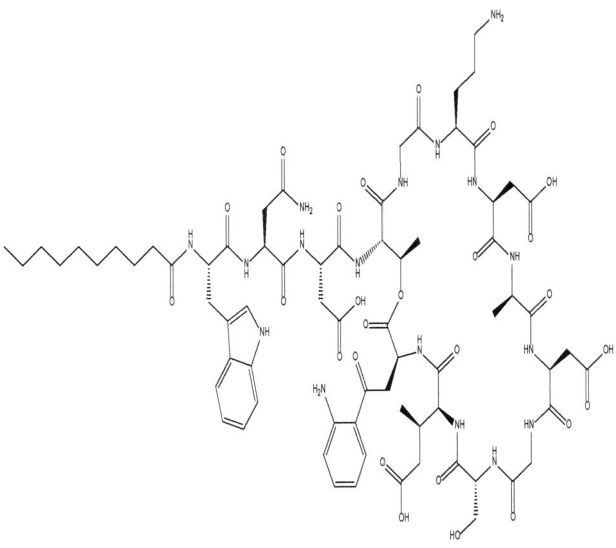	The range of spectrum covers Gram-positive bacteria, including methicillin resistant *Staphylococcus aureus* (MRSA) and vancomycin resistant Enterococci (VRE). Daptomycin disrupts many features of the cell membrane of susceptible bacteria causing a rapid depolarization of the membrane potential leading to the intracellular inhibition of DNA, RNA and protein synthesis. This action leads to bacterial death (Miller et al. 2016).

Table 13.1 contd. ...

...Table 13.1 contd.

***Streptomyces* spp.**	**Antibiotic**	**Structure**	**Mechanism of action and uses**
S. griseus	Streptomycin	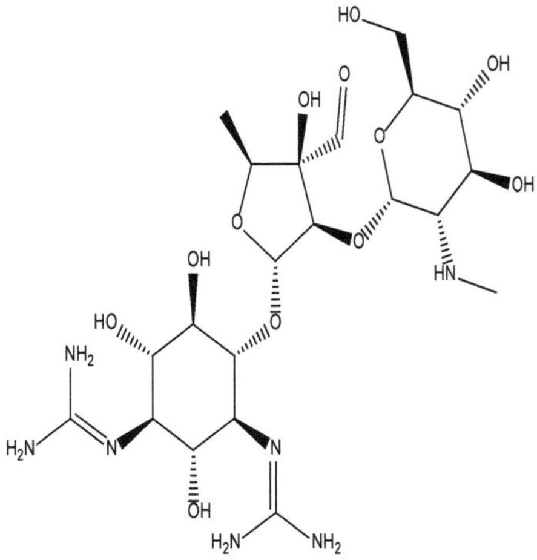	Aminoglycoside antibiotic
S. fradiae	Neomycin	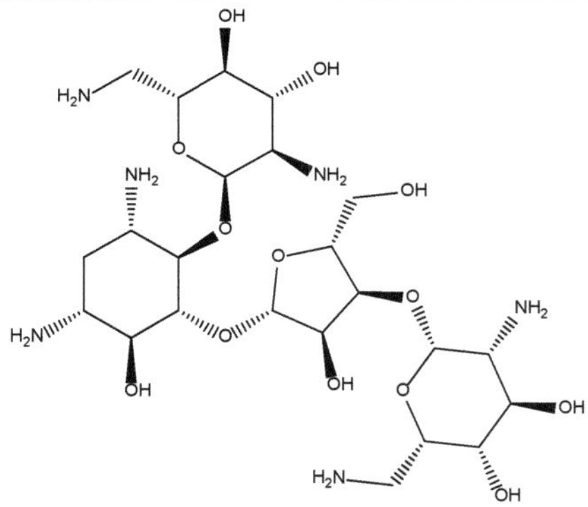	It works by binding the bacterial 30S ribosomal sub-16S unit's rRNA and S12 protein allowing the inhibition of protein synthesis initiation. Neomycin also leads to misinterpretation and premature ending of mRNA prototype leading to cell death.
S. ribosidificus	Ribostamycin	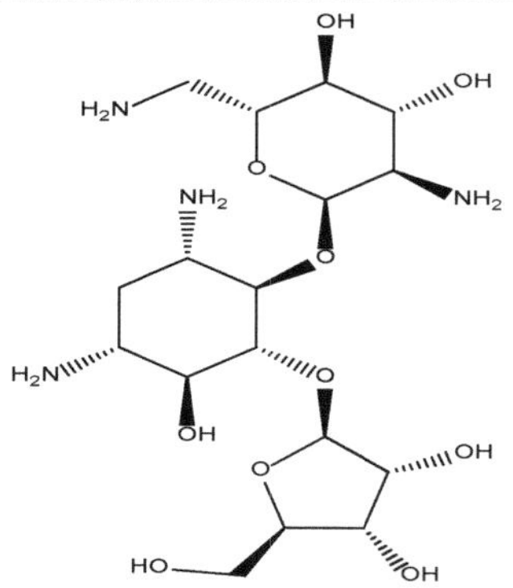	Broad spectrum antibiotic. Active against Gram-positive, Gram-negative bacteria, and mycobacteria (Sanchez and Demain 2011).

Table 13.1 contd. ...

...Table 13.1 contd.

***Streptomyces* spp.**	**Antibiotic**	**Structure**	**Mechanism of action and uses**
S. kanamyceticus	Kanamycin	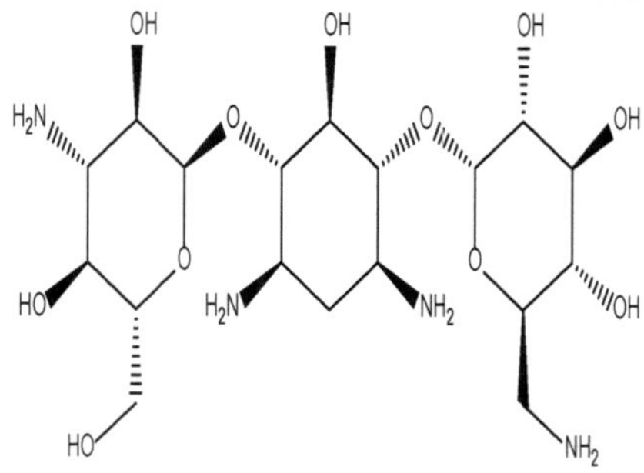	Kanamycin A is the major component of the kanamycin complex, an aminoglycoside antibiotic.
S. stenebrarius	Tobramycin	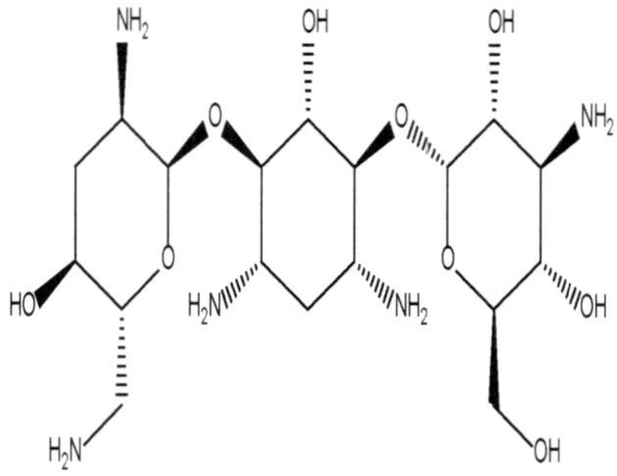	Aminoglycoside antibiotic
S. spectabilis	Spectinomycin	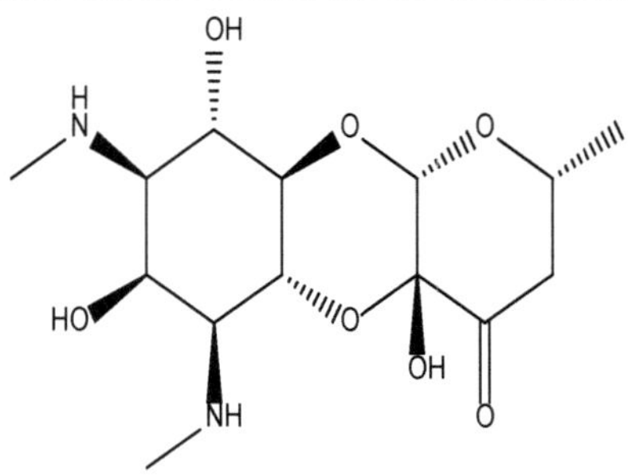	Active against Gram-negative bacteria and used for the treatment of gonorrhea.
S. clavuligerus	Clavulanic acid	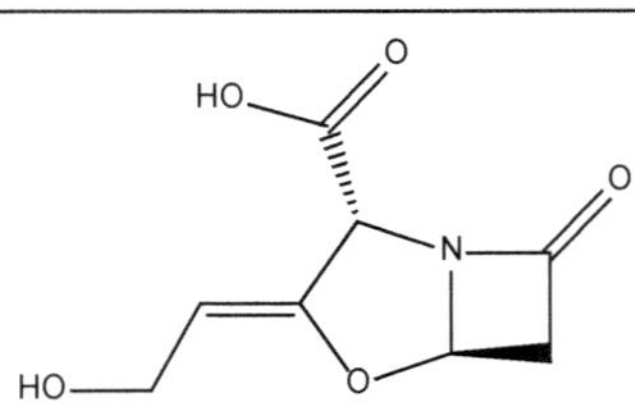	Potent β-lactamases inhibitor that responsible for the hydrolysis of β-lactam antibiotics (Neto et al. 2005).

Table 13.1 contd. ...

...Table 13.1 contd.

***Streptomyces* spp.**	**Antibiotic**	**Structure**	**Mechanism of action and uses**
S. clavuligerus	Cephamycin C	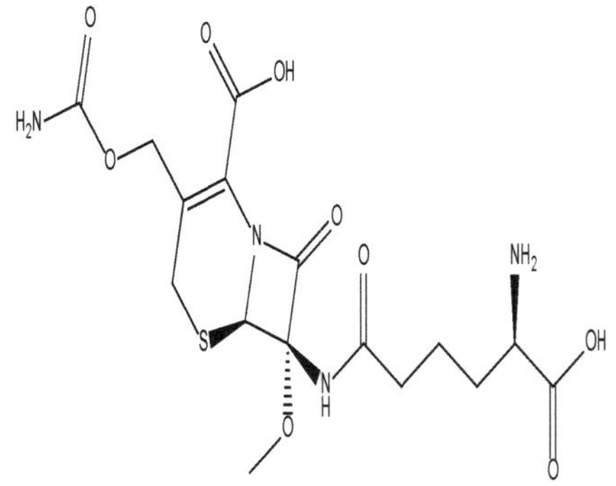	Broad Gram-negative activity (Jacks et al. 1980; Antonio et al. 2012)
S. parvulus, S. sindenensis	Actinomycin D (dactinomycin)	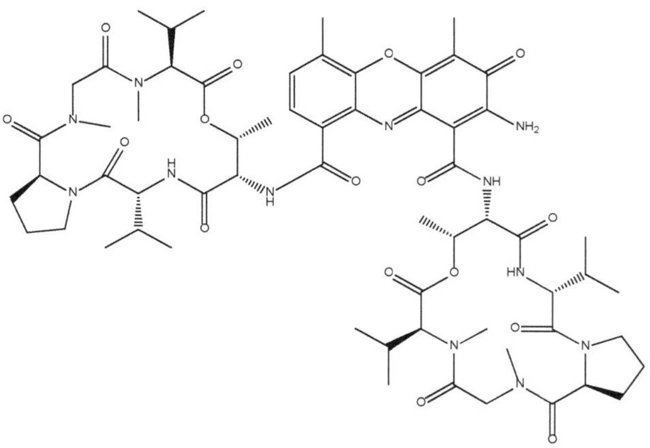	It inhibits DNA-directed RNA synthesis by intercalating into DNA between adjacent G-C base pairs that can be used as antineoplastic agent (Praveen et al. 2008).
S. fradiae	Dekamycin (neomycin B)	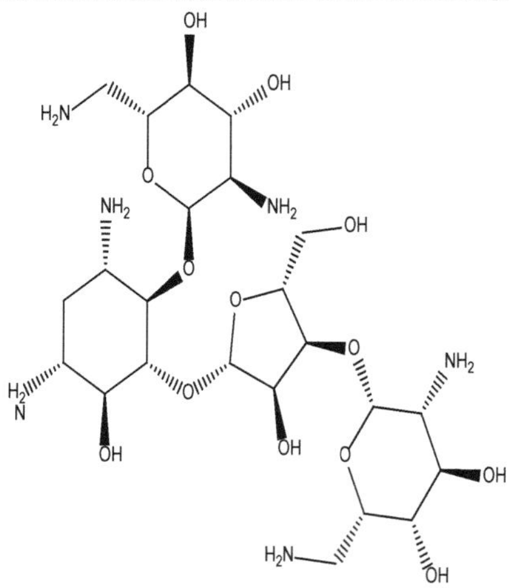	Broad spectrum aminoglycoside antibiotic (Van et al. 1988).

Table 13.1 contd. ...

...Table 13.1 contd.

***Streptomyces* spp.**	**Antibiotic**	**Structure**	**Mechanism of action and uses**
S. nodosus	Amphotericin B	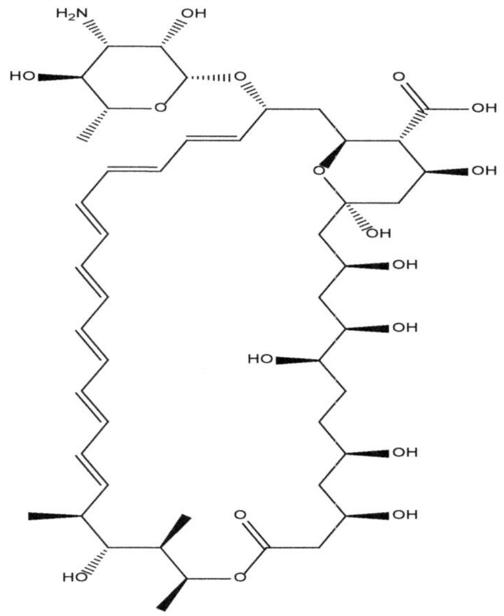	It works by binding with ergosterol, an integral part of the fungal cell membrane. This results in the depolarity and permeability of the cell membrane allowing the leakage of intra-cell materials, cell rupture and finally cell death so it is widely used as antifungal agent (Zhang et al. 2020).
S. noursei	Nystatin	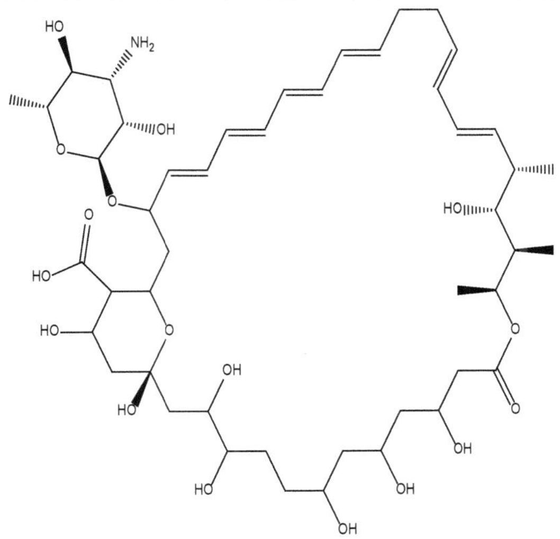	Similar to amphotericin b (Fjærvik and Zotchev 2005).
S. fradiae	Fosfomycin	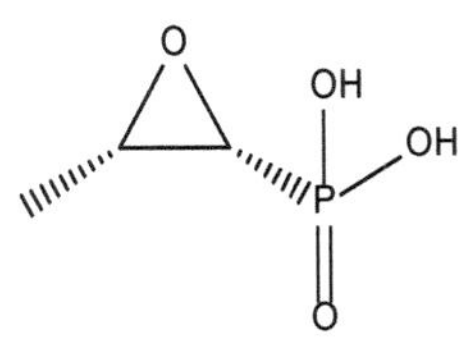	It blocks cell wall synthesis by inhibiting the synthesis of peptidoglycans. It has good activity against staphylococci, pneumococci, and Gram-negative bacteria (Horton 2014).

Table 13.1 contd. ...

...Table 13.1 contd.

***Streptomyces* spp.**	**Antibiotic**	**Structure**	**Mechanism of action and uses**
S. mediterranei	Rifampin	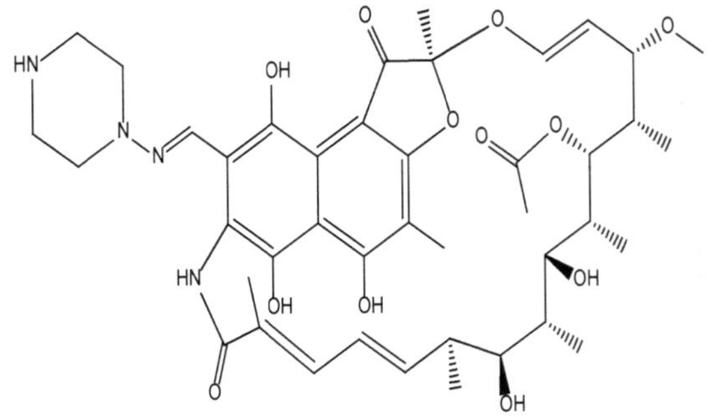	Rifamycin-derived semi-synthetic antibiotic belonging to macrocyclic class antibiotic. Active against diverse Gram-positive and negative bacteria. It is also involved in the fight against TB and most Mycobacterium species. It inhibits DNA-dependent RNA polymerase (Kumar 2017).
S. vinaceus, S. puniceus	Viomycin	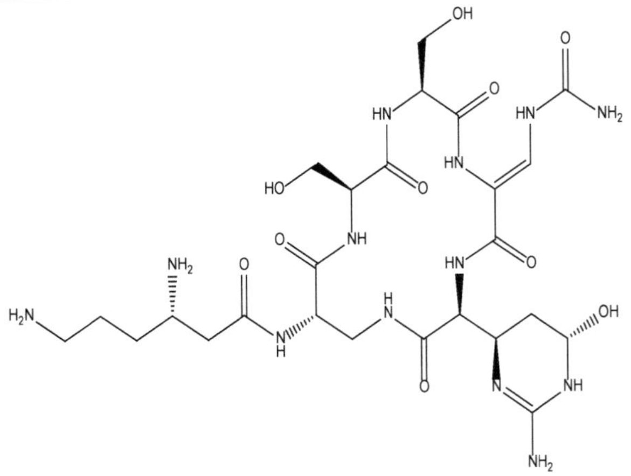	Cyclic polypeptide used to treat tuberculosis by blocking bacterial ribosomal translocation.
S. pristinaespiralis, S. virginiae	Virginiamycin	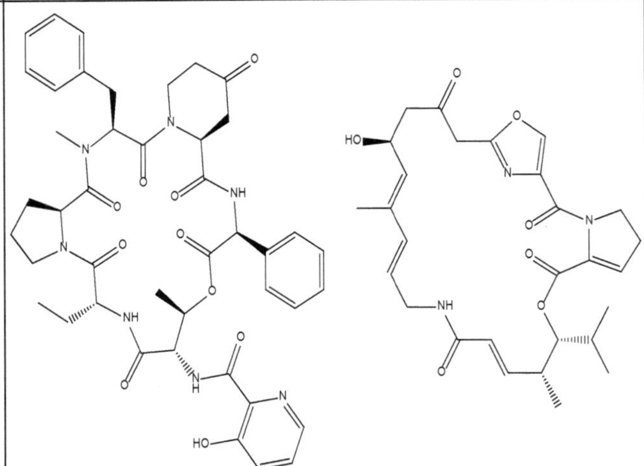	A mixture of two components named virginiamycin M1 and virginiamycin S1. They inhibit protein synthesis by binding to and inhibition of ribosomal assembly (Rath et al. 2010).

Table 13.1 contd. ...

...Table 13.1 contd.

***Streptomyces* spp.**	**Antibiotic**	**Structure**	**Mechanism of action and uses**
S. erythraeus	Erythromycin	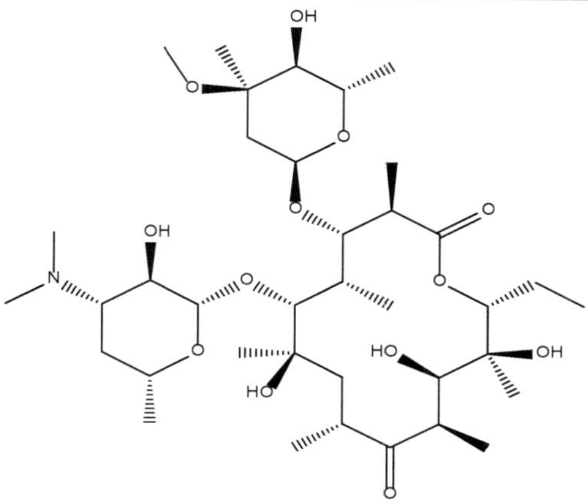	A macrolide antibiotic. It blocks elongation of proteins by binding to the 50S subunit inhibiting translocation during protein synthesis. It is also active in Gram-positive, including *Neisseria, Bodetella, Brucella, Campylobacter, Legionella, Treponema, Chlamydia, Mycoplasma*, and certain Gram-negative bacteria.
S. lincolnensis	Lincomycin	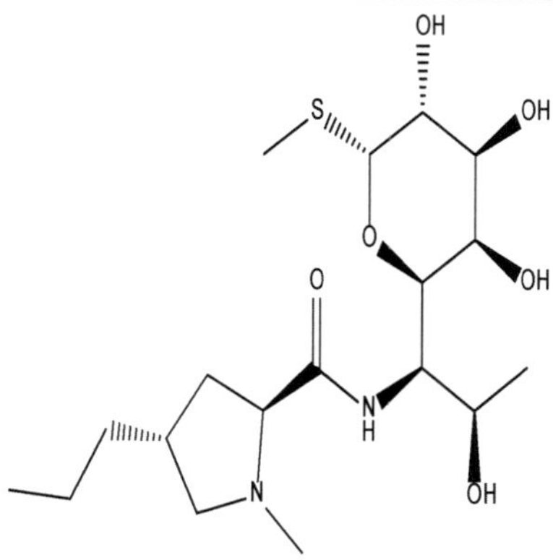	It inhibits protein synthesis through binding to the bacterial 50s ribosomal subunit. It is active against Gram-positive and anaerobic bacteria.
S. garyphalus or S. orchidaceus	cycloserine	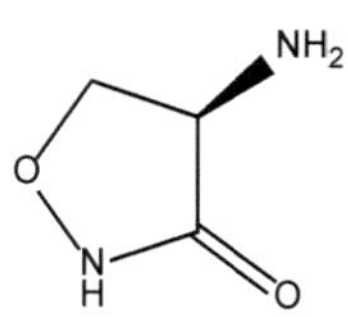	It works as a competitive L-alanine racemase and D-alanine ligases inhibitor by interfering with the bacterial cell wall synthesis via the impairment of the shape of the peptidoglycan required for synthesis of a bacterial cell wall.

Table 13.1 contd. ...

...Table 13.1 contd.

***Streptomyces* spp.**	**Antibiotic**	**Structure**	**Mechanism of action and uses**
S. orientalis	Vancomycin	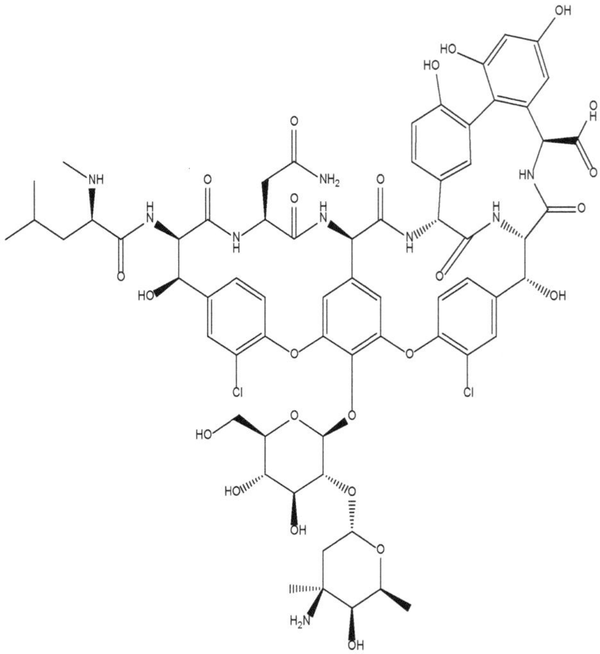	It exerts bactericidal effect by the inhibition of peptidoglycans polymerization in the bacterial cell wall. It is used for the treatment and prevention of many bacterial infections caused by Gram-positive bacteria including methicillin-resistant *Staphylococcus aureus* (MRSA). The spectrum of activity covers also infections caused by streptococci, enterococci, and methicillin-susceptible *Staphylococcus aureus* (MSSA).
S. niveus	Novobiocin (albamycin, cathomycin)	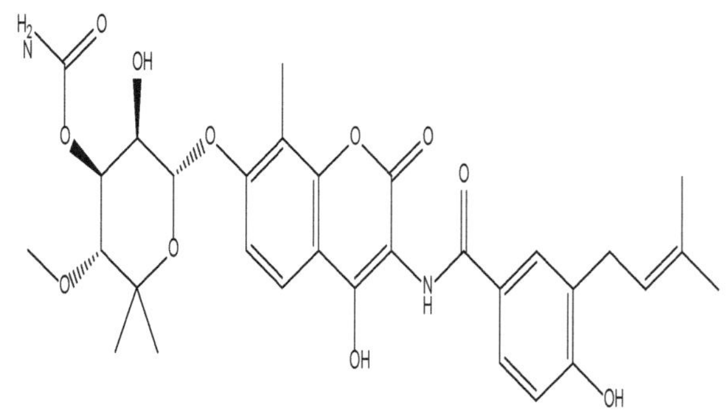	An aminocoumarin antibiotic. Targeting DNA gyrase and the associated DNA topoisomerase IV leads to inhibition of bacterial DNA synthesis is the main mechanism of action of novobiocin.

Table 13.1 contd. ...

...Table 13.1 contd.

Streptomyces spp.	Antibiotic	Structure	Mechanism of action and uses
S. mediterranei	Rifamycin	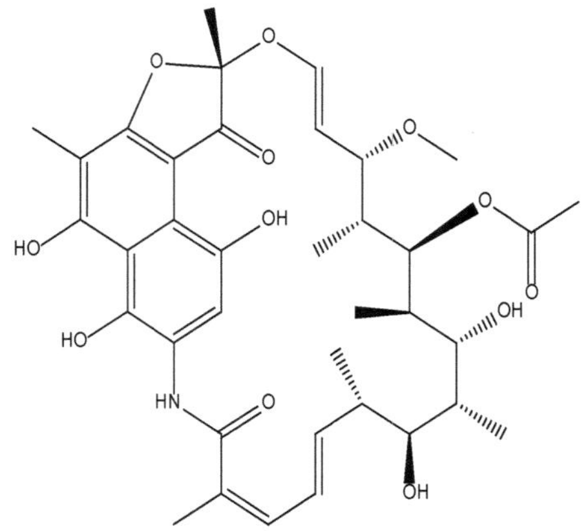	Rifamycin prevents the induction of transcription of RNA polymerase, thus inhibiting prokaryotic RNA synthesis and protein synthesis. It is active against both Gram-positive and negative bacteria, but the main use against *Mycobacterium* spp.
S. alboniger	Puromycin	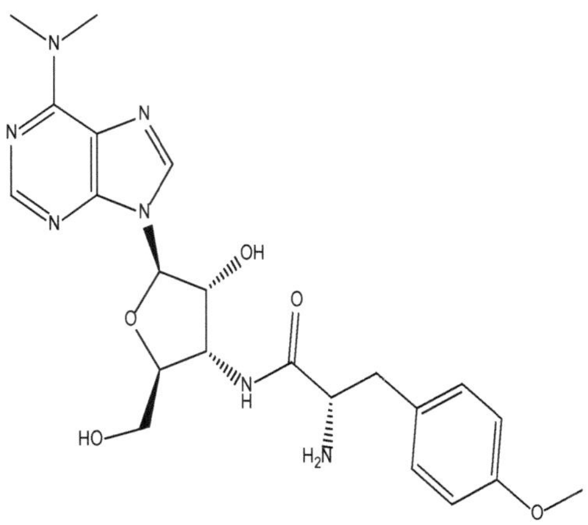	It has dual mechanisms; the first involves the development of nascent peptide by targeting peptidyl-tRNA in the P site as an acceptor substrate. The second mechanism is to compete with aminoacyl-tRNA for binding to the A′ site within the peptidyl transferase nucleus, which is known as the binding site for the 3′ end of aminoacyl-tRNA (Specific bonding of puromycin to full-length protein at the C-terminus). Therefore, it prevents cell growth by blocking the growth of polypeptide chains, effectively suppressing cell growth for both prokaryote and eukaryote cells (Rath et al. 2010).
S. lincolnensis	Clindamycin	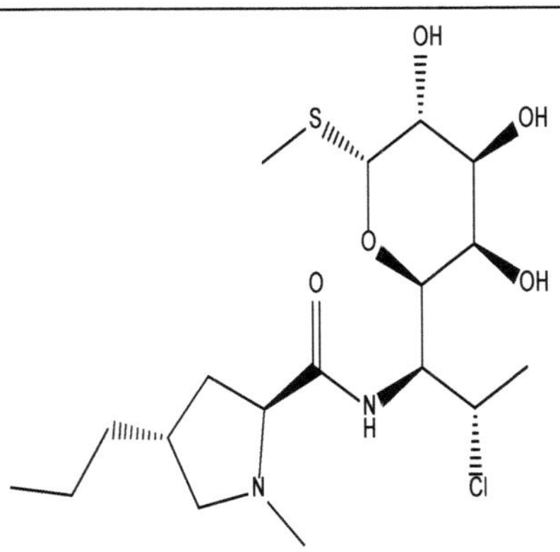	Semisynthetic antibiotic made by modifying the parent compound lincomycin. It prevents bacteria from producing proteins by separating peptidyl-tRNA from the bacterial ribosome.

Novel approaches for antibiotic discovery from *Streptomyces*

Methods for searching for new bacteria/compounds

Investigating different environments

The discovery of new *Streptomyces* is supported by diverse ecology. They have been isolated from the Himalayan mountains, numerous coasts, the Atacama desert, Antarctica and a variety of extremophile environments (Manteca and Yagüe 2019), it could be found even in many volcanic environments, and deep-sea waters at high salt concentrations are the most extreme places on the planet (Arayes et al. 2021). Surpassing the rainforests and the coral reefs, the deep sea is the place with the greatest abundance in biodiversity. Due to the harsh conditions in the deep-sea world, which include the lack of light, a low percentage of oxygen and incredibly high pressures, species that live there must make many biochemical and physiological changes to deal with such a harsh environment, creating distinct chemical entities with diverse biological activities. Until now, the South China Sea has been a prolific producer of *Streptomyces* strains, from which several possible natural compounds have been isolated (Sivalingam et al. 2019). Due to the extreme climatic conditions, the desert is thought to be a haven for rare possible *Streptomyces* with novel bioactive compounds. Arid environments such as high UV radiation, elevated temperatures, high salinity, the occurrence of inorganic oxidants, low amounts of organic material and physical instability caused by heavy winds describe the desert climate. With their novel bioactive compounds, *Streptomyces* strains isolated from the desert are studied to neutralize new drug-resistant infectious diseases and cancer (Sivalingam et al. 2019). New active compounds from new *Streptomyces* strains separated from different environments are described in Table 13.2.

Symbiotic relationships

Symbiosis can be defined as any kind of close or interaction between two dissimilar organisms shaped by selective pressures. One organism (usually a bacterium or fungus) produces small bioactive molecules that affect the host in symbiosis (possibly other symbionts). Mutualistic, commensalism and parasitic symbioses exist. The genetic machinery available to the small molecule creator is essential for symbiosis (Seipke et al. 2012).

Bacteria and plant roots interact with *Streptomyces* genera, creating natural protection against fungal infections. Bacteria and insects are involved in other essential *Streptomyces* interactions. *Streptomyces*, for example, protects the larval cocoon of Beewolf digger wasps from diseases by developing a combination of nine antibiotic compounds. Many new bioactive compounds, including natamycin A and the potent anti-proliferative agent geldanamycin, have been isolated from *Streptomyces* colonies of the South African termite *Macrotermes natalensis*. Another *Streptomyces* extract from the same termite produces the new potent antifungal cyclic analog of dentigerumycin. Finally, a third *Streptomyces* isolates yielded many glycosylated isoflavonoids, including termisoflavone A (Manteca and Yagüe 2019). *Streptomyces* strains isolated from attine ants were shown to be effective against *Escovopsis*, a fungal pathogen that affects insects, and *Leishmania donovanii*, a protozoan parasite that affects humans (Ortega et al. 2019).

Nonculturable microorganisms

Nonculturable microorganisms make up many microbial species and are thought to be a source of new secondary metabolites. The isolation chip (iChip), a system for extracting bacterial cultures in a natural environment, and co-cultivation of actinomycetes with other

Table 13.2. A summary of the non-specific methods and some successful examples of their enforcement.

Methods	Microorganism	Natural product
Media manipulation	*S. roseosporus*	Daptomycin
Stress Response	*S. venezuelae*	Jadomycin B
	S. hygroscopicus	Validamycin A
	S. parvulus	Manumycin family
	S. coelicolor	Ectoine, 5-hydroxyectoine
		Methylenomycin
One Strain Many Compounds (OSMAC)	*S. parvulus*	20 cryptic compounds
Random Mutagenesis	*S. clavuligerus*	Clavulanic acid
	S. hygroscopicus	Rapamycin
	S. coelicolor	Actinorhodin, Undecylprodigiosin
Ribosomal Engineering	*S. coelicolor*	Actinorhodin
Engineering Global Regulation	*S. coelicolor*	Actinorhodin, Prodigiosin, Calcium-Dependent Antibiotic
	S. griseus	Streptomycin
	S. griseochromogenes	Blasticidin S
Elicitors	*S. coelicolor*	Actinorhodin
	S. pristinaespiralis	Desferrioxamine B/E
	S. peucetius	Doxorubicin, Baumycin
	S. coelicolor	Actinorhodin, Undecylprodigiosin
	S. lividans	Prodiginine
	S. griseus	Streptomycin
	S. natalensis	Pimaricin
	29 strains	Cryptic compounds
Metabolic Engineering	*S. clavuligerus*	FK606
	S. coelicolor	Actinorhodin
	S. rimosus	Oxytetracycline
Co-cultures	*S. rimosus* MY02	Antifungal activity
	S. coelicolor	Actinorhodin
	S. fradiae 007	Phenolic polyketides
Conditioning Morphology (PCD + MII)	*S. cattleya*	Tienamycin
	S. cinereoruber	Rodomycin
	Saccharopolyspora erythraea	Erithromycin
	S. coelicolor	Actinorhodin

bacteria are two techniques that have been successfully used to enhance cultivation-based methods. The combination, as mentioned above, of iChip and unexplored environments or symbiotic relationships is also exciting (Manteca and Yagüe 2019).

Methods for assessment the improvement in a known microorganism (unselective)

One Strain Many Compounds Principal (OSMAC)

Triggering the expression of "silent" gene clusters that produce secondary metabolites could have fascinating and novel medical and biotechnological implications. Under laboratory settings, these clusters remain silent (Romano et al. 2018). The "one strain, many compounds" (OSMAC) concept has been used to conceptualize the ideas underlying cultivation-based methods, which highlights how a single microbial strain can produce diverse secondary metabolites when grown under different environmental conditions (Romano et al. 2018).

Alteration in the nutrient regimen

For the production of natural products, secondary metabolism can be influenced. This is accomplished by managing and modifying carbon, nitrogen, phosphorus, sulfur and trace element supplies (Romano et al. 2018).

Changing physical parameters

Standard practice monitors optimum growth conditions, specifically temperature, salinity, and pH, during the taxonomic characterization of newly isolated microorganisms. This and other causes unrelated to the nutritional content of the media may have a significant effect on the growth mechanisms of bacteria and the formation of secondary metabolites. The effect of pH on secondary metabolites production has been documented by Sakar and colleagues, with biofilm formation influencing antibiotic production in a marine *Streptomyces* spp. When the antimicrobial activity was tested in culture broths, it was discovered that no action was shown in an acidic medium (pH 4.0), but activity was noticed up to pH 10.0, with pH 9 being the optimum (Romano et al. 2018).

Co-cultivation and other environmental cues

The Secondary Metabolites (SMs) profiles were changed when *Streptomyces fradiae* 007 was co-cultured with *Penicillium* spp. WC-29-5 to supply two additional polyketides with anticancer properties. This method of co-cultivation was deemed to be the most effective. Bacteria can react to the presence of fungal strains by producing different SMs. Indeed, co-cultivation of *Aspergillus fumigatus* MR2012 and two hyper-arid deserts *Streptomyces leeuwenhoekii* strains resulted in dual induction of bacterial and fungal metabolites, resulting in the discovery of the new bioactive compounds luteoride D and pseurotin G (Romano et al. 2018).

Addition of chemical elicitors

Elicitors are small diffusible molecules that have been shown to increase the synthesis of secondary metabolites and cause cryptic pathways (Manteca and Yagüe 2019).

Stress response

Microorganisms create and export bioactive compounds while they are stressed. This classic approach includes heat and ethanol shock therapies, high salt conditions, elevated hydrostatic pressure, acidic pH shock and feed limitations.

Random mutagenesis

Random mutagenesis technology is a valuable method for generating enzymes, proteins, whole metabolic processes or entire genomes with required or modified characteristics using an iterative mechanism involving recombinant production (Labrou 2010). Chemical, physical and transposon mutagenesis, are effective ways of generating randomized gene mutations in bacterial genomes, are examples of different mutagenesis methods (Manteca and Yagüe 2019).

Ribosomal engineering

One of the rational engineering approaches is to manipulate regulatory genes genetically. Overexpression of the pathway-specific regulatory gene sanG, for example, resulted in a significant increase in nikkomycin production. More recently, evidence has emerged that *Streptomyces* signaling mechanisms are critical for secondary metabolite biosynthesis. Overexpression of the -butyrolactone signaling molecule synthase gene afsA-gR in *Streptomyces fradiae* resulted in a 1.7-fold rise in neomycin development compared to the wild form (Li et al. 2019).

Differentiation in a liquid culture based on morphology

Antibiotic production on a large scale is often done in liquid cultures. The macroscopic morphology of the mycelium is almost universally recognized as being linked to the development of secondary metabolites. Differentiation and secondary metabolism followed the development of pellets and clumps. As a result, new methods for observing macroscopic morphology have been established. Pellet size was measured using laser diffraction. The pellet size distribution of culture populations was determined using flow cytometry. A useful algorithm for characterizing the morphology of filamentous microorganisms in liquid cultures was recently created as a plug-in for the open-source program ImageJ. The activity of *Streptomyces* liquid cultures has been predicted using mathematical models focused on pellet/clump morphology (Manteca and Yagüe 2018).

Programmed cell death and MII differentiation

In liquid and solid cultures, Programmed Cell Death (PCD) is a crucial step that triggers the separation of antibiotic producer mycelium (MII). Modifying culture conditions and morphology can influence PCD and MII differentiation. Secondary metabolite production improves as developmental factors promote PCD and MII separation (Manteca and Yagüe 2019). Table 13.2 shows new active compounds extracted from *Streptomyces* strains with nonselective methods (Manteca and Yagüe 2018).

Methods for improvement in a known compound (selective)

Regulatory engineering

The addition of previously synthesized precursor molecules of the compound of interest can improve the production of certain compounds through the overexpression of its positive regulators. On the contrary, knocking out the suppressors gene are examples of methods used in regulatory engineering (Manteca and Yagüe 2019).

Heterologous expression

Heterologous expression of SMs gene pathways is used to increase the amount of desired compounds generated, trigger cryptic gene clusters, exploit clusters from genetically unamenable strains, extract bioactive compounds from uncultivable organisms and develop new unusual pathways (Ahmed et al. 2020)

Genome mining

It is widely accepted that any bioinformatics investigation used to detect the biosynthetic mechanism of bioactive natural products and their potential functional and chemical associations are referred to as "genome mining." Genome mining entails discovering previously unknown secondary metabolites biosynthetic gene clusters within the genomes of sequenced species and sequence analysis of the enzymes encoded by these gene clusters, and laboratory recognition of the gene clusters' properties (Albarano et al. 2020). The integration of genome-mining and metabolomics plays an important role in the discovery of microbial natural products by the aid of bioinformatics (Sukmarini 2021).

Combinatorial biosynthesis

Combinatorial biosynthesis can be defined as metabolic engineering and mutasynthesis to generate new "unnatural" natural products (Manteca and Yagüe 2019). There has not been much work carried out to explore such artificial architecture of the biosynthetic pathways in order to produce a plethora of a structurally designed antibiotics that can be used efficiently to combat for severe infections.

Conclusion

The genus *Streptomyces* is distributed worldwide. It has an amazing ability to survive in harsh environmental conditions such as the deep sea, arid desert with elevated temperature and lack of humidity, high latitude mountains and even volcanic environment as well. This genus provides a gold mine for researchers since decades to explore its power to produce many bioactive secondary metabolites, which play an important role in the pharmaceutical industry as antimicrobial and immune-suppressant agents. These secondary metabolites can be used as antibiotics, antiprotozoal, antibacterial, antifungal and antiviral agents, anthelmintic, herbicides/pesticides, plant growth promoting rhizobia, anticancer drugs and immune modulators. The *Streptomyces* species produce about two thirds of the known naturally occurring antibiotics. This chapter has shed light on the natural antibiotics produced by the genus Streptomyces starting from the discovery of Streptothricin in 1942, discussing their mechanism of action and uses. However, more efforts are still required based on the successful approaches for antibiotic discovery from *Streptomyces* that include searching for new strains of the genus, screening already known micro-organism for new bioactive compounds, and selective production improvement of a known compound. These efforts may represent the only hope for finding new antibiotics that can overcome the multidrug-resistance microbes that represent a great challenge for many health care systems responsible for the death in many developing countries.

References

Ahmed, Y., Rebets, Y., Estévez, M.R., Zapp, J., Myronovskyi, M., Luzhetskyy, A. (2020). Engineering of *Streptomyces lividans* for heterologous expression of secondary metabolite gene clusters. Microb Cell Fact 19(1): 5. https://doi.org/10.1186/s12934-020-1277-8.

Albarano, L., Esposito, R., Ruocco, N., Costantini, M. (2020). Genome mining as new challenge in natural products discovery. Mar Drugs 18(4): 199. https://doi.org/10.3390/md18040199.

Aminov, R.I. (2010). A brief history of the antibiotic era: lessons learned and challenges for the future. Front Microbiol 1: 134. https://doi.org/10.3389/fmicb.2010.00134.

Anderson, A.S., Wellington, E.M.H. (2001). The taxonomy of *Streptomyces* and related genera. Int J Syst Evol Microbiol 51(3): 797–814. https://doi.org/10.1099/00207713-51-3-797.

Antonio, T., Bellão, C., Corrêa, T., Cavallieri, A.P., Badino, A.C., Araujo, M.L.G.D.C. (2012). Evaluation of different media for the production of cephalosporins by *Streptomyces clavuligerus* ATCC 27064. Braz Arch Biol Techn, 55(6): 819–825. https://doi.org/10.1590/S1516-89132012000600003.

Arai, T. (1997). What are Actinomycetes? Atlas of Actinomycetes', the Society for Actinomycetes Japan (SAJ).

Arayes, M.A., Mabrouk, M.E.M., Sabry, S.A., Abdella, B. (2021). Diversity and characterization of culturable haloalkaliphilic bacteria from two distinct hypersaline lakes in northern Egypt. Biologia 76(2): 751–761. https://doi.org/10.2478/S11756-020-00609-5.

Bhattacharyya, B.K., Pal, S.C., Sen, S.K. (1998). Antibiotic production by *Streptomyces hygroscopicus* D1.5: Cultural effect. Revist Microbiol 29(3): 167–169. https://doi.org/10.1590/S0001-37141998000300003.

Bibb, M.J. (2005). Regulation of secondary metabolism in *Streptomycetes*. Curr Opin Microbiol 8(2): 208–215. https://doi.org/10.1016/j.mib.2005.02.016.

Brook, I. (2016). Spectrum and treatment of anaerobic infections. J Infect Chemother 22(1): 1–13. https://doi.org/10.1016/j.jiac.2015.10.010.

Chopra, I., Roberts, M. (2001). Tetracycline antibiotics: mode of action, applications, molecular biology, and epidemiology of bacterial resistance. Microbiol Mol Biol R 65(2): 232–260. http://doi.org/10.1128/MMBR.65.2.232-260.2001.

de Lima Procópio, R.E., da Silva, I.R., Martins, M.K., de Azevedo, J.L., de Araújo, J.M. (2012). Antibiotics produced by *Streptomyces*. Braz J Infect Dis 16(5): 466–471. https://doi.org/10.1016/j.bjid.2012.08.014.

Dinos, G.P., Athanassopoulos, C.M., Missiri, D.A., Giannopoulou, P.C., Vlachogiannis, I.A., Papadopoulos, G.E., Papaioannou, D., Kalpaxis, D.L. (2016). Chloramphenicol derivatives as antibacterial and anticancer agents: historic problems and current solutions. Antibiotics 5(2): 20. http://doi: 10.3390/antibiotics5020020.

Dominguez-Godinez, C., Carracedo, G., Pintor, J. (2018). Diquafosol delivery from silicone hydrogel contact lenses: Improved effect on tear secretion. J Ocul Pharmacol Ther 34(1–2): 170–176. https://doi.org/10.1089/jop.2016.0193.

Fjærvik, E., Zotchev, S.B. (2005). Biosynthesis of the polyene macrolide antibiotic nystatin in *Streptomyces noursei*. Appl Microbiol Biot 67(4): 436-443. https://doi.org/10.1007/s00253-004-1802-4.

Gaynes, R. (2017). The discovery of penicillin—new insights after more than 75 years of clinical use. Emerg Infect Dis 23(5): 849–853. https://doi.org/10.3201/eid2305.161556.

Holt, J., Krieg, N., Sneath, P. (1994). Bergey's Manual of Determinative Bacterology.

Horton, J.M. (2015). 36 Urinary Tract Agents: Nitrofurantoin, Fosfomycin, and Methenamine.

Hutchings, M., Truman, A., Wilkinson, B. (2019). Antibiotics: past, present and future. Curr Opin Microbiol 51: 72–80. https://doi.org/10.1016/j.mib.2019.10.008.

Jacks, T.M., Schleim, K.D., Judith, F.R., Miller, B.M. (1980). Cephamycin C treatment of induced enterotoxigenic colibacillosis (scours) in calves and piglets. Antimicrob Agents Ch 18(3): 397–402. http://doi: 10.1128/aac.18.3.397.

Jose, P.A., Maharshi, A., Jha, B. (2021). Actinobacteria in natural products research: Progress and prospects. Microbiol Res 246: 126708. https://doi.org/10.1016/J.MICRES.2021.126708.

Kämpfer, P. (2006) The Family Streptomycetaceae, Part I: Taxonomy. *In*: Dworkin, M., Falkow, S., Rosenberg, E., Schleifer, K.H., Stackebrandt, E. (eds.). The Prokaryotes. Springer, New York, NY. https://doi.org/10.1007/0-387-30743-5_22.

Kämpfer, P., Huber, B., Buczolits, S., Thummes, K., Grün-Wollny, I., Busse, H.J. (2008). *Streptomyces specialis* sp. nov. Int J Syst Evol Micr 58(11): 2602–2606. https://doi.org/10.1099/ijs.0.2008/001008-0.

Kariminik, A., Baniasadi, F. (2010). Pageantagonistic activity of actinomycetes on some gram negative and gram positive bacteria. World Appl Sci J 8(7): 828–832.

Kieser, T., Bibb, M.J., Buttner, M.J., Chater, K.F., Hopwood, D.A. (2000). General introduction to actinomycete biology. Practical Streptomyces Genetics.

Krause, K.M., Serio, A.W., Kane, T.R., Connolly, L.E. (2016). Aminoglycosides: An overview. Cold Spring Harb Perspect Med 6(6): a027029. https://doi.org/10.1101/cshperspect.a027029.

Kumagai, T., Koyama, Y., Oda, K., Noda, M., Matoba, Y., Sugiyama, M. (2010). Molecular cloning and heterologous expression of a biosynthetic gene cluster for the antitubercular Agent D-cycloserine produced by *Streptomyces lavendulae*. Antimicrob Agents Chemother 54(3): 1132–1139. https://doi.org/10.1128/AAC.01226-09.

Kumar, P. (2017). Pharmacology of specific drug groups. Pharmacology and Therapeutics for Dentistry, 7th ed.; Elsevier: Amsterdam, The Netherlands, 457–487.

Labrou, N. (2010). Random mutagenesis methods for *in vitro* directed enzyme evolution. Curr Protein Pept Sci 11(1): 91–100. https://doi.org/10.2174/138920310790274617.

Li, C., Zhang, F., Kelly, W.L. (2011). Heterologous production of thiostrepton A and biosynthetic engineering of thiostrepton analogs. Molecul BioSyst 7(1): 82–90. https://doi.org/10.1039/c0mb00129e.

Li, D., Zhang, J., Tian, Y., Tan, H. (2019). Enhancement of salinomycin production by ribosome engineering in *Streptomyces albus*. Sci China Life Sci 62(2): 276–279. https://doi.org/10.1007/s11427-018-9474-7.

Liras, P., Demain, A.L. (2009). Enzymology of β-lactam compounds with cephem structure produced by Actinomycete. *In*: Methods in Enzymology 458: 401–429. Academic Press. https://doi.org/10.1016/S0076-6879(09)04816-2.

Madigan, M., Martinko, J., J, P. (2000). Brock Biology of Microorganisms. Prentice Hall. 1552.

Maleki, H., Dehnad, A., Hanifian, S., Khani, S. (2013). Isolation and molecular identification of streptomyces spp. with antibacterial activity from Northwest of Iran. BioImpacts 3(3): 129–134. https://doi.org/10.5681/bi.2013.017.

Mann, J. (2001). Natural products as immunosuppressive agents. Nat Prod Rep 18(4): 417–430. https://doi.org/10.1039/b001720p.

Manteca, Á., Yagüe, P. (2018). *Streptomyces* differentiation in liquid cultures as a trigger of secondary metabolism. Antibiotics 7(2): 41. https://doi.org/10.3390/antibiotics7020041.

Manteca, Á., Yagüe, P. (2019). *Streptomyces* as a source of antimicrobials: novel approaches to activate cryptic secondary metabolite pathways. *In*: Antimicrobials, Antibiotic Resistance, Antibiofilm Strategies and Activity Methods. IntechOpen. https://doi.org/10.5772/intechopen.81812.

Martens, E., Demain, A.L. (2011). Platensimycin and platencin: promising antibiotics for future application in human medicine. J Antibiot 64(11): 705–710. https://doi.org/10.1038/ja.2011.80.

Miller, W.R., Bayer, A.S., Arias, C.A. (2016). Mechanism of action and resistance to daptomycin in *Staphylococcus aureus* and enterococci. CSH Perspect Med 6(11). http://doi:10.1101/cshperspect.a026997a026997.

Mohanraj, G., Sekar, T. (2013). Isolation and screening of Actinomycetes from marine sediments for their potential to produce antimicrobials. Int J Life Sci Biotechnol Pharm Res 2(3): 115–126.

Morin, J.B., Sello, J.K. (2010). Efficient synthesis of a peculiar vicinal diamine semiochemical from streptomyces natalensis. Org Lett 12(15): 3522–3524. https://doi.org/10.1021/ol1013763.

Murphy, B., Anderson, K., Borissow, C., Caffrey, P., Griffith, G., Hearn, J., Ibrahim, O., Khan, N., Lamburn, N., Lee, M., Pugh, K., Rawlings, B. (2010). Isolation and characterisation of amphotericin B analogues and truncated polyketide intermediates produced by genetic engineering of *Streptomyces nodosus*. Org Biomolecul Chem 8(16): 3758. https://doi.org/10.1039/b922074g.

Neto, A.B., Hirata, D.B., Cassiano Filho, L.C.M., Bellão, C., Badino Júnior, A.C., Hokka, C.O. (2005). A study on clavulanic acid production by *Streptomyces clavuligerus* in batch, fed-batch and continuous processes. Braz J Chem Eng 22: 557–563.

Ortega, H.E., Ferreira, L.L.G., Melo, W.G.P., Oliveira, A.L.L., Alvarenga, R.F.R., Lopes, N.P., Bugni, T.S., Andricopulo, A.D., Pupo, M.T. (2019). Antifungal compounds from *Streptomyces* associated with attine ants also inhibit *Leishmania donovani*. PLoS Negl Trop Dis 13(8). https://doi.org/10.1371/journal.pntd.0007643.

Praveen, V., Tripathi, C.K.M., Bihari, V., Srivastava, S.C. (2008). Production of actinomycin-D by a new isolate, *Streptomyces sindenensis*. Ann Microbiol 58(1): 109–114. https://doi.org/10.1007/BF03179453.

Rath, C.M., Scaglione, J.B., Kittendorf, J.D., Sherman, D.H. (2010). NRPS/PKS hybrid enzymes and their natural products. https://doi.org/10.1016/B978-008045382-8.00725-5.

Reza Dehnad, A., Yeganeh, P., Bakhshi, R., Mokhtarzadeh, A., Soofiani, S., Monadi, A.R., Gasanova, S., Abusov, R. (2010). Investigation antibacterial activity of *Streptomycetes* isolates from soil samples, West of Iran. African J Microbiol Res 4(16): 1685–1693.

Ribeiro da Cunha, Fonseca, Calado. (2019). Antibiotic discovery: where have we come from, where do we go? Antibiotics 8(2): 45. https://doi.org/10.3390/antibiotics8020045.

Robinson, R. (2014). Encyclopedia of Food Microbiology. 2nd edition.

Romano, S., Jackson, S.A., Patry, S., Dobson, A.D. (2018). Extending the "One Strain Many Compounds" (OSMAC) principle to marine microorganisms. Mar Drugs 16(7): 244. https://doi.org/10.3390/md16070244.

Sadhasivam, S., Shanmugam, P., Yun, K.S. (2010). Biosynthesis of silver nanoparticles by Streptomyces hygroscopicus and antimicrobial activity against medically important pathogenic microorganisms. Colloids Surf B: Biointerfaces 81(1): 358–362. https://doi.org/10.1016/j.colsurfb.2010.07.036.

Sanchez, S., Demain, A.L. (2011). Secondary metabolites. (1.12.2.2 Aminoglycosides), in Comprehensive Biotechnology (Second Edition) 1: 155–167. https://doi.org/10.1016 B978-0-08-088504-9.00018-0.

Sanglier, J.J., Wellington, E.M.H., Behal, V., Fiedler, H.P., Ellouz Ghorbel, R., Finance, C., Hacene, M., Kamoun, A., Kelly, C., Mercer, D.K., Prinzis, S., Trigo, C. (1993). Novel bioactive compounds from *Actinomycetes*. Res Microbiol 144(8): 661–663. https://doi.org/10.1016/0923-2508(93)90071-9.

Schatz, A., Bugle, E., Waksman, S.A. (1944). Streptomycin, a substance exhibiting antibiotic activity against Gram-positive and Gram-negative bacteria. Experiment Biol Med 55(1): 66–69. https://doi.org/10.3181/00379727-55-14461.

Seipke, R.F., Kaltenpoth, M., Hutchings, M.I. (2012). *Streptomyces* as symbionts: an emerging and widespread theme? FEMS Microbiol Rev 36(4): 862–876. https://doi.org/10.1111/j.1574-6976.2011.00313.x.

Sivalingam, P., Hong, K., Pote, J., Prabakar, K. (2019). Extreme environment *Streptomyces*: potential sources for new antibacterial and anticancer drug leads? Int J Microbiol 2019: 5283948. https://doi.org/10.1155/2019/5283948.

Song, Q., Huang, Y., Yang, H. (2012). Optimization of fermentation conditions for antibiotic production by Actinomycetes YJ1 Strain against *Sclerotinia sclerotiorum*. J Agric Sc 4(7). https://doi.org/10.5539/jas.v4n7p95.

Straits, F., Brown, R., Elion, G. (2009). Women chemists in the national inventors' Hall of Fame: their remarkable lives and their award-winning research. Bull Hist Chem 34(1).

Sukmarini, L. (2021). Recent advances in discovery of lead structures from microbial natural products: genomics- and metabolomics-guided acceleration. Molecules 26(9): 2542. https://doi.org/10.3390/MOLECULES26092542.

Van, T.D., Do Khac, D., Fetizon, M., Tabet, J.C., Le Beyec, Y., Quyen, T.C. (1988). Identification of dekamycin antibiotics. J Chromatogr A 440: 71–85. https://doi.org/10.1016/S0021-9673(00)94512-6.

Vickers, N.J. (2017). Animal communication: when I'm calling you, will you answer too? Curr Biol 27(14): R713–R715. https://doi.org/10.1016/j.cub.2017.05.064.

Watve, M.G., Tickoo, R., Jog, M.M., Bhole, B.D. (2001). How many antibiotics are produced by the genus *Streptomyces*? Arch Microbiol.176(5): 386–390. https://doi.org/10.1007/s002030100345.

Williams, S.T., Goodfellow, M., Alderson, G., Wellington, E.M., Sneath, P.H., Sackin, M.J. (1983). Numerical classification of *Streptomyces* and related genera. J General Microbiol 129(6): 1743–1813. https://doi.org/10.1099/00221287-129-6-1743.

Zhang, B., Zhou, Y.T., Jiang, S.X., Zhang, Y.H., Huang, K., Liu, Z.Q., Zheng, Y.G. (2020). Amphotericin B biosynthesis in *Streptomyces nodosus*: quantitative analysis of metabolism via LC–MS/MS based metabolomics for rational design. Microb Cell Fact 19(1): 1–12. https://doi.org/10.1186/s12934-020-1290-y.

14

Impact of Endosymbionts on Antimicrobial Properties of Medicinal Plants

Flávia Figueira Aburjaile, José Ribamar Costa Ferreira-Neto, Thamara de Medeiros Azevedo, Juan Carlos Ariute, Jéssica Barboza da Silva, Roberta Lane de Oliveira Silva, Valesca Pandolfi and *Ana Maria Benko-Iseppon**

Introduction

Plants are cornerstones for supporting life on Earth. Together with some protists (algae and euglena) and bacteria, they represent photosynthetic organisms: i.e., transform solar energy and CO_2 into glucose, which is the primary energy source for each living organism. Additionally, they provide O_2 (a by-product of photosynthesis) for the consumption of heterotrophs. Such plants burn fuel (sugars and fatty acids) in their cells to produce energy from that element. In addition to this *'sine qua non'* importance to the maintenance of life, plants have a diversified use, acting as food, firewood, timber, fibers, dyes, pesticides, oils, rubber and now medicines (Hodge 2010).

Regarding medicine, the therapeutic properties of these organisms have been used since ancient times. According to Kelly (2009), the oldest written evidence of medicinal plants used to prepare drugs was found on a Sumerian clay slab from Nagpur, India dated approximately 5000 years ago. Plants are currently considered the largest 'biochemical and pharmaceutical stores' on the planet. These 'living stores' generate a myriad of bioactive compounds.

The plants' bioactive compounds are species-specific. They may vary due to several factors, such as (Maggini et al. 2017): genetics, cultivation techniques, plant material, isolation methods, plant tissue treatment, physiological and phytosanitary status. Recently, attention has been focused on the plant microbiome, especially on plant microbiota, and its roles in producing specialized metabolites, and its influence on the host's biochemistry. The term 'plant microbiome' refers to the entire habitat, including the microorganisms (bacteria, archaea, lower and higher eukaryotes and viruses), their genomes (i.e., genes), and the surrounding environmental conditions (Marchesi and Ravel 2015). 'Plant microbiota', in turn, means the set of endophytic organisms that are systematically associated with a

Laboratório de Genética e Biotecnologia Vegetal, Depto. Genética, Universidade Federal de Pernambuco, Recife, PE, Brazil.
* Corresponding author: ana.iseppon@gmail.com

given host plant (Lemanceau et al. 2017). This microbial community can originate from the following niches (Compant et al. 2016):

- the phyllosphere (i.e., the aerial plant habitat or the leaf surface concerning the external environment; Fig. 14.1A);
- the laimosphere (i.e., microbiologically enriched zone of soil that surrounds below-ground portions of plant stems; Fig. 14.1B);
- the rhizosphere (the narrow region of soil that is directly influenced by root secretions and associated soil microorganisms; Fig. 14.1C);
- the anthosphere (the external environment of flowers; Fig. 14.1D);
- the caulosphere (i.e., the region immediately around the stem surface; Fig. 14.1E);
- the carposphere (i.e., the external fruit surface; Fig. 14.1F);
- and the spermosphere (the exterior of germinated seed; Fig. 14.1G), among others.

From different external areas, endophyte microbiota can penetrate the inner of plant tissues (Fig. 14.1F) and establish an intimate partnership—frequently based on mutual benefits—with their hosts. Endophytes have been found in every plant species examined to date. According to Strobel and Daisy (2003), of the nearly 300,000 plant species on this planet, each plant is the host to one or more associated microbe, in some cases species-specific.

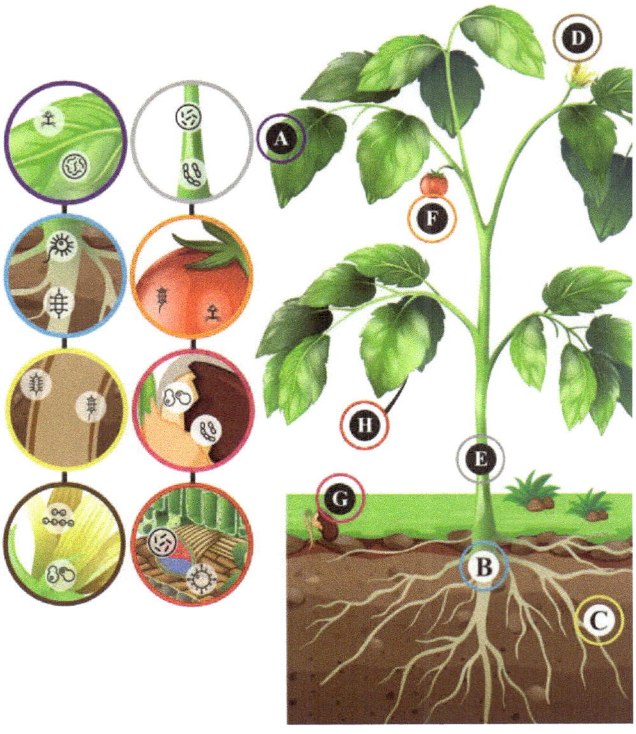

Figure 14.1. Different niches/origins of endophytes in plants: the phyllosphere 'A', the laimosphere 'B', rhizosphere 'C', the anthosphere 'D', the caulosphere 'E', the carposphere 'F', and the spermosphere 'G'; besides the plant inner tissues localization 'H'.

Due to the positive impact of this mutualistic endophytism, one modern approach to focus on this topic is to consider the combination 'plant host-microbiota' as a "holobiont". The definition of the term 'holobiont' was initially employed to designate a 'simple biological entity involving a host and a single inherited symbiont' (Margulis 1991); subsequently, it was adapted to 'individual host and its [entire] microbial community' (Theis et al. 2016). Considering the plant and its host organisms as holobionts is more true and realistic. Evolution acts on the phenotypic properties, which result from the integration of the holobiont organism (e.g., a plant) and its microbial guests. Therefore, it is reasonable to consider that evolution acts on the 'hologenome', i.e., 'the collective genomic content of all the individual bionts within the holobiont' (Rosenberg et al. 2007; Zilber-Rosenberg and Rosenberg 2008).

The influence of plant-associated microbiota in traditional herbal medicine has been discussed (Gunatilaka 2006; Köberl et al. 2013; Huang et al. 2018), and it is clear that this association plays an essential ethnobotanical and ethnomedicinal factor. The active properties of some medicinal plants may be due to the presence of a given microbial host, often also associated with a specific environment or, of the synergistic action of both plant and the associated microorganisms. In some cases, the antimicrobial properties of plants have been proposed to rely exclusively on a given endophytic. For example, Myo et al. (2020) identified endophytic bacteria from *Catharanthus roseus* and *Boscia variabilis* with potent antimicrobial activity against pathogenic microbes, with potential for use in medicine and agriculture. In other cases, only the synergistic action of the plant and the microbial community may achieve the complete phytotherapeutic effect. Nguyen et al. (2019) evaluated the synergistic effects of the Crude Ethyl Acetate Extract (CEAE) from an endophytic actinomycete (MPT42) and the Essential Oil (EO) of the host plant *Litsea cubeba*. The combinations of CEAE with EO led to significant synergistic antimicrobial effects against microbes, with a 4–16 fold reduction in MIC (Minimum Inhibitory Concentration) values when compared to their single-use.

Considering the intimate 'plant-microbiota' relationship and the effects appearing from this interaction, the present work aims to explore fundamental aspects of this topic, focusing on species considered medicinal and the microbiota-associated influence on plant bioactive compounds studied. An overview of this research area will also be presented, highlighting (i) significant examples of association, (ii) methodologies used to study the medicinal plants and their microbiota (iii) microorganisms commonly associated with medicinal plants; and (iv) importance of plant-microbiota association in the 'one health' (human, animal and environmental) context.

Factors affecting medicinal plants and their microbiome

The presence of microorganisms found within a plant was defined by the German botanist Heinrich Anton de Bary (1866), who also introduced the term 'epiphytes' and 'endophytes' (referring to those microorganisms living on the surface or living inside the plant tissue, respectively) (de Bary 1866). However, the first endophyte discovery was made only 38 years later (1904) in the Eurasian darnel ryegrass (*Lolium temulentum*) (Freeman 1904). Since then, the use of the word 'endophyte' has been expanded to invoke all those microorganisms (bacteria, fungi, archaea, etc.) that inhabit the intra- or intercellular parts of the plant, for at least a period of its life cycle, without causing apparent damage (no symptoms of disease) to its host (Wilson 1995; Hardoim et al. 2015; Gupta et al. 2020).

Plant-microorganism interactions are critical to the adaptation and survival of both the partners in any environment. Due to a long period of co-evolution, microbial communities inhabiting the plant have established a highly complex and dynamic relationship with their host's (Andreote et al. 2014; Mitter et al. 2016), including mutualism, antagonism, and rarely parasitism (Owen and Hundley 2004; Nair and Padmavathy 2014).

In symbiotic association, the host plant protects and feeds the endophyte with products derived from its metabolism (Singh et al. 2017). On the other hand, the endophytes may also increase host fitness and competitive abilities by increasing growth rate through evolving biochemical pathways by production of metabolites that result in increased growth and development, enhanced tolerance to biotic and abiotic stresses (Gunatilaka 2006; Turner et al. 2013; White et al. 2014; Prakash et al. 2015; Berg et al. 2017; Huang et al. 2018; Tian et al. 2020). In this case, a microorganism may offer new opportunities for enhancement of quantity and quality of bioactive metabolites of the host plant, by either increasing plant yield and/or modulating a given biosynthetic pathway (Gupta et al. 2020).

The plant microbiota also influences the composition of bioactive compounds in medicinal plants (Köberl et al. 2013; Kilam et al. 2017), many of which can be considered potential sources for the discovery of new drugs (Egamberdieva and Jabborova 2020).

Medicinal plants are known as a source for biologically active metabolites with therapeutic potential since ancient times and have been used for the treatment of a large number of diseases (Chen et al. 2016; Egamberdieva and Jabborova 2020). Phytotherapy is one of the oldest forms of medicine practiced since the primordia of human history. Reports based on archaeological findings, ancient manuscripts, for example, provide evidence that Egypt, India, Greek, Roman, Summaria, Babylon and China used medicinal plants to treat health disorders and prevent diseases (Makarska-Białokoz 2019). The knowledge of their medicinal properties has been transmitted over the centuries within and among human communities. Studies have revealed that almost 80% of the population in developing countries use plant-derived medicines for their wellness (Chen et al. 2016). The ability of some endophytes to increase the therapeutic potential of the host plant turned these microorganisms into a focus for bioprospecting new bioactive compounds and biotechnological implications (Köberl et al. 2013; Berg et al. 2017; Tanvir et al. 2017). Thus, studies involving medicinal plants' microbiome in their natural environment provide an interesting contrast to those with domesticated plant species, especially in discovering a new spectrum of microorganisms able to enhance the production of new medicinal compounds (Huang et al. 2018).

The microbiota greatly influences the growth, quality and health of the medicinal plants through microbial metabolisms and host interactions. One of the beneficial effects of endophytes is stress tolerance to biotic and abiotic factors. Plants are usually challenged by various kinds of stress in the environment, both abiotic (i.e., drought, heat, cold, salinity, heavy metal toxicity, etc.) and biotic (e.g., herbivory and pathogen attack). Thus, the survival of plants depends on their ability to rapidly adjust their physiology and growth to escape or mitigate the impacts of these stresses (Fraire-Velázquez et al. 2011). The strength of several plant species to adapt to different environments can be determined by their association with certain microorganisms (Zelicourt et al. 2013), and the distribution and diversity of microbes that inhabit these plants are also affected (Turner et al. 2013).

Plant metabolism is altered under stress, consequently affecting their microbiome—due to changes in plant metabolism (Hacquard et al. 2017). In response, many endophytes (which mostly belong to bacteria and fungi) may significantly influence the metabolic products in plants (Jia et al. 2016), inducing known and yet unknown biosynthetic pathways, leading to diverse alterations in the plant metabolome (Etalo et al. 2018).

Studies have shown that various compounds (antioxidant enzymes, hormones, metabolites, etc.) are produced by the plant under stress. Similarly, endophytes help in stress-tolerant gene expression. However, the kind of mechanism used depends on the plant and the microorganism, as well as the stress type (Prakash et al. 2015). The microbial communities together with the plant species and soil type comprise important factors determining the types and quantity of biologically active metabolites in medicinal plants (Jia et al. 2016). Furthermore, the content of secondary metabolites from the same medicinal plant taxon can be different, depending on their geographic location and specific tissue (Li et al. 2018).

According to Berg et al. (2017), each plant has to be considered as a metaorganism containing many more microbial cells than plant cells. Thus, the intense interplay between plants and their microbiome is that it maintains the functioning of the holobiont. Advances in multi-omics (comprising genomics, transcriptomics, proteomics, metabolomics and phenomics), and high-throughput sequencing technologies have added valuable knowledge on medicinal plant–microbiome interactions and their external environment (Hao and Xiao 2017), discussed below.

Methodologies for the microbiome study of medicinal plants

Research with medicinal plants initially focused on their compounds, but in recent years, it has covered the structure and function of the microbiome associated with the host plant (Köberl et al. 2013). In addition to the study of general microbial diversity, the recognition that endophytes of medicinal plants are involved in producing bioactive metabolites equal to or similar to that of the host has stimulated research groups to explore endophytic communities (Tanvir et al. 2017). Different methodologies have been applied to investigate the structure and composition of cultivable or non-cultivable endophytic microbial communities associated with different tissues of medicinal plants (Hong et al. 2018; Lu et al. 2020; Castronovo et al. 2020). Studies have compared the effectiveness of culture-dependent and independent methods for characterizing microbial diversity associated with medicinal plants, demonstrating that joint and multiphasic analysis using these approaches is essential for a more detailed investigation of the microbial communities (Pei et al. 2017; Lestari et al. 2018). Table 14.1 presents the main methodologies used in recent studies of the endophytic microbiome of medicinal plants, applied in an isolated or combined manner according to the biological question of the research.

Cultivation-based techniques are generally used in the study of medicinal plant endophytes to isolate strains (Swapna et al. 2019). Among the different strategies for the isolation of endophytes, the method commonly used for their detection and enumeration consists of sterilization of the host plant surfaces and the tissue of interest (Gouda et al. 2016). This method follows three main steps: surface sterilization of the collected plant parts, followed by the fractionation of the plant material into small pieces or homogenization through maceration and transfer to specific culture media for isolation of each endophyte (Golinska et al. 2015).

The sterilization step is essential to guarantee isolation of the endophytic (internal) microbial community. This process is usually carried out by washing the plant tissue in running water followed by treatment with sterilizing reagents for a certain period, followed by a rinse with sterile water (Qin et al. 2011). The most used disinfectant agents for this purpose are: ethanol (70–95%), sodium hypochlorite (2–5%), mercuric chloride (0.1%), hydrogen peroxide, Triton X-100, and Tween 80 (Joe et al. 2016; Kekuda 2016; Sushma et al. 2018). After surface sterilization, the plant tissues are transferred to a specific isolation medium and kept in incubation for a particular period and temperature. Media used for isolating bacterial endophytes include nutrient agar, Luria-Bertani agar and Tryptic Soy

Table 14.1. Main methodologies employed in studies of endophytic microbiomes associated with medicinal plants (2015–2020).

Methodology	Targets	Host Medicinal Plant	Tissue	Main Microorganisms	References
Traditional cultivation-dependent	Isolation and identification of cultured microorganisms using morphological or biochemical characteristics	*Simarouba glauca*	Leaves	*Bacillus cereus*	Kavya and Madhu (2019)
		Ricinus communis	Leaves, stems and seeds	*Aspergillus* spp.; *Penicillium* spp.	Jain and Sharma (2016)
		Basella rubra	Leaves and stems	*Curvularia* sp.; *Cladosporium* sp.; *Aspergillus* sp.; *Cercospora* sp.; *Chaetomium indicum*	Hema et al. (2015)
		Dioscorea composita	Rhizome	*Fusarium; Alternaria*	Gupta et al. (2015)
		Ocimum sanctum, Aegle marmelos, Cinnamomum cassia, Azadirachta indica, Calotropis procera and *Rauwolfia serpentina*	Leaves, stems and petiole	*Bacillus; Pseudomonas* sp.	Panigrahi et al. (2018)
		Kadsura angustifolia	Roots and stems	*Verticillium; Penicillium; Phomopsis; Fusarium; Botrytis; Ceratocystis*	Huang et al. (2015)
		Melia azedarach	Leaves, stems, flowers and fruits	*Chaetomium* sp.; *Aspergillus*	Kanaga et al. (2018)
Amplification and sequencing of rRNA genes and ITS (Metabarcoding)	Molecular identification and phylogenetic analysis of endophytic microbial isolates	*Moringa oleifera*	Stem bark	*Bacillus; Cocus*	Ilmi et al. (2018)
		Thymus vulgaris	Leaves, stems and roots	*Bacillus; Microbacterium; Enterobacter*	Abdelshafy Mohamad et al. (2020)
		Paeonia suffruticosa 'Luoyang Hong'; *P. ostii* 'Luoyang Feng Dan'; *P. ostii* 'Feng Dan'	Leaves, stem and roots	*Alternaria; Fusarium; Pestalotiopsis/Fusarium; Phoma; Alternaria; Paraconiothyrium/Phomopsis; Nigrospora; Alternaria; Pestalotiopsis*	Yang et al. (2018)
		Aloe vera	Roots	*Proteobacteria; Firmicutes; Bacteroidetes; Actinobacteria*	Silva et al. (2019)
		Ligusticum chuanxiong	Rhizome	*Bacillus subtilis*	Yin et al. (2018)
		Teucrium polium	Leaves	*Bacillus; Penicillium*	Hassan (2017)
		Carapa guianensis	Leaves	*Colletotrichum* sp.; *Diaporthe* cf. *mayteni; Pestalotiopsis* sp.	Ferreira et al. (2015)

Table 14.1 contd. ...

...Table 14.1 contd.

Methodology	**Targets**	**Host Medicinal Plant**	**Tissue**	**Main Microorganisms**	**References**
Chromatographic and spectroscopic approaches	Identification of bioactive compounds produced by endophytic microorganisms	*Arnica montana*	Roots	*Streptomyces*	Wardecki et al. (2015)
		Ocimum sanctum	Leaves and stems	Ascomycota	Chowdhary and Kaushik (2015)
		Sinopodophyllum emodi	Roots and rhizome	*Alternaria tenuissima*	Liang et al. (2016)
		Conyza blinii	Leaves and stems	*Chaetomium globosum*; *Xylariaceae*	Tang et al. (2020)
		Paeonia lactiflora	Leaves, roots and stems	*Aspergillus*; *Alternaria*; *Penicillium*	Cheng et al. (2018)
		Dendrobium nobile	Stems	*Trichoderma longibrachiatum*	Sarsaiya et al. (2020)
		Zingiber officinale	Rhizome	*Acremonium* sp.; *Gliocladiopsis* sp.; *Fusarium* sp.; *Colletotrichum* sp.; *Aspergillus* sp.; *Phlebia* sp.; *Earliella* sp.; *Pseudolagarobasidium* sp.	Anisha and Radhakrishnan (2017)
Molecular methods	Identification of the structure and diversity of endophytic microbial communities	*Alpinia officinarum*	Roots, rhizomes, stems and leaves	*Bacillus*; *Methylobacterium*	Li et al. (2015)
		Tinospora crispa	Roots, leaves and stems	*Williamsia*; *Streptomyces*	Primanita et al. (2015)
		Pseudowintera colorata	Roots and stems	*Angustibacter peucedani*; *Streptomyces* sp.	Purushotham et al. (2018)
		Rhododendron zoelleri	Leaves and stems	*Agromyces*; *Gordonia*, *Microbacterium*; *Micromonospora*; *Propionibacterium*, *Saccharomonospora*; *Streptomyces*	Lestari et al. (2018)
		Glycyrrhiza inflata	Roots, stems and leaves	*Microbacterium oxydans*; *Micromonospora saelicesensis*; *Jishengella endophytica*; *Saccharopolyspora* sp.; *Pseudonocardia* sp.	Zhao et al. (2018)
		Ficus deltoidea	Roots, stems, fruits and leaves	*Streptomyces*; *Verrucocispora*; *Kineospora*; *Rhodococcus*; *Intrasporangium*; *Actinomadura*	Janatiningrum et al. (2018)
		Centella asiatica	Roots, stolons and leaves	*Streptomyces*; *Micromonospora*; *Verrucosispora*; *Actinoplanes*; *Couchioplanes*; *Gordonia*	Ernawati et al. (2016)

High throughput sequencing (amplicon and shotgun)	Identification of the structure and diversity of endophytic microbial communities	*Magnolia officinalis*	Stem bark	*Klebsiella; Enterobacter; Enterococcus; Actinomycetospora; Bacillus; Capnobotryella; Devriesia; Devriesia; Candida; Aspergillus; Devriesia*	Wu et al. (2019)
		Aloe vera	Roots, stems and leaves	*Proteobacteria; Firmicutes; Actinobacteria* and *Bacteriodetes*	Akinsanya et al. (2015)
		Polygonum cuspidatum	Roots	*Stenotrophomonas*	Zhang et al. (2020)
		Paris polyphylla var. *yunnanensis*	Roots, stems and leaves	*Cyanobacteria; Rhizobium; Flavobacterium; Sphingobium; Bacillus; Pseudomonas*	Liu et al. (2020)
		Eruca sativa	Rhizosphere and Phyllosphere	*Thaumarchaeota; Euryarchaeota*	Taffner et al. (2019)
		Paullinia cupana	Roots and leaves	*Acidothermus; Porphyromonas*	Liotti et al. (2019)
		Panax notoginseng	Roots	*Enterobacter; Bacillus*	Zhang et al. (2020a)
		Emilia sonchifolia	Whole plant	*Pseudomonas; Stenotrophomonas, Cronobacter; Lysinibacillus; Pantoea; Kluyvera; Pseudorhodoferax*	Urumbil et al. (2020)
		Dendrobium officinale	Roots, stems and leaves	*Ascomycota; Fusarium; Glomerella; Cladosporium*	Liu et al. (2017)

Agar (TSA) (Ramalashmi et al. 2018; Mamangkey et al. 2019; Cardoso et al. 2020), while for fungal endophytes, most used media consider Potato Dextrose Agar, Potato + Carrot Agar and Malt Yeast Extract Agar (Manoharachary and Nagaraju 2016; Hassanein et al. 2016). To identify endophytes, morphological characteristics of the cultures (such as colony color, morphological characteristics of spores and mycelia) can be used, together with biochemical characterization (Martinez-Klimova et al. 2017). These characteristics are often used for a preliminary identification of isolated endophytes. They have been reported, for example, in the identification of microorganisms of the medicinal plant *Emblica officinalis, Terminalia chebula, T. arjuna, Murraya koenigii, Rauwolfia serpentina, Azadirachta indica,* and *Tinospora cordifolia* (Gohain et al. 2019; Duhan et al. 2020). Although this approach allows preliminary microbiota identification, they are sometimes not enough to differentiate between microbial genera and still less at the species level.

The advent of molecular methods brought a new set of reliable and reproducible tools for the systematic characterization of microbial isolates (Barman and Bhattacharjee 2019). Studies based on endophytic microbial cultivation are frequently followed by extraction of genomic DNA from pure cultures and amplification of ribosomal RNA genes (16S e 18S) and ITS (Internal Transcribed Spacer) by PCR (Polymerase Chain Reaction) using specific primers, followed by sequencing of the PCR amplicon(s). The sequences are used for molecular identification and reconstruction of the microbial phylogeny of the endophytic members associated with the medicinal plant of interest (Li et al. 2018; Al-Rashdi et al. 2020).

Taxonomic identities of endophytic bacterial isolates were attributed by analyzing the 16S rRNA sequences in studies with plants such as *Camellia sinensis* (Shan et al. 2018) and *Ficus deltoidea* (Janatiningrum et al. 2018). Through the ITS sequences analysis, taxonomic identities of endophytic fungal isolates have been identified in studies with plants as *Achillea millefolium* (Satari et al. 2018) and *Huperzia serrata* (Wang et al. 2016).

Culture-dependent methods are additionally used to obtain isolates and perform functional tests. In this context, the antimicrobial activity of endophytic isolates obtained from tissues of medicinal plants is usually tracked by different culture-dependent methods, including: (i) cross-streak assays to screen for antibacterial activity (Chandrakar and Gupta 2017); (ii) double culture assays to test action against phytopathogens (Purushotham et al. 2020), and (iii) deferred antagonist assay to assess antibacterial activity (Bibi 2017). Besides, earlier research highlights the potential of endophytes from medicinal plants to produce new bioactive compounds, resulting in increased interest in prospecting these microorganisms as a source of natural products, with promising applications in the agricultural, pharmaceutical and industrial sectors (Rai et al. 2014; Chandrakar and Gupta 2019). To carry out analysis in this context, studies with medicinal plants have adopted the methodology of subjecting isolated endophytes to fermentation for the production of secondary metabolites through their inoculation and incubation in a specific culture medium, followed by extraction and analysis of extracts for identification of bioactive compounds using different techniques, including HPLC (High-Performance Liquid Chromatography) (Gos et al. 2017; Eze et al. 2019), GC-MS (Gas Chromatography-Mass Spectrometry) (Hateet 2017; Chatterjee et al. 2019) and LC-MS/MS (Liquid Chromatography-Tandem Mass Spectrometry) (Jasim et al. 2016; Ghiasvand et al. 2020).

Despite the importance of culture-dependent methods in the study of endophytes, especially for the isolation of endophytes producing bioactive compounds (Shubin et al. 2014), analysis of the microbial biodiversity of endophytic communities of medicinal plant tissues using this approach are sometimes limited (Jin et al. 2014). Microbial isolation using these methods can be affected by the culture medium, growth conditions or manipulation of the inoculated plant tissue (Yang et al. 2017), in addition to the impossibility of recovering the most considerable fraction of microbial populations in culture media (Alain and Querellou 2009; Nadkarni et al. 2009). As a result, molecular tools, based on DNA analysis extracted directly from samples of plant material, are widely used to study medicinal plant microbiomes (Jin et al. 2014; Liu et al. 2016).

Molecular methods have been, in general, dependent on the PCR amplification of DNA barcoding sequences that correspond to entire phylogenetic groups, 16S rRNA for screening bacteria and 18S rRNA or ITS for fungi, followed by the analysis of amplicons generated through molecular DNA fingerprinting approaches (El-Badry and Sheir 2017; Johnston-Monje and Lopez Mejia 2020). The mentioned sequences (16S, 18S, and ITS) are useful as barcodes to identify genera and sometimes species. Therefore, this approach is generally called 'Metabarcoding' in cases where the sequencing applies to the whole PCR product.

Among the most widespread techniques in the analysis of the diversity profile of endophytic communities are DGGE (Denaturing Gradient Gel Electrophoresis), T-RFLP (Terminal Restriction Fragment Length Polymorphism) and ARDRA (Amplified Ribosomal DNA (rDNA) Restriction Analysis) (Zhou et al. 2014; Zhao et al. 2018; Barman and Dkhar 2020). Among these, the DGGE of the 16S and 18S rRNA genes has been a popular choice among publications to elucidate the profile and diversity of bacterial and fungal communities of medicinal plants in different tissues, including the phyllosphere and endosphere of plant parts such as roots, stems, leaves and fruits (Bogas et al. 2015; Wicaksono et al. 2018; Janatiningrum et al. 2018).

Due to methodological advances, 'Metagenomic' studies are actually the best choice. This approach involves DNA extraction of whole environmental samples (e.g., from the soil, from the roots of a plant individual or the gut of an animal). In the case of plant endophytes, it can involve any of the niches illustrated in Fig. 14.1. After extraction, purification and quantification the whole sample is submitted to High-Throughput Sequencing (HTS, also called Next-Generation Sequencing, NGS). Such an inference can provide the sequencing depth necessary to cover complex microbial communities (Shendure and Ji 2008).

Therefore, the association of metagenomics with NGS data allows a precise estimation of the diversity of microbial communities associated with medicinal plants (Pei et al. 2017). The method has benefited by lower costs and higher productivity, being increasingly applied to studies of the structure and composition of the microbiome of a given plant individual (Akinsanya et al. 2015; Yang et al. 2017; Wang et al. 2020). Currently, NGS sequencing of rRNA amplicons (16S, 18S, and ITS, i.e., metabarcoding) has been widely used to explore the diversity and composition of endophytic communities (Dong et al. 2018; Bai et al. 2020).

High-throughput shotgun metagenomic sequencing has been estimated as a complementary approach to amplicon sequencing due to its ability to provide a deeper understanding of the structure and function of the microbiome associated with medicinal plants (Hao et al. 2018). As a result, publications in recent years have adopted HTS technology followed by bioinformatics analysis as an independent and successful approach to characterize and elucidate the diversity and composition of endophytic bacterial and fungal communities of medicinal plants (Li et al. 2017; Lu et al. 2020; Chen et al. 2020). Some recent publications have also adopted a combined methodology between culture-dependent methods or molecular approaches with metagenomics or metabarcoding approaches, to integrate and compare information provided in different ways. This is the case of studies on the endophytic microbiomes of medicinal plants *Passiflora incarnata* (Goulart et al. 2019), *Astragalus membranaceus*, *A. mongholicus* (Sun et al. 2017), and *Paris polyphylla* (Liu et al. 2016) (Table 14.1).

Microorganisms associated with medicinal plants and their diversity

Microorganisms establish essential relationships with plants, whether they are harmful or beneficial to the host plant. Those that develop symbiotic relationships with microorganisms such as fungi and bacteria have benefits, such as (i) resistance to toxic compounds present in the soil (Siciliano et al. 2001), (ii) increase in phytoremediation potential (Ryan et al. 2008), and (iii) increased resistance against several phytopathogens (Strobel 2003). Many bioactive compounds with medicinal potential are secondary metabolites, such as terpenoids, alkaloids and various aromatic compounds. These metabolites are sometimes produced by the mutualistic-symbiotic organisms themselves, regulated by the plant in the presence of a particular tissue, or even only made after the symbiotic process, whereas they can be used by the plant during the defense process (Bacon and White 2000; Mehanni and Safwat 2010; Maggini et al. 2017).

The process of HGT (Horizontal Gene Transfer) between microbionts-holobionts has only been proven for endophytic bacteria, such as in the case of the association of *Burkholderia cepacia* and poplar (*Populus*) (Taghavi et al. 2005). The symbiont-host interaction can be so intimate that in some cases both organisms can synthesize the same compound (e.g., taxol), suggesting that the genes responsible for this process may have the hologenome coming from the microbiont and their generations (host DNA) (Rosenberg and Zilber-Rosenberg 2018).

An example of a compound produced by both holobionts and symbionts is taxol, a low toxicity substance discovered around the 1960s in extracts of the western yew (*Taxus brevifolia*). Taxol has high anticarcinogenic and antitumoral potential, being used for the treatment of breasts and the uterus affected by the disease (Wani et al. 1971). Interestingly, subsequent studies revealed that up to 20 species of endophytic fungi of the genera *Fusarium* and *Alternaria* were also capable of synthesizing taxol, this compound being found even in other plant species (Zhou et al. 2010). Since the yew is a slow-growing and rare tree, discoveries like these were revolutionary because, associated with fermentation, the *in vitro* cultivation of taxol-synthesizing fungi was economically more effective than the previous method of removing the natural compound from the tree wood (Lin et al. 2003). However, it is worth noting that the sequences of the genes that synthesize taxol in fungal and plant organisms showed low homology to each other, indicating that the two metabolic pathways evolved independently (Heinig et al. 2013). In the public domain sequence bank UniProt, seven cured taxol (taxane) sequences are registered assigned to species of the genus *Taxus* (The Uniprot Consortium 2015).

In general, symbionts can be classified into two major groups, according to the tissue in which the host plant is colonized: the endophytes, that is, those that develop totally or partially within the host plant, and that generally do not establish negative relationships, even sometimes being optional saprophytes or pathogens, depending on the internal conditions (Bacon and White 2000); or rhizospheric, which inhabit the soil and interact with the roots of the host plant, influencing, for example, plant nutrition through the acquisition of inorganic iron from the soil, cell leakage in times of high turbidity and also cellular metabolism of new roots (Yang and Crowley 2000).

Once they are among the internal plant tissues, endophytic microorganisms trigger a series of biochemical reactions in the plant that can directly stimulate the host's defense against possible pathogens through metabolites that attack other organisms interacting competitively; or indirect, releasing substances that induce the plant's innate immune system or even induce organ growth (Alvin et al. 2014).

Some endophytic bacteria of the genera *Pseudomonas, Burkholderia* and *Bacillus*, are known to produce substances with an antibiotic, antifungal, antiviral, insecticide, and even immunosuppressive potential, such as ecomycin (Ryan et al. 2008). In Brazil, a plant called 'arnica-do-campo' (*Pseudobrickellia brasiliensis*) has known antimicrobial potential, hosts bacteria of the genera *Pseudomonas* and *Bacillus*, which are good candidates in biotechnology for the production of hydrolytic enzymes and other bactericidal compounds (Cardoso et al. 2020).

Reviews with well-known domesticated plants such as *Arabidopsis thaliana*, rice (*Oryza sativa*), potato (*Solanum tuberosum*) and sweet potato (*Ipomea batatas*) revealed a greater diversity of endophytic bacteria belonging to the phylum of Proteobacteria (about 50%), followed by Actinobacteria, Firmicutes and Bacteriodetes (about 10% each), especially in the root tissues (Fig. 14.2) (Liu et al. 2017). It is suggested that this proportion is also valid in medicinal plants. The herbaceous *Gentianella weberbaueri* and *Valeriana pycnantha*, wild medicinal plants from the microbiome of the *Cordillera Branca*, in the Andes, were associated with species of the genera *Rahnella, Pseudomonas, Serratia* (Proteobacteria) and *Bacillus* (Firmicutes), with high diversity in both species (Ulloa-Munoz et al. 2020).

As an external factor, the environment also contributes to the distribution and persistence of these bacteria in specific hosts. An example is the White mountain range, a region of difficult access, low temperatures, and mostly frozen soil, but which houses a great diversity of medicinal plants, including 28 species with antimicrobial and antioxidant properties (Tamariz-Angeles et al. 2018). As it is a challenging environmental condition, the microorganisms present in these plants may be related to the acquisition of resources for survival in the hostile climate and poor soil, allowing nutrient uptake such as iron, nitrogen and phosphorus, helping to fight possible pathogens (Liu et al. 2017). It is also worth noting that global climate changes can also put these little explored and characterized species at risk of extinction since environmental conditions can also influence interactions with microorganisms and cause damage to the relationships established by the host plant (Bacon and White 2000; Ulloa-Munoz et al. 2020).

Endophytic fungi predominantly belong to the group of ascomycetes in both angiosperms and gymnosperms (Sieber 2007), but basidiomycetes, oomycetes and

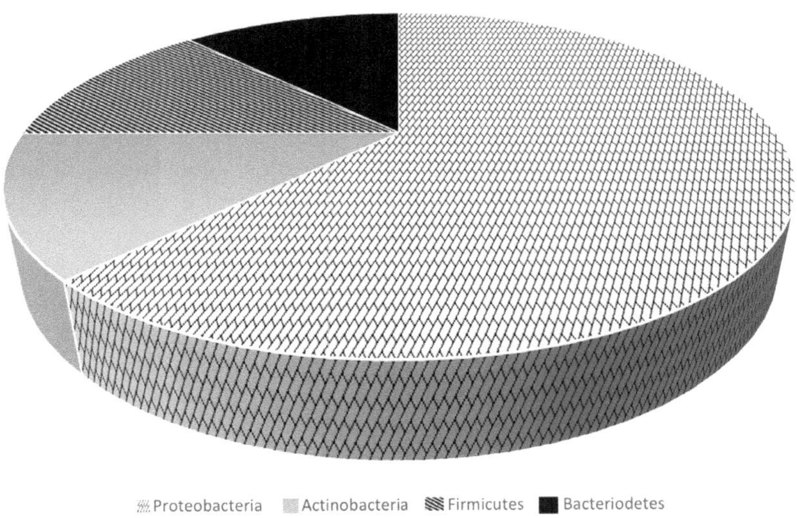

Figure 14.2. Diversity of endophytic bacteria found in the root tissues. It is possible to notice that proteobacteria is the dominant class, followed by actinobacteria, firmicutes and bacteriodetes.

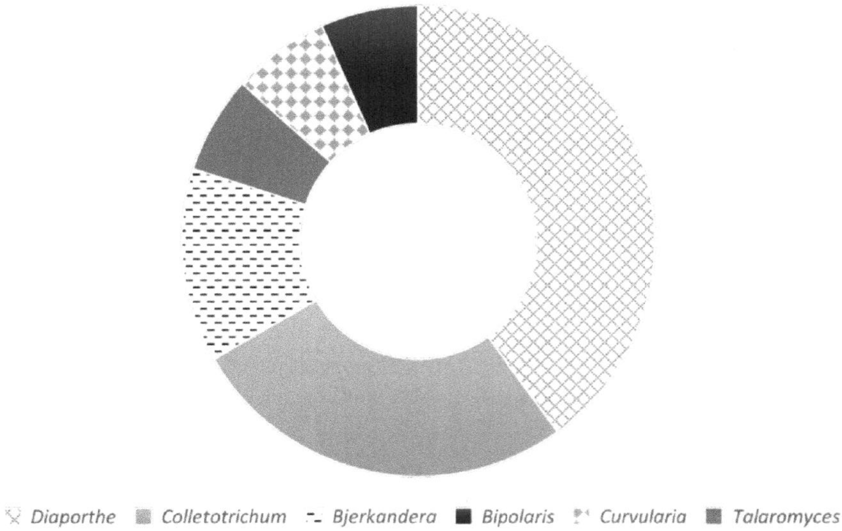

Figure 14.3. Several genera present in Brazilian medicinal plants. *Diaporthe* was the genus with the highest wealth of isolates, followed by *Colletotrichum*, both Ascomycetes (Santos et al. 2020).

zygomycetes are also found (Gupta et al. 2020). A study involving four Brazilian medicinal plants, *Campomanesia xanthocarpa*, *Guazuma ulmifolia*, *Lafoensia pacari* and *Siparuna guianensis*, revealed that the presence of certain endophytic fungi increased the anti-inflammatory potential. In contrast, fungi isolated of the genus *Talaromyces* had anti-inflammatory activity, almost 90% (Santos et al. 2020). It was possible to identify six genera of endophytic fungi, of which five were ascomycetes, and only one was a basidiomycete, *Bjerkandera*, obtained from leaf and stem fragments (Fig. 14.3).

Impacts of the microbiota on medicinal plants

Microenvironments associated with plants are abundantly colonized by several microorganisms that can affect the health and growth of these plants in a beneficial, harmful or neutral way (Müller et al. 2016; Köberl et al. 2013). In this sense, plant microbiomes have been considered important indicators for evaluating the quality and productivity of plants, and studies involving this subject have recently gained prominence (Huang et al. 2018; Turner et al. 2013). Additionally, functional microbiome assays have also allowed investigating biogeochemical environments, complex ecosystems related to hosting organisms, metabolism and their interactions, revealing new perspectives on the role of microbiota associated with ecosystems (Chakraborty et al. 2019).

In general, the structure of communities of bacteria and fungi associated with plants is not composed at random and varies mainly depending on the type of soil, season, plant species or tissue and environmental factors (Hassani et al. 2018; Toju et al. 2014; Bulgarelli et al. 2013; Bulgarelli et al. 2012; Davison et al. 2011; Gottel et al. 2011). The bacterial microbiota is dominated by three main phyla (proteobacteria, actinobacteria and bacteroidetes). In turn, the phylum ascomycota and basidiomycota are the most prevalent among the fungal microbiota present in plant tissues below and above ground. In the roots, the phylum glomeromycota stands out as the main target of studies. However, endophytic fungi also represent an essential fraction of the fungal microbiota present in this tissue and have gained attention in recent years (Hassani et al. 2018; Hacquard et al. 2015; Bonito et al. 2014; Bulgarelli et al. 2013; Toju et al. 2013).

Currently, it is evident that plant phenotypes result from several multiple interactions and that plants constitute an ecosystem that must be considered from the perspective of systems biology (Müller et al. 2016). Thus, systematic approaches accompanied by advanced computational tools regard a strategy of great relevance to elucidate comprehensive biochemical and environmental interactions in host plants and the associated microbial community (Chakraborty et al. 2019). Besides, this approach promotes a broad understanding of the various interactions occurring at the systemic level and offers advantages for possible applications in several areas, particularly sustainable agriculture (Müller et al. 2016).

Medicinal plants represent a valuable source of bioactive compounds with potential use in medical and agronomic applications (Köberl et al. 2013). However, little is known about their microbiomes and the impact caused by this interaction. Köberl et al. (2011) analyzed the microbiome of three medicinal plants (*Matricaria chamomilla, Calendula officinalis* and *Solanum distichum*) grown under organic management on a farm in an Egyptian desert. The author verified the presence of particular bacterial communities in the roots of the investigated plants, with native *Bacillus* strains and *Paenibacillus* showing promising antagonistic properties against a wide range of soil phytopathogens. On the other hand, several extremophilic bacterial groups have decreased or disappeared from the soil after agricultural use. These authors also reported that the bacterial communities in the soil were strongly driven by abiotic factors (water availability and pH) and that *Bacillus* strains were able to promote growth and increase the production of flavonoids in plants.

In a study conducted by Chen et al. (2018), the seed microbiome of *Salvia miltiorrhiza*, an important medicinal plant used for the treatment of coronary and cerebrovascular diseases in Asian countries, was evaluated using 16S rDNA and ITS sequencing, uncovering the presence of 17 phyla and 39 classes of bacteria. The prediction analysis of functional pattern and enrichment indicated functions related to the metabolism of terpenoids, polyketides and biosynthesis of secondary metabolites, suggesting that the identified microbiome may contribute with additional metabolic capacities to those encoded in the *S. miltiorrhiza* genome, thus constituting a genetic resource distinct for the host plant.

The production of bioactive compounds with antimicrobial potential from endophytic bacteria isolated from four medicinal plants (*Dodonaea viscosa, Fagonia indica, Caralluma tuberculata* and *Calendula arvensis*) was reported by Iqrar et al. (2019). After initial screening for antimicrobial activity, nine species belonging to seven distinct genera were selected and identified through 16S rRNA sequencing. Tests for antimicrobial activity revealed that *Pseudomonas taiwanensis* has vigorous anti-*Phytophthora* activity and the volatile compounds produced by the bacterium inhibited the growth of *P. parasitica* by more than 80%. Extracts obtained from other bacteria (*S. alboniger, P. taiwanensis, E. hormaechei, B. tequilensis* and *D. lacustris*) showed potent inhibition *in vitro* against *P. parasitica* at different concentrations (4 to 400 μg/mL).

Bacterial extracts with higher bioactivity (> 80% inhibition *in vitro*) were selected for testing against *P. parasitica* on tobacco leaves, with a reduction in the size and frequency of lesions (68 to 81%) caused by the pathogen. Iqrar and collaborators (2019) concluded that bacterial endophytes have potential as biological control agents and can be applied to formulate biopesticides for use in sustainable agriculture. In a previous study, Kumar et al. (2016) isolated and characterized, through morphological, biochemical analysis and 16S rRNA sequencing, 14 endophytic bacteria from the rhizome of *Curcuma longa*, a medicinal plant grown in India and Middle Eastern countries, and found that the endophytic strains produced phytohormones (indole acetic acid – IAA), solubilized phosphate and siderophores during plant growth. In addition, some strains were resistant to the antimicrobial polymyxin B (specific for gram-negative bacteria), as well as effectively inhibited the growth of *Escherichia coli*, *Klebsiella pneumoniae*, *Fusarium solani*, and *Alternaria alternata*, besides increasing tolerance to salinity (up to 8% NaCl).

Recently, Köberl et al. (2019) carried out a comparative evaluation of the phyllosphere microbiome of two medicinal plants (*Matricaria chamomilla* and *Calendula officinalis*), concerning colonization with Lactic Acid Bacteria (LAB), to identify whether the LABs could be enriched through a material fermentation process. The authors reported significant changes in the plant's microbiome during the implemented fermentations and suggested that the combination of phytotherapeutic properties with a prebiotically active microbiome after fermentation may have a beneficial health effect when applied to food production. Besides, the identified LABs also represent an excellent alternative to synthetic chemical pesticides in sustainable agriculture, combining advantages for human and animal consumers.

In another study, Khan et al. (2020) reported 121 fungal OTUs and 3662 bacterial OTUs in the rhizosphere of three medicinal plants (*Adenium obesum*, *Aloe dhufarensis* and *Cleome austroarabica*) from the Arabian peninsula, grown under different climatic conditions. The microbiome analysis showed that microbial diversity varied significantly depending on the host, location and soil and that the presence of extracellular enzymes can help maintain plant growth during adverse environmental conditions.

Microbiota-medicinal plants interaction

Over the last few years, several studies have demonstrated that endophytes are responsible for the antimicrobial activity of medicinal plants. A new perspective on how the role of the endophytic community influences the chemical composition of medicinal plant species has created interest among researchers. In this context, studies have focused on the modulation and biosynthesis of phytoconstituents and biomolecules carried out by the microbial community. It has been suggested that endophytes can still favor the health of the host plant, as they can compete and produce antimicrobial compounds. These compounds are often promising for the industry since they can act directly on the membrane lysis of the pathogen present in these species of medicinal plants (Castronovo et al. 2021). Studies in *Hypericum perforatum* and *Ziziphora capitata* showed that endophytic bacteria could provide promising antagonistic activity against bacterial and fungal pathogens that cause damage in the agricultural industry, such as *Fusarium oxysporum* (Egamberdieva et al. 2017). Furthermore, studies in *Fagonia indica* point to the production of therapeutic compounds from endophytic bacteria. These compounds can prevent several fungal and bacterial diseases, as they showed significant antagonism in *Klebisiella pneumoniae*, *Stenotrophomonas maltophilia*, *Enterobacter hormaechei*, *Bacillus tequilensis*, *Erwinia* sp. and *Aspergillus niger* (Rahman et al. 2017).

Most drugs used to treat various diseases have been obtained from secondary metabolites or bioactive compounds isolated from medicinal plants due to their complete availability and effectiveness, which have led to the development of several studies, with significant contributions to public health worldwide (Agampodi et al. 2018; Huang et al. 2018). These compounds produced by endophyte have an antagonistic potential effect for different pathogens (e.g., bacteria, fungi or virus). Lai et al. (2016) investigated the effects of an aqueous extract of *Blechnum orientale* (a fern traditionally used to treat skin wounds) in a hydrogel formulation aimed at healing diabetic ulcers. They found that one of the extract fractions exhibited strong antioxidant and antibacterial activity against methicillin-resistant *Staphylococcus aureus* (MRSA), at 0.25 mg/mL. Hydrogels loaded with the extract exhibited the ability to close wounds in approximately 12 days, while healing in controls was not effective or was only achieved after 14 days, proving its potential therapeutic use. In another study, Sharma et al. (2018) evaluated the effect of curcumin (a polyphenolic compound isolated from the medicinal plant *Curcuma longa*) conjugated to Hyaluronic Acid (HA) and demonstrated that the tests carried out on human keratinocyte cells revealed improvements in cell proliferation, decreased oxidative damage induced by H_2O_2 and migration of cells in wounds when compared to treatment with natural curcumin. Also, the topical application of the conjugate in wounds of diabetic mice led to better healing compared to treatment with curcumin without HA and HA alone, suggesting that the conjugation of these compounds is synthesized by endophytic microorganisms that may represent a promising formulation for pharmacological and agro-industrial applications, for example antimicrobials.

The domestication process of the plants allowed the selection of several beneficial characteristics applicable to food and human health and favoring the sustainability of agriculture. The identification of specific microorganisms and their association with host plants has been extensively studied in recent years, mainly after the implementation of cutting-edge technologies, such as NGS, omics and bioinformatics approaches (Chakraborty et al. 2019). Such efforts have generated valuable discoveries that increasingly highlight the potential of medicinal plants as a source of secondary metabolites and bioactive compounds with pharmaceutical, industrial and agronomic applications. The advance in the development of systematic research on microbiomes associated with medicinal plants will expand the knowledge about the role of microorganisms in the modulation of vital metabolic pathways and may help the production of new classes of compounds with medical applications (Huang et al. 2018).

Plant, microbiota and the one health concept (human, animal, and environmental)

The understanding of biological systems has benefited from integrative studies derived from omics sciences. Additionally, the 'One Health' concept advocates for collaborative efforts involving multiple disciplines and collaborations at the local, national and global levels to achieve optimal health for people, animals and our environment (Amuasi et al. 2020). In turn, nutrigenomics is a field aiming to understand specific pathologies, trying to correlate food, genome and external factors (behavioral and environmental). Each gene can modulate its function through expression in a specific situation, modulating the microbial community concerning the host. Thus, understanding the structure and function of the plant microbiome is essential for good health, especially for disease control, in potential cases of identifying pathogens in the environment (Nyholm et al. 2020; Zilber-Rosenberg and Rosenberg 2008; Rosenberg and Zilber-Rosenberg 2019).

Regarding the one health effort, one knows the importance of intestinal homeostasis in studies promoting human health, since everything one eats, including vegetables, can affect the way one absorbs nutrients and changes the composition of the intestinal microbial community, which in turn can change the external environment (environment and animals; Nyholm et al. 2020).

Hologenomics aims to understand this interaction and how molecular mechanisms may have helped in the adaptation and evolution over time to the generation of the many existing species, including the global study of the genomes of the hosts and their microbiota (microbiomes). Among the main events, the horizontal gene transfer, microbial amplification and acquisition of new symbionts (Douglas et al. 2016; Zilber-Rosenberg and Rosenberg 2008) can be mentioned. The latter is causally linked to the pan and core microbiome, which are essential genes shared between these microbial communities and are transferred and can modulate several crucial functions in the organism. Studies show that part of the microbiome (core microbiome) is always transferred from generation to generation, emphasizing that they are species or population-specific and may play an essential adaptive role.

For the maintenance of homeostasis of holobionts, the concept of 'one health' is applied, which depends on good health, which means that for a right balance of the system, not only the host (e.g., a plant) but also the microbiome must be intact and in equilibrium to allow the necessary interactions. If there is any interruption in this system, it is possible that the plant, for example, is more susceptible to pathogens present in its environment, showing the influence of its microbiome (Zilber-Rosenberg and Rosenberg 2008; Zilber-Rosenberg and Rosenberg 2019).

A significant part of the beneficial human gut microbiota comes from raw plant ingestion and the same applies to the animals used for human consumption. Thus, considering the concept 'you are what you eat' (e.g., Zmora et al. 2019) plants and their microbiota are in the core of the 'one health' concept, with large evidence that human and animal health are greatly dependent on the uptake of healthy raw plant biomass.

Conclusion and future directions

In plants, microbial inhabitants collaborate and shape the functioning of each niche, often with advantages for the host plant and even providing adaptive advantages and new phenotypic aptitudes, including antimicrobial activity. Evidently, the pros are much bigger than the cons when it comes to the coexistence of the host plant vs. microbiota, and also considering medicinal plants.

In this regard, it is interesting to note that in the case of several plants, the active principle has not only plant origin, but derives from a 'plant-microorganism-environment' conjunction, or in some cases it may have its main origin in the microbiota. This is a frontier research area which needs to be explored.

The difference between a wild plant and a crop relies not only on its genotype but also on the microbial community it hosts. This finding helps to explain the dependence that monocultures have on the protection by humans with their agro-defensives. Between the 1960s and 1970s, intending to eliminate hunger in the world, the so-called 'green revolution' took place, which involved a process of modernization of agriculture to increase food productivity, including the process of planting, irrigation, application of pesticides and mechanization of the harvesting. In fact, the success of this initiative was unquestionable. But every day the perception grows that defensive agents have deleterious cumulative effects that negatively impact the plant microbiota and the environment, besides harmful effects on animal and human health. The United States, Brazil, Japan and Ukraine, among others, are among the largest buyers of pesticides manufactured on European soil but banned for use in the European Union (EU) and the UK. The EU authorizes the export of pesticides that it considers too dangerous for application in its agriculture but tolerates the import of food grown with these products in other continents. It is the so-called 'poison cycle', including bactericides, fungicides, insecticides with known cytotoxicity on various organisms. Besides, the soil and the groundwater are contaminated and cleaning will take many decades.

The expanded study of endophytes through different omics approaches opens up perspectives for using this microbial community in various biotechnological and industrial applications, such as the production of inoculants for biological control to the detriment of pesticides. Besides, more effective natural herbal compounds, with increased and standardized effects can be achieved if the medicinal plant is associated with the most adequate microbiota.

Like other areas of life sciences, the perception of the role, breadth of functions and the importance of the plant microbiome, especially of endophytes, has been growing gradually. This perception that 'the whole' reflects in 'each one' is inexorable and irreversible. This applies not only to plants but to all organisms including humans, and that the concept that 'nobody is an island' applies to living beings as a whole, at their deepest core and essence. There is no question that this concept applies also to plants, where each individual comprises a megacity with millions of microscopic inhabitants, distributed unevenly across their surfaces and the interior of their tissues.

Many discoveries about these processes will still be uncovered in the next decades. They will revolutionize the production of foods, herbal medicines and raw materials of plant origin.

Acknowledgments

The authors would like to thank Coordenação de Aperfeiçoamento de Pessoal de Nível Superior (CAPES), Conselho Nacional de Desenvolvimento Cientifico e Tecnológico (CNPq) for fellowships and financial support.

References

Abdelshafy Mohamad, O.A., Ma, J.-B., Liu, Y.-H., Zhang, D., Hua, S., Bhute, S., Hedlund, B.P., Li, W.-J., Li, L. (2020). Beneficial endophytic bacterial populations associated with medicinal plant *Thymus vulgaris* alleviate salt stress and confer resistance to *Fusarium oxysporum*. Front Plant Sci 11.

Agampodi, V.A., Collet, C., Collet, T. (2018). A review of antibacterial properties of medicinal plants in the context of wound healing. Unit J Pharmacol Ther 1: 1–15.

Akinsanya, M.A., Goh, J.K., Lim, S.P., Ting, A.S.Y. (2015). Metagenomics study of endophytic bacteria in *Aloe vera* using next-generation technology. Genomics Data 6: 159–163.

Alain, K., Querellou, J. (2009). Cultivating the uncultured: limits, advances and future challenges. Extremophiles 13: 583–594.

Al-Rashdi, F.K.H., Al-Sadi, A.M., Al-Riyamy, B.Z., Maharachchikumbura, S.S.N., Al-Sabahi, J.N., Velazhahan, R. (2020). Endophytic fungi from the medicinal plant *Aloe dhufarensis* Lavranos exhibit antagonistic potential against phytopathogenic fungi. S Afr J Bot 00: 1–8.

Alvin, A., Miller, K.I., Neilan, B.A. (2014). Exploring the potential of endophytes from medicinal plants as sources of antimycobacterial compounds. Microbiol Res 169: 483–495.

Amuasi, J.H., Lucas, T., Horton, R., Winkler, A.S. (2020). Reconnecting for our future: The Lancet one health commission. The Lancet 395: 1469–1471.

Andreote, F.D., Gumiere, T., Durrer, A. (2014). Exploring interactions of plant microbiomes. Sci Agri 71: 528–539.

Anisha, C., Radhakrishnan, E.K. (2017). Metabolite analysis of endophytic fungi from cultivars of *Zingiber officinale* Rosc. identifies myriad of bioactive compounds including tyrosol. 3 Biotech 7: 146.

Bacon, C.W., White, J.F. (2000). Microbial Endophytes. Dekker M, New York 3–29.

Bai, L., He, L., Yu, P., Luo, J., Yang, M., A, X., Wei, X. (2020). Molecular characterization of Mycobiota and *Aspergillus* species from *Eupolyphaga sinensis* walker based on high-throughput sequencing of ITS1 and CaM. J Food Qual e1752415: 1–7.

Barman, D., Bhattacharjee, K. (2019). Endophytic bacteria associated with medicinal plants: The treasure trove of antimicrobial compounds. *In*: Medically Important Plant Biomes. Source of Secondary Metabolites. Springer Nat 15: 153–187.

Barman, D., Dkhar, M.S. (2020). Seasonal variation influence endophytic actinobacterial communities of medicinal plants from tropical deciduous forest of Meghalaya and characterization of their plant growth-promoting potentials. Curr Microbiol 77: 1689–1698.

Bary, A. de (1866). Morphologie und physiologie der Pilze, Flechten und Myxomyceten. W. Engelmann, Leipzig, 1831–1888.

Berg, G., Köberl, M., Rybakova, D., Müller, H., Grosch, R., Smalla, K. (2017). Plant microbial diversity is suggested as the key to future biocontrol and health trends. FEMS Microbiol Ecol 93: 1–9.

Bibi, F. (2017). Diversity of antagonistic bacteria isolated from medicinal plant *Peganum harmala* L. Saudi J Biol Sci 24: 1288–1293.

Bogas, A.C., Ferreira, A.J., Araújo, W.L., Astolfi-Filho, S., Kitajima, E.W., Lacava, P.T., Azevedo, J.L. (2015). Endophytic bacterial diversity in the phyllosphere of Amazon *Paullinia cupana* associated with asymptomatic and symptomatic anthracnose. Springer Plus 4: 1–13.

Bonito, G., Reynolds, H., Robeson, M.S., Nelson, J., Hodkinson, B.P., Tuskan, G., Schadt, C.W., Vilgalys, R. (2014). Plant host and soil origin influence fungal and bacterial assemblages in the roots of woody plants. Mol Ecol 23: 3356–3370.

Bulgarelli D., Rott M., Schlaeppi K., Ver Loren van Themaat, E., Ahmadinejad, N., Assenza, F., Rauf, P., Huettel, B., Reinhardt, R., Schmelzer, E., Peplies, J., Gloeckner, F.O., Amann, R., Eickhorst, T., Schulze-Lefert, P. (2012). Revealing structure and assembly cues for Arabidopsis root-inhabiting bacterial microbiota. Nat 488: 91–95.

Bulgarelli, D., Schlaeppi, K., Spaepen, S., van Themaat, E.V.L., Schulze-Lefert, P. (2013). Structure and functions of the bacterial microbiota of plants. Annu Rev Plant Biol 64: 807–838.

Cardoso, V.M., Campos, F.F., Santos, A.R.O., Ottoni, M.H.F., Rosa, C.A., Almeida, V.G., Grael, C.F.F. (2020). Biotechnological applications of the medicinal plant *Pseudobrickellia brasiliensis* and its isolated endophytic bacteria. J Appl Microbiol 129: 926–934.

Castronovo, L.M., Calonico, C., Ascrizzi, R., Del Duca, S., Delfino, V., Chioccioli, S., Vassallo, A., Strozza, I., De Leo, M., Biffi, S., Bacci, G., Bogani, P., Maggini, V., Mengoni, A., Pistelli, L., Lo Nostro, A., Firenzuoli, F., Fani, R. (2020). The cultivable bacterial microbiota associated to the medicinal plant *Origanum vulgare* L.: from antibiotic resistance to growth-inhibitory properties. Front Microbiol 11: 1–17.

Castronovo, L.M., Vassallo, A., Mengoni, A., Miceli, E., Bogani, P., Firenzuoli, F., Fani, R., Maggini, V. (2021). Medicinal plants and their bacterial microbiota: a review on antimicrobial compounds production for plant and human health. Pathogens 10: 1–17.

Chakraborty, A., Kundu, S., Mukherjee, S., Ghosh, B. (2019). Endophytism in Zingiberaceae: Elucidation of beneficial impact. pp. 1–26. *In*: Mérillon, J.-M., Ramawat, K.G. (eds.). Bioactive Molecules in Food. Springer Cham.

Chandrakar, S., Gupta, A.K. (2017). Antibiotic potential of endophytic Actinomycetes of medicinal herbs against human pathogenic bacteria. Proc Natl Acad Sci India Sect B Biol Sci 87: 905–915.

Chandrakar, S., Gupta, A.K. (2019). Actinomycin-producing endophytic *Streptomyces* parvulus associated with root of *Aloe vera* and optimization of conditions for antibiotic production. Probiotics & Antimicro Prot 11: 1055–1069.

Chatterjee, S., Ghosh, R., Mandal, N.C. (2019). Production of bioactive compounds with bactericidal and antioxidant potential by endophytic fungus *Alternaria alternata* AE1 isolated from *Azadirachta indica* A. Juss. PLoS One 14: 1–18.

Chen, H., Wu, H., Yan, B., Zhao, H., Liu, F., Zhang, H., Sheng, Q., Miao, F., Liang, Z. (2018). Core microbiome of medicinal plant Salvia miltiorrhiza seed: a rich reservoir of beneficial microbes for secondary metabolism? Int J Mol Sci 19: 672.

Chen, S.L., Yu, H., Luo, H.M., Wu, Q., Li, C.F., Steinmetz, A. (2016). Conservation and sustainable use of medicinal plants: problems, progress, and prospects. Chin Med 11: 1–10.

Chen, W.H., Wu, S.J., Sun, X.L., Feng, K.M., Rahman, K., Tan, H.Y., Yu, L.Y., Li, T.Q., Xu, L.C., Qin, L.P., Han, T. (2020). High-throughput sequencing analysis of endophytic fungal diversity in *Cynanchum* sp. S Afr J Bot 134: 349–358.

Cheng, X., Wei, Z., Pu, S., Xiang, M.-, Yan, A., Zhang, Y., Wang, X. (2018). Diversity of endophytic fungi of *Paeonia lactiflora* Pallas and screening for fungal paeoniflorin producers. FEMS Microbiol Lett 365.

Chowdhary, K., Kaushik, N. (2015). Fungal endophyte diversity and bioactivity in the indian medicinal plant *Ocimum sanctum* Linn. PLOS ONE 10: e0141444.

Compant, S., Saikkonen, K., Mitter, B., Campisano, A., Mercado-Blanco, J. (2016). Editorial special issue: soil, plants and endophytes. Plant Soil 405: 1–11.

Davison, J., Öpik, M., Daniell, T.J., Moora, M., Zobel, M. (2011). Arbuscular mycorrhizal fungal communities in plant roots are not random assemblages: Selectivity in AMF-plant associations. FEMS Microbiol Ecol 78: 103–115.

de Zelicourt, A., Al-Yousif, M., Hirt, H. (2013). Rhizosphere microbes as essential partners for plant stress tolerance. Mol Plant 6: 242–245.

Dong, L., Cheng, R., Xiao, L., Wei, F., Wei, G., Xu, J., Wang, Y., Guo, X., Chen, Z., Chen, S. (2018). Diversity and composition of bacterial endophytes among plant parts of *Panax notoginseng*. Chin Med 13: 1–9.

Douglas, A.E., Werren, J.H. (2016). Holes in the Hologenome: Why host-microbe symbioses are not holobionts. mBio 7: e02099–15.

Duhan, P., Bansal, P., Rani, S. (2020). Isolation, identification and characterization of endophytic bacteria from medicinal plant *Tinospora cordifolia*. S Afr J Bot 134: 43–49.

Egamberdieva, D., Wirth, S., Behrendt, U., Ahmad, P., Berg, G. (2017). Antimicrobial activity of medicinal plants correlates with the proportion of antagonistic endophytes. Front Microbiol 8: 1–11.

Egamberdieva, D., Jabborova, D. (2020). Plant microbiome: source for biologically active compounds. pp. 1–9. *In*: Biodiversity and Biomedicine. Elsevier.

El-Badry, M.A., Sheir, D.H. (2017). Microbial fungal community fingerprint of medicinal plant rhizosphere collected form Sinai, Egypt. Merit Res J Agric Sci Soil Sci 5: 201–207.

Ernawati, M., Solihin, D.D., Lestari, Y. (2016). Community structures of endophytic actinobacteria from medicinal plant *Centella asiatica* L. Urban based on metagenomic approach. Int J Pharm Pharm Sci 8: 292–297.

Etalo, D.W., Jeon, J.S., Raaijmakers, J.M. (2018). Modulation of plant chemistry by beneficial root microbiota. Nat Prod Rep 35: 398–409.

Eze, P.M., Nnanna, J.C., Okezie, U., Buzugbe, H.S., Abba, C.C., Chukwunwejim, C.R., Okoye, F.B.C., Esimone, C.O. (2019). Screening of metabolites from endophytic fungi of some Nigerian medicinal plants for antimicrobial activities. The Euro Biotech J 3: 10–18.

Ferreira, M.C., Vieira, M. de L.A., Zani, C.L., Alves, T.M. de A., Junior, P.A.S., Murta, S.M.F., Romanha, A.J., Gil, L.H.V.G., Carvalho, A.G. de O., Zilli, J.E., Vital, M.J.S., Rosa, C.A., Rosa, L.H. (2015). Molecular phylogeny, diversity, symbiosis and discover of bioactive compounds of endophytic fungi associated with the medicinal Amazonian plant *Carapa guianensis* Aublet (Meliaceae). Biochem Syst Ecol 59: 36–44.

Fraire-Velazquez, S., Rodriguez-Guerra, R., Sanchez-Caldero, L. (2011). Abiotic and biotic stress response crosstalk in plants. pp. 1–26. *In*: Shanker, A. (eds.). Abiotic Stress Response in Plants—Physiological, Biochemical and Genetic Perspectives. InTechOpen Book Series.

Freeman, E.M. (1904). The seed fungus of *Lolium temulentum* L., the darnel. Phil Trans R Soc Lond B 196: 1–27.

Ghiasvand, M., Makhdoumi, A., Matin, M.M., Vaezi, J. (2020). Exploring the bioactive compounds from endophytic bacteria of a medicinal plant: *Ephedra foliata* (Ephedrales: Ephedraceae). Adv Tradit Med 20: 61–70.

Gohain, A., Sarma, R.K., Debnath, R., Saikia, J., Singh, B.P., Sarmah, R., Saikia, R. (2019). Phylogenetic affiliation and antimicrobial effects of endophytic actinobacteria associated with medicinal plants: prevalence of polyketide synthase type II in antimicrobial strains. Folia Microbiol 64: 481–496.

Golinska, P., Wypij, M., Agarkar, G., Rathod, D., Dahm, H., Rai, M. (2015). Endophytic actinobacteria of medicinal plants: Diversity and bioactivity. Antonie van Leeuwenhoek 108: 267–289.

Gos, F.M.W.R., Savi, D.C., Shaaban, K.A., Thorson, J.S., Aluizio, R., Possiede, Y.M., Rohr, J., Glienke, C. (2017). Antibacterial activity of endophytic Actinomycetes isolated from the medicinal plant *Vochysia divergens* (Pantanal, Brazil). Front Microbiol 8: 1–17.

Gottel, N.R., Castro, H.F., Kerley, M., Yang, Z., Pelletier, D.A., Podar, M., Karpinets, T. (2011). Distinct microbial communities within the endosphere and rhizosphere of *Populus deltoides* roots across contrasting soil types. Appl Environ Microbiol 77: 5934–5944.

Gouda, S., Das, G., Sen, S.K., Shin, H.-S., Patra, J.K. (2016). Endophytes: atreasure house of bioactive compounds of medicinal importance. Front Microbiol 7: 1–8.

Goulart, M.C., Cueva-Yesquén, L.G., Martinez, K.J.H., Attili-Angelis, D., Fantinatti-Garboggini, F. (2019). Comparison of specific endophytic bacterial communities in different developmental stages of *Passiflora incarnata* using culture-dependent and culture-independent analysis. Microbiol Open 8: 1–16.

Gunatilaka, A.A.L. (2006). Natural products from plant-associated microorganisms: Distribution, structural diversity, bioactivity, and implications of their occurrence. J Am Chem Soc Am Soc Pharmacogn 69: 509–526.

Gupta, S., Kaul, S., Dhar, M.K. (2015). Exploring the antimicrobial potential of fungal endophytes of Dioscorea composita: a north-west Himalayan medicinal plant. ENVIS Bull Himalayan Ecol 23: 113–118.

Gupta, S., Chaturvedi, P., Kulkarni, M.G., Van Staden, J. (2020). A critical review on exploiting the pharmaceutical potential of plant endophytic fungi. Biotechnol Adv 39: 1–123.

Hacquard, S., Garrido-Oter, R., González, A., Spaepen, S., Ackermann, G., Lebeis, S., McHardy, A.C., Dangl, J.L., Knight, R., Ley, R., Schulze-Lefert, P. (2015). Microbiota and host nutrition across plant and animal kingdoms. Cell Host Microbe 17: 603–616.

Hacquard, S., Spaepen, S., Garrido-Oter, R., Schulze-Lefert, P. (2017). Interplay between innate immunity and the plant microbiota. Annu Rev Phytopathol 55: 565–589.

Hao, D., Xiao, P. (2017). Rhizosphere microbiota and microbiome of medicinal plants: from molecular biology to omics approaches. Chin Herb Med 9: 199–217.

Hao, D.C., Zhang, C.-R., Xiao, P.G. (2018). The first Taxus rhizosphere microbiome revealed by shotgun metagenomic sequencing. J Basic Microbiol 58: 501–512.

Hardoim, P.R., van Overbeek, L.S., Berg, G., Pirttilä, A.M., Compant, S., Campisano, A., Döring, M., Sessitsch, A. (2015). The hidden world within plants: Ecological and evolutionary considerations for defining functioning of microbial endophytes. Microbiol Mol Biol Rev 79: 293–320.

Hassan, S.E.-D. (2017). Plant growth-promoting activities for bacterial and fungal endophytes isolated from medicinal plant of *Teucrium polium* L. J Adv Res 8: 687–695.

Hassanein, N., El-Gendy, M., Abdelhameed, N. (2016). Endophytic fungi of some medicinal plants in Egypt. Egypt Acad J Biol Sci G Microbiol 8: 65–78.

Hassani, M.A., Durán, P., Hacquard, S. (2018). Microbial interactions within the plant holobiont. Microbiome 6: 1–17.

Hateet, R.R. (2017). Isolation and identification of three bioactive compounds from endophytic fungus *Trichoderma* sp. J Al-Nahrain Univ 20: 108–113.

Heinig, U., Scholz, S., Jennewein, S. (2013). Getting to the bottom of Taxol biosynthesis by fungi. Fungal Div 60: 161–170.

Hema, P., Murali, M., Thriveni, M.C., Prathibha, M., Jayaramu, S.C., Amruthesh, K.N. (2015). Phytochemical analysis and antibacterial activity of endophytic fungi isolated from *Basella rubra* L.-A medicinal plant. J Pure Appl Microbiol 9: 2971–2978.

Hodge, J. (2010). Plants we Use. Benchmark Educ Co, Pelham, NY 1–32.

Hong, C.E., Jo, S.H., Jo, I.H., Park, J.M. (2018). Diversity and antifungal activity of endophytic bacteria associated with *Panax ginseng* seedlings. Plant Biotechnol Rep 12: 409–418.

Huang, Q., An, H., Song, H., Mao, H., Shen, W., Dong, J. (2015). Diversity and biotransformative potential of endophytic fungi associated with the medicinal plant *Kadsura angustifolia*. Res Microbiol 166: 45–55.

Huang, W., Long, C., Lam, E. (2018). Roles of plant-associated microbiota in traditional herbal medicine. Trends Plant Sci 23: 559–562.

Ilmi, N., Jekti, D.S.D., Zulkifli, L. (2018). Molecular identification of endophytic bacteria from the stem's bark of moringa plant and their antibacterial activities. IOSR J Biotechnol Biochem 4: 21–30.

Iqrar, I., Shinwari, Z.K., El-Sayed, A., Ali, G.S. (2019). Bioactivity-driven high throughput screening of microbiomes of medicinal plants for discovering new biological control agents. BioRXiv Microbiol 1–26.

Jain, P., Sharma, P. (2016). Isolation and preliminary screening of endophytic fungi of *Ricinus communis* for their antimicrobial potential. JMBFS 05: 230–233.

Janatiningrum, I., Solihin, D.D., Meryandini, A., Lestari, Y. (2018). Comparative study on the diversity of endophytic actinobacteria communities from *Ficus deltoidea* using metagenomic and culture-dependent approaches. Biod J Biol Diversity 19: 1514–1520.

Jasim, B., Sreelakshmi, K.S., Mathew, J., Radhakrishnan, E.K. (2016). Surfactin, Iturin, and Fengycin biosynthesis by endophytic *Bacillus* sp. from *Bacopa monnieri*. Microb Ecol 72: 106–119.

Jia, M., Chen, L., Xin, H.L., Zheng, C.J., Rahman, K., Han, T., Qin, L.P. (2016). A friendly relationship between endophytic fungi and medicinal plants: A systematic review. Front Microbiol 7: 1–14.

Jin, H., Yang, X.-Y., Yan, Z.-Q., Liu, Q., Li, X.-Z., Chen, J.-X., Zhang, D.-H., Zeng, L.-M., Qin, B. (2014). Characterization of rhizosphere and endophytic bacterial communities from leaves, stems and roots of medicinal *Stellera chamaejasme* L. Syst Appl Microbiol 37: 376–385.

Joe, M.M., Devaraj, S., Benson, A., Sa, T. (2016). Isolation of phosphate solubilizing endophytic bacteria from *Phyllanthus amarus* Schum & Thonn: evaluation of plant growth promotion and antioxidant activity under salt stress. J Appl Res Med Aromat Plants 3: 71–77.

Johnston-Monje, D., Lopez Mejia, J. (2020). Botanical microbiomes on the cheap: inexpensive molecular fingerprinting methods to study plant-associated communities of bacteria and fungi. Appl Plant Sci 8: 1–12.

Kanaga, S., Ramamurthi, D., Ajamanikandan, K.C.P., Prasanth, D.A. (2018). Isolation and identification of endophytic fungi from *Melia azedarach* L. of Kalvarayan Hills, Eastern Ghats, Tamil Nadu, India. IJAM 21: 58–68.

Kavya, P.S., Madhu, P.C. (2019). Isolation and characterization of l-asparaginase producing endophytic bacteria from *Simarouba gluaca*. Int J Animal Sci Technol 3: 1–6.

Kekuda, T.R.P. (2016). Isolation, characterization and antimicrobial potential of endophytic Actinomycetes. Int J Curr Microbiol App Sci 5: 100–116.

Kelly, K. (2009). History of Medicine. New York, Facts on File, 29–50.

Khan, A.L., Asaf, S., M. Abed, R.M., Ning Chai, Y., N. Al-Rawahi, A., Mohanta, T.K., Al-Rawahi, A., Schachtman, D.P., Al-Harrasi, A. (2020). Rhizosphere microbiome of arid land medicinal plants and extra cellular enzymes contribute to their abundance. Microorganisms 8: 1–19.

Kilam, D., Sharma, P., Agnihotri, A., Kharkwal, A., Varma, A. (2017). Microbial symbiosis and bioactive ingredients of medicinal plants. pp. 283–302. *In*: Varma, A., Prasad, R., Tuteja, N. (eds.). Mycorrhiza-eco-physiology, Secondary Metabolites, Nanomaterials. Springer Cham 1.

Köberl, M., Müller, H., Ramadan, E.M., Berg, G. (2011). Desert farming benefits from microbial potential in arid soils and promotes diversity and plant health. PLoS One 6: 1–9.

Köberl, M., Schmidt, R., Ramadan, E.M., Bauer, R., Berg, G. (2013). The microbiome of medicinal plants: diversity and importance for plant growth, quality and health. Front Microbiol 4: 1–9.

Köberl, M., Erschen, S., Etemadi, M., White, R.A., El-Arabi, T.F., Berg, G. (2019). Deciphering the microbiome shift during fermentation of medicinal plants. Sci Rep 9: 1–11.

Kumar, A., Singh, R., Yadav, A., Giri, D.D., Singh, P.K., Pandey, K.D. (2016). Isolation and characterization of bacterial endophytes of *Curcuma longa* L. 3 Biotech 6: 60.

Lai, J.C., Lai, H.Y., Nalamolu, K.R., Ng, S.F. (2016). Treatment for diabetic ulcer wounds using a fern tannin optimized hydrogel formulation with antibacterial and antioxidative properties. J Ethnopharmacol 189: 277–89.

Lemanceau, P., Blouin, M., Muller, D., Moënne-Loccoz, Y. (2017). Let the core microbiota be functional. Trends Plant Sci 22: 583–595.

Lestari, Y., Murdini, L.A., Solihin, D.D. (2018). The existence of endophytic Actinobacteria from *Rhododendron zoelleri* revealed by culture-dependent and culture-independent approaches. HAYATI J Biosci 25: 54–62.

Li, L., Mohamad, O.A.A., Ma, J., Friel, A.D., Su, Y., Wang, Y., Musa, Z., Liu, Y., Hedlund, B.P., Li, W. (2018). Synergistic plant–microbe interactions between endophytic bacterial communities and the medicinal plant *Glycyrrhiza uralensis* F. Antonie van Leeuwenhoek 111: 1735–1748.

Li, O., Xiao, R., Sun, L., Guan, C., Kong, D., Hu, X. (2017). Bacterial and diazotrophic diversities of endophytes in *Dendrobium catenatum* determined through barcoded pyrosequencing. PLoS One 12: 1–21.

Li, S.B., Huang, J., Zhou, R.C., Li, Z., Xu, S., Ruan, T., Pang, Q. (2015). Diversity and plant growth-promoting potential of bacterial endophytes of Alpinia officinarum Hance, a famous south-China medicinal plant. Acta Ecologica Sinica 35: 3.

Liang, Z., Zhang, J., Zhang, Xuan, Li, J., Zhang, Xiaoqian, Zhao, C. (2016). Endophytic fungus from *Sinopodophyllum emodi* (Wall.) Ying that produces podophyllotoxin. J Chromatogr Sci 54: 175–178.

Lin, F., Liu, X., Wang, H., Zhang, C. (2003). Recent research and prospect on taxol and its producing fungi. Acta Microbiol Sinica 43: 534–538.

Liotti, R.G., da Silva, R.C., Pinheiro, D.G., da Silva, J.M., da Silva, G.F., de Mendonça, E.A.F., Soares, M.A. (2019). Characterization and comprehensive analysis of the ecological interaction networks of bacterial communities in *Paullinia cupana* var. *sorbilis* by 16S rRNA gene metabarcoding. World J Microbiol Biotechnol 35: 182.

Liu, H., Carvalhais, L.C., Crawford, M., Singh, E., Dennis, P.G., Pieterse, C.M.J., Schenk, P.M. (2017). Inner plant values: Diversity, colonization and benefits from endophytic bacteria. Front Microbiol 8: 1–17.

Liu, T., Greenslade, A., Yang, S. (2016). Levels of rhizome endophytic fungi fluctuate in *Paris polyphylla* var. *yunnanensis* as plants age. Plant Divers 39: 60–64.

Liu, T.H., Zhou, Y., Tao, W.-C., Liu, Y., Zhang, X.-M., Tian, S.-Z. (2020). Bacterial diversity in roots, stems, and leaves of chinese medicinal plant *Paris polyphylla* var. *yunnanensis*. Pol J Microbiol 69: 91–97.

Liu, W., Zhou, Z., Liu, Y., Hu, X., Guo, Y., Li, J. (2017). Application of high-throughput internal transcribed spacer rRNA metagenomics analysis in deciphering endophytic fungi diversity of *Dendrobium officinale*. J Biobased Mater Bioenergy 11: 106–118.

Lu, Y., Zhang, E., Hong, M., Yin, X., Cai, H., Yuan, L., Yuan, F., Li, L., Zhao, K., Lan, X. (2020). Analysis of endophytic and rhizosphere bacterial diversity and function in the endangered plant *Paeonia ludlowii*. Arch Microbiol 202: 1–12.

Maggini, V., De Leo, M., Mengoni, A., Gallo, E.R., Miceli, E., Reidel, R.V.B., Biffi, S., Pistelli, L., Fani, R., Firenzuoli, F., Bogani, P. (2017). Plant-endophytes interaction influences the secondary metabolism in *Echinacea purpurea* (L.) Moench: an *in vitro* model. Sci Rep 7: 1–8.

Makarska-Białokoz, M. (2019). History and significance of phytotherapy in the human history 1. Phytotherapy in ancient times. Arch Physiother Glob Res 23: 15–19.

Mamangkey, J., Suryanto, D., Munir, E., Lutfia, A., Hartanto, A., Huda, M.K. (2019). First report of plant growth promoting endophytic bacteria from medicinal invasive plants (*Chromolaena odorata*). IOP Conf Ser Earth Environ Sci 305: 1–10.

Manoharachary, C., Nagaraju, D. (2016). Endophytic hyphomycetous fungi associated with some medicinal plants from Telangana state, India. Indian Phytopathol 69: 169–172.

Marchesi, J.R., Ravel, J. (2015). The vocabulary of microbiome research: A proposal. Microbiome 3: 31–40.

Margulis, L. (1991). Symbiosis as a source of evolutionary innovation: speciation and morphogenesis. pp. 1–14. *In*: Cambridge Mass MIT Press. Symbiogenesis and Symbionticism.

Martinez-Klimova, E., Rodríguez-Peña, K., Sánchez, S. (2017). Endophytes as sources of antibiotics. Biochem Pharmacol 134: 1–17.

Mehanni, M.M., Safwat, M.S.A. (2010). Endophytes of medicinal plants. Acta Hortic 854: 31–40.

Mitter, B., Pfaffenbichler, N., Sessitsch, A. (2016). Plant-microbe partnerships in 2020. Microb Biotechnol 9: 635–640.

Müller, D.B., Vogel, C., Bai, Y., Vorholt, J.A. (2016). The plant microbiota: Systems-level insights and perspectives. Annu Rev Genet 50: 211–234.

Myo, E.M., Maung, C.E.H., Mya, K.M., Khai, A.A. (2020). Characterization of bacterial endophytes from Myanmar medicinal plants for antimicrobial activity against human and plant pathogens. Braz J Pharm Sci 56: 1–8.

Nadkarni, M.A., Martin, F.E., Hunter, N., Jacques, N.A. (2009). Methods for optimizing DNA extraction before quantifying oral bacterial numbers by real-time PCR. FEMS Microbiol Lett 296: 45–51.

Nair, D.N., Padmavathy, S. (2014). Impact of endophytic microorganisms on plants, environment and humans. Sci World J 2014: 1–11.

Nguyen, Q.H., Nguyen, H.V., Vu, T.H.N., Chu-Ky, S., Vu, T.T., Hoang, H., Quach, N.T., Bui, T.L., Chu, H.H., Khieu, T.N., Sarter, S., Li, W.J., Phi, Q.T (2019). Characterization of endophytic *Streptomyces griseorubens* MPT42 and assessment of antimicrobial synergistic interactions of its extract and essential oil from host plant *Litsea cubeba*. Antibiot 8: 197–211.

Nyholm, L., Koziol, A., Marcos, S., Botnen, A.B., Aizpurua, O., Gopalakrishnan, S., Limborg, M.T., Gilbert, M.T.P., Alberdi, A. (2020). Holo-Omics: Integrated host-microbiota multi-omics for basic and applied biological research. iSci 23: 1–16.

Owen, N.L., Hundley, N. (2004). Endophytes—the chemical synthesizers inside plants. Sci Prog 87: 79–99.

Panigrahi, S., Dash, D., Rath, C.C. (2018). Characterization of endophytic bacteria with plant growth promoting activities isolated from six medicinal plants. J Exp Biol Agric Sci 6: 782–791.

Pei, C., Mi, C., Sun, L., Liu, W., Li, O., Hu, X. (2017). Diversity of endophytic bacteria of *Dendrobium officinale* based on culture-dependent and culture-independent methods. Biotechnol Biotechnol Equip 31: 112–119.

Prakash, O., Sharma, R., Rahi, P., Karthikeyan, N. (2015). Role of microorganisms in plant nutrition and health. pp. 125–161. *In*: Rakshit, A., Singh, H.B., Sen, A. (eds.). Nutrient Use Efficiency: From Basics to Advances. Springer India, New Delhi.

Primanita, M., Wahyudi, A.T., Lestari, Y. (2015). 16S rRNA-based metagenomic analysis of endophytic actinomycetes diversity from *Tinospora crispa* L. Miers. Microbiology Indonesia 9: 25–34.

Purushotham, N., Jones, E., Monk, J., Ridgway, H. (2018). Community structure of endophytic actinobacteria in a New Zealand native medicinal plant *Pseudowintera colorata* (Horopito) and their influence on plant growth. Microb Ecol 76: 729–740.

Purushotham, N., Jones, E., Monk, J., Ridgway, H. (2020). Community structure, diversity and potential of endophytic bacteria in the primitive New Zealand medicinal plant *Pseudowintera colorata*. Plants (Basel) 9: 1–14.

Qin, S., Xing, K., Jiang, J.H., Xu, L.H., Li, W.J. (2011). Biodiversity, bioactive natural products and biotechnological potential of plant-associated endophytic actinobacteria. Appl Microbiol Biotechnol 89: 457–473.

Rahman, Lubna, Shinwari, Z.K., Iqrar, I., Rahman, Lutfur, Tanveer, F. (2017). An assessment on the role of endophytic microbes in the therapeutic potential of *Fagonia indica*. Ann Clin Microbiol Antimicrob 16: 1–12.

Rai, M., Agarkar, G., Rathod, D. (2014). Multiple applications of endophytic *Colletotrichum* species occurring in medicinal plants. pp. 227–236. *In*: Gurib-Fakim, A. (ed.). Novel Plant Bioresources. John Wiley & Sons, Ltd, Chichester, UK.

Ramalashmi, K., Prasanna, V.K., Magesh, K., Sanjana, R., Siril, J.S., Ravibalan, K. (2018). A potential surface sterilization technique and culture media for the isolation of endophytic bacteria from *Acalypha indica* and its antibacterial activity. J Med Plant Stud 6: 181–184.

Rosenberg, E., Koren, O., Reshef, L., Efrony, R., Zilber-Rosenberg, I. (2007). The role of microorganisms in coral health, disease and evolution. Nat Rev Microbiol 5: 355–362.

Rosenberg, E., Zilber-Rosenberg, I. (2018). The hologenome concept of evolution after 10 years. Microbiome 6: 78.

Rosenberg, E., Zilber-Rosenberg, I. (2019). The hologenome concept of evolution: medical implications. Rambam Maimonides Med J 10: e0005.

Ryan, R.P., Germaine, K., Franks, A., Ryan, D.J., Dowling, D.N. (2008). Bacterial endophytes: recent developments and applications. FEMS Microbiol Lett 278: 1–9.

Santos, T.F.B., dos Santos Carvalho, C., de Almeida, M.A., Delforno, T.P., Duarte, I.C.S. (2020). Endophytic fungi isolated from Brazilian medicinal plants as potential producers of antioxidants and their relations with anti-inflammatory activity. Biotech 10: 1–8.

Sarsaiya, S., Jain, A., Fan, X., Jia, Q., Xu, Q., Shu, F., Zhou, Q., Shi, J., Chen, J. (2020). New insights into detection of a dendrobine compound from a novel endophytic *Trichoderma longibrachiatum* strain and its toxicity against phytopathogenic bacteria. Front Microbiol 11.

Satari, A.H., Zargar, M.I., Shah, W.A., Bansal, R., Bhat, M.F. (2018). Isolation, molecular identification, phytochemical screening and *in vitro* anti-oxidant activity of endophytic fungi from *Achilea millefolium* Linn. J Pharmacogn Phytochem 7: 87–92.

Shan, W., Zhou, Y., Liu, H., Yu, X. (2018). Endophytic Actinomycetes from tea plants (*Camellia sinensis*): Isolation, abundance, antimicrobial, and plant-growth-promoting activities. Biomed Res Int 2018: 1–12.

Sharma, M., Sahu, K., Singh, S.P., Jain, B. (2018). Wound healing activity of curcumin conjugated to hyaluronic acid: *In vitro* and *in vivo* evaluation. Artif Cells Nanomed Biotechnol 46: 1009–1017.

Shendure, J., Ji, H. (2008). Next-generation DNA sequencing. Nat Biotechnol 26: 1135–1145.

Shubin, L., Juan, H., RenChao, Z., ShiRu, X., YuanXiao, J. (2014). Fungal endophytes of *Alpinia officinarum* rhizomes: insights on diversity and variation across growth years, growth sites, and the inner active chemical concentration. PLoS One 9: 1–21.

Siciliano, S.D., Fortin, N., Mihoc, A., Wisse, G., Labelle, S., Beaumier, D., Ouellette, D., Roy, R., Whyte, L.G., Banks, M.K., Schwab, P., Lee, K., Greer, C.W. (2001). Selection of specific endophytic bacterial genotypes by plants in response to soil contamination. Appl Environ Microbiol 67: 2469–2475.

Sieber, T.N. (2007). Endophytic fungi in forest trees: are they mutualists? Fungal Biol Rev 21: 75–89.

Silva, C.F., Vitorino, L.C., Mendonça, M.A.C., Araújo, W.L., Dourado, M.N., Albuquerque, L.C., Soares, M.A., Souchie, E.L. (2019). Screening of plant growth-promoting endophytic bacteria from the roots of the medicinal plant *Aloe vera*. South African Journal of Botany.

Singh, M., Kumar, A., Singh, R., Pandey, K.D. (2017). Endophytic bacteria: A new source of bioactive compounds. Biotech 7: 1–14.

Strobel, G., Daisy, B. (2003). Bioprospecting for microbial endophytes and their natural products. Microbiol Mol Biol Rev 67: 491–502.

Strobel, G.A. (2003). Endophytes as sources of bioactive products. Microbes Infect 5: 535–544.

Sun, Haifeng, Kang, B., Chai, Z., Sun, Huanhuan, Du, H., Gao, J., Feng, Q., Zhang, C., Cao, Q., Guo, L. (2017). Characterization of root-associated microbiota in medicinal plants *Astragalus membranaceus* and *Astragalus mongholicus*. Ann Microbiol 67: 587–599.

Sushma, K.s., Jayashankar, M.A.k.V., Saeed, M.A. (2018). Identification of endophytic fungi from the medicinal plants of Biligirirangana hill, Karnataka. J Appl Nat Sci 10: 1156–1161.

Swapna, P.S., Savita, S.K., Nagaraju, Y. (2019). Potentials of endophytic bacteria in sustainable agriculture and the use of genomic approach in plant-endophytic interactions. J Pharmacogn Phytochem 8: 2166–2168.

Taffner, J., Cernava, T., Erlacher, A., Berg, G. (2019). Novel insights into plant-associated archaea and their functioning in arugula (*Eruca sativa* Mill.). J Adv Res 19: 39–48.

Taghavi, S., Barac, T., Greenberg, B., Borremans, B., Vangronsveld, J., van der Lelie, D. (2005). Horizontal gene transfer to endogenous endophytic bacteria from poplar improves phytoremediation of toluene. Appl Environ Microbiol 71: 8500–8505.

Tamariz-Angeles, C., Olivera-Gonzales, P., Santillán-Torres, M. (2018). Antimicrobial, antioxidant and phytochemical assessment of wild medicinal plants from Cordillera Blanca (Ancash, Peru). Boletín Latinoamericano y del Caribe de Plantas Medicinales yAromáticas 17: 270–285.

Tang, Z., Wang, Y., Yang, J., Xiao, Y., Cai, Y., Wan, Y., Chen, H., Yao, H., Shan, Z., Li, C., Wang, G. (2020). Isolation and identification of flavonoid-producing endophytic fungi from medicinal plant *Conyza blinii* H.Lév that exhibit higher antioxidant and antibacterial activities. PeerJ 8: e8978.

Tanvir, R., Javeed, A., Bajwa, A.G. (2017). Endophyte bioprospecting in South Asian medicinal plants: An attractive resource for biopharmaceuticals. Appl Microbiol Biotechnol 101: 1831–1844.

The UniProt Consortium. UniProt: a hub for protein information. (2015) Nucleic Acids Res. 43 (Database issue): D204-12.

Theis, K.R., Dheilly, N.M., Klassen, J.L., Brucker, R.M., Baines, J.F., Bosch, T.C.G., Cryan, J.F., Gilbert, S.F., Goodnight, C.J., Lloyd, E.A., Sapp, J., Vandenkoornhuyse, P., Zilber-Rosenberg, I., Rosenberg, E., Bordenstein, S.R. (2016). Getting the hologenome concept right: An eco-evolutionary framework for hosts and their microbiomes. mSystems 1: e00028–16.

Tian, L., Lin, X., Tian, J., Ji, L., Chen, Y., Tran, L.-S.P., Tian, C. (2020). Research advances of beneficial microbiota associated with crop plants. Int J Mol Sci 21: 1–18.

Toju, H., Guimarães, P.R., Olesen, J.M., Thompson, J.N. (2014). Assembly of complex plant–fungus networks. Nat Commun 5: 1–7.

Turner, T.R., James, E.K., Poole, P.S. (2013). The plant microbiome. Genome Biol 14: 1–10.

Ulloa-Muñoz, R., Olivera-Gonzales, P., Castañeda-Barreto, A., Villena, G.K., Tamariz-Angeles, C. (2020). Diversity of endophytic plant-growth microorganisms from *Gentianella weberbaueri* and *Valeriana pycnantha*, highland Peruvian medicinal plants. Microbiol Res 233: 1–11.

Urumbil, S.K., Kumar, M.A. (2020). Diversity analysis of endophytic bacterial microflora in *Emilia sonchifolia* (Linn.) DC on Illumina Mi Seq platforms. J Pure Appl Microbiol 14: 679–687.

Wang, Y., Lai, Z., Li, X.-X., Yan, R.-M., Zhang, Z.-B., Yang, H.-L., Zhu, D. (2016). Isolation, diversity and acetylcholinesterase inhibitory activity of the culturable endophytic fungi harboured in *Huperzia serrata* from Jinggang Mountain, China. World J Microbiol Biotechnol 32: 20.

Wang, Y., Wang, H., Cheng, H., Chang, F., Wan, Y., She, X. (2020). Niche differentiation in the rhizosphere and endosphere fungal microbiome of wild *Paris polyphylla* Sm. Peer J 8: 1–21.

Wani, M.C., Taylor, H.L., Wall, M.E., Coggon, P., McPhail, A.T. (1971). Plant antitumor agents. VI. Isolation and structure of taxol, a novel antileukemic and antitumor agent from *Taxus brevifolia*. J Am Chem Soc 93: 2325–2327.

Wardecki, T., Brötz, E., De Ford, C., von Loewenich, F.D., Rebets, Y., Tokovenko, B., Luzhetskyy, A., Merfort, I. (2015). Endophytic Streptomyces in the traditional medicinal plant *Arnica montana* L.: secondary metabolites and biological activity. Antonie van Leeuwenhoek 108: 391–402.

White, J.F., Torres, M.S., Johnson, H., Irizarry, I., Tadych, M. (2014). A functional view of plant microbiomes: endosymbiotic systems that enhance plant growth and survival. pp. 425–439. *In*: Verma, V.C., Gange, A.C. (eds.). Adv Endophytic Res. Springer India, New Delhi.

Wicaksono, W.A., Sansom, C.E., Eirian Jones, E., Perry, N.B., Monk, J., Ridgway, H.J. (2018). Arbuscular mycorrhizal fungi associated with *Leptospermum scoparium* (mānuka): Effects on plant growth and essential oil content. Symbiosis 75: 39–50.

Wilson, D. (1995). Endophyte: the evolution of a term, and clarification of its use and definition. Oikos 73: 274–276.

Wu, Q., Wei, D., Dong, L., Liu, Y., Ren, C., Liu, Q., Chen, C., Chen, J., Pei, J. (2019). Variation in the microbial community contributes to the improvement of the main active compounds of *Magnolia officinalis* Rehd. et Wils in the process of sweating. Chin Med 14.

Yang, R., Liu, P., Ye, W. (2017). Illumina-based analysis of endophytic bacterial diversity of tree peony (*Paeonia Sect. Moutan*) roots and leaves. Braz J Microbiol 48: 695–705.

Yang, C.H., Crowley, D.E. (2000). Rhizosphere microbial community structure in relation to root location and plant iron nutritional status. Appl Environ Microbiol 66: 345–351.

Yang, G., Li, P., Meng, L., Xv, K., Dong, F., Qiu, Y., He, L., Lin, L., Yang, G., Li, P., Meng, L., Xv, K., Dong, F., Qiu, Y., He, L., Lin, L. (2018). Diversity and communities of culturable endophytic fungi from different tree peonies (geoherbs and non-geoherbs), and their biosynthetic potential analysis. Braz J Microbiol 49: 47–58.

Yang, R., Liu, P., Ye, W. (2017). Illumina-based analysis of endophytic bacterial diversity of tree peony (*Paeonia Sect. Moutan*) roots and leaves. Braz J Microbiol 48: 695–705.

Yin, D.D., Yang, M., Wang, Y.L., Yin, D.K., Liu, H.K., Zhou, M., Li, W., Chen, R., Jiang, S.S., Ou, M.F., Xu, F. (2018). High tetramethylpyrazine production by the endophytic bacterial *Bacillus subtilis* isolated from the traditional medicinal plant *Ligusticum chuanxiong* Hort. AMB Expr 8: 193.

Zhang, C., Ma, X., Zhu, R., Liu, Z., Gu, M., Zhang, J., Li, Y., Xu, Y., Zhu, D. (2020a). Analysis of the endophytic bacteria community structure and function of *Panax notoginseng* based on high-throughput sequencing. Curr Microbiol 10: 2745–2750.

Zhang, Y., Zheng, L., Zheng, Y., Xue, S., Zhang, J., Huang, P., Zhao, Y., Hao, X., He, Z., Hu, Z., Zhou, C., Chen, Q., Liu, J., Wang, G., Sang, M., Sun, X., Wang, X., Xiao, X., Li, C. (2020b). Insight into the assembly of root-associated microbiome in the medicinal plant *Polygonum cuspidatum*. Ind Crops Prod 145: 112163.

Zhao, K., Li, J., Zhang, Xiaoyue, Chen, Q., Liu, M., Ao, X., Gu, Y., Liao, D., Xu, K., Ma, M., Yu, X., Xiang, Q., Chen, J., Zhang, Xiaoping, Penttinen, P. (2018). Actinobacteria associated with *Glycyrrhiza inflata* Bat. are diverse and have plant growth promoting and antimicrobial activity. Sci Rep 8.

Zhou, R.-C., Huang, J., Li, Z.-E., Li, S.-B. (2014). Diversity and tissue distribution of fungal endophytes in *Alpinia officinarum*: an important south-China medicinal plant. Zhongguo Zhong Yao Za Zhi 39: 3023–3029.

Zhou, T., Wang, Y., Chen, J.-Q., Araki, H., Jing, Z., Jiang, K., Shen, J., Tian, D. (2004). Genome-wide identification of NBS genes in japonica rice reveals significant expansion of divergent non-TIR NBS-LRR genes. Mol Genet Genomics 271: 402–415.

Zhou, X., Zhu, H., Liu, L., Lin, J., Tang, K. (2010). A review: recent advances and future prospects of taxol-producing endophytic fungi. Appl Microbiol Biotechnol 86: 1707–1717.

Zilber-Rosenberg, I., Rosenberg, E. (2008). Role of microorganisms in the evolution of animals and plants: The hologenome theory of evolution. FEMS Microbiol Rev 32: 723–735.

Zmora, N., Suez, J., Elinav, E. (2019). You are what you eat: diet, health and the gut microbiota. Nat Rev Gastroenterol Hepatol 16: 35–56.

15

Potential of Mycochemicals in the Prevention and Control of Microbial Diseases

V. O. Oyetayo[1,*] and *C. O. Ogidi*[2]

Introduction

Microorganisms can be divided into the following groups: bacteria, protozoa, fungi and viruses. They are ubiquitous; hence their presence has beneficial, unapparent and harmful effects on the environment where they exist. Society benefits from microorganisms in many ways. They are involved in the production of indigenous African fermented foods such as "garri", "ogi", "pupuru". Microorganisms also play important roles in the production of bread, cheese, beer, antibiotics, probiotics, vaccines, vitamins, enzymes and many other important products. Hence, they are useful in food, pharmaceutical, waste treatment and in the generation of electricity.

While useful as microorganisms, they are also the causative agents of wide range of infectious diseases of humans and domestic animals. The human race and microbial entities have been involved in the war of survival from time immemorial. Microbes have shown incredible ability to adapt, re-adapt, survive and challenge human ingenuity in the last two to three decades (Fonkwo 2008). Hence, pathogens constantly change their genetic make-up, which normally results to ineffectiveness of vaccines and other chemotherapeutic agents used in fighting pathogenic microorganisms and microorganisms responsible for the spoilage of foods. This genetic flexibility allows many infectious agents to mutate or evolve into more deadly strains against which humans have little or no resistance against (Fonkwo 2008). Diseases such as plague, tuberculosis, anthrax and gonorrhea are caused by bacteria. Protozoa cause diseases including malaria, sleeping sickness, dysentery and toxoplasmosis. Microscopic fungi cause diseases such as ringworm, candidiasis and histoplasmosis. Pathogenic viruses cause diseases such as influenza, yellow fever and AIDs. The noxious effects of microorganisms are not limited to animals alone as they are also important pathogens responsible for various diseases of crop plants. Plant and crop diseases such as maize leaf rust, wheat stem rust, powdery

[1] Department of Microbiology, Federal University of Technology, Akure, Nigeria.
[2] Biotechnology Unit, Department of Biological Sciences, Kings University, PMB 555, Odeomu, Nigeria.
* Corresponding author: ovonew67@gmail.com

mildew, leaf spot, and crown gall, and so on are caused by microbial pathogens in crops and plants. The great Irish famine of the 1840s as a result of potato blight was caused by a fungal plant pathogen, Oomycete, *Phytophthora infestans* (Kinealy 1994).

While microorganisms are legitimately trying to survive and procreate, unknowingly to these little organisms; their survival has a lot of detrimental effect to humans and other organisms around them. Specifically, the human race and other animals have survived these many centuries because they are equipped with various defensive mechanisms against the pestiferous effects of microorganisms (Amos 1981). Some of these mechanisms are general in nature. This serves to protect against many types of harmful agents and are on this basis referred to as non-specific immunity. A good example of this is phagocytosis of bacteria by specialized cells. Other defensive mechanisms are specific in that each is effective against a certain noxious agent and no other, and are therefore referred to as specific immunity (Stewart and Beswicvk 1977). The vertebrate host does not depend solely on its immune system to protect it from the agent of disease. Other defense mechanisms, which could be mechanical such as the unbroken skin, and to a lesser extent, mucous membranes provide a physical barrier to the entrance of most pathogens. The effectiveness of these barriers is often enhanced by the presence of chemical barriers as well (Stewart and Beswick 1977). Tears, saliva, and the secretions of the nasopharynx contain lysozome, an enzyme that digest the peptidoglycan of bacterial cell walls. Ecological controls which involve the normal body flora exist in a balance state and thereby protect the host from invasion by pathogens (Van Der Waaij et al. 1982).

Additionally, man has developed chemotherapeutic agents to checkmate the noxious effects of microorganisms. These agents have been in use since the 1950s until it was found to result in the development of resistant populations of bacteria, which made subsequent administration of antibiotics for therapy difficult (Fuller 1989). The use of antibiotics as therapeutic agents had also been found to result in intestinal disorders, which often follow oral treatment with these compounds. Looking at the myriads of demerits associated with the use of antibiotics, it is therefore necessary to find an alternative source of antimicrobial agent that will be effective, safe to the user and also eco-friendly.

The search for safe and efficient pharmacological agents against resistant microorganisms, free radicals generated from oxidation and inflammation and other ailments in the body have increased in the last three decades. Commonly used pharmacological agents which are mostly synthetic compounds have been found to have some side effects (Veglioglu et al. 1998). Hence, there is urgent need for safe and effective antimicrobial agents. Mushrooms have been found to be a reservoir of new natural bioactive compounds with potential application in medicine (Oyetayo and Oyetayo 2020). The use of mushrooms by man has been from ancient times and in some cases connected with mysticism. The first report of a particular species of mushroom used as a hallucinogenic agent was credited to the Yoruba tribe of Nigeria in Africa (Griensven 2009). According to Samorini (1992), this dates back to the Paleolithic period (7000–9000 years ago). However, in the Far East, especially China and Japan the knowledge on the use of edible and medicinal mushrooms had been passed on from one generation to the other in documented form. The earliest Chinese material medical book, *Shennong Bencao Jing*, and other succeeding Chinese medical book had been in use for over 2,500 years to record and depict many medicinal mushrooms (Zhu 2009).

They are known to produce a wide range of secondary metabolites with high therapeutic value (Demain 1999). These secondary metabolites produced by mushrooms have been reported to have nutraceutical properties such as antioxidant, antimicrobial, anticancer, cholesterol lowering and immunostimulatory effects (Mau et al. 2004; Oyetayo et al. 2013;

Ogidi et al. 2018). Myconutraceutical is a term used to describe nutraceutical substances obtained mainly from edible and medicinal mushrooms. Several species of mushrooms grow in the environment by withstanding the deteriorative effect of microorganisms which are ubiquitous. This shows that mushrooms may be producing antibacterial and antifungal compounds to survive in their natural environment (Lindequist et al. 2005). They may therefore be rich source of natural antibiotics.

Mushrooms are laden with pharmacologically active components that do not cause any harm nor place additional stress to the body (Wasser 2002). They have become attractive as functional foods and as a source of physiologically beneficial medicines, while being devoid of undesirable side-effects (Sagakami et al. 1991). Chang (2009) stated that the safety of myconutraceuticals is enhanced by the following facts.

i. Edible mushrooms have been used as part of human diets from time immemorial.
ii. The overwhelming majority of mushrooms used for production of mushroom dietary supplements or nutraceuticals are cultivated commercially and not gathered in the wild. This ensures proper identification, pure and unadulterated products. This also guarantees genetic uniformity.
iii. Mushrooms are easily propagated vegetatively and thus keep to one clone. The mycelium can be stored for a long time.
iv. Many edible/medicinal mushrooms are capable of growing in the form of mycelia biomass in submerged cultures. This is an area one needs to explore since a lot of new bioactive compounds with myconutraceutical properties can be derived from cultural manipulation in submerged cultures.

The functional properties of mushrooms have been attributed to the presence of some bioactive compounds in them. The major bioactive compounds found in mushrooms are biologically active compounds, such as polysaccharides, vitamins, terpenes, steroids, amino acids and trace elements (Lorenze and Anke 1998; Wasser and Weis 1999; Mizuno 1999; Zaidman et al. 2005). Some of these substances are also present in broth used in submerged cultivation. It has been reported that a higher level and number of different polysaccharides are extracted from fruit bodies than from the other cultural sources (Reshetnikov et al. 2001).

Macrofungi extracts and their bioactive constituents have all-encompassing nutritional and therapeutic applications. Certain macrofungi regulate gut microbiota by acting as a prebiotic. The prebiotic characteristics of macrofungi have greatly expanded their applications in developing innovative food and health care products with enhanced market value (Shirur and Shivalingegowda 2015). Several studies indicated that high consumption of edible mushrooms rich in bioactive compounds with antioxidants, vitamins, phenolic compounds have health benefits on humans and lessen the risk of numerous diseases, such as cancer, heart disease, stroke, Alzheimer's, diabetes, cataracts, and age-related functional decadence (Xu and Beelman 2015; Lee et al. 2019). Natural antimicrobial agents from macrofungi now need to be explored for greater uses/gaining an increasing importance as potential alternatives to synthetic preservatives. Mushrooms have strong bioactivity, pleasant and unique flavors, macrofungi serve as nutrient-rich foods, commercial nutraceuticals and pharmaceutical products (Thu et al. 2020).

The bioactive compounds derived from mushrooms function as Biological Response Modifiers (BRM), which are substances that stimulate the body's response to infection and disease. The body is known to produce these substances but not in appreciable quantity hence, exogenous supplies through diet or dietary supplements are needed (Oyetayo

2008). Mushroom nutraceuticals may unarguably be the source of this exogenous supply (Oyetayo 2020). The bioactive compounds sourced from mushrooms are known to be effective, safe and eco-friendly.

Mycochemicals in mushrooms with antimicrobial properties

Most pharmaceutical industries are in search of new antibiotics and other therapeutics such as antihypertensives, anticancer agents, antiparasitics and immunosuppressants (Gupta et al. 2018). Secondary metabolites which are a heterogeneous group of low molecular weight or high molecular weight produced by variety of microorganisms are known to possess therapeutic properties (Table 15.1).

Some fungi species in the class; Basidiomycetes and Ascomycetes, are called medicinal and edible macrofungi as a result of the therapeutical properties they possess. They are producers of high- and low-molecular-weight bioactive compounds such as alkaloids, lectins, lipids, peptidoglycans, phenolics, polyketides, polysaccharides, proteins, polysaccharide-protein/peptides, ribosomal and non-ribosomal peptides, steroids, alkaloids, terpenoids polyphenolics, Fungal Immunomodulatory Proteins (FIPs), terpenoids include mono and sesquiterpenoids, ergosterols and volatile organic compounds such as ceramides (Hyde et al. 2019). Mushrooms are capable of producing bioactive metabolites that have extensive uses in the medicinal field. Some of these important bioactive compounds are highlighted below.

Phenolic compounds

Phenolic compounds are aromatic hydroxylated compounds with one or more aromatic rings and one or more hydroxyl groups. Phenolic compounds include phenolic acids, oxidized polyphenols, hydroxybenzoic acids, flavonoids, tannins, hydroxycinnamic acids, stilbenes and lignans both benzoic (protocatechuic, 4-OH-benzoic, vanillic, syringic), cinnamic acid derivatives (caffeic, p-coumaric, ferulic), flavonoids, hydroxybenzoic acids, hydroxycinnamic acids, lignans, tannins, stilbenes and oxidized polyphenols (Kumar and Goel2019). Phenolic compounds are excellent sources of antioxidants and possess anti-tumor, cardiovascular, anti-viral, anti-microbial, anti-inflammatory and anti-allergenic properties (Cateni et al. 2021). They also help to balance blood sugar level and support the body's detoxification mechanisms. Synthetic phenolic; butylated hydroxyanisole (BHA), butylated hydroxytoluene (BHT), propyl gallate, tert-butylhydroquinone (TBHQ), ethoxyquin (EQ) are effective in inhibiting oxidation. However, some synthetic phenolic may cause adverse toxic effects under certain conditions. BHA, which is very often used as an additive in food industry, can have negative effects on the regulation of the activity of Mitogen-Activated Protein Kinase (MAPK) depending on the dosage. Phenolic compounds such as quercetin, quercetin-3-O-rutinoside, myricetin, naringenin, kaempferol (+)-catechin, (-)-epicatechin, ascorbic acid, β-carotene, lycopene, 4-hydroxybenzoic, ferulic, p-coumaric, protocatechuic, t-cinnamic and vanillic acids and naringenin were detected in mushrooms (Gąsecka et al. 2016). Mushroom extracts of *Agaricus bisporus, Boletus edulis, Calocybe gambosa, Cantharellus scibarius, Craterellus cornucopioides, Hygrophorus marzuolus, Lactarius deliciosus* and *Pleurotus ostreatus* contain caffeic acid, catechin, chlorogenic acid, p-coumaric acid, ferulic acid, gallic acid, gentisic acid, p-hydroxybenzoic acid, homogentisic acid, myricetin, protocatechuic acid, and pyrogallol (Palacios et al. 2011). Phenolic compounds are effective antimicrobial agents against some microorganisms that are resistant to commercial antibiotics. The presence of carboxylic acid (COOH), two hydroxyl (OH) groups in para and ortho-positions

Table 15.1. Bioactive compounds in Medicinal Mushrooms.

Bioactive compounds	Mushrooms	Bioactivities	References
Phenolic acids, monoterpenes, Volatile organic compounds	*Ophiocordyceps sinensis, Ganoderma lucidum, Grifola frondosa, Hericium erinaceus, Antrodia camphorata, Flammulina velutipes*	Antibacterial Antifungal	Pennerman et al. (2015); Fukushima-Sakuno (2020)
Phenolic compounds	*Amanita* spp., *Lactarius Volemus, Russula* spp., *Termitomyces* spp., *Tricholoma crissum, Volvariella volvacea, Astraeus hygriometricus, Auricularia auricula-judae, Cantharellus cibarius, Craterellus aureus, Lentinus* spp.	Antibacterial Antifungal	Butkhup et al. (2018)
Polysaccharides	*Lentinus edodes, Schizophyllum commune, Ganoderma lucidum, Grifola frondosa* and *Pleurotus abalones., Trametes versicolor, Calvatia gigantea, Phellinus linteus, Pleurotus pulmonarius, Agaricus bisporus, Lentinula edodes, Schizophyllum commune, Agaricus subrufescens, Ganoderma lucidum Amanita citrine, Amanita muscaria, Amanita pantherina, Armillaria mellea, Bjerkandera adusta, Clavicorona pyxidata, Clavulina cinerea, Clitocybe gibba, Clitocybe nebularis, Coprinus micaceus, Cortinarius cinnamomeus, Cortinarius sanguineus, Daedaleopsis confragosa, Fomes fomentarius, Fomitopsis pinicola, Ganoderma applanatum, Gymnopilus penetrans, Gymnopilus dryophilus, Heterobasidion annosum, Hygrophoropsis aurantiaca, Hyphodontia paradoxa, Laccaria amethystina, Laccaria laccata, Lactarius aurantiacus, Lactarius helvus, Lactarius resimus, Lactarius rufus, Lactarius vellereus, Laetiporus sulphureus, Leccinum scabrum, Lepista flaccida, Lycoperdon perlatum* Pers.: *Macrolepiota procera, Marasmius oreades, Morchella conica, Panellus stipticus, Paxillus involutus, Pholiota mutabilis, Piptoporus betulinus, Pseudoclitocybe cyanthiformis, Psilocybe capnoides, Psilocybe fascicularis, Psilocybe lateritia, Rhodocollybia maculata, Russula fragilis, Scleroderma citrinum, Sparassis crispa, Suillus bovinus, Suillus variegatus, Trametes hirsuta, Trametes versicolor, Tubaria furfuracea* and *Xerocomus badius*	Antibacterial Antiviral Prebiotic	Sánchez (2017); Nowak et al. (2018)
Fatty acids, amino acids, tocopherol content, carotenoids (β-carotene, lycopene), flavonoids, ascorbic acid, and anthocyanidins	*Agaricus* spp., *Cantharellus cibarius, Conocybe tenera, Gymnopilus junonius, Hygrocybe* spp., *Inocybe splendens, Lactarius pubescens, Laccaria laccata, Laccaria laccata, Lepista nuda, Lentinus cladopus, Pleurotus cystidiosus, Russula* spp.	Antibacterial Antifungal	Sharma and Gautam (2015)

Table 15.1 contd. ...

...*Table 15.1 contd.*

Bioactive compounds	**Mushrooms**	**Bioactivities**	**References**
Protein and peptide (pleurostrin) Cordymin,	*P. ostreatus, Russula paludosa, Agrocybe cylindracea, Cordyceps sinensis, Cordyceps militaris, Xylaria hypoxylon, Pholiota adipose, Russula lepida, Hericium erinaceum, Russula lepida, Hypsizygus marmoreus, Clitocybe maxima. Ganoderma tsugae, Clitocybe maxima, Astraeus hygrometricus, Termitomyces clypeatus, Pleurotus florida, Calocybe indica, V. volvacea, Tricholoma giganteum, Clitocybe sinopica. Agaricus* spp., *Flammulina velutipes, Letinus edodes, Pleurotus* spp.	Antibacterial Antifungal and Antiviral	Sánchez (2017); González et al. (2020)
Ergosterol, Clavilactones A–C, meroterpenoids, named clavipines A–C	*Agaricus silvaticus, Ampulloclitocybe clavipes, Butyriboletus roseoflavus, Cantharellus cibarius, Craterellus cornucopioides, Craterellus odoratus, Fistulina hepatica, Hydnum repandum, Laccaria amethystina, Lactarius hatsudake, Lepista sordida, Lycoperdon pyriforme, Neolentinus lepideus, Phlebopus portentosus, Polyozellus multiplex, Ramaria botrytis, Rugiboletus extremiorientalis, Russula virescens, Sarcodon imbricatus, Termitomyces albuminosus, Termitomyces eurhizus, Termitomyces heimii, Termitomyces microcarpus, Thelephora ganbajun, Volvariella bombycina*, and *Ganoderma lucidum*	Antiplasmodial Antibacterial Antifungal	Oluba (2019)
Phenolic compounds	*Amanita citrina, Amanita muscaria, Amanita pantherina, Amanita porphyria, Bjerkandera adusta, Clavicorona pyxidata, Cortinarius armillatus, Cortinarius sanguineus, Daedaleopsis confragosa, Fomes fomentarius, Fomitopsis pinicola, Gymnopilus penetrans, Heterobasidion annosum, Hygrophoropsis aurantiaca, Hyphodontia paradoxa, Lactarius aurantiacus, Lactarius helvus, Lactifluus vellereus, Lenzites betulina, Panellus stipticus, Pseudoclitocybe cyathiformis, Psilocybe fascicularis, Psilocybe lateritia, Rhodocollybia maculata, Russula fragilis, Scleroderma citrinum, Stereum hirsutum, Thelephora terrestris, Trametes hirsuta, Trichaptum fusco-violaceum, Tubaria furfuracea*	Antibacterial Antifungal Antiparasitic	Nowacka et al. (2015)
Terpenes, Terpenoids, steroids	*Ganoderma amboinense, G. annulare, G. applanatum, G. australe, G. boninense, G. capense, G. carnosum, G. cochlear, G. colossum, G. concinna, G. fornicatum, G. hainanense, G. lipsiense, G. mastoporum, G. neo-japonicum, G. orbiforme, G. pfeifferi, G. resinaceum, G. sinense, G. theaecolum, G. tropicum, G. tsugae, G. lucidum*	Antibacterial Antifungal Antiplasmodial	Baby et al. (2015) Castellano and Torrens (2015) Zhou et al. (2015)
Grifolin, Neogrifolin Confluentin, Orellanine	*Barssia oregonensis, Albatrellus flettii, Amanita augusta, Auricularia fuscosuccinea, Cantharellus cibarius, Chroogomphus Tomentosus, Clavulina cinerea, Coprinus comatus*	Antimicrobial Anti-proliferative Immuno-stimulatory Antituberculosis, Anti-inflammatory	Zeb and Lee (2021)

Polysaccharides, terpenoids, Ganodermanondiol, Lentinan, polypeptide, β-glucan proteoglycan	*Agaricus bisporus, Ganoderma lucidum, Hericium erinaceus, Pleurotus ostreatus, Flammulina velutipes, Lentinula edodes, Trametes* spp., *Russulales* spp., *Hericium erinaceus, Grifola frondosa, Coriolus versicolor, Lentinula edodes*	Antiallergic, Antibacterial, Antifungal, Anti-inflammatory, Antioxidative, Antiviral, Cytotoxic, Neuroprotective, Nephroprotective, Osteoprotective, and Hypotensive, Antidepressive, Antihyperlipidemic, Antidiabetic	Badalyan and Rapior (2021) Venturella et al. (2021)

of the benzene ring and also a methoxyl (OCH_3) group in the meta-position seems to be important for anti-MRSA activity (Alves et al. 2013). Phenolic compounds such as grifolin and its derivatives have been isolated from the edible mushroom; *Albatrellus ovinus* and showed antimicrobial activity. *Hericium erinaceus*, an edible and medicinal mushroom in traditional Chinese medicine contains phenol analogous compounds (hericenones C, D, E, F, G, H) isolated (Thongbai et al. 2015; Corana et al. 2019). Phenols, flavonoids and anthocyanin were greater in amount in *Termitomyces robustus* than *Lentinus squarrosulus* (Ugbogu et al. 2020). Bioactive components; total phenolics, tannins, flavonoids, vitamin C, phytic acid, lycopene, β-carotene, trypsin inhibition and haemagglutinin were reported in *Amanita* spp.; nonconventional edible wild mushroom (Greeshma et al. 2018).

Polysaccharides

Polysaccharides, including polysaccharide-protein complexes are major bioactive constituents of edible and medicinal mushrooms. Polysaccharides from higher basidiomycetes consist of functional components of β-glucans, galactose and mannose with different branching patterns and conformations. Polysaccharides may contain either one monomer unit called homoglycans or two or more monomer units called heteroglycans. Polysaccharides obtained from mushrooms are known to have anti-tumor and immunomodulatory actions. Mushrooms polysaccharides possess large antimicrobial activities against pathogenic microorganisms (Friedman 2016).

Polysaccharides from basidiomycetes have been exploited as functional food substances, and examples include lentinan from *Lentinus edodes*, schizophyllan from *Schizophyllum commune*, pleuran from *Pleurotus* species, calocyban from *Calocybe indica* and ganoderan from *Ganoderma lucidum* (Villares et al. 2012; Zeb and Lee 2021). They were considered as the outstanding representatives of D-glucans with the common (1→3) or (1→6) β-linked glucose backbones and distinguished by different patterns and degrees of branches. These have several health benefits such as mitigating blood cholesterol and glucose levels. Fungal exopolysaccharides produced by species of *Phellinus linteus, Ganoderma lucidum, Fusarium* sp., *Pleurotus* spp., *Inonotus obliquus* and *Aureobasidium pullulans* are of great uses in food, cosmetics, pharmaceutical industries due to their preservatives, bioherbicides and microbicides activities (Mahapatra and Banerjee 2013).

Terpenes and terpenoids

Terpenes and terpenoids are generally known as terpenes in the subclass of prenyllipids. Terpenoids result from detaching and moving methyl groups and adding oxygen atoms in terpenes. Isoprene units are the building blocks of terpenes. Terpenes obtained from basidiomycetes are termed as terpenoids or isoprenoids, whereas those obtained from plants are present in resin and essential oils (Benkeblia 2015). Terpenoids are categorized on the basis of the number of carbons units in their building blocks. Figure 15.1 shows a typical structure of monoterpene.

Terpenoids are by far the largest class of natural products, but only four are currently approved as antimicrobial drugs (tiamulin, valnemulin, retapamulin and lefamulin), and these are all semisynthetic derivatives of the fungal antibiotic pleuromutilin (Mendes et al. 2016).

Terpenes are the largest group of anti-inflammatory compounds in mushrooms. Several terpenes have been isolated from *Ganoderma lucidum*. These are non-polar metabolites comprised of the following groups: volatile mono and sesquiterpenes oils (C10 and C15),

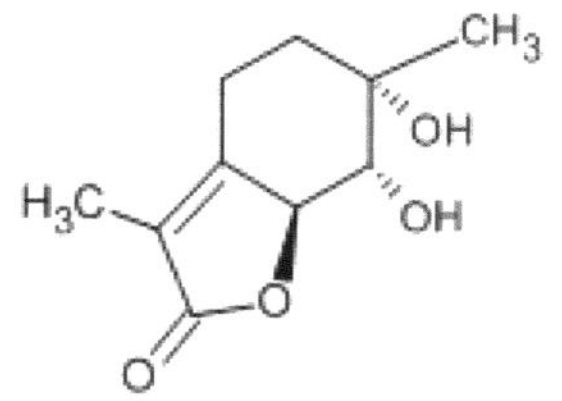

Figure 15.1. Structure of monoterpene.

less volatile diterpenes (C20), involatile triterpenoids and sterols (C30) and the carotenoid pigments (C40). Triterpene chemical structures are based on lanosterol. It is an important intermediate for their synthesis. Stereochemical re-arrangement of these compounds among triterpenoids results in their structural diversity (Paliya et al. 2014; Sánchez 2017). Chen et al. (2017) isolated 40 triterpenoids from the fruiting body of *G. lucidum*. About 380 terpenoids and 30 steroids like lanostane triterpenes as ganoderic and lucidenic acids, meroterpenoids, pentacyclic triterpenes, prenylhydroquinone sesquiterpenoids, steroids, as well as alcohols, aldehydes, alkaloids, esters, glycosides, ketones and lactones with significant bioactivities were isolated from 22 *Ganoderma* species (Baby et al. 2015; Badalyan et al. 2019). Isaka et al. (2017) isolated lanostane triterpenoids, astraeusins M–Q (1–5), 26-epi-artabotryol C1 (6) and 26-epiastrasiaone from the edible mushroom *Astraeus asiaticus*. Among the extracted compounds, astraeusins M exhibits anti-malarial activity and several lanostoids from the same species show anti-bacterial properties (Isaka et al. 2017). *Ganoderma* spp. contain the main active constituents of triterpenoids, they were identified as ganopsoreric acid A, ganoderic acid B, ganoderic acid C_1, ganoderic acid E, ganodermanontriol, lucidenic acids and lanostane-type triterpenic acids (Ma et al. 2011). The bioactive compounds in *Ganoderma* spp. show various bioactivities like anti-HIV-1 protease, anti-tumor and anti-complement activities and thus, used for various medicinal purposes (El-Mekkawy et al. 1998). *Inonotus obliquus* (chaga mushroom) contains various sterols and triterpenes such as inotodiol, trametenolic acid, ergosterol and ergosterol peroxide, which act as functional bioactive ingredients (Elkhateeb et al. 2019).

Basidiomycetes are a rich source of complex, structurally diverse, bioactive sesquiterpenoids and some protoilludene-type sesquiterpene aryl esters produced by the genus Armillaria (class Agaricomycetes, family Physalacriaceae) are promising leads for the development of new drugs (Engels et al. 2021). Chemical structure of a sesquiterpenoid is depicted in Fig. 15.2.

Basidiomycetes fungi also contain diterpenes; metabolites derived from isoprene precursors. Mushrooms possess several active ingredients that exhibit a wide range of anti-fungal and anti-bacterial properties. In particular, crinipellins A, diterpenoid isolated

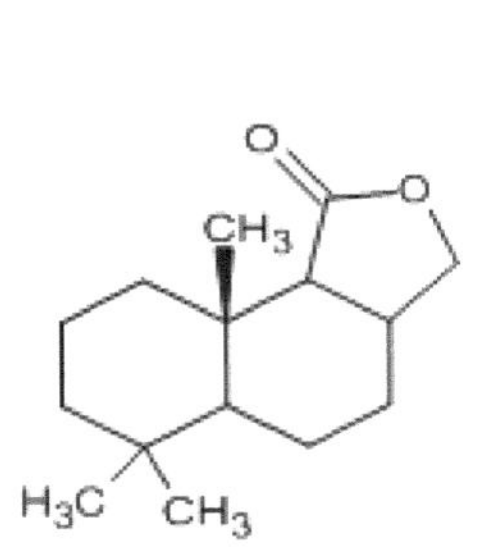

Figure 15.2. Structure of Drimenin, a sesquiterpenoid.

from *Crinipellis rhizomaticola*, exhibits antifungal activity against natural fungicides (Han et al. 2018; Dudekula et al. 2020).

Protein, peptides, volatile organic compounds and others

Mushrooms produce many bioactive proteins and peptides such as lectins, which have no enzymatic activity. Mushrooms also produce bioactive proteins, which possess enzymatic activity such as Fungal Immunomodulatory Proteins (FIPs), Ribosome-Inactivating Proteins (RIPs) and laccases (Xu et al. 2011). Proteins are an essential macronutrient to the growth of the human body and maintenance due to its important physiological functions, such as vital performance of hormones and enzyme action. Mushroom proteins usually have a complete essential amino acid profile, which may cover the dietetic requirements, as well as may have certain economic advantages as compared to animal and plant sources (González et al. 2020). Antifungal peptide (pleurostrin) from *P. ostreatus* exhibited antifungal activity, peptide from *Russula paludosa* showed antiviral properties, agrocybin (an antifungal peptide) from fresh fruiting bodies of *Agrocybe cylindracea.*

Antimicrobial activities of mushrooms bioactive compounds against pathogenic microorganisms

Antimicrobial properties of mushroom extracts are associated with some bioactive compounds; including low- and high-molecular weight compounds. Secondary metabolites, such as sesquiterpenes and other terpenes, steroids, anthraquinones, benzoic acid derivatives and quinolines, and also primary metabolites such as oxalic acid, peptides and proteins present in wild and edible mushrooms are known to possess as antimicrobial property (Alves et al. 2012). Bioactive compounds isolated from cultural broth of *Collybia butyracea, Entoloma nubigenum, Inocybe geophylla, Psathyrella* species and *Protostropharia semiglobata* exhibited antifungal and antibacterial activities against *B. subtilis, Ceratocystis pilifera, E. coli, E. faecalis, Fusarium oxysporum, Penicillium notatum, P. aeruginosa, Rhizoctonia solani* and *S. aureus* (Reinoso et al. 2013). *Lentinus edodes* is the most studied species and seems to have a broad antimicrobial action against both Gram-positive and Gram-negative bacteria. Plectasin peptide, obtained from *Pseudoplectania nigrella*, is the isolated compound with the highest antimicrobial activity against gram-positive bacteria, while 2-aminoquinoline, isolated from *Leucopaxillus albissimus*, presents the highest antimicrobial activity against gram-negative bacteria (Alves et al. 2012). Phenolic acids in the fruiting bodies constitute a good source of natural antimicrobial agents against Gram-positive bacteria like *S. epidermidis, S. aureus, B. subtilis, M. luteus* and Gram-negative bacteria such as *E. coli, K. pneumoniae, P. aeruginosa, P. mirabilis* (Nowacka et al. 2015). *Phellinus linteus* is a medicinal mushroom that is widely used in China, Korea, Japan and other Asian countries. *P. linteus* comprises various bioactive components such as polysaccharides, triterpenoids, phenylpropanoids and furans, and has proven to be an effective therapeutic agent in traditional Chinese medicine for the treatment and the prevention of various diseases (Chen et al. 2019). Methanolic extracts obtained from some fruiting bodies revealed antibacterial and antifungal effects against *Bacillus subtilis, Micrococcus flavus, Staphylococcus aureus, Escherichia coli, Pseudomonas aeruginosa* as well as fungal pathogens like *Aspergillus fumigatus, Candida albicans, C. krusei, Microsporum gypseum, Mucor* sp., and *Trichophyton mentagrophytes* (Al-Fatimi et al. 2013).

The presence of caprylic acid, stearic acid, tetradecanoic acid, methyl-11-octadecenoate, oleic acid and 4-methyl-2-propyl-1-pentanol in *Lenzites quercina* extracts contributed to its antimicrobial properties against *Staphylococcus aureus* (ATCC 29213), *Pseudomonas aeruginosa* (ATCC 27853), *Escherichia coli* (ATCC 35218), Methicillin Resistant *Staphylococcus aureus* (MRSA), *Salmonella Typhi, Bacillus cereus, Enterococcus faecalis, Candida albicans* and *Aspergillus niger* (Ogidi et al. 2015). Terpenoids such as microporenic acids (A–G) isolated from polypore of *Microporus* species showed antibacterial activity against biofilm formation by Gram-positive *S. aureus* as well as antifungal activity against *Candida tenuis* and *C. albicans* (Chepkirui et al. 2018). Extracts obtained from wild mushrooms; *Fistulina hepatica, Lepista nuda, Leucopaxillus giganteus, Mycenarosea* and *Russula delica* displayed antimicrobial activities against biofilm formation by multi-resistant clinical isolates of *Acinetobacter baumannii, E. coli, Proteus mirabilis* and *P. aeruginosa* (Alves et al. 2014). Ethanolic extracts from *A. bisporus, A. brasiliensis, Agrocybeaegerita, C. comatus, C. militaris, F. velutipes, Ganoderma lucidum, G. frondosa, Hypsizygus marmoreus, H. erinaceus, L. edodes, Phellinus igniarius, Pleurotus eryngii* and *P. ostreatus* exhibited concentration-dependent bacteriostatic and bacteriocidic against strain of *Helicobacter pylori* known to cause ulcers (Shang et al. 2013). Ethanolic extract of *Lenzites quercina, Ganoderma lucidum* and *Rigidoporus ulmarius* exhibited a wide range of mycelia inhibition at 50 mg/ml against dermatophytes such as *Epidermophyton floccosum, Trichophyton yaoundei, Microsporum ferrugineum, M. nanum, M. gypseum, T. mentagrophytes, T. verrucosum* and *T. rubrum* (Ogidi and Oyetayo 2016). Ethyl-acetate extracts of genera from Basidiomycetes fungi; *Inonotus, Fomitiporia, Ganoderma, Skeletocutis, Perenniporia, Favolaschia, Hexagonia, Polyporus, Antrodia* and *Echinochaete* inhibited *Bacillus subtilis, Escherichia coli, Mucor plumbeus* and *Candida albicans* with minimum inhibitory concentration (MIC) values of 4.69 µg/ml and < 2.34 µg/ml (Sum et al. 2019).

Mushrooms are considered the producers of different bioactive compounds with antiviral effects (Seo and Choi 2021). Antiviral effect of protein fractions isolated from mycelium of *Lignosus rhinocerotis, Pleurotus giganteus, Hericium erinaceus, Schizophyllum commune* and *Ganoderma lucidium* against human papillomavirus and dengue virus type-2 (DENV-2) strain was reported by Abdullah et al. (2013) and Ellan et al. (2019). Several triterpenes like ganoderiol F, ganodermanontriol, ganoderic acid B from *Ganoderma lucidum* are active as antiviral agents against Human Immunodeficiency Virus type 1 (HIV-1) and HIV-1 protease (El-Mekkawy et al. 1998; Min et al. 1998). In literature collected by Fukushima-Sakuno (2020), lentinamycin produced by *Lentinula edodes* displayed pronounced antibacterial activity against *Alcaligenes faecalis, Bacillus subtilis, Escherichia coli, Flavobacterium aquatile, Burkholderia (Pseudomonas) cepacia, Pseudomonas fluorescens, Salmonella Typhimurium, Staphylococcus aureus,* while antifungal activity was reported against *Aspergillus niger, Aspergillus oryzae, Mucor mucedo, Moniliella tomentosa, Penicillium chrysogenum, Ustilago maydis, Pyricularia setariae, Colletotrichum horii, Bipolarisoryzae (Helminthosporium oryzae), Trichophyton rubrum, Chaetomium globosum,* yeast; *Saccharomyces cerevisiae, Cyberlindnera jadinii, Diutina rugose*. The findings of Shen et al. (2017) revealed different mushroom genera; *Agaricus, Agrocybe, Coprinus, Cordyceps, Dictyophora, Hygrophorus, Lactarius, Leucoagaricus, Ganoderma, Pleurotus* and *Phellinus* with antimicrobial activities against food borne pathogens like *Acinetobacter baumannii, Aeromonas* spp., *Bacillus* spp., *Campylobacter* spp., *C. perfringens, Escherichia coli* O157:H7, *Klebsiella pneumoniae, L. monocytogenes, Pseudomonas aeruginosa, Salmonella enteritidis, Shigella* spp., *Staphylococcus* spp., *V. parahaemolyticus, Yersinia* spp., *Aspergillus* spp. and *Penicillium* spp. Medicinal fungi contain antiparasitic secondary metabolites that are effective against protozoan parasites such as *Plasmodium, Trypanosoma* and *Leishmania,* which are vectors of tropical diseases (Lenzi et al. 2018). Ademola and Odeniran (2017) reported *in vivo* trypanocidal effect of *Pleurotus sajor-caju* aqueous extract against *Trypanosoma congolense.* Ganoderic acid AW1, ganoderic acid AW2, ganomycine A, pinellic acid and ergosterol peroxide were isolated from the fungal fruiting bodies of *Ganoderma* spp. and exhibited good antimalarial activity against *Plasmodium falciparum* (Wahba et al. 2019). Ergosterol from fruit bodies of the basidiomycete *Pleurotus salmoneostramineus* suppressed the growth of *Trypanosoma cruzi trypomastigotes* resulting in permeabilization of the plasma membrane as well as depolarization of mitochondrial membrane potential, leading to parasite death (Alexandre et al. 2017). Mushrooms displayed antiparasitic activity against disease causing parasites such as *Plasmodium* spp., *Trypanosoma* spp., *Leishmania* spp., *Schistosoma* spp., *Haemonchus contortus, Ditylenchus dipsaci, Heterodera glycines* and *Pheretima posthuma* with minimum inhibitory concentration (0.06–54 µg/mL) with no toxicity to host cells *in vitro* or *in vivo*. The bioactive compounds in mushroom extracts (secondary metabolites) targeted cell membrane and lysing cells, caused cell arrest, disrupted mitochondrion function and ribonucleic acid, damaged DNA, disrupted phospholipids (Abugri et al. 2019).

Strategies for producing mushrooms with more effective antimicrobial agents

Large numbers of wild, edible and medicinal mushrooms are cultivated for human uses. Mushrooms are responsible for the breaking down of organic matters; most especially agro-industrial wastes (lignocellulolytic biomass) and thus have an important role in the continual changes that take place in the ecosystem. There is significant and increased interest in mushrooms cultivations due to their nutritional value and medicinal properties (Lu et al. 2020). Many mushrooms have the ability to grow on agro-industrial waste, as submerged cultures, reaching high yields in a short period of time (Wu et al. 2019). Several edible mushrooms are cultivated worldwide on fermented substrates, which could be fermented-pasteurized substrates and a mixture of raw materials that have been steam sterilized (Carrasco et al. 2018). Submerged fermentation and solid state fermentation have been viewed as a promising technology for the efficient production of their valuable compounds, but the approach to improve the production of metabolites using modern biotechnology will improve the yield of secondary metabolites (bioactive compounds). Genetic engineering/metabolic engineering offers innovative and promising strategies to improve the yield of secondary metabolites of mushrooms.

Future perspectives

Over the years, etiological agents responsible for microbial diseases have developed resistance against common microbiostatic and microbiocidal agents used in preventing and controlling them. Microorganisms have developed various mechanisms to evade or neutralize the effects of these antimicrobial agents. Therefore, there is a renewed interest to search for safe and more effective microbiostatic and microbiocidal agents. Edible and medicinal mushrooms are known to be source of safe and effective bioactive agents that can be used to check the negative activities of microorganisms responsible for microbial diseases. Mushrooms can be cultivated on cellulosic waste from which they can acquire various bioactive compounds. The growth medium can also be spiced with various mineral elements that can enhance the effectiveness of the bioactive compound as an antimicrobial agent.

Mushrooms absorb mineral elements, and are able to bio-accumulate them as functional organic compounds during growth. Mushrooms can therefore be used as vehicles to supply effective bioactive compounds that can be used in checking microbial diseases whose causative agent have developed resistance to commonly used antimicrobial agent. This will be a definite strategy to solve the problem of resistance of microorganisms to commonly used antimicrobial agent. Therefore, artificial cultivation of mushrooms on rationally prepared growth medium may help to produce novel bioactive compounds that can help in solving the present problem of ineffectiveness of currently known synthetic antimicrobial agents. Moreover, molecular farming through recombinant DNA technology can be used to insert the gene for the production of effective antimicrobial agent into mushrooms. This may enhance such genetically modified mushrooms to produce bioactive compounds that will be effective against etiological agents responsible for microbial diseases.

Consumption of fortified or genetically mushroom fruit bodies above may help to boost the immune system of individuals thereby preventing or controlling microbial diseases. The modified mushroom fruit bodies can also be prepared apprise about the dietary supplement that can be taken from time to time. It is therefore important that a concerted effort be made to cultivate mushrooms on rationally formulated growth medium that will allow the fruit bodies to absorb an effective and safe bioactive agent that are known to be effective antimicrobial agent.

Conclusion

Microorganisms are major causative agents of diseases in plants and animals. They are also responsible for the spoilage of foods and other organic materials that are of importance to man. The present means of checkmating the pestiferous effect of these microbial agents are becoming ineffective. It is therefore expedient to find effective and safe methods to prevent and control these organisms in order to check their negative social and economic effects. Mushrooms contain unsaturated fatty acids, amino acids phenolics, carotenoids, ascorbic acid, tocopherols, anthocyanidins, polysaccharides, proteins and peptides. These are functional ingredients that could be commercially explored as nutraceuticals and used as nutritional therapy to maintain and promote a healthy life and longevity. Mycochemicals, which are known to be safe, effective and eco-friendly, have the potential of being used to control also microbial diseases. Moreover, higher basidiomycetes can be made to produce novel bioactive compounds through cultural and genetic manipulations. Conscious effort towards harnessing and producing these mycochemicals through appropriate funding of research is therefore important. This will lead to development of effective and eco-friendly microbiostatic/microbiocidal compounds that will be available to treat microbial diseases.

Abbreviations

AIDS	Acquired immune deficiency syndrome
ATCC	American type culture collection
BHA	Butylated hydroxyanisole
BHT	Butylated hydroxytoluene
BRM	Biological response modifier
DENV-2	Dengue virus type-2
DNA	Deoxyribonucleic acid
EQ	Ethoxyquin
FIPs	Fungal immunomodulatory proteins
HIV-1	Human immunodeficiency virus type 1
MAPK	Mitogen-activated protein kinase
MIC	Minimum inhibitory concentration

MRSA	Methycillin resistant *Staphylococcus aureus*
PG	Propyl gallate
RIPs	Ribosome-inactivating proteins
TBHQ	Tert-butylhydroquinone

References

Abdullah, N.M.Z.D.H., Lau, B.F., Annuar, M.S.M. (2013). Domestication of a wild medicinal sclerotial mushroom, *Lignosus rhinocerotis* (Cooke) Ryvarden. Ind Crop Prod 47: 256–261. https://doi.org/10.1016/j.indcrop.2013.03.012.

Abugri, D.A., Ayariga, J.A., Tiimob, B.J., Yedjou, C.G., Mrema, F., Witola, W.H. (2019). Medicinal mushrooms as novel sources for new antiparasitic drug development. *In*: Agrawal, D., Dhanasekaran, M. (eds.). Medicinal Mushrooms. Springer, Singapore. https://doi.org/10.1007/978-981-13-6382-5_9.

Adams, M., Christen, M., Plitzko, I., Zimmermann, S., Brun, R., Kaiser, M., Hamburger, M. (2010). Antiplasmodial lanostanes from the *Ganoderma lucidum* mushroom. J Nat Prod 73(5): 897–900. doi: 10.1021/np100031c. PMID: 20384295.

Ademola, I.O., Odeniran, P.O. (2017). Novel trypanocide from an extract of *Pleurotus sajor-caju* against *Trypanosoma congolense*. Pharm Biol 55(1): 132–138. doi: 10.1080/13880209.2016.1230878.

Alexandre, T.R., Lima, M.L., Galuppo, M.K., Mesquita, J.T., do Nascimento, M.A., Dos Santos, A.L., Sartorelli, P., Pimenta, D.C., Tempone, A.G. (2017). Ergosterol isolated from the basidiomycete *Pleurotus salmoneostramineus* affects *Trypanosoma cruzi* plasma membrane and mitochondria. J Venom Anim Toxins Incl Trop Dis 23: 30. https://doi.org/10.1186/s40409-017-0120-0.

Al-Fatimi, M., Schröder, G., Kreisel, H., Lindequist, U. (2013). Biological activities of selected basidiomycetes from Yemen. Pharmazie 68: 221–226. https://doi.org/10.1691/ph.2013.2729.

Alves, M., Ferreira, I., Lourenço, I., Eduardo Costa, Martins, A., Pintado, M. (2014). Wild mushroom extracts as inhibitors of bacterial biofilm formation. Pathogens 3: 667–679. https://doi.org/10.3390/pathogens3030667.

Alves, M.J., Ferreira, I.C., Dias, J., Teixeira, V., Martins, A., Pintado, M. (2012). A review on antimicrobial activity of mushroom (Basidiomycetes) extracts and isolated compounds. Planta Med 78(16): 1707–18.

Alves, M.J., Ferreira, I.C., Froufe, H.J., Abreu, R.M., Martins, A., Pintado, M. (2013). Antimicrobial activity of phenolic compounds identified in wild mushrooms, SAR analysis and docking studies. J Applied Microbiol 115(2): 346–357. https://doi.org/10.1111/jam.12196.

Amos, W.M.G. (1981). Basic Immunology. Butterworths and Co. pp. 115–121.

Baby, S., Johnson, A.J., Govindan, B. (2015). Secondary metabolites from *Ganoderma*. Phytochem 114: 66–101. https://doi.org/10.1016/j.phytochem.2015.03.010.

Badalyan, S.M., Barkhudaryan, A., Rapior, S. (2019). Recent progress in research on the pharmacological potential of mushrooms and prospects for their clinical application. *In*: Agrawal, D., Dhanasekaran, M. (eds.). Medicinal Mushrooms. Springer, Singapore. https://doi.org/10.1007/978-981-13-6382-5_1.

Badalyan, S.M., Barkhudaryan, A., Rapior, S. (2019). Recent progress in research on the pharmacological potential of mushrooms and prospects for their clinical application. pp. 1–70. Agrawal, D.C., Dhansekaran, M. (eds.). Medicinal Mushrooms: Recent Progress in Research and Development, Springer Nature.

Badalyan, S., Rapior, S. (2021). Agaricomycetes mushrooms (Basidiomycota) as potential neuroprotectants. Italian J Mycol 50: 30–43.

Benkeblia, N. (2015). *Ganoderma lucidum* polysaccharides and terpenoids: profile and health benefits. J Food Nutr Diet 1: 1–6. doi: 10.19104/jfnd.2015.101.

Butkhup, L., Samappito, W., Jorjong, S. (2018). Evaluation of bioactivities and phenolic contents of wild edible mushrooms from Northeastern Thailand. Food Sci Biotechnol 27: 193–202. https://doi.org/10.1007/s10068-017-0237-5.

Carrasco, J., Zied, D., Pardo, J.E., Preston, G.M., Pardo-Giménez, A. (2018). Supplementation in mushroom crops and its impact on yield and quality. AMB Expr 8: 146.

Castellano, G., Torrens, F. (2015). Information entropy-based classification of triterpenoids and steroids from Ganoderma. Phytochem 116: 305–313. https://doi.org/10.1016/j.phytochem.2015.05.008.

Cateni, F., Gargano, M.L., Procida, G., Venturella, G., Cirlincione, F., Ferraro, V. (2021). Mycochemicals in wild and cultivated mushrooms: nutrition and health. Phytochem Rev https://doi.org/10.1007/s11101-021-09748-2.

Chang, S.T. (2009). Overview of mushroom cultivation and utilization as functional foods. pp. 1–33. *In*: Peter, C.K. Cheung (ed.). Mushrooms as Functional Foods.

Chen, S., Li, X., Yong, T., Wang, Z., Su, J., Jiao, C., Xie, Y., Yang, B.B. (2017). Cytotoxic lanostane-type triterpenoids from the fruiting bodies of *Ganoderma lucidum* and their structure-activity relationships. Oncotarget 8(6): 10071–10084. https://doi.org/10.18632/oncotarget.14336.

Chen, W., Tan, H., Liu, Q., Zheng, X., Zhang, H., Liu, Y., Xu, L. (2019). A review: the bioactivities and pharmacological applications of *Phellinus linteus*. Molecules 24: 1888.

Chepkirui, C., Yuyama, K.T., Wanga, L.A., Decock, C., Matasyoh, J.C., Abraham, W.R., Stadler, M. (2018). Microporenic acids A-G, biofilm inhibitors, and antimicrobial agents from the basidiomycete *Microporus* Species. J Nat Prod 27: 81(4): 778–784. doi: 10.1021/acs.jnatprod.7b00764.

Corana, F., Cesaroni, V., Mannucci, B., Baiguera, R.M., Picco, A.M., Savino, E., Ratto, D., Perini, C., Kawagishi, H., Girometta, C.E., Rossi, P. (2019). Array of metabolites in Italian *Hericium erinaceus* mycelium, primordium, and sporophore. Molecules 24(19): 3511. https://doi.org/10.3390/molecules24193511.

Demain, A.L. (1999). Pharmaceutically active secondary metabolites of microorganisms. Appl Microbiol Biotechnol 52(4): 455–63.

Dudekula, U.T., Doriya, K., Devarai, S.K. (2020). A critical review on submerged production of mushroom and their bioactive metabolites. 3 Biotech 10: 337. https://doi.org/10.1007/s13205-020-02333-y.

Elkhateeb, W.A., Daba, G.M., Thomas, P.W., Wen, T.C. (2019). Medicinal mushrooms as a new source of natural therapeutic bioactive compounds. Egypt Pharmaceut J 18: 88–101.

Ellan, K., Thayan, R., Raman, J., Hidari, K.I.P.J., Ismail, N., Sabaratnam, V. (2019). Anti-viral activity of culinary and medicinal mushroom extracts against dengue virus serotype 2: an *in-vitro* study. BMC Complement Altern Med 19(1): 260. doi:10.1186/s12906-019-2629-y.

El-Mekkawy, S., Meselhy, M.R., Nakamura, N., Tezuka, Y., Hattori, M., Kakiuchi, N., Shimotohno, K., Kawahata, T., Otake, T. (1998). Anti-HIV-1 and anti-HIV-1-protease substances from *Ganoderma lucidum*. Phytochem 49(6): 1651–7. doi: 10.1016/s0031-9422(98)00254-4.

Engels, B., Heinig, U., McElroy, C., Meusinger, R., Grothe, T., Stadler, M., Jennewein, S. (2021). Isolation of a gene cluster from *Armillaria gallica* for the synthesis of armillyl orsellinate-type sesquiterpenoids. Appl Microbiol Biotechnol 105(1): 211–224. https://doi.org/10.1007/s00253-020-11006-y.

Fonkwo, P.N. (2008). Pricing infectious disease: The economic and health implications of infectious diseases. Science and Society, EMBO Reports 9: S13.

Friedman, M. (2016). Mushroom polysaccharides: chemistry and antiobesity, antidiabetes, anticancer, and antibiotic properties in cells, rodents, and humans. Foods 5(4): 80. https://doi.org/10.3390/foods5040080.

Fukushima-Sakuno, E. (2020). Bioactive small secondary metabolites from the mushrooms *Lentinula edodes* and *Flammulina velutipes*. J Antibiot 73(10): 687–696. https://doi.org/10.1038/s41429-020-0354-x.

Fuller, R. (1989). Probiotics in man and animals—A review. J Appl Bacteriol 66: 365–378.

Gąsecka, M., Mleczek, M., Siwulski, M., Niedzielski, P. (2016). Phenolic composition and antioxidant properties of *Pleurotus ostreatus* and *Pleurotus eryngii* enriched with selenium and zinc. Eur Food Res Technol 242: 723–732.

González, A., Cruz, M., Losoya, C., Nobre, C., Loredo, A., Rodríguez, R., Contreras, J., Belmares, R. (2020). Edible mushrooms as a novel protein source for functional foods. Food Funct 11(9): 7400–7414.

Greeshma, A.A., Sridhar, K.R., Pavithra, M., Tomita-Yokotani, K. (2018). Bioactive potential of nonconventional edible wild mushroom Amanita. *In*: Gehlot, P., Singh, J. (eds.). Fungi and Their Role in Sustainable Development: Current Perspectives. Springer, Singapore. https://doi.org/10.1007/978-981-13-0393-7_37.

Griensven, L.V. (2009). Mushrooms, must action be taken? Proceeding of The 5th International Medicinal Mushroom Conference, Nantong, China. pp. 407–412.

Gupta, S., Summuna, B., Gupta, M., Annepu, S.K. (2018). Edible mushrooms: cultivation, bioactive molecules, and health benefits. *In*: Mérillon, J.M., Ramawat, K. (eds.). Bioactive Molecules in Food. Reference Series in Phytochemistry. Springer, Cham. https://doi.org/10.1007/978-3-319-54528-8_86-1.

Han, J.W., Oh, M., Lee, Y.J., Choi, J., Choi, G.J., Kim, H. (2018). Crinipellins A and I, two diterpenoids from the basidiomycete fungus *Crinipellis rhizomaticola*, as potential natural fungicides. Molecules 23: 1–11. https://doi. org/10.3390/molecules23092377.

Hyde, K.D., Xu, J., Rapior, S., Jeewon, R., Lumyong, S., Niego, A.G.T., Abeywickrama, P.D., Aluthmuhandiram, J.V.S., Brahamanage, R.S., Brooks, S., Chaiyasen, A., Chethana, K.W.T., Chomnunti, P., Chepkirui, C., Chuankid, B., de Silva, N.I., Doilom, M., Faulds, C., Gentekaki, E., Gopalan, V., Kakumyan, P., Harishchandra, D., Hemachandran, H., Hongsanan, S., Karunarathna, A., Karunarathna, S.C., Khan, S., Kumla, J., Jayawardena, R.S., Liu, J.K., Liu, N., Luangharn, T., Macabeo, A.P.G., Marasinghe, D.S., Meeks, D., Mortimer, P.E., Mueller, P., Nadir, S., Nataraja, K.N., Nontachaiyapoom, S., O'Brien, M., Penkhrue, W., Phukhamsakda, C., Ramanan, U.S., Rathnayaka, A.R., Sadaba, R.B., Sandargo, B., Samarakoon, B.C., Tennakoon, D.S., Siva, R., Sriprom, W., Suryanarayanan, T.S., Sujarit, K., Suwannarach, N., Suwunwong, T., Thongbai, B., Thongklang, N., Wei, D., Wijesinghe, S.N., Winiski, J., Yan, J., Yasanthika, E., Stadler, M. (2019). The amazing potential of fungi:

50 ways we can exploit fungi industrially. Fungal Divers 97: 1–136. https://doi.org/10.1007/s13225-019-00430-9.

Isaka, M., Palasarn, S., Sommai, S., Veeranondha, S., Srichomthong, K., Kongsaeree, P., Prabpai, S. (2017). Lanostane triterpenoids from the edible mushroom *Astraeus asiaticus*. Tetrahed 73: 1561–1567. https://doi.org/10.1016/j.tet.2017.01.070.

Kinealy, C. (1994). This Great Calamity. Gill & Macmillan. pp. xv and assim. ISBN 978-0-7171-1881-6.

Kumar, N., Goel, N. (2019). Phenolic acids: Natural versatile molecules with promising therapeutic applications. Biotechnol Rep 24: e00370. https://doi.org/10.1016/j.btre.2019.e00370.

Lee, W., Fujihash, A., Govindarajulu, M., Ramesh, S., Deruiter, J., Majrashi, M., Almaghrabi, M., Rishi, M., Moore, T., Agrawal, D.C., Dhanasekaran, M. (2019). Role of mushrooms in neurodegenerative diseases. *In*: Agrawal, D., Dhanasekaran, M. (eds.). Medicinal Mushrooms. Springer, Singapore. http://doi-org-443.webvpn.fjmu.edu.cn/10.1007/978-981-13-6382-5_8.

Lenzi, J., Costa, T.M., Alberton, M.D., Goulart, J., Tavares, L. (2018). Medicinal fungi: a source of antiparasitic secondary metabolites. Appl Microbiol Biotechnol 102: 5791–5810.

Lindequist, U., Niedermeyer, T.H.J., Julich, W.-D. (2005). The pharmacological potential of mushrooms—Review. Evid Based Complement Alternat Med 2(3): 285–99.

Lorenzen, K., Anke, T. (1998). Basidiomycetes as a source for new bioactive natural products. Curr Org Chem 2: 329–64.

Lu, H., Lou, H., Hu, J., Liu, Z., Chen, Q. (2020). Macrofungi: a review of cultivation strategies, bioactivity, and application of mushrooms. Comp Rev Food Sci Food Saf 19(5): 2333–2356.

Ma, B., Ren, W., Zhou, Y., Ma, J., Ruan, Y., Wen, C.N. (2011). Triterpenoids from the spores of *Ganoderma lucidum*. North Am J Med Sci 3(11): 495–498. https://doi.org/10.4297/najms.2011.3495.

Mahapatra, S., Banerjee, D. (2013). Fungal exopolysaccharide: Production, composition and applications. Microbiol Insights 6: 1–16.

Mau, J.L., Chang, C.N., Huang, S.J., Chen, C.C. (2004). Antioxidant properties of methanolic extracts from *Grifola frondosa*, *Morchella esculenta* and *Termitomyces albuminosus* mycelia. Food Chem 87: 111–118.

Mendes, R.E., Farrell, D.J., Flamm, R.K., Talbot, G.H., Ivezic-Schoenfeld, Z., Paukner, S., Sader, H.S. (2016). *In vitro* activity of lefamulin tested against *Streptococcus pneumoniae* with defined serotypes, including multidrug-resistant isolates causing lower respiratory tract infections in the United States. Antimicrob Agents Chemother 60: 4407–4411. https://doi.org/10.1128/AAC.00627-16.

Min, B.S., Nakamura, N., Miyashiro, H., Bae, K.W., Hattori, M. (1998). Triterpenes from the spores of *Ganoderma lucidum* and their inhibitory activity against HIV-1 protease. Chem Pharm Bull 46(10): 1607–12. doi: 10.1248/cpb.46.1607.

Mizuno, T. (1999). The extraction and development of antitumor active polysaccharides from medicinal mushrooms in Japan—review. Int J Med Mushrooms 1: 9–30.

Nowacka, N., Nowak, R., Drozd, M., Olech, M., Los, R., Malm, A. (2015). Antibacterial, antiradical potential and phenolic compounds of thirty-one polish mushrooms. PloS One 10(10): e0140355. https://doi.org/10.1371/journal.pone.0140355.

Nowak, R., Nowacka-Jechalke, N., Juda, M., Malm, A. (2018). The preliminary study of prebiotic potential of Polish wild mushroom polysaccharides: the stimulation effect on *Lactobacillus* strains growth. Eur J Nutr 57(4): 1511–1521. https://doi.org/10.1007/s00394-017-1436-9.

Ogidi, C.O., Oyetayo, V.O., Akinyele, B.J. (2015). *In vitro* evaluation of antimicrobial efficacy of extracts obtained from raw and fermented wild macrofungus, *Lenzites quercina*. Int J Microbiol 1–7.

Ogidi, C.O., Oyetayo, V.O. (2016). Phytochemical property and assessment of antidermatophytic activity of some selected wild macrofungi against pathogenic dermatophytes. Mycol DOI: 10.1080/21501203.2016.1145608.

Ogidi, C.O., Oyetayo, V.O., Akinyele, B.J., Ogbole, O.O., Adeniji, A., Oluremi, B. (2017). Molecular identity and cytotoxicity of extracts of a macrofungus; *Lenzites quercina* toward cancer cell lines. Biotechnologia 98(1): 25–32.

Ogidi, C.O., Oyetayo, V.O., Akinyele, B.J. (2018). Estimation of total phenol, flavonoid content and free radical scavenging activity of a wild macrofungus, *Lenzites quercina*. Curr Res Environ Applied Mycol 8(4): 425–437.

Oluba, O.M. (2019). *Ganoderma* terpenoid extract exhibited anti-plasmodial activity by a mechanism involving reduction in erythrocyte and hepatic lipids in *Plasmodium berghei* infected mice. Lipids Health Dis 18: 12 https://doi.org/10.1186/s12944-018-0951-x.

Oyetayo, V.O. (2008). Mushrooms indigenous to Nigeria as potential source of myconutraceuticals. Curr Trends Biotechnol Pharm 2(4): 471–477.

Oyetayo, V.O., Nieto-Camacho, A., Baldomero, E.R., Jimenez, M. (2012). Assessment of anti-inflammatory, lipid peroxidation and acute toxicity of extracts obtained from wild macrofungi collected from Akure Southwest Nigeria. Int J Med Mushrooms 14(6): 573–578.

Oyetayo, V.O., Nieto-Camacho, A., Ramırez-Apana, M.T., Baldomero, E.R., Jimenez, M. (2013). Total phenol, antioxidant and cytotoxic properties of wild macrofungi collected from Akure Southwest Nigeria. Jordan J Biol Sci 6(2): 5–10.

Oyetayo, V.O. (2020). Health Promotion: The Probiotic and Myconutraceutical Approach. 118th Inaugural Lecture of The Federal University of Technology Akure. 106p.

Oyetayo, V.O., Oyetayo, F.L. (2020). Mushrooms in Nigeria and their nutraceutical potentials. pp. 157–184. *In*: Chaurasia, P.K., Bharati, S.L. (eds.). Research Advances in the Fungal World: Culture, Isolation, Identification, Classification, Characterization, Properties and Kinetics. Nova Science Publishers, Inc., USA. Chapter 9.

Palacios, I., Lozano, M., Moro, C., D'Arrigo, M., Rostagno, M.A., Martínez, J.A., García-Lafuente, A., Guillamón, E., Villares, A. (2011). Antioxidant properties of phenolic compounds occurring in edible mushrooms. Food Chem 128(3): 674–678.

Paliya, B.S., Verma, S., Chaudhary, H.S. (2014). Major bioactive metabolites of the medicinal mushroom: *Ganoderma lucidum*. Int J Pharm R 6(1): 12–24.

Pennerman, K.K., Yin, G., Bennett, J.W. (2015). Health effects of small volatile compounds from East Asian medicinal mushrooms. Mycobiol 43(1): 9–13. https://doi.org/10.5941/MYCO.2015.43.1.9.

Perveen, S. (2018). Introductory chapter: terpenes and terpenoids. IntechOpen. rhttp://dx.doi.org/10.5772/intechopen.79683.

Reinoso, R., Cajas-Madriaga, D., Martínez, M., San Martín, A., Pérez, C., Fajardo, V., Becerra, J. (2013). Biological activity of macromycetes isolated from Chilean subantarctic ecosystems. J Chil Chem Soc 58(4): 2016–2019. https://doi.org/10.4067/S0717-97072013000400024.

Reshetnikov, S.V., Wasser, S.P., Tan, K.K. (2001). Higher Basidiomycetes as a source of antitumour and immunostimulating polysaccharides (Review). Int J Med Mushrooms 3: 361–394.

Sagakami, H., Aohi, T., Simpson, A., Tanuma, S. (1991). Induction of immunopotentiation activity by a protein bound polysaccharide, PSK. Anticancer Res 11: 993–1000.

Samorini, G. (1992). The oldest representations of hallucinogenic mushrooms in the world (Sahara desert, 9000–7000 B.P.). Integration 2(3): 69–78.

Sánchez, C. (2017). Bioactives from mushroom and their application. *In*: Puri, M. (eds.). Food Bioactives. Springer, Cham. https://doi.org/10.1007/978-3-319-51639-4_2.

Seo, D.J., Choi, C. (2021). Antiviral bioactive compounds of mushrooms and their antiviral mechanisms: a review. Viruses 13(2): 350. https://doi.org/10.3390/v13020350.

Shang, X., Tan, Q., Liu, R., Yu, K., Li, P., Zhao, G.P. (2013). *In vitro* anti-Helicobacter pylori effects of medicinal mushroom extracts, with special emphasis on the Lion's mane mushroom, Hericiumerinaceus (higher Basidiomycetes). Int J Med Mushrooms 15(2): 165–174.

Sharma, S.K., Gautam, N. (2015). Chemical, bioactive, and antioxidant potential of twenty wild culinary mushroom species. Biomed Res Int Article ID 346508, p 12.

Shen, H.S., Shao, S.Q., Chen, J.C., Zhou, T. (2017). Antimicrobials from mushrooms for assuring food safety. Compr Rev Food Sci F16: 316–329.

Shirur, M., Shivalingegowda, N.S. (2015). Mushroom marketing channels and consumer behaviour: a critical analysis. Mysore J Agric Sci 49(2): 390–393.

Stewart, F.S., Beswick, T.S.L. (1977). Bacteriology, Virology and Immunity for Students of Medicine. 10th edn. Macmillan Publishing Co. Inc., New York. pp. 47–58.

Sudheer, S., Alzorqi, I., Manickam, S., Ali, A. (2019). Bioactive compounds of the wonder medicinal mushroom "*Ganoderma lucidum*". *In*: Mérillon, J.M., Ramawat, K. (eds.). Bioactive Molecules in Food. Reference Series in Phytochemistry. Springer, Cham. https://doi.org/10.1007/978-3-319-78030-6_45.

Sum, W.C., Indieka, S.A., Matasyoh, J.C. (2019). Antimicrobial activity of Basidiomycetes fungi isolated from a Kenyan tropical forest. Afr J Biotechnol 18(5): 112–123.

Thongbai, B., Rapior, S., Hyde, K.D., Stadler, M. (2015). *Hericium erinaceus*, an amazing medicinal mushroom. Mycol Prog 14: 91. https://doi.org/10.1007/s11557-015-1105-4.

Thu, Z.M., Myo, K.K., Aung, H.T., Clericuzio, M., Armijos, C., Vidari, G. (2020). Bioactive phytochemical constituents of wild edible mushrooms from Southeast Asia. Molecules 25(8): 1972. https://doi.org/10.3390/molecules2508197.

Ugbogu, E.A., Emmanuel, O., Salem, A.Z.M., Elghandour, M.M.M.Y. (2020). Nutritional composition of *Termitomyces robustus* (Agaricomycetes) and *Lentinus squarrosulus* (Mont.) singer in South East Nigeria. Agroforest Syst 94: 1291–1300. https://doi.org/10.1007/s10457-018-0323-6.

Van Der Waaij, D., Horstra, H., Wiegersma, N. (1982). Effect of ß-lactam antibiotics on the resistance of the digestive tract to colonisation. J Infect Dis 146: 417–422.

Veglioglu, Y.S., Mazza, G., Gao, L., Oomah, B.D. (1998). Antioxidant activity and total phenolics in selected fruits, vegetables, and grain products. J Agric Food Chem 46: 4113–17.

Venturella, G., Ferraro, V., Cirlincione, F., Gargano, M.L. (2021). Medicinal mushrooms: bioactive compounds, use, and clinical trials. Int J Mol Sci 22(2): 634. doi: 10.3390/ijms22020634.

Villares, A., Mateo-Vivaracho, L., Guillamón, E. (2012). Structural features and healthy properties of polysaccharides occurring in mushrooms. Agriculture 2(4): 452–471.

Wahba, A.E., El-Sayed, A.K.A., El-Falal, A.A., Soliman, E.M. (2019). New antimalarial lanostane triterpenes from a new isolate of Egyptian *Ganoderma* species. Med Chem Res 28: 2246–2251. https://doi.org/10.1007/s00044-019-02450-1.

Wasser, S.P., Weis, A.L. (1999). Therapeutic effects of substances occurring in higher basidiomycetes mushrooms a modern perspective. Cri Rev Immunol 19(1): 65–96.

Wasser, S.P. (2002). Medicinal mushrooms as a source of antitumour and immunostimulating polysaccharides. Applied Microbiol Biotechnol 60: 258–274.

Wu, F., Zhou, L.W., Yang, Z.L., Bau, T., Li, T.H., Dai, Y.C. (2019). Resource diversity of Chinese macrofungi: edible, medicinal and poisonous species. Fungal Divers 98: 1–76. https://doi.org/10.1007/s13225-019-00432-7.

Xu, T., Beelman, R.B. (2015). The bioactive compounds in medicinal mushrooms have potential protective effects against neurodegenerative diseases. Adv Food Technol Nutr Sci Open J 1(2): 62–66. doi: 10.17140/AFTNSOJ-1-110.

Xu, X., Yan, H., Chen, J., Zhang, X. (2011). Bioactive proteins from mushrooms. Biotechnol Adv 29(6): 667–74. doi: 10.1016/j.biotechadv.2011.05.003.

Zaidman, B.Z., Yassin, M., Mahajna, J., Wasser, S.P. (2005). Medicinal mushroom modulators of molecular targets as cancer therapeutics. Appl Microbiol Biotechnol 67: 453–8.

Zeb, M., Lee, C.H. (2021). Medicinal properties and bioactive compounds from wild mushrooms native to North America. Molecules 26(2): 251. https://doi.org/10.3390/molecules26020251.

Zhou, L.W., Cao, Y., Wu, S.H., Vlasák, J., Li, D.W., Li, M.J., Dai, Y.C. (2015). Global diversity of the *Ganoderma lucidum* complex complex (Ganodermataceae, Polyporales) inferred from morphology and multilocus phylogeny. Phytochem 114: 7–15.

Zhu, P. (2009). The present status and prospects of medicinal fungal research and development in China. Proceeding of The 5th International Medicinal Mushroom Conference, Nantong, China. pp. 26–33.

16

Potential Activity of Mushrooms Against Viruses:

Current Trends and Perspectives

Denise Grotto,[1,*] *Erika Leão Ajala Caetano,*[1] *Raquel Andrade Leite Vieira,*[1] *Fernanda Gomes Leite,*[1] *Nathália Roberta Cardoso Mendes Castanho*[1,2] and *Angela Faustino Jozala*[2,*]

Introduction

Fungi represent an extremely diverse kingdom, from morphology to its applications (Blackwell 2011; Chambergo and Valencia 2016). Mushrooms are fungi that produce large and fleshy fruit bodies (sporophore). Edible mushrooms are nutritious, characterized by low carbohydrate, lipid contents and high protein content, thus the cultivation of mushrooms is important for the production of food and medicine. Edible and medicinal mushrooms need to be grown on standardized substrates, and 1 kg of mushroom can generate up to 5 kg of the spent mushroom substrate, which means that 5 kg of residual mass can be applied in other technological sectors, such as the production of biochar.

In addition to their striking flavor and nutritional potential, edible mushrooms have been used to treat and prevent many diseases, including diabetes, hypercholesterolemia, chronic hepatitis, nephritis, liver disease, neurasthenia, arthritis, bronchitis, asthma, gastric ulcer, cancer, infections caused by bacteria among others (Spim et al. 2017; Gargano et al. 2017; Ma et al. 2018), and also can be used against COVID-19 (Hetland et al. 2020).

The pathogenic microorganisms such as viruses and bacteria have become increasingly resistant or that they mutate very quickly, searching for alternative treatments to antibiotics and antivirals has been growing. Particularly with COVID-19, the indiscriminate use of antimicrobials and disinfectants may further increase microorganism resistance, and there is a factual demand for possibilities for treating diseases.

For thousands of years, natural products have had an important position in traditional medicine and still provide the most affordable treatment for diseases in many developing countries. Among the alternatives, mushrooms (medicinal and edible) deserve to be highlighted. These mushrooms belong to the phylum basidiomycota (also known as

[1] LAPETOX – Laboratory of Toxicological Research, University of Sorocaba, Sorocaba/SP, Brazil.
[2] LAMINFE – Laboratory of Industrial Microbiology and Fermentation Process, University of Sorocaba, Sorocaba/SP, Brazil.
* Corresponding authors: angela.jozala@prof.uniso; denise.grotto@prof.uniso

basidiomycetes) and are widely found and consumed; consequently, they are the most studied specimens. Their properties have been extensively studied for several pathologies, including viral diseases.

Mushrooms fit into a good circular economy, since using their fruiting bodies for medicine or food, up to the spent mushroom substrate, residue from fungi culture that can be applied in the treatment of contaminated water. In this chapter, the different applications of mushrooms will be approached for medicinal purposes directly related to antiviral treatments.

Mushrooms as an antiviral alternative to pathogenic viruses

The COVID-19 pandemic brought up a subject that has not been explored yet; although some viral diseases are well known, it has some particularly morbid characteristics. Severe Acute Respiratory Syndrome Coronavirus (SARS-CoV-2), for example, is responsible for developing a serious respiratory syndrome, which can cause death quickly (Di Pierro et al. 2020). However, some viruses, such as the Herpes virus, are easily found in humans who only develop an injury when there is an immuno-compromised (Eo et al. 2000).

Viral diseases are a serious problem in modern medicine, where they can trigger several types of processes, from inflammatory cascades (such as the influenza virus) (Krupodorova et al. 2014) even the development of cancer cells (Papilloma Virus) (Donatini 2014). Although there are some treatments for these viruses, some problems such as toxicity, adverse effects and limitations of medications, in addition to the rapid mutation ability of viruses make them unsuccessful (Ellan et al. 2019). Therefore, finding an alternative that can assist the endogenous immune response process, acting concurrently with antiviral drugs, seems to be the solution to this problem (Shahzad et al. 2020).

Considering the antiinflammatory, antioxidant, immunostimulating and even anti-carcinogenic potential, mushrooms can be considered as nutraceuticals. Even though mushrooms are a widely consumed food in the East, they are still entering the field in the west. Mushrooms are rich in proteins, fibers and polysaccharides, such as β-glucan, and are used completely, from mycelium to the fruiting body (Cardozo et al. 2011; Krupodorova et al. 2014; Teplyakova and Kosogova 2016).

The indirect antiviral effects are the result of the immunostimulating activity of polysaccharides from mushrooms or other complex molecules, such as lentinan and beta glucan from Fig. 16.1A and B. Thus, mushrooms have great potential for use as successful antiviral treatments with a reduced chance of adverse side effects (Chen et al. 2012; Rincão et al. 2012; Linnakoskil et al. 2018).

There are several species of mushrooms found in literature with antiviral potential, among them *Ganoderma lucidum, Grifola frondosa, Agaricus brasiliensis, Pleurotus nebrodensis, Lentinus edodes,* and others mentioned in Table 16.1. In the same table, diseases caused by viruses as well as the results obtained in each study are reported and described. It is important to note that different diseases triggered by viruses have been assessed, from the simplest ones, which theoretically do not induce major damage to the host, to more complex diseases such as cancer.

In addition, especially in *in vitro* studies, the mechanism of action of the antiviral activity of mushrooms is, in large part the inhibition of virus replication, a direct action of the compounds produced by mushrooms on viruses. On the other hand, according to some clinical studies, it is perceived that the beneficial effects of mushrooms are indirect. There was, for example, the normalization of the liver function in patients with viral hepatitis in a

Table 16.1. Mushrooms studied for the treatment of pathogenic viruses, including *in vitro, in vivo* studies and also clinical trials.

Species	Samples	Viruses	Type of study	Outcome	Reference
Ganoderma lucidum	Fruiting body called carpophore - aqueous extract	Herpes simplex (1 and 2)	*In vitro*	Polysaccharides linked to acidic proteins showed high antiviral activity for both serotypes. Polysaccharides still had an inhibitory effect on viral penetration.	Eo et al. 2000
Inonotus hispidus	Aqueous and ethanolic extract of fruit bodies	Influenza (type A and B)	*In vitro*	Ethanol extracts from fruit bodies of *I. hispidus* showed interesting antiviral activity against influenza virus types A and B. This activity may be related to the presence of hispolon and hispidine in the extracts.	Ali et al. 2003
Ganoderma lucidum	Mycelium - aqueous extract (bioactive fraction)	Herpes simplex	*In vitro*	Bioactive fraction showed strong antiviral activity against herpes. The concentrations with effective antiviral property are far from the threshold of cytotoxicity, being a product of high possibility of development.	Lui et al. 2004
Grifola frondosa	Fruiting body - aqueous extract	Herpes simplex	*In vitro* and *In vivo*	The protein from the extract of *G. frondosa* was able to inhibit replication of the herpes simplex virus *in vitro*, and reduced the severity of blepharitis, neovascularization and stromal keratitis in a murine model.	Chang-Qing et al. 2007
Agaricus blazei murill	Dry fungal bodies- Aqueous extract	Hepatitis B	Clinical trial	*A. blazei murill* extract normalized liver function in patients with chronic hepatitis B, showing a decrease in the average level of aspartate aminotransferase and alanine aminotransferase.	Chung-Hua et al. 2008
Pleurotus nebrodensis	Basidiocarp	Human immunodeficiency virus 1	*In vitro*	Nebrodeolysin was an isolated hemolytic compound, which had a suppressive effect on the action of HIV-1 in addition to an antiviral effect. Although it did not help in the production of anti-HIV-1 antibodies.	Lv et al. 2009
Ganoderma, Lentinus, Pleurotus, Laetiporus, Polyporus, Inonotus, Flammulina, Grifola, Trametes	Aqueous extracts and polysaccharide fractions of the mycelium and fruit bodies	Herpes simplex	*In vitro*	Eleven preparations of *Ganoderma, Lentinus* and *Pleurotus* completely inhibited infectious activity at doses of not less than 100 PPU (Herpes simplex type 2). The antiviral activity was probably due to the content of polysaccharides or their derivatives in the composition. It increased with the increase in the amount of the total polysaccharide fraction or its concentration.	Razumovi A. et al. 2010

Table 16.1 contd. ...

...Table 16.1 contd.

Species	Samples	Viruses	Type of study	Outcome	Reference
Agaricus brasiliensis	Mycelium - polysaccharide isolation	Herpes simplex	*In vitro*	The isolated polysaccharide presented interference in several stages of virus replication, mainly in penetration, but also in viral protein expression, hindering intercellular communication and replication.	Cardozo et al. 2011
Lentinula edodes	Aqueous, ethanolic extract and polysaccharide fractions of fruit bodies	Poliovirus type 1	*In vitro*	Extract and polysaccharide fractions of *L. edodes* acted in the initial poliovirus replication processes. Extracts and polysaccharides can be considered a source of potential antiviral substances. However, further studies are needed to better understand the stage(s) of viral replication where inhibition occurs.	Rinção et al. 2012
Agaricus brasiliensis	Mycelium isolated and modified by sulfation	Herpes simplex	*In vitro*	The mycelium of *A. brasiliensis* was not effective in reducing eye disease and skin lesions in topical treatments. However, it was effective in the groups treated orally and vaginally, reducing the disease scores, suggesting that the cure was accelerated, in addition to showing there was no toxicity.	Cardozo et al. 2013
Agaricus brasiliensis	Fruiting body - isolated and sulfated polysaccharides	Herpes simplex	*In vitro*	Sulfated polysaccharides have an interesting anti-herpetic mechanism, inhibiting viral cell-cell adsorption, penetration and propagation. The authors also suggest a beneficial synergistic effect with acyclovir.	Cardozo et al. 2014
Ganoderma lucidum *Trametes versicolor*	Fruiting body biomass powder	Human Papillomavirus (16 and 18)	Clinical trial	The combination of powder from the two mushrooms was effective in the oral presentation of HPV. The authors believe that it is a complementary therapy to vaccination.	Donatini 2014
Auriporia aurea *Flammulina velutipes* *Fomes fomentarius* *Ganoderma lucidum* *Lentinus edodes* *Lyophyllum shimeji* *Pleurotus eryngii* *Pleurotus ostreatus* *Schizophyllum commune* *Trametes versicolor*	Mycelium - aqueous extract	Influenza type A and Herpes simplex 2	*In vitro*	All mushrooms have antiviral activity against the influenza A virus, while *A. aurea*, *F. fomentarius*, *G. lucidum*, and *T. versicolor* have antiviral activity against HSV-2	Krupodorova et al. 2014

Agaricomycetes	-	Influenza, Human immunodeficiency, Herpes, Hepatitis, among others	Narrative review	The antiviral effects of mushrooms involve polysaccharides, proteins, melanins, terpenoids and nucleosides. Considering the complex composition of its compounds, agaricales have a wide spectrum and biotechnological applicability.	Teplyakova and Kosogova 2016
Grifola frondosa	Mycelium	Enterovirus 71	*In vitro*	Purified polysaccharide from *G. frondosa* was effective in inhibiting EV71 replication in infected cells and the expression of viral VP1 protein and genomic RNA synthesis were suppressed. In addition, GFP1 exhibited apoptotic activities by suppressing EV71-induced caspase-3 cleavage and downregulation of IκBα.	Zhao et al. 2016
Agaricus brasiliensis	Fruiting body - aqueous extract	Influenza (subtype H1N1)	*In vitro*	Extract of *A. brasiliensis* had no direct effects on the virus, but exerted antiviral effects through cells, demonstrating that the extract can effectively prevent the spread of the influenza virus.	Eguchi et al. 2017
G. lucidium, H. erinaceus, L. rhinocerotis, P. giganteus, S. commune	Fruiting body - aqueous extract	Dengue (serotype 2)	*In vitro*	Extracts were more effective when added simultaneously to the virus, suggesting that the inhibitory activity is dependent on the time of exposure.	Ellan et al. 2019
Agaricus bitorquis	Dry fruiting body - aqueous extract	Human immunodeficiency virus	*In vitro*	Lectin from *A. bitorquis* has unique properties of molecular mass and subunit structure compared to other lectins of the genus *Agaricus*. It is an inulin specific lectin and prefers rabbit erythrocytes, human type A and AB. It manifests significant inhibitory activity of HIV-1 RT, antiproliferative activity in relation to cancer cells and mitogenic activity in relation to murine splenocytes.	Zhang et al. 2019
Lentinula edodes	Mycelium – α-glucans	SARS-CoV	Narrative review	α-glucan-derived mushroom is described as safe and effective in promoting immune responses to a wide range of viral infections, as well as in healthy humans, although its effectiveness has not been specifically assessed for coronavirus.	Di Pierro et al. 2020

Table 16.1 contd. ...

...Table 16.1 contd.

Species	Samples	Viruses	Type of study	Outcome	Reference
Lentinus edodes	Fruiting body – β-glucans	SARS-CoV	*In vitro*	Glucan showed potential in the treatment of lung injury induced by SARS-CoV, reducing inflammation, and showing immunomodulatory activity.	Murphy et al. 2020
Lentinus edodes *Grifola frondosa* *Ganoderma lucidum* *Inonotus obliquus*	Aqueous extract and ethanolic extract	SARS-CoV	Narrative review	Many mushrooms can be used in several viral infections, especially if they have a high β-glucan and phenolic compounds levels. These components can be natural inhibitors against coronavirus enzymes. In addition, this review highlighted the therapeutic potential of *I. obliquus* as a natural antiviral treatment against SARS-COV-2. Previous studies on this mushroom have laid the foundation for the antiviral capabilities of *I. obliquus*.	Shahzad et al. 2020

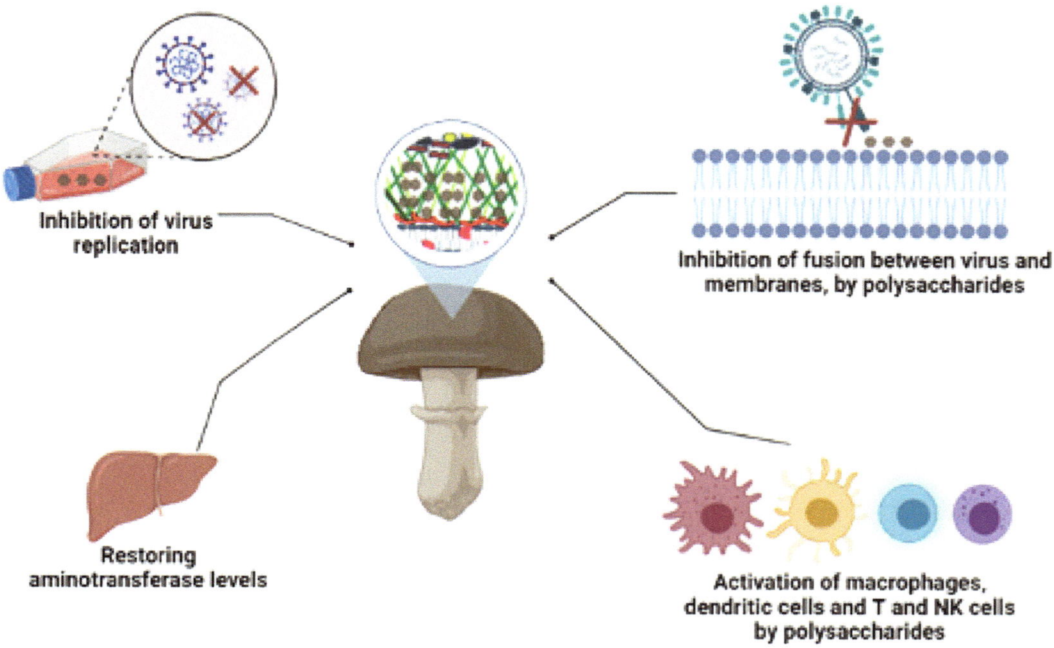

Figure 16.1. Suggestion of mechanisms of action present in mushrooms.

study using aqueous extract of *Agaricus blazei* Murill for restoring aminotransferase levels (Chung-Hua et al. 2008).

As mentioned earlier, one of the main antiviral mechanisms present in mushrooms, is related to the inhibition of viral replication. However, there are numerous mechanisms. Polysaccharides may also be involved in the activation of macrophages, dendritic cells and T cells or even increasing the cytotoxicity of the NK cell and inducing the production of IL-12 (Chang-Qing et al. 2007). Another possible mechanism of action is related to the role of polysaccharides in inhibiting fusion between the pathogen's membranes and mucous cells; polysaccharide-pathogen binding in order to avoid binding to host receptors (Cardozo et al. 2013) can also be mentioned. Figure 16.1 explains the proposals of mechanisms of action.

Future trends and perspectives

Especially in the pandemic and post-pandemic of COVID-19, due to the indiscriminate use of antimicrobials and disinfectants, it is feasible that there is the increase in the number of resistant microorganisms, including bacteria and viruses. In addition to resistance, knowledge gaps arise in cases such as COVID-19, in which a virus that had an animal as a reservoir jumps to humans, and the dynamics of pathogen-human/host animal interaction needs to be unraveled and treatment developed.

However, mushrooms come with good prospects as a treatment, since a protective immune response can make a big difference. In this line, the active ingredients of mushrooms are safe natural derivatives, endowed with immunomodulatory capacity, effective in promoting regulated immune responses to a wide number of viral infections.

Conclusion

Diseases from viruses are a critical challenge in modern medicine. Viruses are responsible for respiratory syndrome until the development of cancer. Although there are some conventional therapies, adverse effects, limitations of medications, and the rapid mutation ability of viruses make the treatments unsuccessful, sometimes. Therefore, mushrooms polysaccharides, lentinan and beta-glucan, are a natural alternative for antiviral treatments, with a reduction of side effects by immunostimulating activity.

References

Ali, N.A., Mothana, R.A.A., Lesnau, A., Pilgrin, H., Lindequist, A. (2003). Antiviral activity of *Inonotus hispidus*. Fitoterapia 74: 483–485.

Anasonye, F., Winquiste, E., Räsänen, M., Kontro, J., Björklöf, K., Vasilyeva, G., Jørgensen, K.S., Steffen, K.T., Tuomela, M. (2015). Bioremediation of TNT contaminated soil with fungi under laboratory and pilot scale conditions. Int Biodeterior Biodegrad 105: 07–12.

Blackwell, M. (2011). The fungi: 1, 2, 3 ...5.1 million species? Am J Bot 98: 426–438.

Blánquez, P., Sarrà, M., Vicente, T. (2008). Development of a continuous process to adaot the textile wastewater treatment by fungi to industrial conditions. Process Biochem 43: 1–7.

Cardozo, F.T.G.S., Cameline, C.M., Mascarello, A., Rossi, M.J., Nunes, R.J., Barardi, C.R.M., Mendonça, M.M., Simões, C.M.O. (2011). Antiherpetic activity of a sulfated polysaccharide from *Agaricus brasiliensis* mycelia. Antivir Res 92(1): 108–114.

Cardozo, F.T., Larsen, I.V., Carballo, E.V., Jose, G., Stern, R.A., Brummel, R.C., Camelini, C.M., Rossi, M.J., Simões, C.M., Brandt, C.R. (2013). *In vivo* anti-herpes simplex virus activity of a sulfated derivative of *Agaricus brasiliensis* Mycelial polysaccharide. Antimicrob Agents Chemother 57: 2541–2549.

Cardozo, F.T., Camelini, C.M., Leal, P.C., Kratz, J.M., Nunes, R.J., Mendonça, M.M., Simões, C.M. (2014). Antiherpetic mechanism of a sulfated derivative of *Agaricus brasiliensis* fruiting bodies polysaccharide. Intervirology 57(6): 375–383.

Chambergo, F.S., Valencia, E.Y. (2016). Fungal biodiversity to biotechnology. Appl Microbiol Biotechnol 100: 2567–2577.

Chang-Qing, G., Li, J.W., Chao, F., Jin, M., Wang, X.W., Shen, Z.Q. (2007). Isolation, identification and function of a novel anti-HSV-1 protein from *Grifola frondosa*. Antiviral Res 7: 250–257.

Chen, S., Xu, J., Liu, C., Zhu, Y., Nelson, D.R., Zhou, S., Li, C., Wang, L., Guo, X., Sun, Y., Luo, H., Li, Y., Song, J., Henrissat, B., Levasseur, A., Qian, J., Li, J., Luo, X., Shi, L., He, L., Xiang, L., Xu, X., Niu, Y., Li, Q., Han, M.V., Yan, H., Zhang, J., Chen, H., Lv, A., Wang, Z., Liu, M., Schwartz, D.C., Sun C. (2012). Genome sequence of the model medicinal mushroom *Ganoderma lucidum*. Nat Commun 3: 913.

Chung-Hua, H.S.U., Kung-Chang, H., Yi-Hsiung, C., Pesus, C. (2008). The mushroom *Agaricus blazei* Murill extract normalizes liver function in patients with chronic hepatitis B. J Altern Complement Med 4: 299–301.

Crini, G., Lichtfouse, E. (2019). Advantages and disadvantages of techniques used for wastewater treatment. Environ Chem Lett 17: 145–155.

Di Pierro, F., Bertuccioli, A., Cavecchia, I. (2020). possible therapeutic role of a highly standardized mixture of active compounds derived from cultured *Lentinula edodes* mycelia (AHCC) in patients infected with 2019 novel coronavirus. Minerva Gastroenterol Dietol 66: 172–176.

Donatini, B. (2014). Control of Oral Human Papillomavirus (HPV) by medicinal mushrooms, *Trametes versicolor* and *Ganoderma lucidum*: A preliminary clinical trial. Int J Med Mushrooms 16: 497–498.

Eguchi, N., Fujino, K., Thanasut, K., Taharaguchi, M., Motoi, M., Motoi, A., Oonaka, K., Taharaguchi, S. (2017). *In vitro* anti-influenza virus activity of *Agaricus brasiliensis* KA21. Biocontrol Sci 22: 171–174.

Ellan, K., Thayan, R., Raman, J., Hidari, K.I.P.J., Ismail, N., Sabaratnam, V. (2019). Anti-viral activity of culinary and medicinal mushroom extracts against dengue virus serotype 2: an *in-vitro* study. BMC Complement Altern Med 19: 1–12.

Eo, S.K., Kim, Y.S., Lee, C.K., Han, S.S. (2000). Possible mode of antiviral activity of acidic protein bound polysaccharide isolated from *Ganoderma lucidum* on herpes simplex viruses. J Ethnopharmacol 72: 475–481.

Gargano, M.L.G., Griensven, L.J.L.D., Isikhuemhen, O.S., Lindequist, U., Venturella, G., Zervakis, S.P.W.G.I. (2017). Medicinal mushrooms: Valuable biological resources of high exploitation potential. Plant Biosyst 151: 548–65.

Gao, L., Sun, Y., Si, J., Liu, J., Sun, G., Sun, X., Cao, L. (2014). *Cryptoporus volvatus* extract inhibits influenza virus replication *in vitro* and *in vivo*. PLoS One 9: 1–17.

Hetland, G., Johnson, E., Bernardshaw, S.V., Grinde, B. (2020). Can medicinal mushrooms have prophylactic or therapeutic effect against COVID-19 and its pneumonic superinfection and complicating inflammation? Journal of Immunology 12937: 1–12.

Jin, Y., Zhang, M., Jin, Z., Wang, G., Li, R., Zhang, X., Liu, X., Qu, J., Wang, H. (2020). Characterization of biochars derived from various spent mushroom substrates and evaluation of their adsorption performance of Cu(II) ions from aqueous solution. Environ Res 1: 51.

Kariuki, Z., Kiptoo, J., Onyancha, D. (2017). Biosorption studies of lead and copper using Rogers mushroom biomass 'Lepiota hystrix'. S Afr J Sci 23: 62–70.

Krupodorova, T., Rybalko, S., Barshteyn, V. (2014). Antiviral activity of Basidiomycete mycelia against influenza type A (serotype H1N1) and herpes simplex virus type 2 in cell culture. Virol Sin 29: 284–290.

Linnakoski, R., Reshamwala, D., Veteli, P., Cortina-Escribano, M., Vanhanen, H., Marjomäki, V. (2018). Antiviral agents from fungi: diversity, mechanisms and potential applications. Frente Microbiol 9: 1–18.

Liu, J., Yang, F., Ye, L.B., Yang, X.J., Timani, K.A., Zheng, Y., Wang, Y.H. (2004). Possible mode of action of antiherpetic activities of a proteoglycan isolated from the mycelia of *Ganoderma lucidum in vitro*. J Ethnopharmacol 95: 265–272.

Lv, H., Yang Kong, Qing Yao, Bo Zhang, Fang-wei Leng, He-jiaoBian, Jan Balzarini, Els Van Damme, Jin-ku Bao. (2009). Nebrodeolysin, a novel hemolytic protein from mushroom *Pleurotus nebrodensis* with apoptosis-inducing and anti-HIV-1 effects. Phytomedicine 16: 198–205.

Ma, G., Yang, W., Zhao, L., Pei, F., Fang, D., Hu, Q. (2018). A critical review on the health promoting effects of mushrooms nutraceuticals. Food Sci Hum Wellness 7: 125–133.

Marco-Urrea, E., Pérez-Trujillo, M., Vicent, T., Caminal, G. (2009). Ability of white-rot fungi to remove selected pharmaceuticals and identification of degradation products of ibuprofen by *Trametes versicolor*. Chemosphere 74: 765–71.

Menk, J.J., do Nascimento, A.I.S., Leite, F.G., de Oliveira, R.A., Jozala, A.F., de Oliveira Junior, J.M., Chaud, M.V., Grotto, D. (2019). Biosorption of pharmaceutical products by mushroom stem waste. Chemosphere 237: 124515.

Murphy, E.J., Masterson, C., Rezoagli, E., O'Toole, D., Major, I., Stack, G.D., Lynch, M., Laffey, J.G., Rowan, N.J. (2020). β-Glucan extracts from the same edible shiitake mushroom *Lentinus edodes* produce differential *in-vitro* immunomodulatory and pulmonary cytoprotective effects—Implications for coronavirus disease (COVID-19) immunotherapies. Sci Total Environ 732: 139330.

Razumov, I.A., Kosogova, T.A., Kazachinskaia, E.I., Puchkova, L.I., Shcherbakova, N.S., Gorbunova, I.A., Mikhaĭlovskaia, I.N., Loktev, V.B., Tepliakova, T.V. (2010). Antiviral activity of aqueous extracts and polysaccharide fractions from mycelium and fruit bodies of higher fungi. Antibiot Khimiote 55: 14–8.

Rincão, V.P., Yamamoto, K.A., Ricardo, N.M., Soares, S.A., Meirelles, L.D., Nozawa, C., Linhares, R.E. (2012). Polysaccharide and extracts from Lentinula edodes: structural features and antiviral activity. Virol J 9: 1–6.

Shahzad, F., Anderson, D., Najafzadeh, M. (2020). The antiviral, anti-inflammatory effects of natural medicinal herbs and mushrooms and SARS-CoV-2 infection. Nutrients 12: 1–13.

Skariyachan, S., Prasanna, A., Manjunath, S.P., Karanth, S.S., Nazre, A. (2016). Environmental assessment of the degradation potential of mush-room fruit bodies of *Pleurotus ostreatus* (Jacq.: Fr.) P. Kumm. towards synthetic azo dyes and contaminating effluents collected from textile industries in Karnataka, India. Environ Monit Assess 188: 121.

Spim, S.R.V., de Oliveira, B.G.C.C., Leite, F.G., Gerenutti, M., Grotto, D. (2017). Effects of Lentinula edodes consumption on biochemical, hematologic and oxidative stress parameters in rats receiving high-fat diet. Eur J Nutr 56: 2255–2264.

Šrédlová, K., Škrob, Z., Filipová, A., Mašín, P., Holecová, J., Cajthaml, T. (2019). Biodegradation of PCBs in contaminated water using spent oyster mushroom substrate and atrickle-bed bioreactor. Water Res 1: 115274.

Teplyakova, T.V., Kosogova, T.A. (2016). Antiviral effect of *Agaricomycetes Mushrooms* (Review). Int J Med Mushrooms 18: 375–386.

Yang, X., Guo, M., Wu, Y., Wu, Q., Zhang, R. (2014). Removal of emulsifed oil from water by fruiting bodies of macro-fungus (*Auricularia polytricha*). PLoS ONE 9: e95162.

Zhang, G.Q., Chen, Q.J., Hua, J., Liu, Z.L., Sun, Y., Xu, X., Han, P. (2019). An inulin-specific lectin with anti-HIV-1 reverse transcriptase, antiproliferative, and mitogenic activities from the edible mushroom *Agaricus bitorquis*. Biomed Res Int 2019: 1–9.

Zhao, C., Gao, L., Wang, C., Liu, B., Jin, Y., Xing, Z. (2016). Structural characterization and antiviral activity of a novel heteropolysaccharide isolated from *Grifola frondosa* against enterovirus 71. Carbohydr Polym 144: 382–9.

Zulfadhly, Z., Mashitah, M.D., Bhatia, S. (2001). Heavy metals removal infixed-bed column by the macro fungus *Pycnoporus sanguineus*. Environ Pollut 112: 463–470.

Index

P

S

T

V

X

Editors' Biography

Dr. Mahendra Rai is presently a visiting Professor at the Department of Microbiology, Nicolaus Copernicus University, Torun, Poland. Formerly he was Professor and Head of department of Biotechnology, SGB Amravati University, Maharashtra state, India. He was a visiting scientist at the University of Geneva, Debrecen University, Hungary; University of Campinas, Brazil; VSB Technical University of Ostrava, Czech Republic, and National University of Rosario, Argentina, and University of Sao Paulo. He has published more than 425 research papers in national and international journals. In addition, he has edited/authored more than 62 books and 6 patents. Recently, he has been featured in Stanford's list of top 2% scientists under Nanoscience and Nanotechnology.

Dr. Chistiane M. Feitosa is a Professor at the Department of Chemistry, Federal University of Piaui, Teresina, Brazil. Her research areas include pharmacology, natural products and their secondary metabolites with the potential for treatment of neurodegenerative diseases and antimicrobials. She has published more than 55 research papers in national and international journals. In addition, she has edited/authored more than 12 books/chapters from eminent publishers like CRC press/ Taylor and Francis and also 5 patents.